IN VIVO OPTICAL IMAGING of BRAIN FUNCTION

SECOND EDITION

FRONTIERS IN NEUROSCIENCE

Series Editors
Sidney A. Simon, Ph.D.
Miguel A.L. Nicolelis, M.D., Ph.D.

Published Titles

Apoptosis in Neurobiology
Yusuf A. Hannun, M.D., Professor of Biomedical Research and Chairman, Department of Biochemistry and Molecular Biology, Medical University of South Carolina, Charleston, South Carolina
Rose-Mary Boustany, M.D., tenured Associate Professor of Pediatrics and Neurobiology, Duke University Medical Center, Durham, North Carolina

Neural Prostheses for Restoration of Sensory and Motor Function
John K. Chapin, Ph.D., Professor of Physiology and Pharmacology, State University of New York Health Science Center, Brooklyn, New York
Karen A. Moxon, Ph.D., Assistant Professor, School of Biomedical Engineering, Science, and Health Systems, Drexel University, Philadelphia, Pennsylvania

Computational Neuroscience: Realistic Modeling for Experimentalists
Eric DeSchutter, M.D., Ph.D., Professor, Department of Medicine, University of Antwerp, Antwerp, Belgium

Methods in Pain Research
Lawrence Kruger, Ph.D., Professor of Neurobiology (Emeritus), UCLA School of Medicine and Brain Research Institute, Los Angeles, California

Motor Neurobiology of the Spinal Cord
Timothy C. Cope, Ph.D., Professor of Physiology, Wright State University, Dayton, Ohio

Nicotinic Receptors in the Nervous System
Edward D. Levin, Ph.D., Associate Professor, Department of Psychiatry and Pharmacology and Molecular Cancer Biology and Department of Psychiatry and Behavioral Sciences, Duke University School of Medicine, Durham, North Carolina

Methods in Genomic Neuroscience
Helmin R. Chin, Ph.D., Genetics Research Branch, NIMH, NIH, Bethesda, Maryland
Steven O. Moldin, Ph.D., University of Southern California, Washington, D.C.

Methods in Chemosensory Research
Sidney A. Simon, Ph.D., Professor of Neurobiology, Biomedical Engineering, and Anesthesiology, Duke University, Durham, North Carolina
Miguel A.L. Nicolelis, M.D., Ph.D., Professor of Neurobiology and Biomedical Engineering, Duke University, Durham, North Carolina

The Somatosensory System: Deciphering the Brain's Own Body Image
Randall J. Nelson, Ph.D., Professor of Anatomy and Neurobiology, University of Tennessee Health Sciences Center, Memphis, Tennessee

The Superior Colliculus: New Approaches for Studying Sensorimotor Integration
William C. Hall, Ph.D., Department of Neuroscience, Duke University, Durham, North Carolina
Adonis Moschovakis, Ph.D., Department of Basic Sciences, University of Crete, Heraklion, Greece

New Concepts in Cerebral Ischemia
Rick C. S. Lin, Ph.D., Professor of Anatomy, University of Mississippi Medical Center, Jackson, Mississippi

DNA Arrays: Technologies and Experimental Strategies
Elena Grigorenko, Ph.D., Technology Development Group, Millennium Pharmaceuticals, Cambridge, Massachusetts

Methods for Alcohol-Related Neuroscience Research
Yuan Liu, Ph.D., National Institute of Neurological Disorders and Stroke, National Institutes of Health, Bethesda, Maryland
David M. Lovinger, Ph.D., Laboratory of Integrative Neuroscience, NIAAA, Nashville, Tennessee

Primate Audition: Behavior and Neurobiology
Asif A. Ghazanfar, Ph.D., Princeton University, Princeton, New Jersey

Methods in Drug Abuse Research: Cellular and Circuit Level Analyses
Dr. Barry D. Waterhouse, Ph.D., MCP-Hahnemann University, Philadelphia, Pennsylvania

Functional and Neural Mechanisms of Interval Timing
Warren H. Meck, Ph.D., Professor of Psychology, Duke University, Durham, North Carolina

Biomedical Imaging in Experimental Neuroscience
Nick Van Bruggen, Ph.D., Department of Neuroscience Genentech, Inc.
Timothy P.L. Roberts, Ph.D., Associate Professor, University of Toronto, Canada

The Primate Visual System
John H. Kaas, Department of Psychology, Vanderbilt University
Christine Collins, Department of Psychology, Vanderbilt University, Nashville, Tennessee

Neurosteroid Effects in the Central Nervous System
Sheryl S. Smith, Ph.D., Department of Physiology, SUNY Health Science Center, Brooklyn, New York

Modern Neurosurgery: Clinical Translation of Neuroscience Advances
Dennis A. Turner, Department of Surgery, Division of Neurosurgery, Duke University Medical Center, Durham, North Carolina

Sleep: Circuits and Functions
Pierre-Hervé Luoou, Université Claude Bernard Lyon, France

Methods in Insect Sensory Neuroscience
Thomas A. Christensen, Arizona Research Laboratories, Division of Neurobiology, University of Arizona, Tuscon, Arizona

Motor Cortex in Voluntary Movements
Alexa Riehle, INCM-CNRS, Marseille, France
Eilon Vaadia, The Hebrew University, Jerusalem, Israel

Neural Plasticity in Adult Somatic Sensory-Motor Systems
Ford F. Ebner, Vanderbilt University, Nashville, Tennessee

Advances in Vagal Afferent Neurobiology
Bradley J. Undem, Johns Hopkins Asthma Center, Baltimore, Maryland
Daniel Weinreich, University of Maryland, Baltimore, Maryland

The Dynamic Synapse: Molecular Methods in Ionotropic Receptor Biology
Josef T. Kittler, University College, London, England
Stephen J. Moss, University College, London, England

Animal Models of Cognitive Impairment
Edward D. Levin, Duke University Medical Center, Durham, North Carolina
Jerry J. Buccafusco, Medical College of Georgia, Augusta, Georgia

The Role of the Nucleus of the Solitary Tract in Gustatory Processing
Robert M. Bradley, University of Michigan, Ann Arbor, Michigan

Brain Aging: Models, Methods, and Mechanisms
David R. Riddle, Wake Forest University, Winston-Salem, North Carolina

Neural Plasticity and Memory: From Genes to Brain Imaging
Frederico Bermudez-Rattoni, National University of Mexico, Mexico City, Mexico

Serotonin Receptors in Neurobiology
Amitabha Chattopadhyay, Center for Cellular and Molecular Biology, Hyderabad, India

Methods for Neural Ensemble Recordings, Second Edition
Miguel A.L. Nicolelis, M.D., Ph.D., Professor of Neurobiology and Biomedical Engineering, Duke University Medical Center, Durham, North Carolina

Biology of the NMDA Receptor
Antonius M. VanDongen, Duke University Medical Center, Durham, North Carolina

Methods of Behavioral Analysis in Neuroscience
Jerry J. Buccafusco, Ph.D., Alzheimer's Research Center, Professor of Pharmacology and Toxicology, Professor of Psychiatry and Health Behavior, Medical College of Georgia, Augusta, Georgia

***In Vivo* Optical Imaging of Brain Function, Second Edition**
Ron Frostig, Ph.D., Professor, Department of Neurobiology, University of California, Irvine, California

IN VIVO OPTICAL IMAGING of BRAIN FUNCTION

SECOND EDITION

Edited by
Ron D. Frostig
University of California, Irvine
California

CRC Press
Taylor & Francis Group
Boca Raton London New York

CRC Press is an imprint of the
Taylor & Francis Group, an **informa** business

CRC Press
Taylor & Francis Group
6000 Broken Sound Parkway NW, Suite 300
Boca Raton, FL 33487-2742

First issued in paperback 2019

ISBN-13: 978-1-4200-7684-4 (hbk)
ISBN-13: 978-0-367-38565-1 (pbk)

Visit the Taylor & Francis Web site at
http://www.taylorandfrancis.com

and the CRC Press Web site at
http://www.crcpress.com

Contents

Preface

These are exciting times for the field of optical imaging of brain function. This field has continued to rapidly advance in both its theoretical and technical domains, and its contributions to brain research continue to considerably advance our understanding of brain function. Thus, the description of these techniques and their applications should be attractive to both basic and clinical researchers who are interested in brain function. The comparison between this edition and the previous edition of the book (CRC 2002) serves as a good indicator for the continuing rapid growth of the field. In contrast to the previous edition consisting of 9 chapters, the current edition contains 15 chapters—9 new and 6 thoroughly revised. Further, some of the imaging techniques described in the new chapters did not even exist when the previous edition was published!

This book brings together a description of the wide variety of optical techniques available for the specific study of neuronal activity in the living brain and their application for animal and human functional imaging research. These in vivo techniques can vary by their level of temporal resolution (milliseconds to seconds), spatial resolution (microns to millimeters), degree of invasiveness to the brain (removal of the skull above the imaged area to complete noninvasiveness), use of signals intrinsic to the brain versus use of external probes, and their current application to basic and clinical studies. Besides their usefulness for the study of brain function, these optical techniques are also appealing because they are typically based on compact, mobile, and affordable equipment, and can therefore be more readily integrated into a typical neuroscience, cognitive science, or neurology laboratory, as well as a hospital operating room. The chapters of this book describe the theory, setup, analytical methods, and examples of experiments that highlight the advantages of the different techniques. With the growing sophistication of the hardware, there is a parallel growing sophistication of the software. Therefore, with this edition, I have included a chapter dedicated to the description of principles of a software system, in the same way that other chapters describe the principles behind hardware, and I foresee that such chapters becoming more prevalent in the future.

Because this book focuses on in vivo imaging techniques, examples of the applications are focused on living brains of animals and humans rather than tissue cultures or brain slices. In addition, this book focuses on optical imaging techniques and consequently does not include nonoptical techniques of functional brain imaging such as fMRI, PET, and MEG techniques.

I have no doubt that the study of brain function using optical imaging techniques, as described in this book, will continue to expand and flourish in the future. The interest in these methods coupled with a continual rapid pace in relevant theoretical and technological advances should ensure that these techniques not only remain but also increase their dominance as major players of future brain research.

Ron Frostig, Ph.D.
Irvine, California 2008

The Editor

Ron Frostig, Ph.D., is a professor in the department of neurobiology and behavior, and in the department of biomedical engineering, at the University of California–Irvine (UCI). He received his B.Sc. in biology and his M.Sc. in neurobiology from Hebrew University, Jerusalem, Israel. He received his Ph.D. in neurobiology from the University of California–Los Angeles (UCLA), and was a postdoctoral researcher at the Rockefeller University in New York. In 1990 he joined the department of neurobiology and behavior at UCI, where he served as the director of graduate studies and currently serves as the vice chair of the department.

Dr. Frostig's main research interests are the functional organization and plasticity of sensory cortex, and he published research on these topics in leading scientific journals.

Dr. Frostig is a member of the American Society for Neuroscience and the American Physiological Society. For his teaching at UCI, he received an Excellence in Teaching Award. For his research he received the Alexander Heller Prize and the Arnold and Mabel Beckman Young Investigator Award, in addition to research grants from National Institute of Health and National Science Foundation.

Contributors List

David Abookasis
Laser Microbeam and Medical Program (LAMMP)
Beckman Laser Institute and Medical Clinic
University of California, Irvine
Irvine, California

David A. Boas
A.A. Martinos Center for Biomedical Imaging
Massachusetts General Hospital
Charlestown, Massachusetts

K.C. Brennan
Laboratory of Neuro Imaging
Department of Neurology
School of Medicine
University of California, Los Angeles
Los Angeles, California

Gang Chen
Department of Neuroscience
University of Minnesota
Minneapolis, Minnesota

Cynthia H. Chen-Bee
Department of Neurobiology and Behavior
Department of Biomedical Engineering
The Center for the Neurobiology of Learning and Memory
University of California, Irvine
Irvine, California

David J. Cuccia
Laser Microbeam and Medical Program (LAMMP)
Beckman Laser Institute and Medical Clinic
University of California, Irvine
Irvine, California

Earl M. Dolnick
Department of Physics
University of California, San Diego
La Jolla, California

Jonathan Driscoll
Department of Physics
University of California, San Diego
La Jolla, California

Timothy J. Ebner
Department of Neuroscience
University of Minnesota
Minneapolis, Minnesota

Monica Fabiani
Beckman Institute and Psychology Department
University of Illinois at Urbana-Champaign
Urbana, Illinois

Isabelle Ferezou
Laboratoire de Neurobiologie et Diversité Cellulaire
Ecole Supérieure de Physique et de Chimie Industrielles
Paris, France

Maria Angela Franceschini
A.A. Martinos Center for Biomedical Imaging
Massachusetts General Hospital
Charlestown, Massachusetts

Ron D. Frostig
Department of Neurobiology and Behavior
Department of Biomedical Engineering, and
The Center for the Neurobiology of Learning and Memory
University of California, Irvine
Irvine, California

Wangcai Gao
Department of Neuroscience
University of Minnesota
Minneapolis, Minnesota

John S. George
Biological and Quantum Physics Group MS-D454
Los Alamos National Laboratory
Los Alamos, New Mexico

Gabriele Gratton
Beckman Institute and Psychology Department
University of Illinois at Urbana-Champaign
Urbana, Illinois

Ronald M. Harper
Department of Neurobiology
University of California, Los Angeles
Los Angeles, California

Fritjof Helmchen
Department of Neurophysiology
Brain Research Institute
University of Zurich
Zurich, Switzerland

Ryuichi Hishida
Department of Neurophysiology
Brain Research Institute
Niigata University
Niigata, Japan

Theodore J. Huppert
Department of Radiology
University of Pittsburgh Medical Center
Pittsburgh, Pennsylvania

T. Robert Husson
Committee on Computational Neuroscience
University of Chicago
Chicago, Illinois

Naoum P. Issa
Department of Neurobiology
University of Chicago
Chicago, Illinois

Valery A. Kalatsky
Department of Electrical and Computer Engineering
University of Houston
Houston, Texas

Hiroki Kitaura
Department of Neurophysiology
Brain Research Institute
Niigata University
Niigata, Japan

David Kleinfeld
Department of Physics and Graduate Program in Neurosciences
University of California, San Diego
La Jolla, California

Thomas Knöpfel
Laboratory for Neural Circuit Dynamics
RIKEN Brain Science Institute
Saitama, Japan

Yamato Kubota
Department of Neurophysiology
Brain Research Institute
Niigata University
Niigata, Japan

Ferenc Matyas
Laboratory of Sensory Processing
Brain Mind Institute
Ecole Polytechnique Federale de Lausanne (EPFL)
Lausanne, Switzerland

Quoc-Thang Nguyen
Department of Physics
University of California, San Diego
La Jolla, California

Carl C.H. Petersen
Laboratory of Sensory Processing
Brain Mind Institute
Ecole Polytechnique Federale de Lausanne (EPFL)
Lausanne, Switzerland

David M. Rector
Department of VCAPP
Washington State University
Pullman, Washington

Katsuei Shibuki
Department of Neurophysiology
Brain Research Institute
Niigata University
Niigata, Japan

Kuniyuki Takahashi
Department of Neurophysiology
Brain Research Institute
Niigata University
Niigata, Japan

Arthur W. Toga
Laboratory of Neuro Imaging
Department of Neurology
School of Medicine
University of California, Los Angeles
Los Angeles, California

Manavu Tohmi
Department of Neurophysiology
Brain Research Institute
Niigata University
Niigata, Japan

Bruce J. Tromberg
Laser Microbeam and Medical Program (LAMMP)
Beckman Laser Institute and Medical Clinic
Department of Neurobiology and Behavior
University of California, Irvine
Irvine, California

Philbert S. Tsai
Department of Physics
University of California, San Diego
La Jolla, California

Matt Wachowiak
Department of Biology
Boston University
Boston, Massachusetts

Xincheng Yao
Department of Biomedical Engineering
University of Alabama at Birmingham
Birmingham, Alabama

1 Optical Imaging of Brain Activity In Vivo Using Genetically Encoded Probes

Matt Wachowiak and Thomas Knöpfel

CONTENTS

1.1 INTRODUCTION

Genetically encoded reporters of neural activity have great promise as tools for imaging brain function in vivo. Reproducible labeling of specific cell types and the continuous presence of indicators for long-term experiments are the most prominent advantages of this new methodology. It is becoming increasingly acknowledged that genetically encoded optical reporters, combined with advanced methods for recording these optical signals, will be the tool of choice for monitoring neural activity in the intact animal. Since the appearance of groundbreaking and influential descriptions of genetically encodable reporters of neuronal activity[1–4] intense efforts have been made to apply these new probes under in vivo conditions. To date, however, a relatively small number of optical reporters have proven useful for in vivo brain imaging or even in vitro neural preparations, and only a subset of these have proven useful for imaging in the intact mammalian brain. The amount of time that it has taken for these tools to mature is due to serious technical challenges in developing optical reporters that generate a sufficient signal in the intact brain, function correctly at mammalian body temperature, and throughout the life of the neuron are expressed at sufficient levels in the desired neurons without significantly altering their function. Nonetheless, many of these hurdles have been overcome in recent years, and several classes of genetically encoded activity reporters work robustly when expressed in a variety of different systems. Continued effort in developing probe and cell type–specific expression systems also promises to increase the utility and number of these reporters in the very near future.

In theory, the use of genetically encoded optical reporters of neural activity in vivo should be similar to (or less demanding than) that of classical, synthetic optical indicators. Indeed, like their synthetic counterparts, the major classes of genetically encoded reporters sense calcium or voltage, with a third class (the pHluorins) reporting synaptic vesicle cycling. Because genetically encoded reporters are typically derivatives of naturally expressed proteins such as calmodulin or ion channel subunits, the potential for interaction with endogenous proteins and enzyme substrates adds another level of complexity to experimental design and data interpretation. Thus, while many of the same technical issues seen with synthetic optical reporters apply to the use of genetically encoded probes, additional factors dependent on the specific design and mechanism of each type of indicator must also be considered when using these probes to measure and interpret neural activity in vivo.

In this chapter, we will review the design principles underlying three major classes of genetically encoded indicators—calcium sensors, reporters of transmitter release, and voltage sensors—as well as strategies for expressing these indicators in vivo. We will focus on the use of these probes in the mammalian brain, where their implementation has been the most challenging, although work in nonmammalian systems (i.e., zebrafish, Drosophila) and in in vitro preparations will be discussed in cases where this work has yielded important insights. We will then present examples using two of these sensors—synaptopHluorin (spH) and GCaMP2—to monitor sensory coding and postsynaptic processing in the mouse olfactory bulb. GCaMP2 and spH work via very different mechanisms and report distinct—though related—aspects of neural activity; consequently, each probe presents different advantages and difficulties when monitoring brain function in vivo. While the work presented here has been done primarily in the olfactory system, each of these probes has proven robust in its ability to report activity in a variety of neuronal systems, and so the principles discussed in this chapter should be generally applicable to imaging elsewhere in the brain.

1.2 DESIGN PRINCIPLES OF DIFFERENT GENETICALLY ENCODED REPORTER TYPES

Optical imaging of neuronal activity using fluorescent probes became a widely used method following the development of the calcium indicator fura-2.[5] In brief, fura-2 combined the properties of a calcium-binding molecule (EGTA) with a bright fluorescent dye (stilbene chromophore) in a manner that caused a change in fluorescence output upon binding of calcium. Fluorescent protein (FP)-based sensors follow this inspiration. They typically involve the molecular fusion of a sensor protein that undergoes conformational transitions in response to fluctuations in a physiologic parameter (e.g., calcium concentration or membrane voltage) with a conformation-sensitive FP-based reporter protein. This approach subsequently became feasible for in vivo application with the development of bright, spectrally and optically improved variants of *Aequorea victoria* green FP and similar proteins from corals.[6] FP sensors have two potential advantages: they can be both genetically engineered and genetically targeted. The first feature is perhaps more practical in nature, while the ability to genetically target a fluorescent probe (that is, to stain a genetically defined cell type) is of great importance and promise for in vivo imaging.

1.2.1 Genetically Engineered FP Calcium Sensors

Förster resonance energy transfer (FRET) describes an energy transfer mechanism between two dye molecules. A donor chromophore in its excited state can transfer energy by a nonradiative, long-range dipole–dipole coupling mechanism to an acceptor chromophore in close proximity (typically <10 nm). The use of FRET to monitor distances and movements at the molecular level is now well established in biology, and the demonstration of FRET between mutational variants of GFP (Mitra et al. 1996) provided a widely used principle to transduce conformational transitions of a sensor protein in to FP optical output. The first FP calcium sensors based on FRET used the calcium-dependent interaction of the protein calmodulin with the M13 peptide as the sensor, together with pairs of blue/green or cyan/yellow FPs as reporters[4,7] (Figure 1.1A). A more recent, but similar design used the skeletal muscle calcium sensor troponin C (TnC) which, unlike calmodulin, is not expressed in neurons (Figure 1.1B). The advantage is that TnC-based sensors are proposed to avoid interactions of the sensor calmodulin with endogenous cellular mechanisms.

A second principle to "sensitize" the reporter FP to conformational alterations is *circular permutation* (cp). In cpFPs, the sequential order of the N- and C-terminal parts of the amino sequence are swapped, with the effect that the amino acids where the initial sequence is divided form the new N- and C-termini. Importantly, positions have been identified in the amino acid sequence of several FPs where permutation conserves the ability of the protein to form a fluorescent chromophore. The principle behind using cpFPs is that the fluorescent properties of a FP are more sensitive to the impact of a fused sensor, following structural weakening by circular permutation (Baird et al., 1999; Akemann et al., 2001) (Figures 1.1C,D). The original cpFP-based Ca^{2+} sensors suffered from pH sensitivity and inefficient protein folding.[8–10] However, optimizing folding efficacy of the fluorescent proteins as well as the Ca^{2+} affinity and dynamic range of the sensing mechanisms has subsequently yielded several generations of probes for each design principle (reviewed in Reference 11).

Characterization of FP calcium sensors in the living mammalian (in all cases rodent) brain remains sparse, but to date CerTN-L15 (a troponin C based FRET

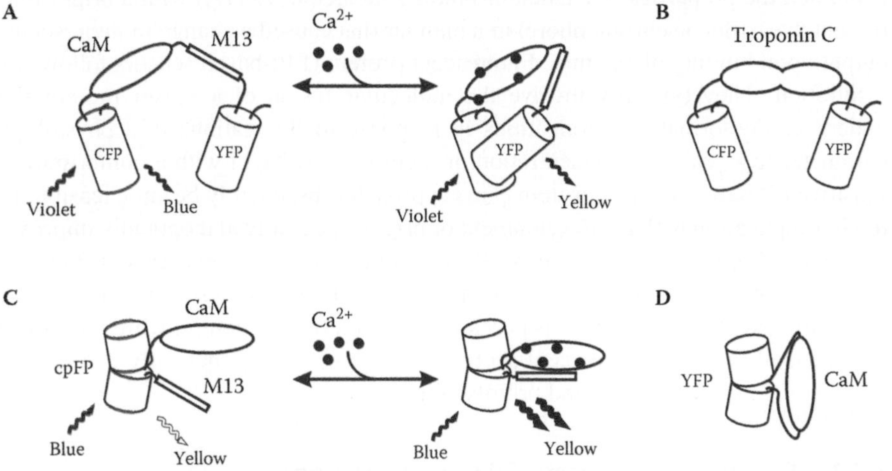

FIGURE 1.1 Genetically encoded calcium sensors: design and use for in vivo imaging. Schematics showing mechanisms of action of four genetically encoded calcium sensors: (A) yellow cameleon (from Miyawaki, A. et al., Nature, 388: 882–887, 1997); (B) TN-L15 (from Heim, N. et al. Improved calcium imaging in transgenic mice expressing a troponin C-based biosensor. Nat Methods, 4: 127–129, 2007); (C) GCaMP/Pericam (from Nagai, T. et al. Proc Natl Acad Sci USA, 98: 3197–3202, 2001; Nakai, J. et al. A high signal-to-noise Ca(2+) probe composed of a single green fluorescent protein. Nat Biotechnol, 19: 137–141, 2001); (D) Camgaroo (from Baird, G.S. et al., Proc Natl Acad Sci USA, 96: 11241–11246, 1999). Evoked neuronal activity imaged in vivo with GCaMP2.

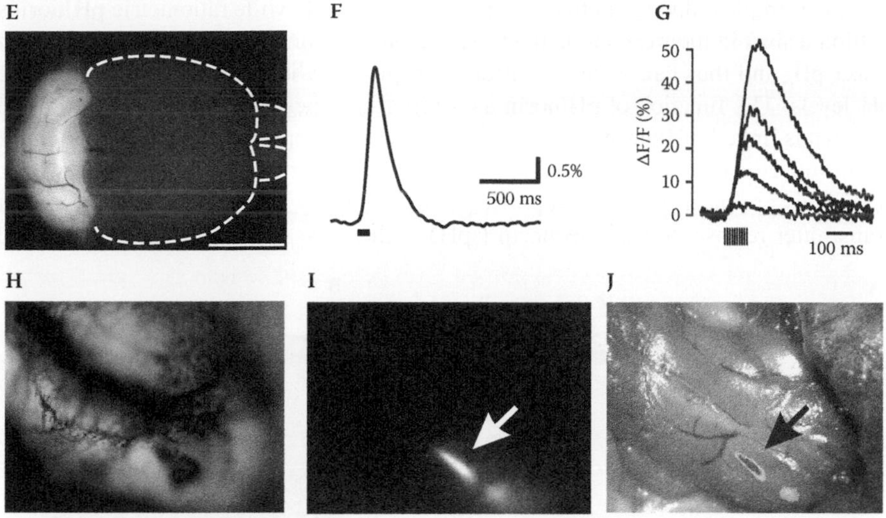

FIGURE 1.1 (continued) Genetically encoded calcium sensors: design and use for in vivo imaging. Schematics showing mechanisms of action of four genetically encoded calcium sensors: (E) whole-brain fluorescence image of a mouse expressing GCaMP2 under control of the Kv 3.1 promoter (line #846HB, showing bright cerebellar fluorescence; scale bar, 3 mm); (F) fluorescence signal evoked by electrical stimulation (10 pulses at 100 Hz, horizontal bar) of the molecular layer of the cerebellum (units are ΔF/F); (G) fluorescence signals imaged in vivo with two-photon laser scanning microscopy (line-scan mode) in response to electrical stimulation, as in (F). Each trace shows the response to an increasing number of pulses (from 2–10) delivered at 100 Hz. Each trace is the average of 10 trials. (H) Resting fluorescence image of the cerebellar surface viewed transcranially (vasculature appears as dark areas). (I) Grayscale map of fluorescence signal evoked by molecular layer stimulation through a small craniotomy in the preparation in (H). (J) Overlay of optical signal map and cerebellar surface after fixation and removal of the bone. Abbreviations: CFP, cyan fluorescent protein; YFP, yellow fluorescent protein; cpFP, circularly permuted fluorescent protein; CaM, Calmodulin complex; M13, M13 polypeptide. ([E–J] From Diez-Garcia, J. et al., Neuroimage, 34: 859–869, 2007.)

sensor), and GCaMP2 (a green cpFP sensor) exhibit the best performance in transgenic mice.[12–14] An example of stimulus-evoked GCaMP2 signals imaged from the cerebellum of a live mouse is shown in Figure 1.1(E–J). Despite their successful use, these sensors still have weaknesses, with the TN class of sensors suffering from slow kinetics and the GCaMP2 class having a relatively low brightness.[15]

1.2.2 Reporters of Transmitter Release: pHluorins

The pHluorins represent a different class of genetically encoded sensors that exploit the pH sensitivity of GFP as a signaling mechanism. "Wild-type" GFP fluorescence has a very weak pH sensitivity; in the pHluorins, this sensitivity has been enhanced through targeted mutations.[2] Several varieties of pHluorins have been generated, and their detailed functional properties are reviewed elsewhere.[16–18] "Ecliptic" pHluorins exhibit a simple reduction in fluorescence at a lower pH, while ratiometric pHluorins exhibit a shift in their excitation maximum from 395 nm at neutral pH to 475 nM at lower pH, and thus can be used—after appropriate calibration—to report absolute pH levels. The function of pHluorin as a reporter depends on the protein target to which it is linked: pHluorins can, in theory, be used to report changes in intracellular pH or movement of proteins between subcellular compartments having different pH. pHluorins are most useful for reporting neural activity when used as indicators of transmitter release. SynaptopHluorin (spH) is the most successful and widely used

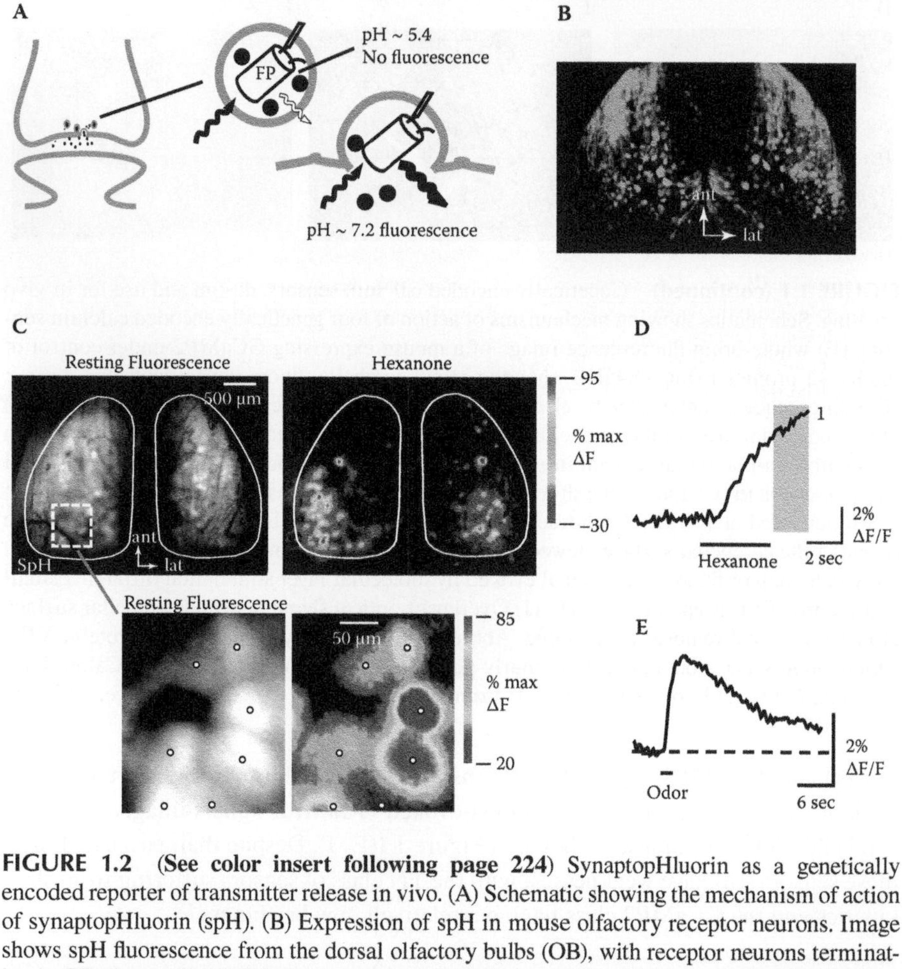

FIGURE 1.2 **(See color insert following page 224)** SynaptopHluorin as a genetically encoded reporter of transmitter release in vivo. (A) Schematic showing the mechanism of action of synaptopHluorin (spH). (B) Expression of spH in mouse olfactory receptor neurons. Image shows spH fluorescence from the dorsal olfactory bulbs (OB), with receptor neurons terminating in OB glomeruli. (C) Odorant-evoked spH signals imaged in vivo: (left) resting fluorescence image (grayscale), (right) pseudocolor map of the odorant-evoked fluorescence change, made by temporally averaging the signal over the time-window indicated in (D). Lower images, higher-magnification imaging of the area indicated by the dashed box showing resolution of individual adjacent glomeruli. (D) Time course of the odorant-evoked fluorescence change imaged from one glomerulus. (E) Longer time course (different preparation) showing the slow recovery of the evoked fluorescence change. Abbreviations: FP, fluorescent protein; spH, synaptopHluorin; ant, anterior; lat, lateral. ([B–D] From Bozza, T. et al., Neuron, 42: 9–21, 2004.)

of the pHluorins, and consists of a "superecliptic" pHluorin linked to the luminal side of the vesicle trafficking protein VAMP-2 (Figure 1.2A).[2,16,17] The low pH inside synaptic vesicles shifts the chromophore of the spH protein to a protonated, nonfluorescent state, causing spH fluorescence to be nearly completely eliminated (importantly, this mechanism "eclipses" photon absorption by the chromophore, and so is not a straightforward fluorescence "quenching"). Vesicle exocytosis associated with transmitter release leads to an increase in fluorescence as the pHluorin is exposed to the neutral pH of the extracellular space. As the vesicle is recycled and reacidified, fluorescence decreases to baseline levels[17,19] (Figure 1.2A).

SpH has been widely used in cell culture and in in vitro preparations for studying the dynamics of vesicle recycling and, to a lesser degree, transmitter release itself.[17,19–21] It has also proven to be among the most robust of the genetically encoded indicators for in vivo or ex vivo imaging of neural activity, being effective in reporting evoked transmitter release in preparations including *Drosophila* olfactory sensory neurons and interneurons,[22] mouse olfactory receptor neurons,[23] mouse neuromuscular junction,[24] and mouse hippocampal mossy fibers.[25] Examples of resting fluorescence and odorant-evoked spH signals imaged from mice expressing spH in olfactory receptor neurons is shown in Figures 1.2(B–D). The robustness of spH across systems and animal models is likely because its mechanism of action depends on a simple change in the surrounding ionic environment rather than a conformational change in its target protein. Limitations of spH include the fact that it specifically reports transmitter release, and thus is a useful indicator of "activity" only at presynaptic terminals. Its efficacy as a reporter will also depend on the reliability of transmitter release at the synapse of interest. Because of the slow time course of vesicle recycling, the temporal resolution of spH in vivo can also be limited (Figure 1.2D,E).

1.2.3 Genetically Engineered FP Voltage Sensors

FP-based probes for membrane voltage follow similar design principles as described for the FP calcium sensors. The sensor protein is derived from a voltage-sensitive protein (entire voltage-gated ion channel or a voltage sensor domain of either an ion channel or a voltage-operated, membrane-associated enzyme) and a FP reporter that is either a FP-FRET pair or a single FP. The first reported FP voltage sensor design was obtained by fusing wtGFP to the C-terminus of the *Drosophila* Shaker potassium channel.[3] This probe was designated FlaSh, for "fluorescent Shaker" and its design suggested that a native (not cp) FP can, in some way be "sensitized" by the surface of the lipid membrane (Figure 1.3A). The second prototypic design was based on the isolated voltage sensor domain from Kv2.1 and was termed VSFP1. Its concept was to exploit the well-established voltage-dependent conformational changes around the fourth transmembrane segment (S4) of the voltage sensor domain, in combination with a FRET FP pair[26] or a cpFP.[27] The third prototype, SPARC,[28] was generated by inserting a FP between domains I and II of the rat skeletal muscle Na^+ channel. Unfortunately, functional optical signals were small or undetectable in either Flare (a Kv1.4 variant of FlaSh), VSFP1, or SPARC.[29] A large, nonresponsive background fluorescence caused by poor membrane targeting of the probes was suggested as the main reason for the

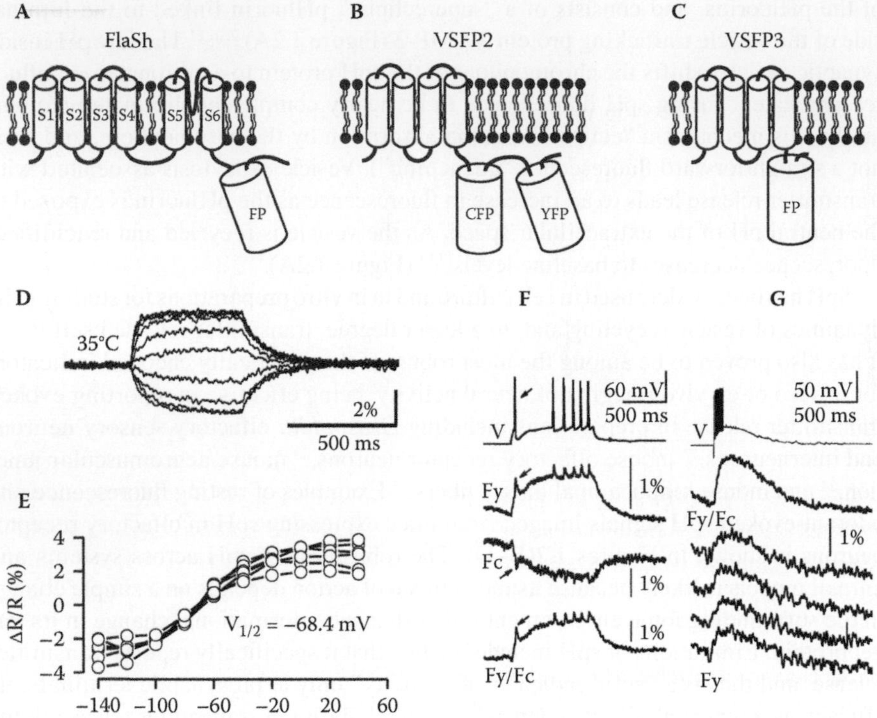

FIGURE 1.3 Genetically encoded optical reporters of membrane potential. (A–C) Schematics showing mechanism of action of three genetically encoded voltage sensors: (A) FLasH (from Siegel, M.S. and Isacoff, E.Y. Neuron, 19: 735–741, 1997), (B) VSFP2 (from Dimitrov, D. et al., PLoS ONE, 2: e440, 2007), and (C) VSFP3. (D–G) Characteristics of VSFP2.1 expressed in vitro. (D) Ratio of yellow/cyan fluorescence during voltage steps (500 ms duration) from a holding potential of –70 mV to potentials ranging from –140 to +40 mV (20 mV increments). (E) Ratio of yellow/cyan fluorescence versus membrane voltage. Plots show data from individual cells. Solid line shows Boltzmann fit to the grand average data. (F,G) Optical signals in PC12 cells expressing VSFP2.1 in response to a voltage trace obtained from a current-clamped mouse olfactory bulb mitral cell. The mitral cell was stimulated to generate a series of action potentials by intracellular injection of a current pulse (F) or by electrical stimulation of the olfactory nerve (G). Traces in (F) are averages of 50 sweeps, upper traces in (G) are the average of 90 sweeps, and the lower four traces in (G) are single sweeps. Fluorescence signals were digitally low-pass filtered (0.2 kHz) and were not corrected for dye bleaching. Recordings were done at 35°C. Abbreviations: FP, fluorescent protein; CFP, cyan fluorescent protein; YFP, yellow fluorescent protein, V, membrane potential; Fy, yellow fluorescence; Fc,cyan fluorescence; Fy/Fc, ratio of yellow and cyan fluorescence. ([D–G] From Dimitrov, D. et al., PLoS ONE, 2: e440, 2007.)

failure of these first-generation FP voltage sensors.[29] A second generation of FP voltage sensors overcame this limitation using the voltage sensor domain of a voltage-sensing nonion channel protein, Ci-VSP (*Ciona intestinalis* voltage-sensor-containing phospatase;[30] Figure 1.3B). The best variant of this second-generation FP voltage sensors is VSFP2.1. The fluorescence-voltage relationship with a $V_{1/2}$

value in the physiological range of neuronal membrane fluctuations, together with relatively fast kinetics, makes VSFP2.1 a suitable candidate for optical measurements of neuronal activity, such as large synaptic potentials, action potential trains, and bistabilities in resting membrane potential (Figures 1.3D–G). In contrast to calcium imaging, VSFP2.1 can also directly monitor membrane hyperpolarization. For resolving single action potentials, a newer series of VSFPs (VSFP3s[31]) with faster kinetics may be advantageous. Generation and characterization of transgenic mice expressing VSFP2s and -3s are under way.

1.2.4 Monochromatic and Dual Color FP Probes

FPs come in different colors and can be utilized in either monochromatic sensors (single FP approaches) or combined to make dual-colored FRET-based sensors. Efficient use of FP probes requires an understanding of the advantages and disadvantages associated with these options. The dual-color types of the previously mentioned FP sensors respond with a fluorescence increase at one wavelength and a decrease at another wavelength when calcium concentration or membrane potential changes. These sensors can, therefore, be used for ratiometric measurements. Theoretically, ratiometric measures allow absolute calibration of the measured signals and are less affected by movements of the object. Absolute calibration in live animals is, however, usually impractical because the calibration procedures require exact subtraction of any offset fluorescence ("background") that is usually inaccessible under conditions of in vivo imaging. Reduced sensitivity to movements is, however, a practical advantage when uncorrected background fluorescence is small. For imaging using a CCD-type camera, dual excitation or emission measurements are technically demanding, and so for this application, monochromatic approaches are preferable. Typically, monochromatic measurements (even with probes that are, in principle, suitable for dual color measurements) result in a better signal-to-noise ratio. One reason for this is that these probes are endowed with a better spectral separation between the excitation and emission wavelengths.

1.3 STRATEGIES FOR EXPRESSION OF GENETICALLY ENCODED REPORTERS IN VIVO

1.3.1 Virus Vectors

Several recombinant virus vectors have been successfully used for expression of FPs in the mature mammalian nervous system. These include adeno-associated virus (AAV), lentivirus, and herpes simplex virus (HSV) amplicon. While attempts to express FP sensors at appropriate levels and cell specificity are ongoing in several laboratories, published results are sparse. It appears that more time is required to establish the important question as to which viral vector is preferable for FP sensor expression.

1.3.2 Transgenic and Gene-Targeting Approaches

Transgenic and gene-targeted mice are time consuming and expensive to generate. However, once available, they are not only the most convenient approaches but are also more reliable reagents for study. They are convenient because such mice can be used immediately for imaging experiments with no need for preceding viral injection and incubation procedures. Specificity of labeling is also best defined in gene-targeted mice, where a specific gene product that defines the cell of interest is replaced by a FP sensor. Despite these obvious advantages, there are some limitations that must be considered: (1) there may be no gene product that is specific to the cell population of interest, (2) many gene products are driven by relative weak promoters resulting in low expression levels, and (3) gene targeting (in the form of standard "knock-in") results in mice which have a normal gene expressed at lower levels (+/– mice) or even fully knocked out (–/– mice). Perhaps one of the most successful examples for application of knock-in techniques are mice that express spH selectively in mouse olfactory receptor neurons.[23] In these mice, the coding sequence for the receptor neuron-specific protein olfactory marker protein was replaced with that of spH via gene targeting. Imaging experiments using these mice are described below.

An excellent alternative are transgenic mice with strong cell type–specific regulatory sequences (promoters). In transgenic mice, typically several or more copies of the transgene are inserted into the genome so that screening for highly expressing lines is often successful. Good examples are mice in which the FP calcium sensor GCaMP2 is expressed under the regulatory sequences of the Kv3.1 potassium channel gene. GCaMP2 mice have proven suitable for imaging of presynaptic calcium transients in cerebellar parallel fibers (axons of granule cells)[13,14] and postsynaptic calcium signals of olfactory bulb mitral and periglomerular cells.[32]

1.3.3 Electroporation In Utero

To overcome difficulties with virus vectors and the limitations of time-consuming and expensive transgenic/gene targeting approaches, electroporation in utero has proved to be a useful alternative in producing rodents expressing FPs in adult neurons.[33,34]

1.3.4 Chemically Inducible Systems for Temporal Control of Reporter Expression

All of the three expression techniques described above can, in principle, be enhanced by including components for inducible expression.[34] In general, such systems consists of two components derived from the *E. coli* tetracycline-resistance operon. The tetracycline (TET) system uses a tetracycline-controlled transactivator that acts on a tetracycline response element (TRE) within the promoter that controls expression of the gene of interest. In principle there are two advantages in the context of FP sensor expression. Inducible systems allow the generation of mice in which FP sensor expression can be "turned on" following normal development and maturation in the absence of transgene expression. This approach has been to chosen to overcome fears that calcium buffering by FP calcium sensors may interfere with normal brain development. The TET system may also allow the use of a comparably weak but highly cell-specific promoter to drive the transactivator, while multiple copies of the TRE within a strong promoter ensure high expression levels. This idea has been pursued for FP calcium sensors, but so far the theoretical advantages have only been partially realized.[35]

1.3.5 Recombinase-Based Systems for Cell Type–Specific Targeting

A second two-component system is based on recombinases such as Cre recombinase.[34] Cre recombinase catalyzes site-specific recombination of DNA between a 34–base pair recognition element (loxP sites). Cre recombinase is used to delete a segment of DNA *flanked* by *LoxP* sites (aka "floxed"). For targeted expression of FP sensors, their genes are delivered in inactive form (e.g., preceeded by a STOP codon flanked by a pair of loxP sites). When Cre recombinase is expressed in a specific cell type (using any of the above methods), the STOP codon is removed in this cell type, and the target gene (FP sensor) is expressed. Thus, this technology permits targeted expression of FP sensors in any subset of neurons for which one of the now many Cre recombinase-mice lines are available. Another promising approach involves placing the sensor gene in a lox-stop-lox cassette, which is then inserted into a virus and expressed under lox recombination as defined by a preexisting transgenic mouse with a FP sensor preceeded by a floxed STOP codon. Furthermore, Cre recombinase may be substituted, or used in combination with, other recombinases (e.g., Flp). These methods are reviewed in more detail elsewhere.[34]

1.4 EXPERIMENTAL SETUP FOR IMAGING WITH GENETICALLY ENCODED REPORTERS

1.4.1 Optical Setup

Design of the experimental setup for in vivo imaging with genetically encoded fluorescent probes follows the same principles as for imaging with conventional, synthetic fluorescent indicators. Because the probes are typically expressed in only a subset of neurons in a particular brain region, resting fluorescence levels are often low relative to those seen after bulk loading of neurons with synthetic indicators such as AM-ester calcium-sensitive dyes or voltage-sensitive dyes. Resting fluorescence is also typically lower than that for the same neurons expressing GFP due to the lower resting fluorescence of the probes themselves. Thus, imaging systems that are optimized for use at low light levels are best suited for imaging with genetically encoded probes.

The authors typically use standard epifluorescence illumination to image optical signals from the surface of the brain in vivo. The system is built on the optics illumination turret of an Olympus BX51WI microscope. The illumination turret is hung from an X95 optical rail so that the entire underlying area is open. This design has proved extremely versatile, in that it allows for surgical setups, electrophysiological equipment, or even apparati for performing behavioral experiments to be positioned underneath the microscope during imaging. Images of two different configurations of this setup are shown in Figure 1.4. Custom-built "macroscopes" may also be used for epifluorescence imaging—typically, at lower cost.[14,36] However, using the standard Olympus illumination turret provides a high degree of flexibility, allowing easy interchange of filter cubes, objectives, and other components, as well as straightforward visual access to the preparation via the eyepieces. The ability to switch between filter cubes for imaging the GFP-based genetically encoded probes versus longer-wavelength synthetic indicators has been extremely useful when initially characterizing signals from novel probes, for example.

FIGURE 1.4 Examples of microscope system used for in vivo imaging in acute, chronic, and awake preparations. (left) The imaging setup configured for acute and chronic imaging in anesthetized animals. The system is built around an optics illumination turret of an Olympus BX51WI microscope that is hung from a horizontal X95 optical rail, leaving the underlying area open for the animal and apparatus for stimulus presentation and electrophysiology. The system is shown with the most commonly used 4× macro objective from Olympus (0.28 numerical aperture, 30 mm working distance). Focusing is achieved with a manual focusing ring (LFM-1, Newport). Custom adapters couple the focusing ring to the objective and allow for different objectives to be used while retaining approximate parfocality. The animal preparation, including custom headholder and heating pad, is mounted on translation stages for XY positioning and a vertical linear stage (MVN80, Newport) for rough height adjustment. The stages are robust enough to accommodate a stereotaxic frame and/or micromanipulators for electrophysiology. This entire assembly is mounted on an optical rail and is slid forward of the microscope for the initial surgical procedures. A stereomicroscope can be swung into place for the surgery (not shown). This arrangement minimizes difficulty in arranging the preparation for imaging. For chronic imaging, alignment rods connect the microscope, camera and preparation for precise repositioning of the preparation at different time points (not shown). (right) Configuration used for awake, head-fixed imaging. The microscope and camera setup is identical. A custom head-fixation bar and restraint chamber is positioned below the microscope along with additional apparatus for behavioral training and monitoring. This assembly is mounted on XY translation stages and a vertical jack (Model 281, Newport) for positioning, and is surrounded by a custom enclosure that allows access for the objective. Focusing is achieved manually from outside of the enclosure.

It is important that the light source used for imaging be low noise ("flutter" of less than 0.1%), especially at frequencies in the range of 1–20 Hz, as the kinetics of many of the stimulus-evoked optical signals recorded in vivo occur at this time-scale. We use a 150-W Xenon arc lamp from Opti-Quip and 150-W short-arc Xenon bulbs or a 150-W Halogen lamp (Moritex). These light sources are typically very stable, although fluctuations at 1–2 Hz and up to several percent of basal intensity can occur if power to the lamp is not properly adjusted or the bulb has aged. LED-based light sources also offer low noise and obviate the need for a shutter; these are now available from several manufacturers (e.g., Cairn Research, United Kingdom). We have found that these produce sufficient light output for use in most in vivo systems. A disadvantage with LED-based sources, however, is that switching to different excitation wavelengths for imaging different probes in the same preparation can be awkward, and only a limited range of wavelengths are available.

A low-noise CCD camera is essential for imaging signals from genetically encoded probes due to the low resting fluorescence of most probes in vivo. We use a 256 × 256 pixel, back-illuminated CCD camera from RedShirt Imaging (NeuroCCD, SM-256) or a cooled, 1600 × 1200 pixel Sensicam (PCO); both cameras have a dynamic range of at least 14 bits and a read noise of less than 30 e- at their highest frame rates (100 Hz and 30 Hz, respectively). In our experience, the pixel resolution of the CCD camera need not be high because light-scattering and biological noise in the mammalian brain limits spatial resolution with epifluorescence imaging. For most in vivo applications, frame rates of 10–100 Hz are sufficient to capture the dynamics of neuronal population activity as reflected by calcium-sensitive probes or by spH. Typical frame rates in our experiments are 25 Hz for GCaMP2 imaging and 7 Hz for spH imaging. The emergence of genetically encoded voltage sensors will hopefully allow for neuronal dynamics to be imaged at a higher temporal resolution, necessitating higher frame rates.

SpH, GCaMP2, and other genetically encoded calcium sensors are also amenable to multiphoton imaging in vivo.[14,32] Most two-photon laser scanning confocal microscopes suitable for in vivo imaging have been custom built to optimize power throughput, light collection, and scanning flexibility. Such a system is described elsewhere in this volume. In our experience it has proved extremely useful to have an epifluorescence/CCD-based imaging system in optical alignment with a two-photon scanning system, so that a relatively large region of brain tissue can be imaged with epifluorescence followed by two-photon imaging in targeted areas within this region.[37] In this way, large-scale functional maps can be acquired in concert with imaging of the underlying architecture at cellular- or subcellular-level resolution.

1.4.2 Physical Setup for In Vivo Imaging

The physical setup for imaging optical signals from genetically encoded probes in vivo is similar to that for imaging with synthetic dyes, with a key concern being minimization of artifacts related to head movement, breathing, and heartbeat. In our experiments in anesthetized animals, we fix the head by attaching a small bar to the skull with cyanoacrylate and dental cement, and fix this bar to a custom holder;

the headholder is mounted to a ball joint to allow positioning of the head at oblique angles—a feature that is useful for imaging more lateral brain areas.

Several steps can be taken to reduce heartbeat and breathing artifacts. If possible, it is desirable to leave the skull intact and image through bone that has been thinned to a depth of ~100 μm using a low-speed (<1000 rpm) dental drill. Placing Ringer's solution above the thinned bone renders it virtually transparent. If the bone and/or dura must be removed in order to perform microelectrode electrophysiology or pharmacology (for example), breathing artifacts can be reduced by elevating the head slightly above the body and by placing a coverslip atop the craniotomy window. We also commonly administer dexamethasone (2 mg/kg, SC) to reduce brain swelling in preparations requiring a craniotomy. Choice of anesthesia can affect the size of heartbeat and breathing artifacts; we have had the most success with pentobarbital, which does not cause an increase in cardiac output and generally suppresses respiratory volume. These effects, however, are accompanied by a narrow safety factor, and it is helpful to monitor EKG during the course of the experiment to detect early signs of hypoxia and to deliver oxygen if necessary.

Choice of anesthesia can also greatly affect the size of stimulus-evoked optical signals imaged with spH or calcium sensors, for several reasons. First, both evoked and spontaneous activity levels in the neurons expressing the reporter can be affected by anesthesia. Second, choice of anesthesia has a significant effect on the magnitude of intrinsic hemodynamic optical signals, the nature of which is discussed in more detail in the section below. In mice, stimulus-evoked hemodynamic signals imaged in the olfactory bulb are smallest with pentobarbital and isoflurane and largest with urethane anesthetics. Ketamine/xylazine has been undesirable as an anesthetic because of spontaneous hemodynamic signals that occur at roughly 0.5–1 Hz—a time course similar to that of many evoked signals seen with spH or GCaMP2.

1.4.3 Repeated, Chronic In Vivo Imaging

The maintained expression of genetically encoded optical probes in select neuronal populations makes these ideal tools for investigating stability and plasticity of brain activity by chronic imaging over long periods. Such imaging requires installation of an optical window over the brain region of interest that remains clear for as long as possible and does not damage the underlying tissue. It is possible to remove the overlying bone and cover the brain tissue with either agarose or silicone elastomer (e.g., Kwik-Sil, WPI) and seal this with a glass coverslip;[38,39] in our hands, however, the craniotomy remains stable and optically clear for only a fraction of windowed animals. An alternative is to thin the overlying bone and coat the thinned bone with cyanoacrylate; the cyanoacrylate renders the bone transparent and itself remains clear after hardening. With this approach, windows stay optically clear for 1–2 weeks, after which they show signs of bone regrowth and become opaque. The window can be rethinned or removed for imaging sessions at later time points. Regardless of the approach, it is important to confirm with independent methods that the windowing has not altered the underlying tissue. Figure 1.5 shows an example of odorant-evoked spH signals imaged from olfactory receptor neurons in a mouse

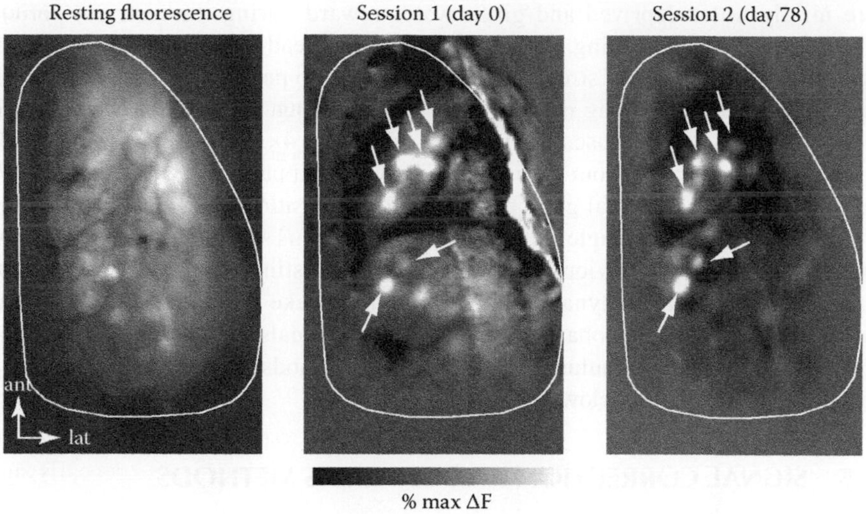

FIGURE 1.5 Long-term, chronic imaging with synaptopHluorin. (left) Resting fluorescence image of the dorsal surface of a mouse expressing spH in all olfactory receptor neurons, viewed through thinned bone. (center) Grayscale spH response map evoked by methyl valerate imaged at the time of chronic window installation. (right) Response map for the same odorant imaged 78 days later. The session 1 response map was from only a single trial and so had a smaller S/N ratio than the session 2 map. Nonetheless individual glomeruli are easily identified across sessions.

at the time of window installation and 78 days later, after rethinning of the window. The response maps remain stable over this time period.

1.4.4 Imaging in Awake, Head-Fixed Animals

Genetically encoded optical probes facilitate the use of imaging to monitor brain activity at high spatial and temporal resolution in awake, behaving animals. Because no labeling procedure is necessary, such imaging can be done without invasive dye-loading procedures that can disrupt behavior. Combining imaging in awake animals with repeated imaging through chronic optical windows has potential as a very powerful approach with which to investigate experience-dependent plasticity among genetically defined neuronal populations and the relationship of this activity to perception and decision making.

We have developed methods for imaging with conventional calcium-sensitive dyes in awake, head-fixed rats as they perform an operant (go/no-go) odor discrimination task;[40] the head-restraint paradigm is well established for both rats and mice[41,42] and can be easily adapted to accommodate imaging.[39] The method is essentially the same in rats or mice. Briefly, a headcap consisting of an attachment base and an inverted screw is chronically implanted onto the skull; for head-fixation, the screw is inserted through a female fitting attached to a small box or tube and the headcap tightened in place with a thumbnut. Mice or rats are mildly water deprived and given water reward during the restraint period; after 1–3 weeks of training, animals will remain head-fixed for 60–90 minutes without showing signs of stress and can be trained to perform operant behavioral tasks (typically by licking or bar press). For imaging, the fixation apparatus is positioned under a microscope objective (we use a 4×, 0.28 n.a., 30 mm working-distance objective from Olympus). The entire apparatus is mounted on XY translation plates and dual goniometers to adjust position and anterior-posterior/medial-lateral rotation angle for optimal imaging. This setup is effective at allowing imaging during behavioral responses to sensory stimuli with minimal motion artifact. Intrinsic hemodynamic signals in the awake animal tend to be more widespread than in pentobarbital-anesthetized animals and occur both spontaneously and during stimulus-evoked activity. Methods for correcting for these artifacts are discussed below.

1.5 SIGNAL CORRECTION AND ANALYSIS METHODS

Transforming optical signals imaged with genetically encoded probes into spatial maps and temporal response patterns reflecting the underlying neural activity requires a moderate degree of signal processing. The key signal processing steps—bleaching correction, elimination of biological artifacts, and data reduction—are the same as those used for data acquired with synthetic probes, though their implementation can differ depending on the nature of the signal generated by a particular probe. At the end of this section, we also discuss methods for estimating patterns of action potential firing from the optical signals obtained in vivo.

1.5.1 PHOTOBLEACHING

Imaging optical signals with epifluorescence optics typically requires some form of correction for bleaching of the probe during the course of the trial. If illumination intensities are low enough, appreciable bleaching can be avoided, obviating the need for correction. However, in our experience, the intensities needed to overcome shot-noise limitations are nearly always high enough to induce at least a moderate degree of bleaching, especially when imaging across relatively large (>1 mm^2) areas using lower-magnification (and thus lower numerical aperture) objectives. The degree of photobleaching can vary with the probe itself, its expression level, its depth in the tissue, and even the anesthetic state of the animal. With spH, using a 4× (0.28 n.a.) objective covering a field of view of ~3 × 3 mm and 25% transmission of light from a 150-W Xenon arc, we typically observe bleaching of 5–10% of resting fluorescence in a single 10-s trial.

Photobleaching effects with in vivo spH imaging are somewhat unusual compared with the behavior of synthetic fluorescent reporters. First, we find that resting spH fluorescence recovers with moderate speed after bleaching, returning nearly to pretrial levels after 90 seconds.[23] Whether such recovery occurs for other genetically encoded sensors (GCaMP2, for example) is unclear, but this may be expected if the sensor protein is being continually replaced by constitutive expression or transport/diffusion from other subcellular compartments. Second, because spH fluorescence is eclipsed at the acidic pH levels within presynaptic vesicles, vesicular spH absorbs little or none of the excitation light and thus is not subject to photobleaching. Instead, bleaching of resting fluorescence arises primarily from the relatively small fraction (estimated to be ~5%) of membrane-bound spH. Thus, a reduction in resting fluorescence induced by photobleaching may not correspond to a reduction in evoked signals.

While these two features slightly complicate corrections for photobleaching—at least when imaging with spH—in practice, we have found that a simple subtraction of a "blank" (i.e., no stimulus) trial is effective at removing bleaching artifacts (Figure 1.6A–C). One concern is limiting the number of blank trials given in an imaging session, especially because the typical length of most trials for collecting in vivo data is at least several seconds. Blank trials that are distributed too sparsely lead to poor correction, since bleaching rate can change slightly during the course of an imaging session. We typically collect a blank trial approximately every 6–10 stimulus trials. The blank signal is subtracted from each stimulus trial on a pixel-by-pixel basis. To reduce the additive effects of shot noise in the blank-subtracted data, the temporally low-pass filtered prior to subtraction.

1.5.2 Hemodynamic Artifacts

Hemoglobin is a strong absorber of photons at the emission and excitation wavelengths used for imaging GFP, which is the basis for all of the current generation of genetically encoded activity probes. As a result, optical interference from blood vessels is one of the most serious limitations to imaging with these sensors in the vertebrate brain in vivo. Large surface blood vessels can obscure signals from underlying tissue. More serious, however, are artifactual signals generated by blood vessel

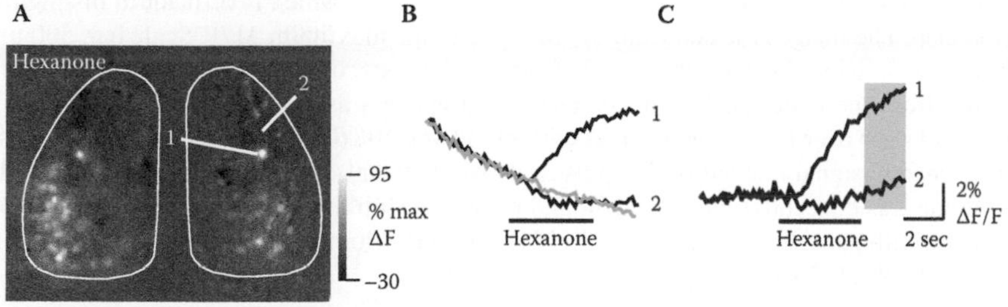

FIGURE 1.6 Bleaching and hemodynamic artifacts in fluorescence signals imaged in vivo. (A) Grayscale map of the change in fluorescence (ΔF) elicited by 2-hexanone (single trial) in a spH-expressing mouse, imaged from the dorsal olfactory bulbs through thinned bone. The response map is scaled from 95% to –30% of the maximal ΔF. Two regions from which the traces in (B) and (C) were derived are indicated. (B) Time course of the fluorescence decrease due to photobleaching (gray trace) taken from a no-stimulus trial, superimposed on the raw traces from two glomeruli (black traces) during a hexanone trial. (C) Time course of the odorant-evoked spH signal from the same two glomeruli after correction for photobleaching (subtraction of gray trace from panel B). Gray box demarcates the image frames used to create the map of activation in panel B. Data were collected at 7 Hz and are presented with no temporal filtering. (From Bozza, T., McGann, J.P., Mombaerts, P., and Wachowiak, M., *Neuron*, 42: 9–21, 2004.)

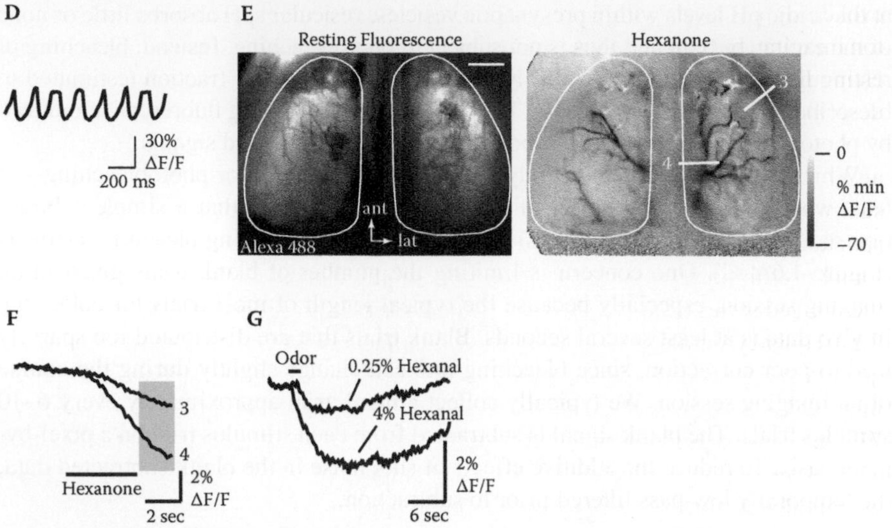

FIGURE 1.6 (continued) Bleaching and hemodynamic artifacts in fluorescence signals imaged in vivo. (D) Fluorescence signal imaged from the dorsal olfactory bulb of a mouse in which olfactory receptor neurons were loaded with Alexa Fluor 488 dextran and the overlying bone removed. Signal is the spatial average of a small (~100 μm) region adjacent to a blood vessel. The large signal fluctuations are due to blood vessel movement and brain pulsation associated with heartbeat. (E) Resting fluorescence and reponse maps on the dorsal olfactory bulbs after loading with Alexa Fluor 488 dextran as in (D). The bone is intact in this preparation. Outlines of the bulbs are indicated by white solid line. The optical signal evoked by hexanone reveals a strong decrease in fluorescence associated with blood vessels (seen as dark lines) and a smaller, spatially diffuse fluorescence decrease which is difficult to distinguish in the map. The image is scaled from 0% to –70% of the maximum ΔF/F. Scale bar, 500 μm. (F) Traces showing fluorescence signals from two regions, one from a blood vessel (4) and another from the underlying bulb (3). The signal appears as a slow fluorescence decrease which is larger over regions associated with blood vessels. (G) Time course of the intrinsic hemodynamic signal elicited by 2-s pulses of two different concentrations of hexanal. The amplitude was concentration dependent and showed a similar time course to the spH signal (compare with panel C). (From Bozza, T., McGann, J.P., Mombaerts, P., and Wachowiak, M., Neuron, 42: 9–21, 2004.)

movement. Heartbeat and respiration are one source of such artifacts, which appear most strongly at the edges of large vessels (Figure 1.6D). Another source, which can be more problematic, is activity-dependent changes in bulk blood flow to the tissue of interest; this often manifests as a widespread darkening of the tissue. This darkening is largest along major blood vessels, but also typically has a diffuse, widespread component (Figure 1.6E–G). A third source of hemodynamic artifact is spontaneous fluctuations in bulk flow that occur at frequencies of 0.2–1 Hz. All of these hemodynamic artifacts are apparent even in tissue labeled with nonactivity-dependent fluorescent markers such as Alexa Fluor 488 or GFP, and appear as fluorescence changes that can be as large as 10% of resting fluorescence—several-fold larger than evoked signals arising from the optical probe. Such artifacts are avoided in two-photon imaging, but can be a significant problem with standard epifluorescence imaging. Hemodynamic artifacts can be minimized with the correct choice of anesthesia, as described above.

Averaging signals across multiple trials can be useful in minimizing heartbeat and respiration artifact, as can triggering trial acquisition from the heartbeat and subtracting blank trials. This approach is of little help, however, in dealing with spontaneous and stimulus-evoked hemodynamic artifacts. An approach that we have found useful, however, involves applying a high-pass spatial filter to each frame of the imaged series (Figure 1.7). We use an algorithm that transforms each frame into the frequency domain and filters using a defined cutoff frequency; the cutoff frequency will vary depending on the spatial dimensions of the signal of interest. For imaging activity in glomeruli of the olfactory bulb that have an average diameter of 75–100 μm, we use a spatial cutoff frequency of 2.6 mm^{-1}; this operation results in a ~20% reduction in glomerular response amplitudes, but a significant increase in overall signal-to-noise ratio. A particular advantage of this spatial filtering strategy is that it filters all widespread signals, regardless of time course. It can thus reduce or eliminate artifacts from photobleaching, hemodynamic signals, and even lamp noise in a single processing step. This strategy, of course, relies on the spatial dimensions of the signal of interest (i.e., that arising from the optical probe) being smaller than that of the artifactual signal—a condition that may not be satisfied in all situations. A final step that can be useful in reducing artifact from more localized hemodynamic signals (near blood vessels, for example) involves manually selecting a background region with no stimulus-evoked signal and subtracting the signal in this region from that in a nearby region of interest (Figure 1.7C).

1.5.3 Green Autofluorescence Signals

Normal brain tissue exhibits "autofluorescence"; that is, fluorescence in the absence of an exogenous dye. The dominant source of green autofluorescence (in the wavelength range of GFP fluorescence excitation and emission) is the chromophore of FAD (flavin adenine dinucleotide). FAD fluorescence increases with neuronal activity and therefore green autofluorescence cannot only add background fluorescence (reducing signal-to-noise ratio) but also can contribute to stimulus-evoked optical signals recorded in the range of 500 to 600 nm. Indeed, such autofluorescence signals can be exploited as a modality of intrinsic signal imaging.[43,44] Thus, when establishing the use of FP reporter mice, it is critical to compare presumed FP signals imaged at wavelengths between 400 and 600 nm with those obtained in wild type control mice. Because evoked FAD-derived fluorescence signals typically have a slow time course, they can be easily separated from faster calcium transients monitored with probes such as GCaMP2.[14] However, contributions from autofluorescence signals must be carefully considered for slower signals with rise times greater than approximately 200 ms (such as spH; see Figure 1.6). Contributions from autofluorescence signals become less of a concern when imaging with probes having a relatively high

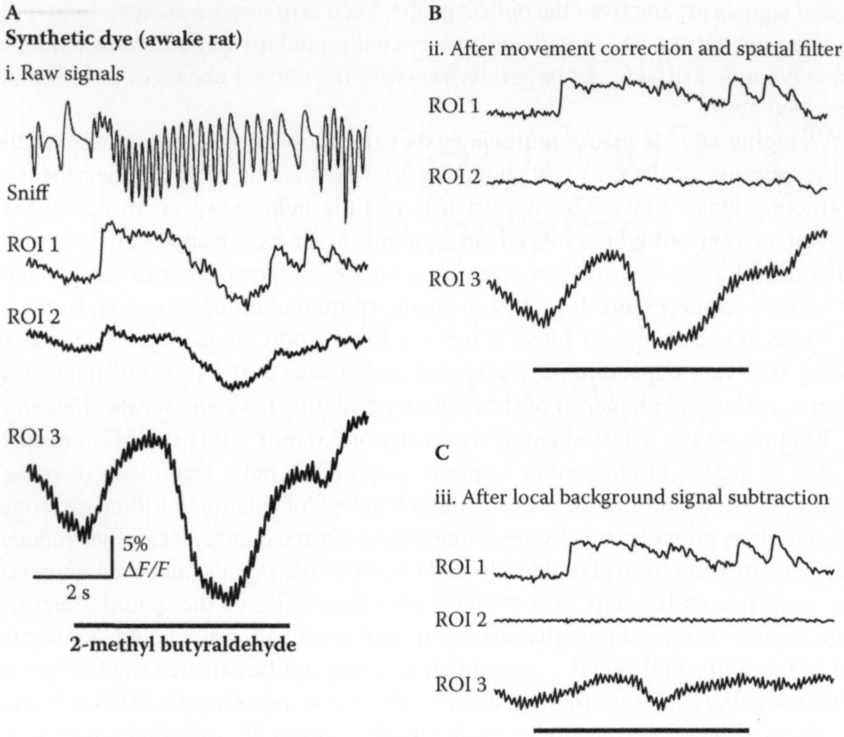

FIGURE 1.7 Strategies for correcting for movement and intrinsic hemodynamic fluorescence signals. (A) Fluorescence signals imaged from the olfactory bulb of an awake, head-fixed rat in which olfactory receptor neurons were loaded with a synthetic calcium-sensitive dye (Calcium Green-1 dextran). Top trace shows the rat's respiratory response ("sniff") to a novel odor stimulus that evokes high-frequency sniffing. Lower traces show raw fluorescence signals from three regions of interest (ROIs) on the dorsal bulb. Signals show slow fluctuations that occur before and during high-frequency sniffing. Smaller heartbeat artifacts are also apparent (e.g., ROI 3). There is no detectable movement artifact. ROI 1 is located over an odorant-activated glomerulus, ROI 2 is over an adjacent, nonactivated area, and ROI 3 is located over a blood vessel. Intrinsic signals and heartbeat artifacts are largest over blood vessels. (B) Same traces after implementing movement correction and high-pass spatial filtering algorithms (see text). The diffuse intrinsic signal and some heartbeat noise is removed. Large intrinsic signals remain over blood vessels. (C) Traces after subtracting signal from "surround" areas adjacent to each ROI to correct for localized intrinsic signals. ROI 2 is used as the "surround" for ROI 1 and for itself. "Surrounds" for ROI 3 is not shown. (From Verhagen, J.V. et al. *Nat Neurosci*, 10: 631–639, 2007.)

resting fluorescence. For example, when imaging odorant-evoked spH signals from the olfactory bulb of OMP-spH mice, resting fluorescence is 4–6 times higher than in wild-type mice; the evoked fractional fluorescence change is thus dominated by a signal from an spH signal rather than from autofluorescence. We have also found that odorant-evoked autofluorescence signals in the olfactory bulb (in wild-type mice) are much less localized than spH or GCaMP2 signals.

1.5.4 Movement Correction

With increasing interest in imaging from awake animals with conventional or genetically encoded optical probes, algorithms for offline correction of movement are becoming increasingly important. Several strategies can be implemented to correct for movement artifacts. For optical signals collected using wide-field epifluorescence, a cross-correlation-based algorithm has proved effective at correcting for relatively small movements that occur largely in the XY plane. First, a reference frame for each trial is constructed by averaging the first five imaged frames. Each subsequent frame is then high-pass filtered by subtraction with a morphologically opened frame (disk radius equals 10 pixels) and upsampled by a factor of 4 via bilinear interpolation; upsampling is necessary to correct for artifacts from movements of less than a full pixel. This step is not necessary if the original pixel resolution is high. Each upsampled frame is then cross-correlated with the reference frame; if the peak of this cross-correlation is different from the origin (indicating movement), the original frame is upsampled further (to 8× original resolution), then shifted according to the peak of its crosscorrelation with the reference. The shifted frame is then down-sampled to the original resolution.

More complex movement-correction algorithms can be implemented to correct for more serious motion artifact. For example, optical signals imaged via two-photon imaging from awake, head-fixed mice are subject to movement both within and out of the plane of focus, and the higher resolution from two-photon imaging presents a more serious challenge for movement correction. To deal with this problem, hidden Markov model-based correction algorithms have been developed that are reasonably effective at correcting for movement.[39]

1.5.5 Estimating Action Potential Firing from Optical Signals

A major goal of imaging from the nervous system with optical probes is to record patterns of action potential firing. However, most genetically encoded probes used in vivo report changes in intracellular calcium concentration, which indirectly reflect action potential firing due to calcium influx through voltage-activated calcium channels. Several methods have been devised for estimating action potential firing patterns from calcium signals. The simplest method involves taking the derivative of the fluorescence signal and thresholding it to identify rapid increases in fluorescence that correspond to single spikes.[45] This approach has proved effective with high-affinity synthetic calcium indicators and in situations when individual cell bodies are visible.

A second approach for estimating action potential firing from calcium signals is not to detect single spikes, but instead to estimate changes in the relative frequency of firing of the imaged neurons by temporal deconvolution of the optical signal.[46] This approach relies on the assumption that the imaged signal is a linear convolution of the optical signal evoked by a single action potential (i.e., the impulse response) and the pattern of action potential firing. Under such conditions, the original firing pattern can be recovered from the optical signal by deconvolution with this impulse response. The impulse response can, in practice, be characterized simply as the decay rate of the calcium signal evoked by a single spike (or even burst of spikes). This decay rate can be determined by imaging from a single cell while evoking controlled spike responses via electrical stimulation or an intracellular recording electrode or while evoking responses to a white-noise-like stimulus.[47] In theory, this decay rate should be determined for each imaged neuron, which would defeat the purpose of using optical imaging as a tool for recording from many neurons. Under the appropriate conditions, however, the deconvolution method is robust enough that small differences in decay rate from neuron to neuron have little effect on estimated firing rate changes, so that the decay need only be determined for a few representative cells.

Temporal deconvolution has proved useful in estimating firing rate changes in multiple neurons imaged simultaneously using synthetic calcium indicators.[46,47] We have also used this approach to estimate temporal patterns of action potential firing at the population level, with the deconvolution applied to each group of receptor neurons converging onto a single glomerulus in the rat olfactory bulb.[40] To date, temporal deconvolution has not been applied to signals imaged with genetically encoded calcium indicators. While there are no theoretical reasons why this approach should not be applicable, limitations of the deconvolution procedure become more serious in the case of the current generation of calcium sensors. First, the response time of these sensors is somewhat slower than that of most synthetic probes due to their different mechanism of action, making it more difficult to distinguish spike-evoked calcium transients based on rise-time. Second, temporal deconvolution is valid only when the onset and decay of the optical signal varies linearly with firing rate, and there are many situations where this may not be the case—for example, if the probe nears saturation during periods of high-frequency firing, if the probe has a high threshold for detecting firing rate changes, or if intracellular calcium concentration changes independent of spiking. Most genetically encoded calcium indicators characterized to date do saturate at lower spike rates than do synthetic indicators, and are less reliable at reporting calcium influx evoked by single action potentials. Instead, a burst of several spikes occurring at relatively high-frequency is often required to evoke a detectable signal.[15] Future improvements in the design and expression of genetically encoded calcium sensors may overcome some of these limitations.

1.6 EXAMPLE EXPERIMENTS

1.6.1 SpH Imaging of Odorant-Evoked Activity and Presynaptic Modulation of Transmitter Release

SpH has proven a sensitive and reliable reporter of stimulus-evoked activity in mouse olfactory receptor neurons.[23,48,49] Because of its mechanism of action, this probe has proved particularly useful for investigating a question of long-standing interest in olfaction: the presynaptic modulation of transmitter release from receptor neurons. Here, we present example experiments in which we use spH imaging to investigate the functional organization of this inhibition in the glomerular layer and its role in shaping odor representations in vivo. While the focus is on in vivo imaging, we have found it extremely useful to complement in vivo experiments with recordings in olfactory bulb slice preparations, which allow for more controlled stimulation paradigms and more reliable pharmacological manipulations. The slice experiments have also proved indispensable for characterizing the nature of the optical signal reported by spH; examples of these experiments are also presented here. A similar approach is used for characterizing genetically encoded optical probes that report calcium or voltage.

1.6.1.1 SpH as a Reporter of Transmitter Release

In other systems, spH has been used to track the cycling of transmitter vesicles, typically by imaging with relatively low temporal resolution and using prolonged trains of presynaptic action potentials to drive vesicle fusion and subsequent transmitter release.[19, 21, 24] However, under the right experimental conditions, spH is also an effective reporter of transmitter release (and hence synaptic transmission) when imaged at high temporal resolution and using single, brief electrical stimuli in olfactory bulb slices to elicit one or a few presynaptic action potentials. An example of one such experiment is shown in Figure 1.8. A single (0.1 ms) shock delivered to the axons of ORNs expressing spH evokes a fluorescence increase in two discrete glomeruli. The optical signal is confined to the glomerulus and is not detectable in the axons, consistent with the optical signal deriving from transmitter release at the synapse (Figure 1.8A). The shock-evoked signal has a rapid rise and a slow decay, with a time-constant of approximately 30 s (Figure 1.8B). The magnitude of the shock-evoked spH signal can be modulated by changing external calcium concentration (Figure 1.8C) or by blocking presynaptic voltage-sensitive calcium channels pharmacologically (not shown). Importantly, the dependence of the shock-evoked spH response amplitude on external [Ca^{2+}] is nearly identical to that of EPSCs in external tufted cells, which are known to receive excitatory, monosynaptic input from ORNs (Figure 1.8D). Thus, spH is capable of reporting graded changes in the probability of transmitter release across the population of ORNs terminating with a single glomerulus.

Nonetheless, high temporal resolution imaging of shock-evoked spH signals reveals that the rise-time of the response (approximately 30 ms) is considerably slower than the influx of calcium into the presynaptic terminal, which can be imaged in the same experiment using synthetic, long-wavelength calcium-sensitive dyes loaded into ORNs (Figure 1.9A). In addition, close inspection of the evoked spH signal reveals a small and brief *decrease* in fluorescence whose origin is unclear (Figure 1.9B). The explanation for both the prolonged rise-time and the initial fluorescence lies in the mechanism of action of spH, which is to report changes in pH surrounding the fluorophore, regardless of whether or not these occur as a result of vesicle cycling. Because the majority of the spH resting fluorescence arises from the small fraction of protein present in the plasma membrane and *not* sequestered inside vesicles, even small changes in extracellular pH can result in significant fluorescence changes that do not directly reflect transmitter release. In the case of olfactory nerve stimulation, nerve shock elicits synchronous release from many (up to several thousand) axons in the glomerulus, which can transiently acidify the extracellular space. Indeed, increasing the buffering capacity of the extracellular solution by adding 20 mM HEPES eliminates the transient fluorescence decrease (Figure 1.9C). Likewise, activation of glutamate transporters after the response can

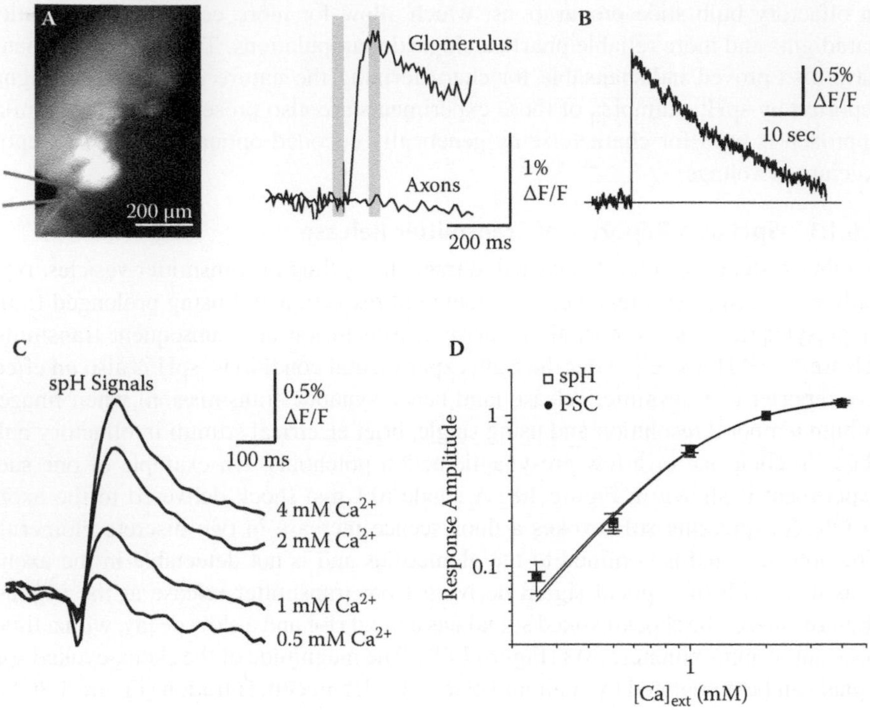

FIGURE 1.8 SynaptopHluorin as a reporter of stimulus-evoked transmitter release: in vitro validation. (A) Resting fluorescence and optical signals imaged from an olfactory bulb slice of a mouse expressing spH in olfactory receptor neurons. The position of the stimulating electrode is indicated in gray. Traces show the stimulus-evoked spH signal measured from the bright glomerulus. Traces are unfiltered signals from a single trial, imaged at 125 Hz. "Flat" trace shows the signal from pixels overlying the axon, averaged across eight trials and low-pass filtered at 20 Hz. Thin vertical line indicates time of nerve shock. (B) Longer record, obtained from a different preparation, demonstrating the slow decay of the evoked spH signal. Trace is average of four trials, imaged at 40 Hz. (C) Shock-evoked spH signals recorded from a glomerulus under different external Ca^{2+} concentrations ($[Ca^{2+}]_{ext}$). Each trace is the average of four trials and is bleach-subtracted and low-pass filtered at 15 Hz. (D) Plots of spH signal (open squares) and external tufted cell EPSC amplitude (closed circles, recorded from different preparations) as a function of $[Ca^{2+}]$ext. Response amplitudes are normalized to the amplitude at 2 mM $[Ca^{2+}{}_{ext}]$. Error bars are s.e.m. Solid curves show fits of each dataset to the Hill equation. The fits are nearly identical for the spH and EPSC recordings. ([A,B] From McGann, J.P. et al., Neuron, 48: 1039–1053, 2005; [C,D] from Wachowiak, M. et al., J Neurophysiol, 94: 2700–2712, 2005.)

transiently alkalinize the extracellular space because protons are cotransported during glutamate reuptake,[50] causing a slower transient fluorescence *increase*. Blocking glutamate reuptake with TBOA eliminates this increase (Figure 1.9D). Finally, shock-evoked spH signals measured in the presence of both 20 mM HEPES and TBOA reveal fluorescence increases that are monophasic and have a rise time of less than 10 ms, as expected for transmitter release (Figure 1.9E). These experiments

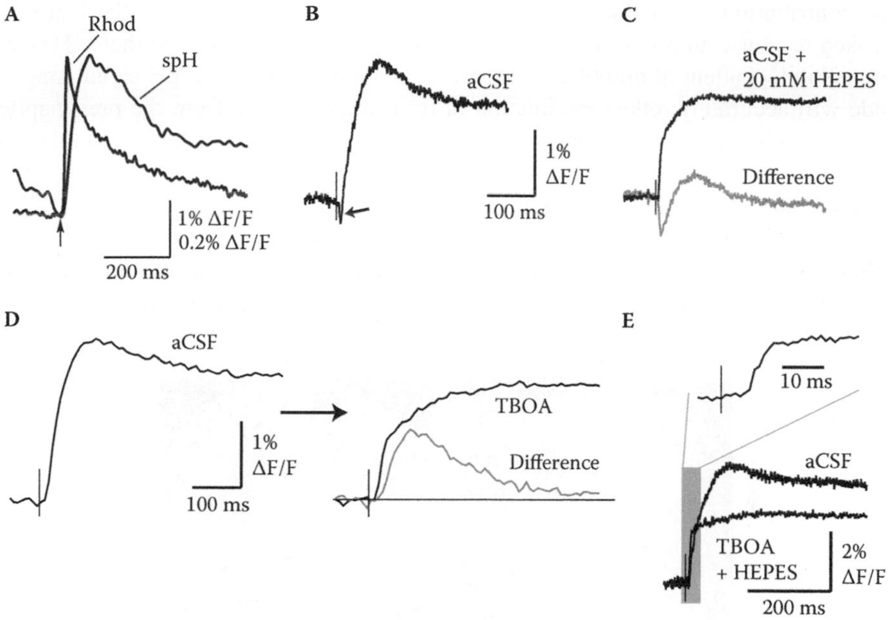

FIGURE 1.9 pH-dependent artifacts affecting stimulus-evoked spH fluorescence signals. (A) Traces showing "high-affinity" rhod dextran signals reflecting shock-evoked presynaptic calcium influx (rhod) and spH signals reflecting transmitter release (spH) acquired in alternate trials from the same glomerulus in an olfactory bulb slice preparation. Both traces are single trials. The spH signal is low-pass filtered at 15 Hz and the rhod signal is low-pass filtered at 50 Hz. The rise-time of the rhod signal is faster than that of spH but the onset times are similar. (B) Trace of the shock-evoked spH signal imaged at 1 kHz frame rate reveals complex signal kinetics. The response begins with a small and rapid decrease in fluorescence (arrow), followed by a slower rise and an initial decay with time constant of ~90 ms. Trace is average of four trials and is unfiltered. (C) spH signal measured from the same glomerulus as in (B) after switching to aCSF containing 20 mM HEPES. The initial fluorescence decrease is eliminated and the rise-time is faster. The difference between the traces in (B) and (C) is shown in gray and reveals the buffering-dependent component of the evoked signal, which consists of a rapid acidification of the synaptic cleft followed by a slower alkalization. (D) A slower alkalization is mediated by glutamate transporters. Left trace shows evoked spH signal imaged in control aCSF (four trial average, imaged at 125 Hz). Right trace shows response in 50 μM TBOA (black) and difference between control and TBOA traces (gray). The alkalization is likely due to cotransport of protons by the glutamate transporter. (E) Shock-evoked spH signals show a rapid and simple fluorescence increase in 20 mM HEPES-buffered aCSF and TBOA (red) without the complex kinetics seen in control aCSF (black). Traces are four trial averages imaged at 2 kHz and unfiltered. Inset (above) shows the evoked signal in HEPES and TBOA on an expanded time-scale. The rise-time of the spH signal is <10 ms. ([A] From Wachowiak, M. et al., J Neurophysiol, 94: 2700–2712, 2005; [B–E] from McGann, J.P. et al., Neuron, 48: 1039–1053, 2005.)

illustrate the importance of thoroughly characterizing the nature of the optical signal being generated by genetically encoded optical probes, especially those that are not direct reporters of membrane potential. Nonetheless, in the case of spH, despite the contribution of "artifactual" sources to the evoked optical signal, the fact that evoked response amplitudes scale linearly with EPSC amplitudes and that spH rise-time is independent of amplitude suggest that changes in evoked spH signal amplitude will accurately reflect modulation in transmitter release from the presynaptic terminal. This relationship appears to approximately hold using odorant stimulation in vivo, as evidenced by a rough correspondence between maps of odorant-evoked spH signals and those imaged with calcium-sensitive dye in the same experiment (Figure 1.10). However, we have not tested the degree to which pH-dependence or other effects may distort the dynamics of responses evoked during odorant stimulation. Thus, in our in vivo experiments, we have limited our interpretation of

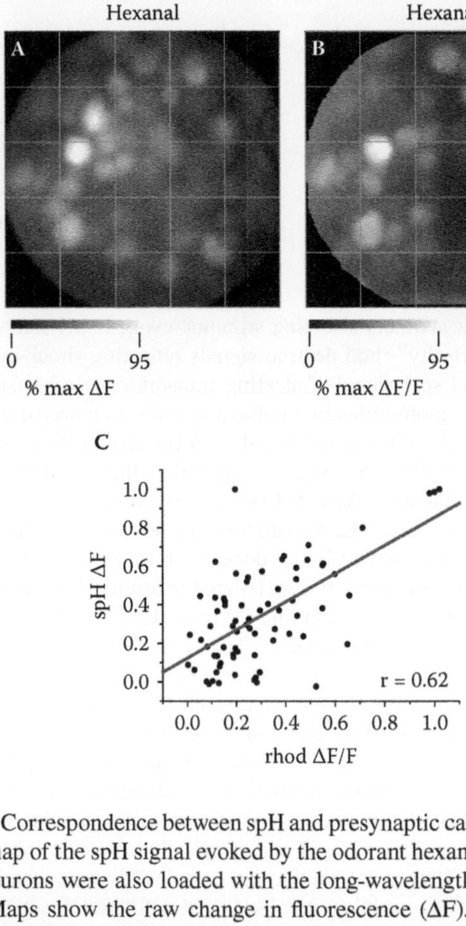

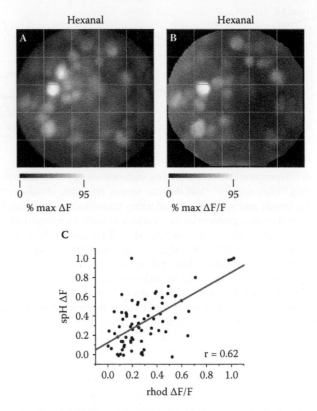

FIGURE 1.10 Correspondence between spH and presynaptic calcium signals imaged in vivo. (A) Grayscale map of the spH signal evoked by the odorant hexanal in a mouse in which olfactory receptor neurons were also loaded with the long-wavelength synthetic calcium indicator rhod dextran. Maps show the raw change in fluorescence (ΔF). Map is normalized to 95% of its maximal ΔF. The overlying grid registers the two images. (B) Response map of odorant-evoked signal using rhod dextran (imaged in a different trial with different optical filter settings), measured as fractional change in fluorescence (ΔF/F). (C) Correlation between the relative amplitudes of the spH and rhod dextran responses in rhod-labeled glomeruli, collected from three preparations. Responses were normalized to the maximal signal amplitude in each preparation before performing the correlation. The correlation coefficient is 0.62. (From Bozza, T., McGann, J.P., Mombaerts, P., and Wachowiak, M. *Neuron*, 42: 9–21, 2004.)

spH signals to focus on modulation of the response by drug application or odorant-evoked inputs to neighboring glomeruli.

1.6.1.2 Using SpH to Probe the Organization of Presynaptic Inhibition In Vivo

We have used spH imaging to probe the organization of presynaptic inhibition in the glomerular layer of the olfactory bulb (OB) and its role in shaping odor representations in vivo. Experiments performed in OB slice preparations using paired electrical stimulation of ORN axons reveal a strong paired-pulse suppression of transmitter release that is partially relieved by blocking ionotropic glutamatergic receptors.[49] These experiments demonstrate that ORNs—which are glutamatergic—activate postsynaptic olfactory bulb neurons that then presynaptically inhibit release in response to subsequent inputs. GABA-ergic interneurons in the glomerular layer play a major role in this inhibition: $GABA_B$ receptors are expressed on the presynaptic terminals of ORNs, and $GABA_B$ antagonists are as effective as glutamatergic antagonists at relieving paired-pulse suppression.[48,49,51] To investigate the role of this $GABA_B$-mediated presynaptic inhibition in odor coding, we have performed several types of experiments using spH signals as measures of odorant-evoked transmitter release from ORNs in vivo.[48] In the first, we applied a $GABA_B$ antagonist (CGP35348) directly to the dorsal surface of the mouse OB while imaging odorant-evoked spH signals. In these experiments, mice were anesthetized with pentobarbital and the dura removed from the dorsal OB prior to data collection. Dura removal (when done carefully) has no negative impact on imaged responses, and in fact slightly improves the quality of imaged response maps. SpH signals remain stable or decrease slightly when control-mouse Ringer's is applied to the OB. In contrast, when $GABA_B$ agonist baclofen is applied, odorant-evoked spH signals are greatly decreased (Figure 1.11A). The effect of baclofen can be reversed by applying CGP35348, a $GABA_B$ receptor antagonist. Applying CGP35348 alone causes a strong increase in odorant-evoked spH signals, with an average increase of approximately 80% in ΔF after 4 s of odorant presentation (Figure 1.11B). This indicates that presynaptic inhibition mediated by $GABA_B$ receptors regulates the magnitude of sensory input to the olfactory bulb.

The localization of the spH signal to the sites of transmitter release (i.e., within glomeruli) allowed us to compare overall spatial patterns of ORN input to glomeruli before and after $GABA_B$ receptor blockade. Normalizing evoked response maps to their peak response magnitude indicates that relative patterns are unchanged by CGP35348—an observation supported by correlation analyses of response maps.[48] In other words, blocking presynaptic inhibition does not recruit input to glomeruli not initially activated by the odorant. This result also provides evidence that the odorant-evoked spH signal responds approximately linearly across much of the dynamic range of receptor inputs to glomeruli; if the spH signal had a significant threshold for reporting presynaptic activity or if it saturated at relatively low levels of activation, increasing transmitter release with CGP35348 would be expected to significantly alter relative response maps imaged with the probe. To more directly test whether presynaptic inhibition alters relative patterns of input to glomeruli—for example, via lateral inhibition between glomeruli—we used odorants chosen to selectively

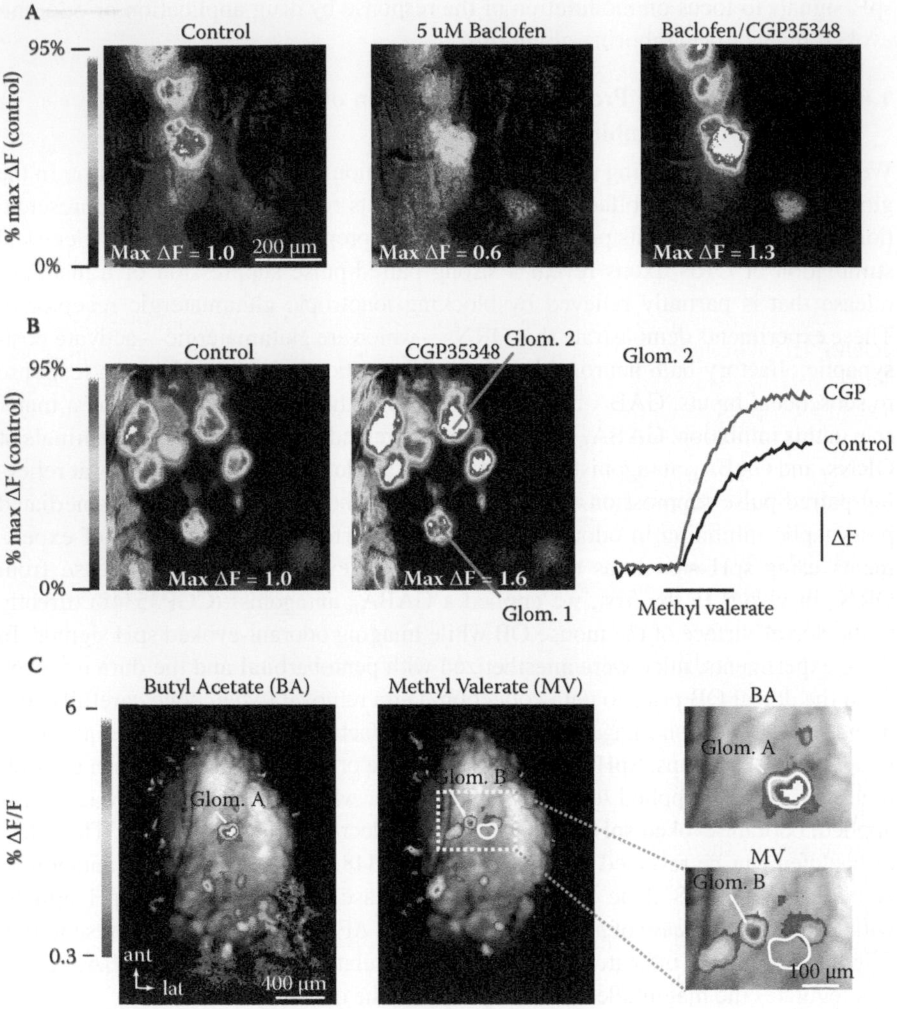

FIGURE 1.11 (See color insert following page 224) Imaging the in vivo regulation of receptor input to the olfactory bulb by $GABA_B$-mediated presynaptic inhibition. (A) Pseudocolor maps of the spH signal evoked by the odorant butyl acetate. The $GABA_B$ receptor agonist baclofen reduces the amplitude of the odorant-evoked response (middle panel) compared to control levels (left panel). The reduction was reversed by coapplication with the $GABA_B$ receptor antagonist CGP35348 (1 mM; right panel). All maps scaled to the maximum of the control response. (B) Maps of the spH signal evoked by methyl valerate before (left) and after (right) application of CGP35348. Maps are scaled to the maximum of the control response. While CGP increased the amplitude of the odorant-evoked responses, the set of responsive glomeruli did not change. Traces (right) show the effect of CGP35348 on the odorant-evoked spH signal in one glomerulus. (C) Pseudocolor map of the spH signal evoked by butyl acetate (BA) or methyl valerate (MV), overlaid on an image of the resting fluorescence of the dorsal olfactory bulb (different preparation from above). Each map is normalized to its own maximum. The region in the dashed box is enlarged in the insets (right). (From McGann, J.P. et al. *Neuron*, 48: 1039–1053, 2005.)

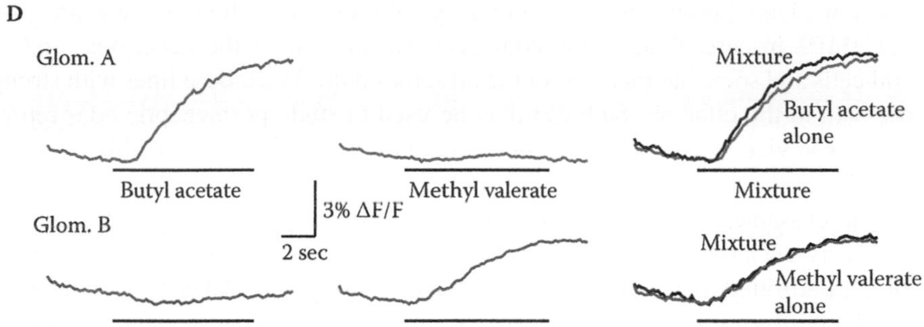

FIGURE 1.11 (continued) Imaging the in vivo regulation of receptor input to the olfactory bulb by $GABA_B$-mediated presynaptic inhibition. (D) Traces showing the spH signal evoked in glomeruli A (top) and B (bottom) by each odorant. Each glomerulus is selectively activated by one of the two odorants. For each glomerulus, the response to the binary mixture is no different from its response to the single component presented alone. (From McGann, J.P. et al. *Neuron*, 48: 1039–1053, 2005.)

activate adjacent glomeruli. The rationale for this experiment is that if lateral interglomerular inhibition modulates ORN input in vivo, then the odorant-evoked spH signal in a glomerulus should be suppressed when its neighbor is coactivated by a different odorant. An example experiment is shown in Figure 1.11C. In this example, butyl acetate evokes strong input to a single glomerulus (the "test glomerulus") in the central dorsal OB and weaker input to additional glomeruli. A different odorant, methyl valerate, evokes strong input to three glomeruli surrounding the test glomerulus, but evokes little or no input to the test glomerulus itself. We then asked whether coactivation of the surrounding glomeruli by methyl valerate suppressed the response of the test glomerulus to butyl acetate by presenting the two odorants as a binary mixture. As shown in Figure 1.11D, the test glomerulus's response to the mixture was identical in both amplitude and time course to the response to butyl acetate alone. These experiments suggest that a major role of presynaptic inhibition in vivo is to regulate the strength of sensory input to the brain while maintaining the spatial maps of glomerular activation that are thought to encode odorant identity. However, several questions about the role of presynaptic inhibition in vivo are not easily addressed using spH—chiefly because of its limited temporal resolution. For example, it is unclear from these experiments whether inputs are modulated by feedback inhibition—which is apparent in slices—or via tonic inhibition under the control of (for example) centrifugal inputs. Presynaptic indicators with higher temporal resolution—such as genetically encoded or synthetic calcium indicators—would be useful in addressing this question.

1.6.2 Imaging Postsynaptic Odor Representations with GCaMP2

To date there have been few examples of genetically encoded calcium sensors used to report neural activity in the in vivo mammalian brain. One successful example is a line of transgenic mice expressing GCaMP2 under the control of the Kv 3.1

potassium channel promoter.[13,14] Certain lines of these mice show stable expression of GCaMP2 in cerebellum, pyramidal neurons throughout the neocortex, and in mitral cells and some interneurons of the olfactory bulb. The mouse lines with strong expression in the olfactory bulb can thus be used to study postsynaptic odor representations in vivo.[32] Here, we show an example of odorant-evoked GCaMP2 signals imaged from the dorsal OB in a pentobarbital-anesthetized mouse (Figure 1.12).

In this example, odorants were applied while artificially "sniffing" air through the nasal cavity via a tube inserted into the nasopharynx (a second tube directed toward the lungs permitted free respiration). Note that the resting fluorescence does not reveal clear glomeruli (Figure 1.12A), as is evident in the OMP-spH mice. This is partially due to the low fluorescence of GCaMP2 at resting calcium concentration and also to the expression in some interneurons whose processes cross glomerular boundaries. Odorant presentation evokes optical signals that have a relatively rapid rise-time as well as a relatively rapid decay; these temporal dynamics are fast enough to allow responses to each individual inhalation to be tracked—at least when presented at 1 Hz (Figure 1.12B). Thus, the temporal resolution of the GCaMP2 signal is higher than that of the spH signal and appears similar to that of synthetic calcium-sensitive dyes. Spatial maps of the evoked signal are spatially organized and appear to consist of discrete glomeruli in locations expected from separate imaging experiments using

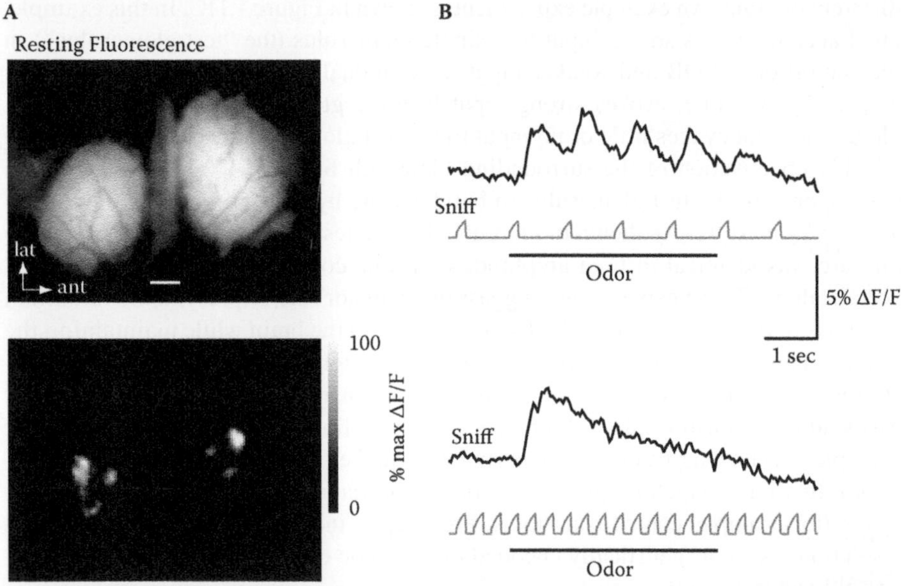

FIGURE 1.12 Postsynaptic odorant-evoked GCaMP2 signals imaged from the olfactory bulb in vivo. (A) Top: Resting fluorescence from the dorsal olfactory bulbs of a mouse expressing GCaMP2 under the control of the Kv 3.1 promoter. Bottom: Grayscale response map evoked by the odorant ethyl butyrate. (B) Optical signal from one glomerulus showing responses evoked during 1 Hz (top) and 3 Hz (bottom) artificial inhalation. The GCaMP2 signals show rapid-onset responses to each sniff of odorant at low inhalation frequencies but show attenuation at higher frequencies.

spH or synthetic dye imaging from ORNs. Finally, responses evoked during high-frequency sniffing of odorant show a lack of modulation by each sniff and instead show clear attenuation of the response during prolonged stimulation (Figure 1.12B). This frequency-dependent attenuation is also consistent with our earlier observations imaging presynaptic calcium signals.[40,52] Thus, GCaMP2 imaging allows the testing hypotheses about how sensory inputs are transformed by synaptic processing. It will, of course, be extremely interesting to combine the postsynaptic GCaMP2 imaging with imaging of presynaptic activity in the same preparation—either using spH or longer-wavelength, synthetic calcium-sensitive dye loaded into ORNs.

1.7 DISCUSSION AND FUTURE DIRECTIONS

The properties of the current generation of the major types of genetically encoded FP-based probes that are most advanced for in vivo applications are summarized in Table 1.1. The spH probes represent the only class that is currently more suitable for in vivo imaging of activity than its conventional ("synthetic," not genetically encoded) analog (FM dyes—the synthetic analog of spH—are not suitable for in vivo experimentation). Synthetic versions of the other two classes—calcium and voltage sensors—have undergone many more person-years of development, and currently surpass genetically encoded sensors in signal-to-noise ratio, response times, linear dynamic range, range of affinities, and range of available imaging wavelengths. In particular, small molecule synthetic calcium indicators are readily applicable in vivo and continue to facilitate many invaluable insights into nervous system function. Nonetheless, genetically encoded activity calcium-sensitive FPs uniquely enable experiments in which cell type–specific and long-lasting imaging approaches are essential. Thus, continued effort in improving probe performance, as well as in developing expression strategies, should remain a high priority.

VSFPs have not been sufficiently explored in vivo to directly compare them with available conventional voltage imaging methods. Because unselective labeling of

TABLE 1.1
Functional Characteristics of Genetically Encoded FP Sensors

		Time Constant		Saturating ΔF/F (ΔR/R)			
Parameter	**Probe**	**ON [ms]**	**OFF [ms]**	**Epifluorescence**	**1P LSM**	**2P LSM**	**Ref.**
Synaptic release	spH	10 ms	~50,000	N/A	—	—	48
Calcium	GCaMP2	40 ms	200–250	3–10%	50–80%	50–100%	14
	CerTN-L15	40 ms?	500–1000	—	—	Up to 65%	12
Voltage	VSFP2.1	20	75	6% (ΔR/R)	—	—	30
	VSFP3.1	4 ms	—	4% (ΔR/R)	—	—	31

Note: Values are at physiological temperature (in vivo) or 35°C (in vitro).

plasma membranes by synthetic voltage probes has prohibitive effects on both the signal to noise ratio and interpretation of signals, VSFPs have the potential to overcome many of the current limitations of voltage imaging based on organic voltage sensitive dyes. Because current VSFPs have relatively modest AF/F values, however, and because S/N_R is directly proportional to $\Delta F/F$, improving this parameter will greatly enhance the prospects of these probes.

One area of improvement that has received relatively little effort to date is the development of longer-wavelength FP probes. The main advantages of excitation with and emission of longer wavelength light are (1) reduced interference with blue and green fluorescence signals from endogenous chromophores (NAD, FAD), (2) reduced interference from activity dependent changes in hemoglobin and cytochrome oxidase absorption, and (3) improved tissue penetration due to an increased "mean free path" (the average distance between scattering events).[53] The increase in signal-to-noise ratio when imaging with longer-wavelength probes in vivo in the mammalian brain—particularly when imaging with epifluorescence—can be an order of magnitude; thus, developing longer-wavelength versions of the current generation of FP sensors could be the most efficient way to increase probe performance. Red-shifted FP calcium and voltage sensors are currently being developed in several labs, but are not yet widely distributed. Additional effort on improving probe performance should focus on increasing the kinetics as well as brightness of genetically encoded probes—in particular, calcium sensors. Fortunately, the increased interest in making genetically encoded probes useful for monitoring activity in the intact brain should help to fulfill the promise of these tools for studying nervous system function in the near future.

Finally, in this review we have focused on genetically encoded probes to optically image brain activity in vivo. During recent years a variety of techniques that allow optical and genetically targeted control of neuronal activity have emerged.[54–56] We envisage that in the near future, genetically encoded tools will allow for optical probing combined with and complemented by optical control of assemblies of specific individual nerve cells in vivo.

ACKNOWLEDGMENTS

The authors would like to thank present and past collaborators for their valuable contributions to the work described here, including J. McGann, N. Pírez, J. Verhagen, R. Carey, M.C. Cheung, D. Wesson, H. Mutoh, D. Dimitrov, W. Akemann, A. Perron, and Y. Iwamoto. We thank L. Cohen for essential guidance in methodological development. Our respective laboratories have been supported by grants from the National Institutes of Health (MW and TK), Boston University (MW), and RIKEN BSI (TK).

REFERENCES

1. Miesenbock, G. and Rothman, J.E. Patterns of synaptic activity in neural networks recorded by light emission from synaptolucins. *Proc Natl Acad Sci USA*, 94: 3402–3407, 1997.

2. Miesenbock, G., De Angelis, D.A., and Rothman, J.E. Visualizing secretion and synaptic transmission with pH-sensitive green fluorescent proteins. *Nature*, 394: 192–195, 1998.
3. Siegel, M.S. and Isacoff, E.Y. A genetically encoded optical probe of membrane voltage. *Neuron*, 19: 735–741, 1997.
4. Miyawaki, A. et al. Fluorescent indicators for Ca2+ based on green fluorescent proteins and calmodulin. *Nature*, 388: 882–887, 1997.
5. Grynkiewicz, G., Poenie, M., and Tsien, R.Y. A new generation of Ca2+ indicators with greatly improved fluorescence properties. *J Biol Chem*, 260: 3440–3450, 1985.
6. Giepmans, B.N., Adams, S.R., Ellisman, M.H., and Tsien, R.Y. The fluorescent toolbox for assessing protein location and function. *Science*, 312: 217–224, 2006.
7. Romoser, V.A., Hinkle, P.M., and Persechini, A. Detection in living cells of Ca2+–dependent changes in the fluorescence emission of an indicator composed of two green fluorescent protein variants linked by a calmodulin-binding sequence. A new class of fluorescent indicators. *J Biol Chem*, 272: 13270–13274, 1997.
8. Baird, G.S., Zacharias, D.A., and Tsien, R.Y. Circular permutation and receptor insertion within green fluorescent proteins. *Proc Natl Acad Sci USA*, 96: 11241–11246, 1999.
9. Nakai, J., Ohkura, M., and Imoto, K. A high signal-to-noise Ca(2+) probe composed of a single green fluorescent protein. *Nat Biotechnol*, 19: 137–141, 2001.
10. Akemann, W., Raj, C.D., and Knöpfel, T. Functional characterization of permuted enhanced green fluorescent proteins comprising varying linker peptides. *Photochem Photobiol*, 74: 356–363, 2001.
11. Knöpfel, T., Diez-Garcia, J., and Akemann, W. Optical probing of neuronal circuit dynamics: Genetically encoded versus classical fluorescent sensors. *Trends Neurosci*, 29: 160–166, 2006.
12. Heim, N. et al. Improved calcium imaging in transgenic mice expressing a troponin C-based biosensor. *Nat Methods*, 4: 127–129, 2007.
13. Diez-Garcia, J. et al. Activation of cerebellar parallel fibers monitored in transgenic mice expressing a fluorescent Ca2+ indicator protein. *Eur J Neurosci*, 22: 627–635, 2005.
14. Diez-Garcia, J., Akemann, W., and Knöpfel, T. In vivo calcium imaging from genetically specified target cells in mouse cerebellum. *Neuroimage*, 34: 859–869, 2007.
15. Mao, T., O'Connor, D.H., Scheuss, V., Nakai, J., and Svoboda, K. Characterization and subcellular targeting of GCaMP-type genetically encoded calcium indicators. *PLoS ONE*, 3: e1796, 2008.
16. Yuste, R., Miller, R.B., Holthoff, K., Zhang, S., and Miesenbock, G. Synapto-pHluorins: Chimeras between pH-sensitive mutants of green fluorescent protein and synaptic vesicle membrane proteins as reporters of neurotransmitter release. *Methods Enzymol*, 327: 522–546, 2000.
17. Sankaranarayanan, S., De Angelis, D., Rothman, J.E., and Ryan, T.A. The use of pHluorins for optical measurements of presynaptic activity. *Biophys J*, 79: 2199–2208, 2000.
18. Miesenbock, G. A Practical Guide: Synapto-pHluorins—Genetically encoded reporters of synaptic transmission. In *Imaging in Neuroscience and Development: A Laboratory Manual*, Ed. R. Yuste and A. Konnerth, Cold Spring Harbor Press, Cold Spring Harbor, New York, 2005, pp. 599–604.
19. Sankaranarayanan, S. and Ryan, T.A. Real-time measurements of vesicle-SNARE recycling in synapses of the central nervous system. *Nat Cell Biol*, 2: 197–204, 2000.
20. Gandhi, S.P. and Stevens, C.F. Three modes of synaptic vesicular recycling revealed by single-vesicle imaging. *Nature*, 423: 607–613, 2003.
21. Li, Z. et al. Synaptic vesicle recycling studied in transgenic mice expressing synaptopHluorin. *Proc Natl Acad Sci USA*, 102: 6131–6136, 2005.
22. Ng, M. et al. Transmission of olfactory information between three populations of neurons in the antennal lobe of the fly. *Neuron*, 36: 463–474, 2002.

23. Bozza, T., McGann, J.P., Mombaerts, P., and Wachowiak, M. In vivo imaging of neuronal activity by targeted expression of a genetically encoded probe in the mouse. *Neuron*, 42: 9–21, 2004.
24. Wyatt, R.M. and Balice-Gordon, R.J. Heterogeneity in synaptic vesicle release at neuromuscular synapses of mice expressing synaptopHluorin. *J Neurosci*, 28: 325–335, 2008.
25. Suyama, S., Hikima, T., Sakagami, H., Ishizuka, T., and Yawo, H. Synaptic vesicle dynamics in the mossy fiber-CA3 presynaptic terminals of mouse hippocampus. *Neurosci Res*, 59: 481–490, 2007.
26. Sakai, R., Repunte-Canonigo, V., Raj, C.D., and Knöpfel, T. Design and characterization of a DNA-encoded, voltage-sensitive fluorescent protein. *Eur J Neurosci*, 13: 2314–2318, 2001.
27. Knöpfel, T., Tomita, K., Shimazaki, R., and Sakai, R. Optical recordings of membrane potential using genetically targeted voltage-sensitive fluorescent proteins. *Methods*, 30: 42–48, 2003.
28. Ataka, K. and Pieribone, V.A. A genetically targetable fluorescent probe of channel gating with rapid kinetics. *Biophys J*, 82: 509–516, 2002.
29. Baker, B.J. et al. Three fluorescent protein voltage sensors exhibit low plasma membrane expression in mammalian cells. *J Neurosci Methods*, 161: 32–38, 2007.
30. Dimitrov, D. et al. Engineering and characterization of an enhanced fluorescent protein voltage sensor. *PLoS ONE*, 2: e440, 2007.
31. Lundby, A., Mutoh, H., Dimitrov, D., Akemann, W., and Knöpfel, T. Engineering of a genetically encodable fluorescent voltage sensor exploiting fast Ci-VSP voltage-sensing movements. *PLoS ONE*, 3:e2514, 2008.
32. Chaigneau, E. et al. The relationship between blood flow and neuronal activity in the rodent olfactory bulb. *J Neurosci*, 27: 6452–6460, 2007.
33. Navarro-Quiroga, I., Chittajallu, R., Gallo, V., and Haydar, T.F. Long-term, selective gene expression in developing and adult hippocampal pyramidal neurons using focal in utero electroporation. *J Neurosci*, 27: 5007–5011, 2007.
34. Luo, L., Callaway, E.M., and Svoboda, K. Genetic dissection of neural circuits. *Neuron*, 57: 634–660, 2008.
35. Hasan, M.T. et al. Functional fluorescent Ca2+ indicator proteins in transgenic mice under TET control. *PLoS Biol*, 2: e163, 2004.
36. Ratzlaff, E.H. and Grinvald, A. A tandem-lens epifluorescence macroscope: Hundred-fold brightness advantage for wide-field imaging. *J Neurosci Methods*, 36: 127–137, 1991.
37. Wachowiak, M., Denk, W., and Friedrich, R.W. Functional organization of sensory input to the olfactory bulb glomerulus analyzed by two-photon calcium imaging. *PNAS*, 101: 9097–9102, 2004.
38. Trachtenberg, J.T. et al. Long-term in vivo imaging of experience-dependent synaptic plasticity in adult cortex. *Nature*, 420: 788–794, 2002.
39. Dombeck, D.A., Khabbaz, A.N., Collman, F., Adelman, T.L., and Tank, D.W. Imaging large-scale neural activity with cellular resolution in awake, mobile mice. *Neuron*, 56: 43–57, 2007.
40. Verhagen, J.V., Wesson, D.W., Netoff, T.I., White, J.A., and Wachowiak, M. Sniffing controls an adaptive filter of sensory input to the olfactory bulb. *Nat Neurosci*, 10: 631–639, 2007.
41. Boyden, E.S. and Raymond, J.L. Active reversal of motor memories reveals rules governing memory encoding. *Neuron*, 39: 1031–1042, 2003.
42. Katz, D.B., Simon, S.A., and Nicolelis, M.A. Dynamic and multimodal responses of gustatory cortical neurons in awake rats. *J Neurosci*, 21: 4478–4489, 2001.
43. Shibuki, K. et al. Dynamic imaging of somatosensory cortical activity in the rat visualized by flavoprotein autofluorescence. *J Physiol*, 549: 919–927, 2003.

44. Coutinho, V., Mutoh, H., and Knöpfel, T. Functional topology of the mossy fibre-granule cell—Purkinje cell system revealed by imaging of intrinsic fluorescence in mouse cerebellum. *Eur J Neurosci*, 20: 740–748, 2004.
45. Smetters, D., Majewska, A., and Yuste, R. Detecting Action potentials in neuronal populations with calcium imaging. *Methods*, 18: 215–221, 1999.
46. Yaksi, E. and Friedrich, R.W. Reconstruction of firing rate changes across neuronal populations by temporally deconvolved Ca(2+) imaging. *Nat Methods*, 3: 377–383, 2006.
47. Ramdya, P., Reiter, B., and Engert, F. Reverse correlation of rapid calcium signals in the zebrafish optic tectum in vivo. *J Neurosci Methods*, 157: 230–237, 2006.
48. McGann, J.P. et al. Odorant representations are modulated by intra- but not interglomerular presynaptic inhibition of olfactory sensory neurons. *Neuron*, 48: 1039–1053, 2005.
49. Wachowiak, M. et al. Inhibition of olfactory receptor neuron input to olfactory bulb glomeruli mediated by suppression of presynaptic calcium influx. *J Neurophysiol*, 94: 2700–2712, 2005.
50. Zerangue, N. and Kavanaugh, M.P. Flux coupling in a neuronal glutamate transporter. *Nature*, 383: 634–637, 1996.
51. Aroniadou-Anderjaska, V., Zhou, F.-M., Priest, C.A., Ennis, M., and Shipley, M.T. Tonic and synaptically evoked presynaptic inhibition of sensory input to rat olfactory bulb via $GABA_B$ heteroreceptors. *J Neurophysiol*, 84: 1194–1203, 2000.
52. Spors, H., Wachowiak, M., Cohen, L.B., and Friedrich, R.W. Temporal dynamics and latency patterns of receptor neuron input to the olfactory bulb. *J Neurosci*, 26: 1247–1259, 2006.
53. Oheim, M., Beaurepaire, E., Chaigneau, E., Mertz, J., and Charpak, S. Two-photon microscopy in brain tissue: Parameters influencing the imaging depth. *J Neurosci Methods*, 111: 29–37, 2001.
54. Miesenbock, G. and Kevrekidis, I.G. Optical imaging and control of genetically designated neurons in functioning circuits. *Annu Rev Neurosci*, 28: 533–563, 2005.
55. Boyden, E.S., Zhang, F., Bamberg, E., Nagel, G., and Deisseroth, K. Millisecond-timescale, genetically targeted optical control of neural activity. *Nat Neurosci*, 8: 1263–1268, 2005.
56. Knöpfel, T. Expanding the toolbox for remote control of neuronal circuits. *Nat Methods*, 5: 293–295, 2008.
57. Nagai, T., Sawano, A., Park, E.S., and Miyawaki, A. Circularly permuted green fluorescent proteins engineered to sense Ca2+. *Proc Natl Acad Sci USA*, 98: 3197–3202, 2001.

2 Two-Photon Functional Imaging of Neuronal Activity

Fritjof Helmchen

CONTENTS

2.1 INTRODUCTION

For centuries scientists have been fascinated by the structure of the brain, all the more since the middle of the 19th century when advances in staining techniques enabled high-resolution light microscopic studies of brain cell morphology. Compared to studying cell morphology, it has been more difficult to experimentally investigate the dynamic nature of brain cells, including their morphological transformations, as well as the molecular and electrical signaling events underlying their specific functions. Still today, it remains challenging to study neural dynamics on the cellular level in intact brains of living animals. For many years, in vivo studies of brain cell dynamics relied solely on electrical recordings of neuronal activity, using either extracellular recordings of neuronal spike patterns or intracellular recordings of membrane potential dynamics. Strong light-scattering of neural tissue generally precluded optical imaging with cellular and subcellular resolution in the intact brain (with few exceptions in favorable cases[1]). A major breakthrough in 1990 was the invention of two-photon excited fluorescence laser scanning microscopy (TPLSM),[2] which enabled optical studies of brain cell morphology and function in vivo.[3,4] The success of TPLSM has been fostered by the parallel development of various novel staining techniques for in vivo labeling of brain cells with fluorescent markers. Expression of variants of fluorescent proteins by genetic means[5,6] complements the in vivo imaging capabilities of TPLSM particularly well. The combination of high-contrast fluorescence labeling of brain cells—neurons as well as glial cells—with high-resolution imaging in vivo has opened new research fields in neuroscience. It is now possible to watch brain cells "at work" so that neuroscientists can directly study cellular and molecular mechanisms underlying both normal brain function as well as brain diseases. Because of its great potential for gaining fundamental insights into how the brain develops, how it computes, and how it can adapt, in vivo two-photon imaging of brain cell dynamics has enormously expanded over the past decade, and it is still growing at a rapid pace.

This chapter provides an overview of two-photon imaging of brain cell dynamics in vivo. Because the principles of two-photon fluorescence excitation and TPLSM technology are treated in detail in Chapter 3, I focus here on the principles of dynamic measurements of single-cell and network activity using state-of-the-art labeling methods and various laser-scanning approaches. A key aspect of "activity" is that it always refers to processes evolving in the temporal domain, be it electrical or chemical signaling, turnover of proteins, or changes in cell morphology. Therefore, imaging modes capable of reading out the "dynamics" of living brain cells over a wide range of time scales will be particularly highlighted. In vivo calcium imaging studies—as the most prominent, current approach for functional measurements—will be summarized, and examples from mammalian brain imaging will be provided on both the single-cell as well as the population level. The chapter ends with an outlook discussing ongoing developments that are expected to become important research directions in the next years.

2.2 IMAGING BRAIN CELL DYNAMICS USING IN VIVO TWO-PHOTON MICROSCOPY

2.2.1 Two-Photon Microscopy

Two-photon microscopy is a laser-scanning technique, in which a laser beam is focused through a microscope objective down to a micrometer-sized light spot in order to excite fluorescent molecules. At each moment in time, laser-scanning microscopes typically collect fluorescence light from a single location in the sample. Spatial information is gathered by moving ("scanning") the laser focus within the tissue. For example, raster-like, line-by-line scanning of a horizontal plane is a standard technique for acquiring two-dimensional (2D) images, and plane-by-plane acquisition of image stacks is a standard method to collect fluorescence data from a sample volume (Figure 2.1A). Exceptions from single-beam, laser-scanning microscopes are multifocal microscopes, in which fluorescence is generated simultaneously in multiple

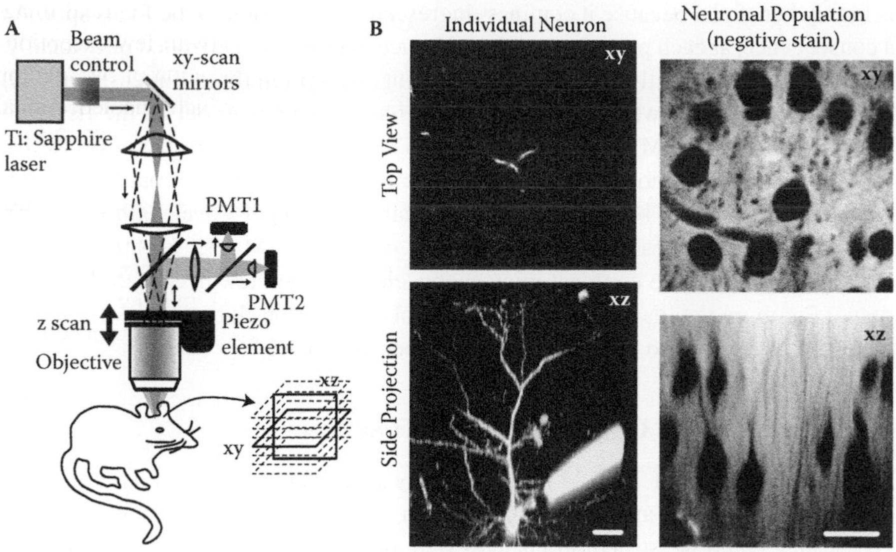

FIGURE 2.1 In vivo two-photon imaging of individual neurons and neuronal populations. (A) Schematic of a two-photon microscope setup for in vivo imaging in the brain. Beam control includes beam expansion and adjustment of laser intensity. Fluorescence can be detected in multiple color channels using photomultiplier tubes (PMTs). Image stacks are acquired using a software-controlled focusing device, for example, a piezoelectric focusing element. (B) The typical view during the experiment is the top view (xy view; top row). To obtain side views (xz or yz views) of neurons, maximum-intensity side projections are computed from stacks of hundreds of images (bottom row). The neuron on the left was filled with a fluorescent calcium indicator through a whole-cell patch pipette. The focal plane of the xy image was just above the main bifurcation of the apical dendrite. The neuronal population on the right was visualized in negative contrast by bulk injection of the bright red fluorescent dye Alexa-594 into the extracellular space. Note that pyramidal cells can be identified by their apical dendritic trunks. Scale bars 20 μm.

laser foci but then needs to be spatially resolved with a camera detector.[7] Two-photon microscopes based on multifocal excitation have been rarely applied in vivo so far; we will not considered them further here.

The key difference of TPLSM compared to confocal laser scanning microscopy is the physical process of light absorption, which is followed by fluorescence emission. In a two-photon absorption process, two photons are absorbed virtually simultaneously (within <1 fs). Because the two photons combine their energies to promote the molecular transition, their individual energies are relatively low. Hence, excitation light is of longer wavelength compared to single-photon absorption, typically in the near-infrared wavelength range (700–1000 nm). The use of near-infrared light is the first key advantage of two-photon excitation because longer-wavelength light is less scattered in biological tissue, enabling larger penetration depths. The second important feature is that the fluorescence signal depends nonlinearly on excitation light intensity (on the square of the intensity in the case of two-photon excitation). This nonlinear dependence (in defining TPLSM as a "nonlinear microscopy technique") is highly beneficial because it confines fluorescence generation to the focus spot. As a consequence, at each point in time fluorescence photons are generated only locally, and they can be correctly assigned to their point of origin in three-dimensional (3D) space irrespective of whether they are scattered on their way out of the tissue to the detector. In TPLSM a confocal pinhole in front of the detector is not necessary to achieve optical sectioning; it should be omitted and as-many-as-possible fluorescence photons should be collected. In summary, the longer-wavelength excitation and the intrinsic optical sectioning property make TPLSM less sensitive to light scattering compared to confocal microscopy, clearing the view deep into the living tissue. For more details about the principles of two-photon fluorescence excitation, the reader is referred to Chapter 3 and a number of reviews.[8–11]

2.2.2 Monitoring Cell Structure and Function

The special features of TPLSM make it ideally suited for the study of dynamic processes on the cellular and subcellular level in the intact nervous system. Dynamic events may comprise structural changes (i.e., morphological changes of individual cells or cell migration, and electrical and biochemical activity of cells, e.g., the generation of action potentials, ion fluxes across the membrane, or enzyme activities). Besides imaging neuronal dynamics TPLSM also has proved beneficial for measuring capillary blood flow[12] (see also Chapter 3) as well as activity in electrically nonexcitable glial cells (for review see References 13,14). A key prerequisite for two-photon imaging is fluorescence labeling of the structures of interest with either an anatomical marker or a functional probe. Labeling everything generally is not very helpful; rather, a sparse, high-contrast label of particular cells is desirable. Meanwhile, several techniques for specific and sparse labeling are available, as will be discussed in Section 2.3.3.

Following the initial demonstration of the advantages of TPLSM for neuroscientific research,[15] the first report on in vivo functional imaging of single neurons[4] was published in 1997. Figure 2.1B shows an image from a similar, more recent experiment. Here, an individual pyramidal neuron in the neocortex of an anesthetized rat was loaded with a fluorescent calcium indicator dye through a whole-cell patch pipette. This type of experiment has been useful for measuring calcium dynamics in neuronal dendrites and to study dendritic excitation.[16–20] Figure 2.1B in addition shows how a cell population can be easily visualized in the living brain using a simple negative-contrast stain. Negative stains can help to identify and to specifically target cells.[21–23] Note the difference between the standard *xy*-view as it typically appears during the experiment and the side view (or "*xz*-view," with *z* being the dimension along the optical axis). Usually the side view is calculated offline from a stack of images; only recently sideways imaging has been implemented as a normal imaging mode in the microscope setup using a fast piezo-electric focusing device.[24,25]

Early on it was noted that TPLSM also is a useful tool for precise photomanipulation.[3] This type of application is complementary to the imaging mode and has provided important results in brain slices during the last decade.[26–28] It still needs to be adapted for in vivo conditions perhaps fostered by the recent developments of light-gated cation channel proteins.[29]

Two-photon microscopy covers the spatial scales between a micron and a millimeter. Its spatial resolution is good enough to resolve synaptic structures such as dendritic spines and axonal boutons in the intact brain. A number of studies have investigated structural dynamics of dendritic spines or axonal boutons,[30–32] especially those modifying the connectivity scheme and thus probably contributing to neural circuit plasticity.[33] In the neocortex, structural dynamics has been found to be particularly high in microglial cells, the immunocompetent cells in the brain; in their "resting state" microglia continually expand and retract their fine processes, on a minutes time scale presumably subserving a homeostatic function.[34,35] Compared to these studies of morphological changes, functional measurements at the single-synapse level in vivo are more difficult; calcium signals in presynaptic terminals or dendritic spines have not been studied in vivo yet. An obvious difficulty is that remaining heartbeat and breathing-induced tissue pulsations make measurements remain on the micron scale difficult, even with good dampening.

On the other end of the spatial scales, the maximal imaging field of TPLSM covers larger and larger volumes. This expansion is driven by special high NA objectives with low magnification[36] and new labeling techniques for measuring activity patterns in distributed cell networks.[37] At present, the side length of typical field-of-views is on the order of a few hundred microns but may reach a millimeter soon. For the future, a major challenge is to cover even larger areas or 3D volumes while maintaining cellular resolution and sufficient temporal resolution. Most likely, such developments will require new technologies.

In summary, TPLSM is a versatile technique to image both structural dynamics and cell function in the living brain (Figure 2.2A). Despite the substantially reduced sensitivity of TPLSM to light scattering, two-photon imaging studies are still limited in terms of depth penetration. The maximal imaging depth that has been reached in the neocortex is about 1 mm, and background fluorescence

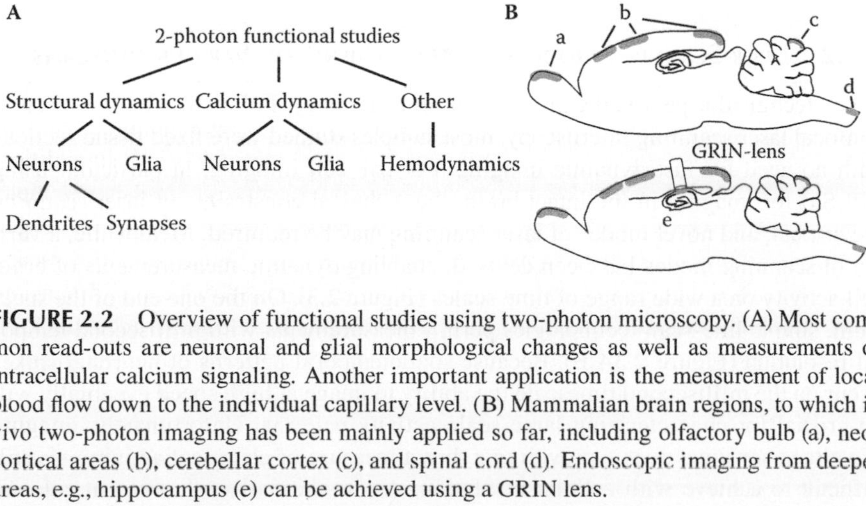

FIGURE 2.2 Overview of functional studies using two-photon microscopy. (A) Most common read-outs are neuronal and glial morphological changes as well as measurements of intracellular calcium signaling. Another important application is the measurement of local blood flow down to the individual capillary level. (B) Mammalian brain regions, to which in vivo two-photon imaging has been mainly applied so far, including olfactory bulb (a), neocortical areas (b), cerebellar cortex (c), and spinal cord (d). Endoscopic imaging from deeper areas, e.g., hippocampus (e) can be achieved using a GRIN lens.

generated during deep imaging sets a fundamental limit to the achievable depth.[38,39] TPLSM, therefore, is restricted to near-surface areas of the brain. So far it has been mainly applied to easily accessible brain areas such as neocortex,[4] cerebellum,[40] olfactory bulb,[18] and, more recently, spinal cord dorsal horn.[41] Special procedures are required to reach deeper brain regions such as the hippocampus. Particularly promising are rodlike lenses (so-called gradient-index (GRIN) lenses), which have diameters below a millimeter and can be targeted to deeper areas for endoscopic imaging (Figure 2.2B). For example, using such an approach, fluorescently stained pyramidal cells have been visualized in the hippocampus of living mice.[42,43] In this chapter we will provide examples from the mammalian neocortex and the cerebellar cortex for illustration.

2.3 MICROSCOPE SETUP AND LABELING METHODS

2.3.1 Two-Photon Microscope Setup

The principle setup of a two-photon microscope is similar to a confocal microscope, and in some aspects even simpler.[9] A main difference is the light source for fluorescence excitation. While wavelength-tunable Ti:sapphire IR lasers as sources of femtosecond laser pulses have developed into user-friendly key-turn instruments, they still are expensive and constitute a major cost for setting up a TPLSM. For several reasons, a number of research labs prefer to custom-build the core microscope part. Besides reducing the overall cost for a TPLSM system, this approach provides more flexibility regarding space issues (fitting an animal underneath the microscope) and scanning modes. New approaches for in vivo imaging often require creative solutions and microscope manufacturers tend to lag behind the rapid developments in the field. Because the technological aspects of setting up an in vivo TPLSM are extensively treated in a different chapter of this book (Chapter 3) I do not provide any further details on instrumentation here but rather focus on special laser-scanning modes and fluorescence labeling techniques.

2.3.2 Laser-Scanning Modes for Measurements on Various Timescales

A key technical aspect is the laser scanning technology. Following the invention of confocal laser scanning microscopy, most samples studied were fixed tissue sections with no need for fast dynamic imaging. For live cell imaging, in particular using TPLSM for imaging in the intact brain, the technical constraints of laser scanning are critical, and novel modes of laser scanning may be required. Meanwhile, a variety of scanning modes has been devised, enabling dynamic measurements of brain cell activity on a wide range of time scales (Figure 2.3). On the one end of the spectrum, simple line-scan technologies permit measurements with millisecond temporal resolution (Figure 2.3A,B). Because spatiotemporal patterns of neuronal spikes occur on the millisecond timescale the ability to image at high speed eventually will be crucial for characterizing large-scale activity patterns. Unfortunately, imaging of voltage-sensitive dyes—providing a direct measure of electrical activity—is still difficult to achieve with cellular resolution. As an alternative, fluorescent calcium

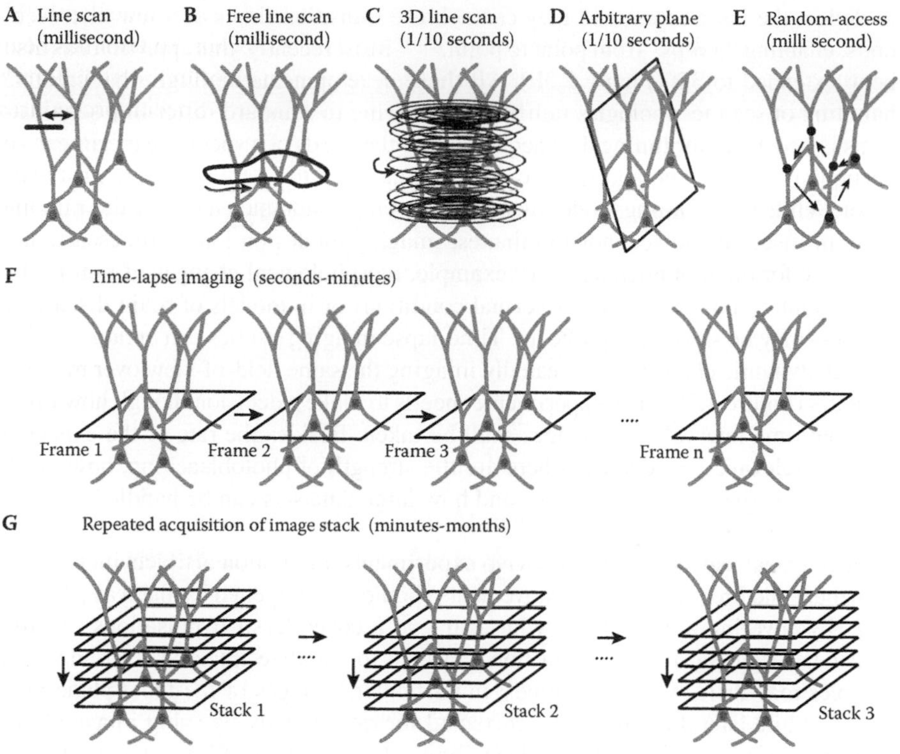

FIGURE 2.3 Laser-scanning modes for imaging on a wide range of timescales. (A) Straight line scans provide measurements from specific cellular structures such as dendrites with millisecond resolution. (B) Free-line scans along closed curved trajectories can be used to obtain fast measurements from several structures of interest. (C) The line scan concept can be extended to 3D by combining free-line scanning in the *xy* plane with continuous fast focusing along the optical axis. Full 3D acquisition rates of 10–20 Hz can be achieved. (D) Mixing of scan signals between *xy* scanners and a rapid *z* focus allows arbitrary orientation of the image plane. (E) In random-access scanning, the laser focus rapidly "jumps" between preselected points of interest. With special nonmechanical scan devices, point-to-point transitions rates of 50–100 kHz can be achieved. (F) Standard time-lapse imaging. A movie is acquired at a particular focal plane. (G) Long-term imaging by repeated sampling of the same tissue volume.

indicators can indirectly probe spiking activity, although the achievable temporal resolution ultimately is limited by their calcium-binding kinetics. It thus remains a challenge to achieve in vivo measurements of large-scale population activity with high temporal accuracy using TPLSM.

Another challenge is to extend existing scan technologies to 3D, since neural circuits operate in 3D space. One approach, which will be exemplified further below, is to combine mechanical scan devices (e.g., galvanometric scan mirrors and a mechanical focusing element) to create 3D scan trajectories[44] (Figure 2.3C,D). Alternatively, nonmechanical scan devices, namely, acousto-optical deflectors (AODs), can be employed. These devices use sound-wave-generated diffraction gratings in crystals to deflect the laser beam, and they can achieve transition times of a few microseconds, enabling "jumps" from point to point.[45–47] Most recently, this approach has also been extended to 3D (Figure 2.3E). All these developments highlight that creative handling of scan technologies rather than adhering to standard raster-like scan patterns is beneficial for tuning data acquisition to the needs of a specific experiment, in particular when spatial and speed considerations need to be traded off against each other. While the scanning modes mentioned so far are suitable on timescales ranging from milliseconds to seconds to minutes, imaging modes on longer timescales are desirable for other applications. For example, morphological changes of brain cells in the adult brain—either under normal conditions or in models of brain disease or injury—may be slow but significant. Time-lapse imaging studies can create a movie of such dynamic changes by repeatedly imaging the same field-of-view over minutes to hours (Figure 2.3F). The experimenter needs to make a decision though how often (at what time intervals) the images shall be taken. Besides the rate of the expected changes relevant considerations here are the strength of photobleaching, how much illumination the tissue can tolerate, and how large data sets can be handled for storage and analysis.

A practical aspect special to in vivo experiments are motion artifacts induced by heartbeat and breathing. Tissue movements should be dampened as much as possible (e.g., by covering a craniotomy with agar and a coverglass.) If pulsations remain, heartbeat triggered image acquisition can help to stabilize imaging.[34] In addition, samples may drift over longer times; in particular, changes of the focal plane may occur. While lateral drifts can be corrected for rather easily by using image alignment routines, changes in focal plane cannot be corrected offline. A safe way to circumvent this problem is to always acquire small image stacks (10–20 planes), with the structures of interest in the center of the image stack (Figure 2.3G). Clearly, such repeated acquisition of image stacks, often referred to as "4D imaging," slows down the measurements, but intervals of <1 min can be easily achieved and are sufficient in many cases. In addition, four-dimensional (4D) imaging can be applied on much longer timescales by repeatedly imaging a particular tissue volume in the same animal over days, weeks, or even months.[30,31,48] For finding the same structures again and again, blood vessels or other coarse landmarks can be utilized. Long-term imaging highlights the in vivo advantage of TPLSM, enabling previously impossible longitudinal studies of brain cell dynamics in the normal and diseased brain.

2.3.3 Traditional Dyes and Fluorescent Proteins

Ideally, in vivo TPLSM is applied to brains with prelabeled cells. In fact, cells contain autofluorescent molecules (e.g., nicotinamide adenine dinucleotide [NADH]), which can be imaged and used to determine a metabolic state.[49] In most cases, however, additional fluorescence labeling is necessary. Negative fluorescence staining has already been mentioned as a simple method to visualize cell bodies and identify at least some cell types.[21–23] For positive staining, a broad toolbox of fluorescent dyes[50] and a spectrum of in vivo labeling techniques are now available. Several classes of fluorescent labels can be distinguished. A first group comprises synthetic organic dyes, often referred to as "traditional" dyes, which need to be introduced into cells by physical or chemical means. Fluorescent proteins establish a second class; they are genetically encoded, and their expression can be induced in specific cell types. A third "mixed" is to attach a fluorophore to a protein of interest by tagging the protein with a specific short peptide sequence (for review see Reference 51). Molecules in all these classes come in different colors with emission spectra ranging from blue to near-infrared. For example, an entire spectrum of fluorescent protein is available, which is continually growing.[52] Multicolor imaging will become increasingly important in the future because it permits simultaneous monitoring of dynamic processes in multiple tissue components, thus allowing a direct view at their interactions. The recent introduction of combinatorial expression of fluorescent proteins in "Brainbow" transgenic animals,[53,54] resulting in nearly a hundred hues for easy distinction of individual cells, most likely is only the beginning of a new multicolor area of in vivo imaging.

Traditional dyes can be loaded into cells using a variety of methods, including uptake from the extracellular space, intracellular recordings, or electroporation. Often these methods lack cell type specificity. An exception is the red fluorescent dye sulforhodamine-101, a derivative of Texas Red, which for unknown reasons specifically labels astroglia in rodent neocortex.[55] A few drops of sulforhodamine-101 briefly applied to the exposed brain surface are sufficient for bright labeling. Combined with tail-vein injection of fluorescent dyes for labeling the microvasculature[12] astroglia staining with sulforhodamine-101 can be recommended as a simple test stain for evaluating and optimizing the in vivo performance of a TPLSM setup. Fluorescent protein expression can be achieved using local methods such as viral transfections[56,57] and targeted single-cell DNA electroporation[23,58] or more globally by generating transgenic animals. A major advantage is that cell type specificity can be achieved if a specific promoter is available. Transgenic animals with cell-specific expression of fluorescent proteins thus make the ideal situation of in vivo imaging of specifically prelabeled cells reality. A further boost of the generation of transgenic animals can be expected driven by the further improvements of activity-dependent fluorescent proteins.

2.3.4 Activity-Dependent Probes

The distinction between "anatomical" and "functional" markers—between markers for cell morphology or protein localization on the one hand and those for particular physiological parameters on the other—is a vague one. Often cell function manifests itself as structural dynamics, either of gross cell morphology or of intracellular rearrangement of molecular components. More precisely one may speak of "activity-dependent probes" in the case of dyes that change their fluorescent properties in response to a specific cellular activity (e.g., in response to ion influx or the activation of an enzyme). Voltage-sensitive fluorescent dyes and fluorescent ion indicators are examples of traditional activity-dependent probes. In addition, ongoing research continually produces new activity-dependent fluorescent proteins and improves their performance for specific purposes. Sometimes, however, an activity-dependent effect may need to be suppressed, as in the case of unwanted pH-sensitivity of some fluorescent proteins. Today, the most commonly applied activity-dependent probes are fluorescent calcium indicators, which will be discussed in more detail in the next section.

2.4 ACTIVITY MEASUREMENTS USING CALCIUM INDICATORS

2.4.1 Calcium Indicator Types and Loading Methods

Intracellular binding of calcium ions to proteins is fundamental to a multitude of cell functions. In particular, considering the role of calcium ions in synaptic transmission, it is the basis for neuronal communication and information processing. Not surprisingly, therefore, fluorescent calcium indicators are the most advanced functional indicators currently applied in neuroscience. The first synthetic fluorescent dyes developed by Roger Tsien around 1980 were based on calcium-chelating molecules such as BAPTA.[59] Over the past decades, a wide palette of synthetic calcium indicators has been created, now comprising dyes with vastly different calcium-binding affinities and different spectral properties, to name just two of the most important dye properties. Different affinities are desirable because calcium concentration changes span a huge range from nanomolar level (e.g., in large cell compartments) to micromolar level (e.g., in highly localized submembraneous calcium domains). For in vivo TPLSM nonratiometric indicators that are single-photon excited around 488 nm (e.g., Calcium Green, Fluo-4, or Oregon Green BAPTA) have proved most suitable. The reason is that they are easily two-photon excited at wavelengths between 800–900 nm and can be loaded into neurons using several methods. But also traditional UV-excitable dyes such as Fura-2 can be employed for in vivo two-photon imaging.[60] Applications of red fluorescent calcium indicators (e.g., Rhod-2) on the other hand are still rare, mostly because far fewer indicator types are available and because labeling is more problematic. Rhod-2 AM, for example, preferentially loads glial cells in vivo, and has not been used for neuronal imaging. Generally synthetic calcium indicators show low photobleaching rates with two-photon excitation and have no obvious pharmacological side effects aside from adding calcium buffering capacity to the cell.

Depending on the research focus, one may want to label an individual cell or entire cell populations. Fortunately, there are now several rather simple methods available, with which one can "titrate" the number of labeled cells (Figure 2.4). In the first single-cell in vivo imaging study,[4] individual neurons were filled with a calcium indicator through an intracellular recording electrode (Figure 2.4A). This approach results in high-contrast labeling of the cell and all its dendrites, and therefore is well suited to study subcellular signaling and dendritic integration (see below). Intracellular loading can be performed through either a sharp electrode or a whole-cell patch pipette.[19] The experimental procedure can be alleviated by visually guiding the pipette tip to particular neurons, using either fluorescence counterstains[61] or negative contrast.[23] Even simpler is the application of brief electrical pulses through a dye-filled micropipette, causing focal electroporation of individual cells[22] or local electroporation of small groups of neurons[62] (Figure 2.4B). This type of labeling should be helpful for

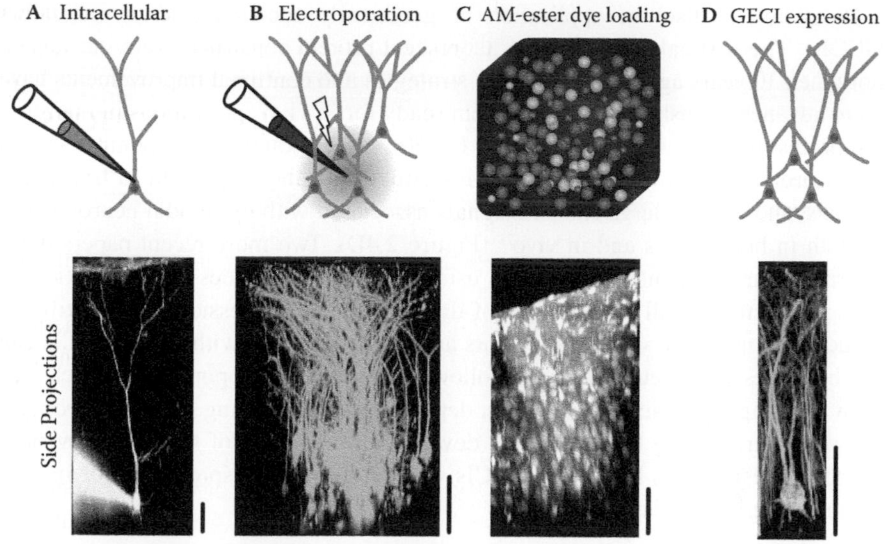

FIGURE 2.4 Methods for in vivo labeling of neurons with calcium indicator. (A) Individual neurons can be filled during an intracellular recording using a sharp electrode or a whole-cell patch pipette. (B) Electroporation through a glass pipette permits dye loading of single cells by visually targeting individual neurons or—as shown here—of groups of neurons by repetitive electroporation in the neuropil. (From Nagayama et al., *Neuron* 53: 789–803, 2007.). (C) Multicell bolus loading of AM-ester calcium indicator dyes permits labeling of all neurons and astrocytes within a local tissue volume. (D) Genetically encoded calcium indicators may be expressed by various means, in particular through generation of transgenic animals. (From Heim et al. *Nat Methods* 4: 127–129, 2007. With permission.) Scale bars 50 μm.

investigating excitation flow through local neural circuits because it permits discrimination of individual cells and zoom in on synaptic structures. Finally, it has been a long-standing goal in neuroscience to achieve functional labeling of entire local populations of neurons with the aim to optically resolve neuronal network dynamics in vivo. A major breakthrough was achieved in 2003 in Arthur Konnerth's group, when they successfully loaded cells in vivo using the membrane-permeable, acetoxymethyl(AM)-ester form of calcium indicator dyes.[37] AM-dyes easily penetrate the cell membrane, but once the ester groups are cleaved by endogenous esterases—reconstituting the original charged indicator molecule—they are trapped in the cytosol. Although AM-ester loading dates back to the early 1980s,[63] only in 2003 Konnerth's group demonstrated that direct injection of a bolus of dissolved AM-ester dye into the brain parenchyma leads to dye uptake in essentially all cells within a circumscribed volume (Figure 2.4C) (the only exception being microglial cells, for which it is unclear in how far they are unlabeled by AM-ester loading). Since then multicell bolus loading (MCBL) has become a widespread method for in vivo investigations of the functional organization of local circuits. Practical tips can be found in a detailed protocol of this method[64] and in reviews.[65,66]

Promising is also the application of genetically encoded calcium indicators (GECIs). The first calcium-sensitive fluorescent protein constructs were introduced more then 10 years ago.[67] Clever design strategies and continual improvements have led to advanced constructs that now seem ready for in vivo applications in mammalian brains (for reviews, see References 68–70). As an example, a troponin-C-based fusion protein of two fluorescent proteins could be stably expressed in transgenic animals, and showed large calcium signals associated with pyramidal neuron activity both in brain slices and in vivo[71] (Figure 2.4D). Two more recent papers demonstrated improved in vivo imaging using GECIs.[72,73] Besides the generation of transgenic animals, all other means of driving protein expression can be utilized, including transfection with viral vectors and electroporation with DNA. GECIs can also be expressed in neuronal groups following in utero electroporation,[72] which may allow labeling of particular cell types, depending on the timing of the electroporation procedure during the embryonal development. In view of the rapid advances in protein design, it is likely that GECIs will soon find widespread use for in vivo calcium imaging studies.

2.4.2 Inferring Neuronal Spiking Activity from Calcium Signals

Calcium indicators are utilized in different ways to investigate various aspects of cellular signaling. Two general applications are (1) the study of the calcium-dependence of particular processes (e.g., of transmitter release or synaptic plasticity) and (2) the indirect inference of electrical excitation and spiking activity from the fluorescence measurement. The second type of application is based on the close linkage between electrical excitation and calcium influx, mediated by voltage-dependent calcium channels. For example, each action potential in a neuron typically leads to opening of calcium channels in the somatic membrane and in at least part of the dendritic tree. With a sensitive calcium indicator, the brief surge of calcium ions can be detected as a rapid fluorescence change, which subsequently decays back to resting fluorescence, reflecting calcium removal from the cytosol (Figure 2.5B). The decay time constant of such a stereotyped "calcium transient" depends on calcium buffering and extrusion mechanisms, and typically is in the range of a few hundred millisecond for somata.[75] Thus, the instantaneous rate of neuronal spikes can be indirectly inferred from the fluorescence measurements,[76] and in the best cases even single action potentials can be detected.[77–79] Clearly, the sensitivity will vary between cell types, and one should also keep in mind that calcium ions might enter the cytosol via other entry routes (e.g., by local release from intracellular stores). Furthermore, the temporal precision of determining spike times is limited by the temporal resolution of the laser scanning technique, which typically is slow compared to electrical recordings. Despite these drawbacks, the optical detection of spiking activity has a special appeal because it promises to enable in vivo measurements of nearly complete spiking patterns (including nonactive neurons) in local neural circuits, perhaps sampling populations up to several hundred cells.[44] Large-scale population calcium imaging complements existing electrophysiological methods and may help to reveal fundamental principles of network dynamics.

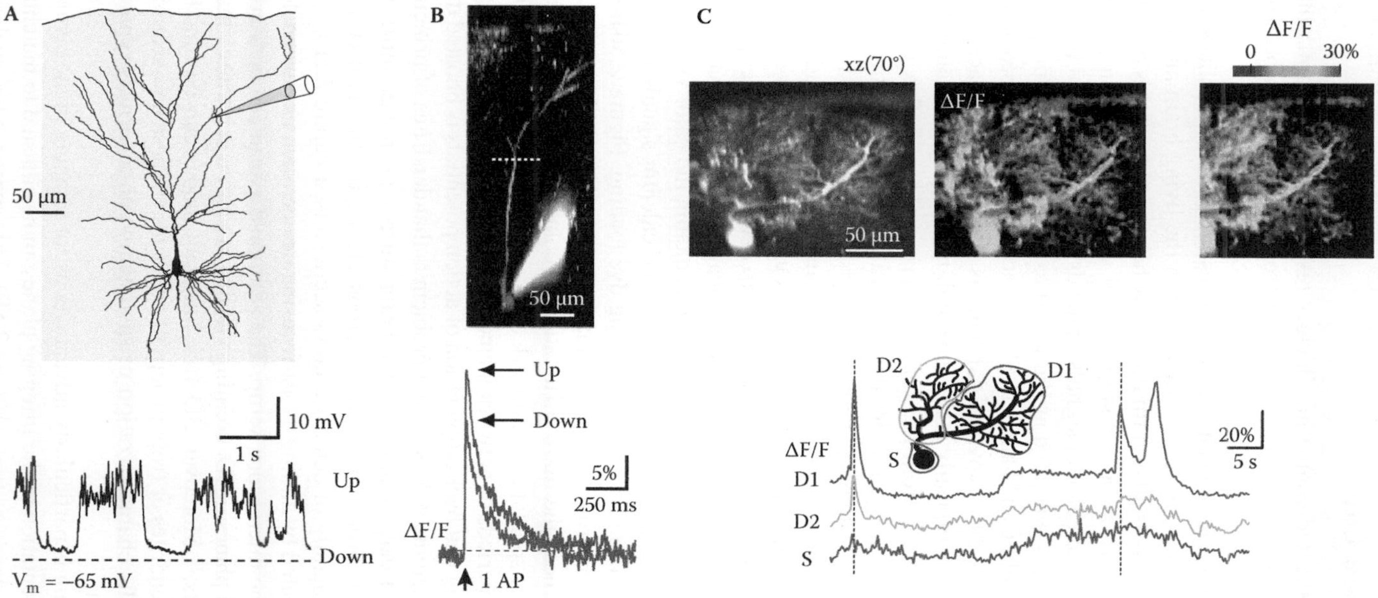

FIGURE 2.5 **(See color insert following page 224)** Examples of in vivo calcium imaging of dendritic excitation. (A) Spontaneous subthreshold membrane potential fluctuation in a neocortical pyramidal cell in an urethane-anesthetized rat revealed by a dendritic whole-cell recording (B). Fast calcium measurement using a line scan near the main bifurcation of the apical dendrite of a different layer 2 neocortical pyramidal cell, which was filled with Oregon Green BAPTA-1 through a somatic whole-cell patch pipette. Bottom: Single action potential-evoked calcium transients were larger if evoked during Up states compared to Down state. (C) Dendritic imaging of Purkinje cell excitation in the cerebellar cortex. A single Purkinje cell was filled with Oregon Green BAPTA-1 via local electroporation. Two large dendritic calcium signals are shown, presumably induced by climbing fiber activation. The time course of fluorescence changes is shown in the lower panel for three regions of interest; ([A,B] Adapted from Waters, J. and Helmchen, F., *J Neurosci* 24: 11127–11136, 2004; [C] from Göbel, W. and Helmchen, F., *J Neurophysiol* 98: 3770–3779, 2007. With permission.)

2.5 EXAMPLES OF IN VIVO IMAGING OF NEURONAL ACTIVITY

2.5.1 Imaging Individual Cells

Single neurons are sophisticated computing devices responsible for synaptic integration. Most dendrites possess nonlinear ("active") properties, which in principle make them capable of generating local regenerative potentials. Although various modes of dendritic integration have been carefully investigated in brain slices, it is still largely unknown which modes are actually used in the living brain and under what specific conditions. In vivo calcium imaging may help to address this question under physiological conditions. As a first step, several studies have taken a closer look at action-potential-evoked calcium signals in dendrites of neocortical pyramidal neurons. Action potentials generated at the initial axon segment backpropagate into the dendritic tree, providing an "I am active" feedback signal to synapses, which is important for synaptic plasticity. In pyramidal neurons action potential backpropagation into the apical dendrite typically is decremental, meaning that the amplitude gradually declines with distance from the soma. Because calcium influx depends on action potential amplitude and duration, dendritic calcium signals indicate how strongly action potentials are attenuated and whether backpropagation might be modulated. Consistent with previous brain slice experiments, in vivo studies have confirmed that backpropagation of fast action potentials varies in different cell types, from nonattenuated in mitral cells of olfactory bulb[18] to decremental in neocortical pyramidal neurons[16,20] to strongly attenuated in cerebellar Purkinje neurons.[25] In the typical experiment, an individual neuron is filled with calcium indicator via a sharp electrode or a patch pipette, and action-potential-evoked calcium signals are measured at variable distances from the soma following dye loading. Figure 2.5 shows an example of a measurement from a neocortical layer 2/3 neuron. A line-scan measurement was performed near the main bifurcation to compare dendritic calcium transients during periods of spontaneous network activity (so-called Up states in the anesthetized animal) and periods of network silence. Calcium transients were larger during Up states, indicating that backpropagation of action potentials is boosted during spontaneous cortical activity.[78] A practical problem is that dendrites of principal neurons often are oriented perpendicular to the brain surface as, for example, the main apical dendrite of neocortical pyramidal neurons. Hence, the standard *xy* view provides a limited view, mainly of dendritic cross sections (see Figure 2.1B). This problem can be circumvented by adding a fast-focusing device, such as a piezoelectric objective mount, to the microscope setup. The additional z-scan can be used to image arbitrarily oriented planes in 3D, creating viewing angles on dendrites similar to brain slice experiments.[25] Moreover, 3D free line-scan modes can be tuned to record along individual dendrites or from groups of dendrites.[25] These new scanning modes should facilitate the characterization of spatiotemporal dendritic excitation profiles in the intact brain.

Arbitrary plane imaging is particularly advantageous for imaging in the cerebellar cortex because the orientation of the imaging plane can be adjusted to match the 2D plane of Purkinje cell dendritic trees (Figure 2.5B). This approach has enabled a direct view of climbing-fiber-evoked dendritic calcium transients[25] and promises to allow a more detailed investigation of the integration of parallel fiber and climbing-fiber inputs. In general, the question of whether, where, and when local dendritic spikes occur in the intact brain is now the focus of dendritic research. Large calcium action potentials have been reported in distal apical dendrites large layer 5 pyramidal neurons in vivo,[17] but their significance for cortical association is still unclear. Layer 2/3 pyramidal neurons apparently have a weaker propensity to generate dendritic spikes.[81] However, the generation of local spikes requires clustered and synchronous synaptic inputs[26] and therefore may crucially depend on behavioral state. High-resolution imaging using TPLSM, eventually in awake animals (see below), will be the key technology to solve these questions.

2.5.2 Imaging Population Activity

Monitoring population activity is the second major application of in vivo TPLSM. The MCBL technique is well suited to label entire neuronal and astroglial networks (Figure 2.6A), making it possible to record from many, if not all, cells in a local area and to analyze the spatial and temporal response patterns. In vivo calcium imaging thus is an ideal tool to study local neuronal network dynamics. During recent years, it has been applied to various brain areas, including the visual cortex of cats and rodents[82–84] and rodent barrel cortex.[76,77] Obvious questions that can be addressed are in how far subassemblies of neurons represent particular stimulus features and how neuronal correlations are distributed throughout the network. A fruitful approach is to combine TPLSM with other imaging modalities. For example, intrinsic optical imaging (Chapter 9) can be used to first obtain a functional map of a particular brain area (e.g., the primary somatosensory cortex) and then target dye loading and two-photon imaging to an identified area.[76,81] Eventually, it would be extremely informative to combine measurements of network function with post hoc high-resolution anatomical analysis of the underlying network connectivity.[83]

One major limitation still is the number of cells that can be recorded from with sufficient temporal resolution. Recently, we have introduced 3D line scanning technology as one possible approach to extend measurements to larger neural networks (Figure 2.3C). Using a mechanically "swinging" objective, this method enables measurements from several hundred cells in 3D with a sampling rate of 10 Hz for the entire population (Figure 2.6B). There still is room for improvements of this purely mechanical approach, and I expect that within a few years 3D measurements at 10 Hz or faster rates will be possible from a thousand or more cells. An alternative approach to this problem is the use of nonmechanical scanners, namely, acousto-optical deflectors (AODs). These devices use sound-wave-generated diffraction gratings in a crystal to deflect the laser beam, and they can achieve point-to-point transition times of a few microseconds.[45–47] Recently, it could be demonstrated that special arrangements of AODs can provide not only fast 2D but also 3D random access scanning by rapidly changing the divergence of the laser beam, which effectively shifts the focal plane (Figure 2.3E).[45] Though currently restricted to rather small volumes, this approach may eventually enable optical imaging of network activity with millisecond time resolution.

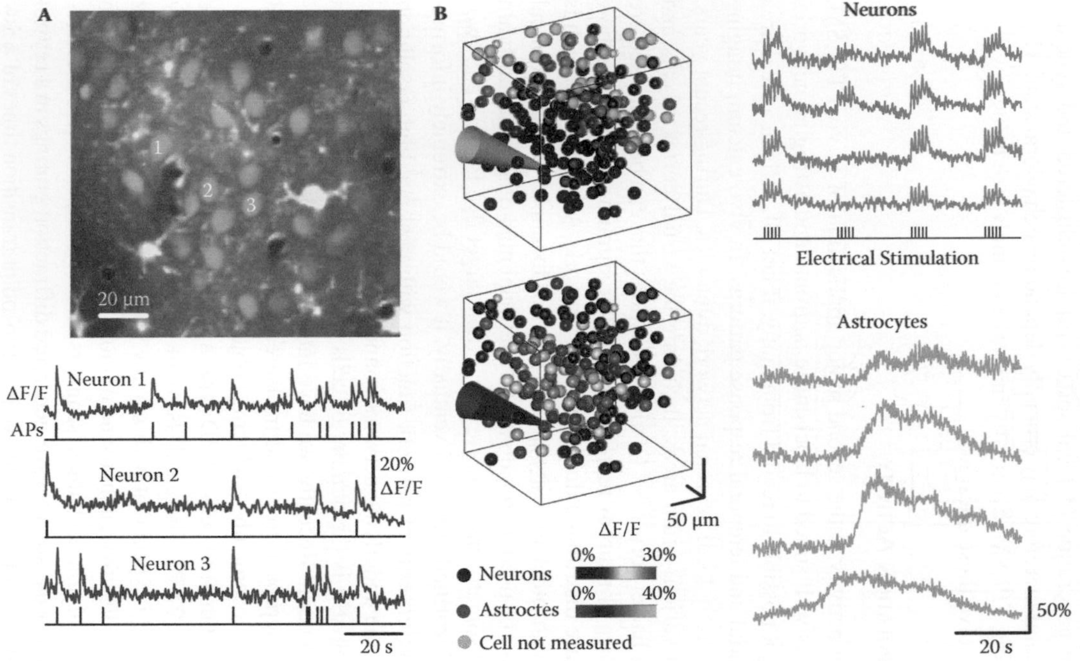

FIGURE 2.6 **(See color insert following page 224)** Example of calcium imaging of population activity. (A) Time-lapse recording of spontaneous activity in neocortical layer 2 neurons following multicell bolus loading with Oregon Green BAPTA-1 AM. Example fluorescence traces from three neurons are depicted together with presumed underlying action potential patterns. Astrocytes were counterstained with sulforhodamine 101 and appear orange. (B) 3D population imaging using the 3D line scan method shown in Figure 2.2. Left: 3D visualization of the activity pattern is based on two different color lookup tables for neurons and astrocytes. Activity patterns are depicted at two time points (top: during an astroglial calcium wave; bottom: during neuronal responses to local electrical stimulation). Right: Example traces from four neurons and four astrocytes. Local electrical stimulation was applied through a glass pipette (yellow, at rest; red, during stimulation). (B) Adapted from Göbel, W. et al. *Nat Methods* 4: 73–79, 2007.

2.6 DISCUSSION

The scope of this chapter was to provide an overview of the new possibilities opened by TPLSM to study brain cell dynamics in vivo. Within only one decade, in vivo TPLSM has proved extremely valuable for addressing fundamental questions of neuronal wiring and single-cell as well as network computation. Although these questions are far from being solved it is clear that, in vivo, TPLSM will be an indispensable tool for their further investigation. Spanning the range from the single synapses to local networks TPLSM builds a bridge between traditional electrophysiological methods, probing the activity in individual neurons or dispersed networks, and larger-scale measurements of brain dynamics (without cellular resolution) such as electroencephalography, fMRI, intrinsic optical imaging (Chapter 9), or voltage-sensitive dye imaging (Chapter 6). Because of its special advantages for deep tissue imaging TPLSM has also expanded in various other research fields beyond neuroscience, for example, to the study of cell dynamics in the immune system, in the kidney, or in tumors.

For the near future, I see several important research fields that are likely to benefit from in vivo two-photon imaging: (1) Repeated imaging of cellular or network activity should enable studies of the postnatal development of neural circuit activity. This goal will require repeated imaging sessions with activity-dependent probes over days and weeks,[72] similar to the approach routinely used for structural imaging. Stable expression of genetically encoded indicators should make this type of measurement feasible. (2) The same approach might also enable in vivo studies of circuit plasticity. The challenge is to perform repeated imaging of the same cellular network (e.g., before and after a sensory deprivation protocol or following brain injury). Again such experiments most likely will rely on genetically encoded markers. Analysis of changes in sensory response probabilities or neuronal cross correlations may then reveal mechanisms underlying circuit plasticity. Data analysis will not be trivial, though, because of the high-dimensionality of functional data sets from large populations. (3) The application of TPLSM to study animal models of brain disease such as Alzheimer's disease[86] or stroke[87] will further expand. Here, the ability to monitor dynamic processes of interacting cell types using multicolor imaging—for example, glial cells and neurons—will be especially important. While most of traditional neuropathology is based on looking for changes in morphology, TPLSM opens the avenue for investigating alterations in cell activity or network dynamics and for probing the effect of potential therapeutic interventions.[88] All these fields will also benefit from further developments of techniques for optical stimulation (or silencing), based on photomanipulation using caged compounds[26–28] or light-activated channels.[86]

Finally, to establish the behavioral relevance of cellular and local network signaling, a key challenge is to perform in vivo imaging experiments in awake rather than anesthetized animals.[90] Two methodological routes toward this goal are possible (Figure 2.7). First, TPLSM can be applied to head-restrained animals that are trained to tolerate head-fixation. This has recently been demonstrated for imaging of cortical network activity following bulk loading of calcium indicators.[91] Although they were head-restrained, mice in this study actually were mobile as they could run on an air-supported styrofoam ball (Figure 2.7A). The advantage of the head-restrained approach is that a standard microscope can be used providing optimal

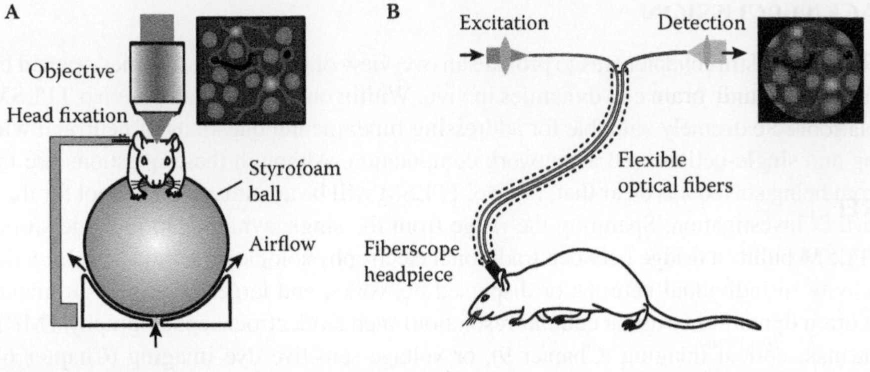

FIGURE 2.7 Routes toward high-resolution two-photon imaging of neural activity in awake behaving animals. (A) Imaging in head-restrained awake mice. The animal is mobile on an air-supported styrofoam ball. (B) Imaging in freely behaving animals using a head-mounted miniaturized microscope. The fiberscope headpiece is connected via optical fibers to the excitation laser beam and the detector system. (B) Adapted from Göbel, W. et al. *Nat Methods* 4: 73–79, 2007. With permission.

imaging conditions. In addition, the special 3D scanning modes for measuring large-scale network activity could possibly be applied to behaving animals. A special problem to be solved is that motion artifacts are more pronounced under awake conditions. Motion correction algorithms will be crucial to correct spatial distortions in the fluorescence measurements as far as possible.[91,92]

The second promising method for high-resolution optical imaging in behaving animals is miniaturization of the microscope so that it can be carried by rats or even mice (Figure 2.7B). The principal idea is to uncouple a small microscope headpiece from all large experimental equipment by guiding the excitation light through an optical fiber. The small headpiece may contain lenses, a small scan device, and a photodetector. Fluorescence light can also be collected through an additional optical fiber and detected remotely. In 2001, imaging of blood vessels in the neocortex of freely moving rats was demonstrated with the first two-photon fiberscope,[93] weighing about 20 g. Since then, various designs of miniaturized two-photon microscopes have been published (for a review see Reference 94). Most recently, we introduced an ultracompact fiberscope design (<1 g) and applied it to imaging calcium signals in Purkinje cell dendrites in anesthetized rats.[91] Using a single photon fiberscope another study recently achieved optical recordings of Purkinje cell activity in freely moving mice.[96] The next major breakthrough—to optically record neural activity with cellular resolution in freely moving animals—is still pending. Both the head-restrained and the freely moving approach should be highly beneficial for relating dendritic signals as well as local network activity to the behavioral state and actual behavior of the animal. For example, these methods may allow proving the in vivo occurrence of local spikes in thin neuronal dendrites and to reveal their functional relevance. A direct correlation of neural activity with animal perception and behavior will be essential to uncover fundamental principles of information processing on the level of local circuits and to formulate models of how large-scale brain activity emerges from the lower-level brain cell dynamics.

ACKNOWLEDGMENTS

Recent work in my laboratory was funded by the Max Planck Society, the University of Zurich, and grants from the Human Frontier Science Program and the Swiss National Science Foundation.

REFERENCES

1. Borst, A. and Egelhaaf, M. In vivo imaging of calcium accumulation in fly interneurons as elicited by visual motion stimulation. *Proc Natl Acad Sci USA* 89: 4139–4143, 1992.
2. Denk, W. et al. Two-photon laser scanning fluorescence microscopy. *Science* 248: 73–76, 1990.
3. Denk W. Two-photon scanning photochemical microscopy: Mapping ligand-gated ion channel distributions. *Proc Natl Acad Sci USA* 91: 6629–6633, 1994.
4. Svoboda, K. et al. In vivo dendritic calcium dynamics in neocortical pyramidal neurons. *Nature* 385: 161–165, 1997.
5. Young, P. and Feng, G. Labeling neurons in vivo for morphological and functional studies. *Curr Opin Neurobiol* 14: 642–646, 2004.
6. Miyawaki, A. Innovations in the imaging of brain functions using fluorescent proteins. *Neuron* 48: 189–199, 2005.
7. Kurtz, R. et al. Application of multiline two-photon microscopy to functional in vivo imaging. *J Neurosci Methods* 151: 276–286, 2006.
8. Zipfel, W.R. et al. Nonlinear magic: Multiphoton microscopy in the biosciences. *Nat Biotechnol* 21: 1369–1377, 2003.
9. Helmchen, F. and Denk, W. Deep tissue two-photon microscopy. *Nat Methods* 2: 932–940, 2005.
10. Svoboda, K. and Yasuda, R. Principles of two-photon excitation microscopy and its applications to neuroscience. *Neuron* 50: 823–839, 2006.
11. Kerr, J.N. and Denk, W. Imaging in vivo: Watching the brain in action. *Nat Rev Neurosci* 9: 195–205, 2008.
12. Kleinfeld, D. et al. Fluctuations and stimulus-induced changes in blood flow observed in individual capillaries in layers 2 through 4 of rat neocortex. *Proc Natl Acad Sci USA* 95: 15741–15746, 1998.
13. Takano, T. et al. Two-photon imaging of astrocytic Ca^{2+} signaling and the microvasculature in experimental mice models of Alzheimer's disease. *Ann N Y Acad Sci* 1097: 40–50, 2007.
14. Helmchen, F. and Kleinfeld, D. In vivo measurements of blood flow and glial cell function with two-photon laser scanning microscopy. *Methods Enzymol,* 444: 231–254, 2008.
15. Denk, W. et al. Anatomical and functional imaging of neurons using 2-photon laser-scanning microscopy. *J Neurosci Methods* 54: 151–162, 1994.
16. Svoboda, K. et al. Spread of dendritic excitation in layer 2/3 pyramidal neurons in rat barrel cortex in vivo. *Nat Neurosci* 2: 65–73, 1999.
17. Helmchen, F. et al. In vivo dendritic calcium dynamics in deep-layer cortical pyramidal neurons. *Nat Neurosci* 2: 989–996, 1999.
18. Charpak, S. et al. Odor-evoked calcium signals in dendrites of rat mitral cells. *Proc Natl Acad Sci USA* 98: 1230–1234, 2001.
19. Helmchen, F. and Waters, J. Ca^{2+} imaging in the mammalian brain in vivo. *Eur J Pharmacol* 447: 119–129, 2002.
20. Waters, J. et al. Supralinear Ca^{2+} influx into dendritic tufts of layer 2/3 neocortical pyramidal neurons in vitro and in vivo. *J Neurosci* 23: 8558–8567, 2003.

21. Euler, T. et al. Directionally selective calcium signals in dendrites of starburst amacrine cells. *Nature* 418: 845–852, 2002.
22. Nevian, T. and Helmchen, F. Calcium indicator loading of neurons using single-cell electroporation. *Pflügers Arch* 454: 675–688, 2007.
23. Kitamura, K. et al. Targeted patch-clamp recordings and single-cell electroporation of unlabeled neurons in vivo. *Nat Methods* 5: 61–67, 2008.
24. Callamaras, N. and Parker, I. Construction of a confocal microscope for real-time x-y and x-z imaging. *Cell Calcium* 26: 271–279, 1999.
25. Göbel, W. and Helmchen, F. New angles on neuronal dendrites in vivo. *J Neurophysiol* 98: 3770–3779, 2007.
26. Losonczy, A. and Magee, J.C. Integrative properties of radial oblique dendrites in hippocampal CA1 pyramidal neurons. *Neuron* 50: 291–307, 2006.
27. Matsuzaki, M. et al. Dendritic spine geometry is critical for AMPA receptor expression in hippocampal CA1 pyramidal neurons. *Nat Neurosci* 4: 1086–1092, 2001.
28. Nikolenko, V. et al. Two-photon photostimulation and imaging of neural circuits. *Nat Methods* 4: 943–950, 2007.
29. Nagel, G. et al. Channelrhodopsins: Directly light-gated cation channels. *Biochem Soc Trans* 33: 863–866, 2005.
30. Trachtenberg, J.T. et al. Long-term in vivo imaging of experience-dependent synaptic plasticity in adult cortex. *Nature* 420: 788–794, 2002.
31. Grutzendler, J. et al. Long-term dendritic spine stability in the adult cortex. *Nature* 420: 812–816, 2002.
32. De Paola, V. et al. Cell type-specific structural plasticity of axonal branches and boutons in the adult neocortex. *Neuron* 49: 861–875, 2006.
33. Chklovskii, D.B. et al. Cortical rewiring and information storage. *Nature* 431: 782–788, 2004.
34. Nimmerjahn, A. et al. Resting microglial cells are highly dynamic surveillants of brain parenchyma in vivo. *Science* 308: 1314–1318, 2005.
35. Davalos, D. et al. ATP mediates rapid microglial response to local brain injury in vivo. *Nat Neurosci* 8: 752–758, 2005.
36. Oheim, M. et al. Two-photon microscopy in brain tissue: Parameters influencing the imaging depth. *J Neurosci Methods* 111: 29–37, 2001.
37. Stosiek, C. et al. In vivo two-photon calcium imaging of neuronal networks. *Proc Natl Acad Sci USA* 100: 7319–7324, 2003.
38. Theer, P. et al. Two-photon imaging to a depth of 1000 μm in living brains by use of a Ti:Al2O3 regenerative amplifier. *Opt Lett* 28: 1022–1024, 2003.
39. Theer, P. and Denk, W. On the fundamental imaging-depth limit in two-photon microscopy. *J Opt Soc America a-Optics Image Science and Vision* 23: 3139–3149, 2006.
40. Sullivan, M.R. et al. In vivo calcium imaging of circuit activity in cerebellar cortex. *J Neurophysiol*, 94: 1636–1644, 2005.
41. Davalos, D. et al. Stable in vivo imaging of densely populated glia, axons and blood vessels in the mouse spinal cord using two-photon microscopy. *J Neurosci Methods* 169: 1–7, 2008.
42. Jung, J.C. et al. In vivo mammalian brain imaging using one- and two-photon fluorescence microendoscopy. *J Neurophysiol* 92: 3121–3133, 2004.
43. Levene, M.J. et al. In vivo multiphoton microscopy of deep brain tissue. *J Neurophysiol* 91: 1908–1912, 2004.
44. Göbel, W. et al. Imaging cellular network dynamics in three dimensions using fast 3D laser scanning. *Nat Methods* 4: 73–79, 2007.
45. Reddy, G.D. et al. Three-dimensional random access multiphoton microscopy for functional imaging of neuronal activity. *Nat Neurosci* 11: 713–720, 2008.

46. Salome, R. et al. Ultrafast random-access scanning in two-photon microscopy using acousto-optic deflectors. *J Neurosci Methods* 154: 161–174, 2006.
47. Vucinic, D. and Sejnowski, T.J. A compact multiphoton 3D imaging system for recording fast neuronal activity. *PLoS ONE* 2: e699, 2007.
48. Christie, R.H. et al. Growth arrest of individual senile plaques in a model of Alzheimer's disease observed by in vivo multiphoton microscopy. *J Neurosci* 21: 858–864, 2001.
49. Kasischke, K.A. et al. Neural activity triggers neuronal oxidative metabolism followed by astrocytic glycolysis. *Science* 305: 99–103, 2004.
50. Giepmans, B.N. et al. The fluorescent toolbox for assessing protein location and function. *Science* 312: 217–224, 2006.
51. O'Hare, H.M. et al. Chemical probes shed light on protein function. *Curr Opin Struct Biol* 17: 488–494, 2007.
52. Shaner, N.C. et al. A guide to choosing fluorescent proteins. *Nat Methods* 2: 905–909, 2005.
53. Livet, J. et al. Transgenic strategies for combinatorial expression of fluorescent proteins in the nervous system. *Nature* 450: 56–62, 2007.
54. Lichtman, J.W. et al. A technicolour approach to the connectome. *Nat Rev Neurosci* 9: 417–422, 2008.
55. Nimmerjahn, A. et al. Sulforhodamine 101 as a specific marker of astroglia in the neocortex in vivo. *Nat Methods* 1: 31–37, 2004.
56. Dittgen, T. et al. Lentivirus-based genetic manipulations of cortical neurons and their optical and electrophysiological monitoring in vivo. *Proc Natl Acad Sci USA* 101: 18206–18211, 2004.
57. Lendvai, B. et al. Experience-dependent plasticity of dendritic spines in the developing rat barrel cortex in vivo. *Nature* 404: 876–881, 2000.
58. Haas, K. et al. Single-cell electroporation for gene transfer in vivo. *Neuron* 29: 583–591, 2001.
59. Tsien, R.Y. Fluorescent probes of cell signaling. *Annu Rev Neurosci* 12: 227–253, 1989.
60. Sohya, K. et al. GABAergic neurons are less selective to stimulus orientation than excitatory neurons in layer II/III of visual cortex, as revealed by in vivo functional Ca^{2+} imaging in transgenic mice. *J Neurosci* 27: 2145–2149, 2007.
61. Margrie, T.W. et al. Targeted whole-cell recordings in the mammalian brain in vivo. *Neuron* 39: 911–918, 2003.
62. Nagayama, S. et al. In vivo simultaneous tracing and Ca^{2+} imaging of local neuronal circuits. *Neuron* 53: 789–803, 2007.
63. Tsien, R.Y. A non-disruptive technique for loading calcium buffers and indicators into cells. *Nature* 290: 527–528, 1981.
64. Garaschuk, O. et al. Targeted bulk-loading of fluorescent indicators for two-photon brain imaging in vivo. *Nat Protoc* 1: 380–386, 2006.
65. Garaschuk, O. et al. Optical monitoring of brain function in vivo: From neurons to networks. *Pflügers Arch* 453: 385–396, 2006.
66. Göbel, W. and Helmchen, F. In vivo calcium imaging of neural network function. *Physiology (Bethesda)* 22: 358–365, 2007.
67. Miyawaki, A. et al. Fluorescent indicators for Ca^{2+} based on green fluorescent proteins and calmodulin. *Nature* 388: 882–887, 1997.
68. Garaschuk, O. et al. Troponin C-based biosensors: A new family of genetically encoded indicators for in vivo calcium imaging in the nervous system. *Cell Calcium*, 2007.
69. Kotlikoff, M.I. Genetically encoded Ca^{2+} indicators: Using genetics and molecular design to understand complex physiology. *J Physiol* 578: 55–67, 2007.
70. Mank, M. and Griesbeck, O. Genetically encoded calcium indicators. *Chem Rev* 108: 1550–1564, 2008.

71. Heim, N. et al. Improved calcium imaging in transgenic mice expressing a troponin C-based biosensor. *Nat Methods* 4: 127–129, 2007.
72. Mank, M. et al. A genetically encoded calcium indicator for chronic in vivo two-photon imaging. *Nat Methods* 5: 805–811, 2008.
73. Wallace, D.J. et al. Single-spike detection in vitro and in vivo with a genetic Ca^{2+} sensor. *Nat Methods* 5: 797–804, 2008.
74. Mao, T. et al. Characterization and subcellular targeting of GCaMP-type genetically encoded calcium indicators. *PLoS ONE* 3: e1796, 2008.
75. Helmchen, F. et al. Ca^{2+} buffering and action potential-evoked Ca^{2+} signaling in dendrites of pyramidal neurons. *Biophys J* 70: 1069–1081, 1996.
76. Yaksi, E. and Friedrich, R.W. Reconstruction of firing rate changes across neuronal populations by temporally deconvolved Ca^{2+} imaging. *Nat Methods* 3: 377–383, 2006.
77. Kerr, J.N. et al. Imaging input and output of neocortical networks in vivo. *Proc Natl Acad Sci USA* 102: 14063–14068, 2005.
78. Kerr, J.N. et al. Spatial organization of neuronal population responses in layer 2/3 of rat barrel cortex. *J Neurosci* 27(48): 13316–13328, 2007.
79. Sato, T.R. et al. The functional microarchitecture of the mouse barrel cortex. *PLoS Biol* 5: e189, 2007.
80. Waters, J. and Helmchen, F. Boosting of action potential backpropagation by neocortical network activity in vivo. *J Neurosci* 24: 11127–11136, 2004.
81. Larkum, M.E. et al. Dendritic spikes in apical dendrites of neocortical layer 2/3 pyramidal neurons. *J Neurosci* 27: 8999–9008, 2007.
82. Ohki, K. et al. Functional imaging with cellular resolution reveals precise micro-architecture in visual cortex. *Nature* 433: 597–603, 2005.
83. Ohki, K. et al. Highly ordered arrangement of single neurons in orientation pinwheels. *Nature* 442: 925–928, 2006.
84. Mrsic-Flogel, T.D. et al. Homeostatic regulation of eye-specific responses in visual cortex during ocular dominance plasticity. *Neuron* 54: 961–972, 2007.
85. Briggman, K.L. and Denk, W. Towards neural circuit reconstruction with volume electron microscopy techniques. *Curr Opin Neurobiol* 16: 562–570, 2006.
86. Bacskai, B.J. et al. Imaging of amyloid-beta deposits in brains of living mice permits direct observation of clearance of plaques with immunotherapy. *Nat Med* 7: 369–372, 2001.
87. Zhang, S. and Murphy, T.H. Imaging the impact of cortical microcirculation on synaptic structure and sensory-evoked hemodynamic responses in vivo. *PLoS Biol* 5: e119, 2007.
88. Busche, M.A. et al. Clusters of hyperactive neurons near amyloid plaques in a mouse model of Alzheimer's disease. *Science* 321: 1686–1689, 2008.
89. Zhang, F. et al. Circuit-breakers: Optical technologies for probing neural signals and systems. *Nat Rev Neurosci* 8: 577–581, 2007.
90. Kleinfeld, D. and Griesbeck, O. From art to engineering? The rise of in vivo mammalian electrophysiology via genetically targeted labeling and nonlinear imaging. *PLoS Biol* 3: e355, 2005.
91. Dombeck, D.A. et al. Imaging large scale neural activity with cellular resolution in awake mobile mice. *Neuron* 56(1): 43–57, 2007.
92. Greenberg, O.S. and Kerr, J.N. Automated correction of fast motion artifacts for two-photon imaging of awake animals. *J Neurosci Meth* 176: 1–15, 2009.
93. Helmchen, F. et al. A miniature head-mounted two-photon microscope. high-resolution brain imaging in freely moving animals. *Neuron* 31: 903–912, 2001.
94. Flusberg, B.A. et al. Fiber-optic fluorescence imaging. *Nat Methods* 2: 941–950, 2005.
95. Engelbrecht, C.J. et al. Ultra-compact fiber-optic two-photon microscope for functional fluorescence imaging in vivo. *Opt Express* 16: 5556–5564, 2008.
96. Flusberg, B.A. et al. High-speed, miniaturized fluorescence microscopy in freely moving mice. *Nat Methods* 5:935–938, 2008.

3 In Vivo Two-Photon Laser Scanning Microscopy with Concurrent Plasma-Mediated Ablation

Principles and Hardware Realization

Philbert S. Tsai and David Kleinfeld

CONTENTS

3.1 INTRODUCTION

Many biological processes of current interest occur below the surface layers of accessible tissue. It is often the case that the surface layers cannot easily be removed without adversely affecting the physiology and function of the deeper layers. A variety of imaging techniques have been developed to perform sectioning deep to the surface using optical, electrical, and magnetic contrast agents and recording methods. Two-photon laser scanning microscopy (TPLSM) with ultrashort (i.e., order 100-fs) pulsed laser light provides optically sectioned images from depths of 500 microns or more below the surface in highly scattering brain tissue.[1–5] This method is unique in that it can provide images with submicrometer lateral resolution and micrometer axial resolution on the millisecond time-scale,[6] as is required for the study of many dynamic biological processes.[1,7–11]

A strength of TPLSM is the ease with which this technique may be combined with electrical measurements of physiological parameters and with other optical techniques. These include one-photon uncaging with gated continuous laser light, two-photon uncaging with ultrashort pulsed laser light and, of particular interest in this work, plasma-mediated ablation through the nonlinear absorption of amplified ultrashort laser pulses.

A nonlinear interaction that is of particular relevance for the ablation of submicrometer volumes of neuronal[12] and vascular[13] tissue deep in the brain is plasma-mediated ablation.[14–16] Here, the optical field of the ultrashort pulse leads to a high rate of tunneling by electrons to form an electron plasma.[17] The density of this plasma rapidly builds up by virtue of field-driven collisions between free electrons and molecules, while the thickness of the plasma approaches a value that is much less than the depth of the focal volume. The absorption of laser light by this thin layer leads to highly localized ablation and the explosive removal of material. One critical feature of plasma-mediated ablation is that light outside of the focal volume has little detrimental effect on the specimens; this limits the thermal build-up and ensuing collateral damage in the targeted tissue.

This chapter provides an overview of the technical considerations relevant to the design and assembly of a system for TPLSM that is particularly suitable for in vivo imaging and histology, either alone or in combination with plasma-mediated ablation. It extends an earlier technical presentation[18] and complements other designs of upright systems.[19–21] The conversion of commercially available confocal microscopes for use in TPLSM, rather than the de novo construction of a system, have been described.[22–26] Commercial systems for TPLSM are available from Prairie (Ultima; www.prairie-technologies.com) and Leica (TPS MP; www.leica.com), and it is likely that the commercial market will continue to mature. Commercial systems that incorporate amplified ultrashort laser pulses do not exist at this time, although it should be simple to adapt current systems to support two beams.

3.2 OVERVIEW

3.2.1 Single Photon Absorption for Imaging through Weakly Scattering Preparations

In traditional fluorescence microscopy, including confocal microscopy, fluorescent molecules that are either intrinsic to the tissue or artificially introduced are excited by a wavelength of light that falls within the linear absorption spectrum of the molecule. A single photon is absorbed by the molecule and promotes it to an excited electronic state. Upon thermal equilibration and relaxation back to its ground state, the molecule will emit a photon that is shifted toward longer wavelengths with respect to the excitation light.

The probability of exciting fluorescence by the absorption of a single photon is linearly proportional to the spatial density of photons (i.e., the light intensity). The total fluorescence is, therefore, constant from any given axial plane (i.e., a plane taken normal to the propagation direction), regardless of the focusing of the incident beam. As one moves toward the focus of a beam, its intensity in a given axial plane increases as the power becomes distributed over a smaller cross-sectional area. At the same time, fewer fluorescent molecules in that plane are exposed to the illumination, so that the lateral extent of the excitation is reduced.

3.2.1.1 Direct Imaging

Biological preparations that do not appreciably scatter or absorb the incident light, such as in vitro slice or cell culture, can be observed by a variety of direct imaging techniques,[27] typically with the use of a charge coupled device array detector. In whole-field fluorescent microscopy, the entire region of interest in the preparation is concurrently illuminated, and all of the collected fluorescent light is recorded. This method has the advantage of a high rate of data collection (e.g., hundreds of frames per second with current array detectors), but suffers from the lack of optical sectioning. The fluorescence generated from planes outside the focal plane contributes in a nonuniform manner to the background level. The result is a lack of axial resolution and a signal-to-noise ratio that is lower than for an image in which fluorescence from outside the focal plane is eliminated.

Axial resolution on the order of the diffraction limit may, in principle, be achieved with direct imaging methods in which light from regions outside the focal plane is systematically eliminated (Table 3.1). One method is to acquire data at multiple depths, by shifting the focus above and below the object plane, and then use the images from out-of-focus planes, in conjunction with a deconvolution procedure, to model and subtract the out-of focus light.[28–31] This method and its extensions have the advantage of experimental simplicity in that only a motorized focus control need be added to a standard microscope. On the other hand, diffraction by the open aperture of the objective leads to a set of dark rings in the transfer function of the objective so that this method is insensitive to the spatial frequencies of these rings. This is the so-called "missing cone" problem in which the cone represents spatial frequencies in radial and axial momentum space (i.e., k_r and k_z). An alternate procedure is to project a line grating into the object plane so that the projected image has spacing on the order of the lateral

TABLE 3.1
Nominal Properties of Techniques for Functional Imaging with Fluorescent Indicators

<table>
<tr><th>Mode</th><th>Technique</th><th>Axial Resolution</th><th>Lateral Resolution</th><th>Frame Period (100 by 100 pixels)</th><th>Depth Penetration</th></tr>
<tr><td rowspan="4">Direct image formation (CCD acquisition)</td><td>Full field</td><td>$\gg\lambda$</td><td rowspan="6">$\sim\lambda$</td><td>~100 Hz</td><td rowspan="3">$\sim\Lambda_{scattering}$</td></tr>
<tr><td>Vibrating grating</td><td rowspan="2">$\sim\lambda$</td><td rowspan="3"><30 Hz</td></tr>
<tr><td>Deconvolution</td></tr>
<tr><td>Interferometric</td><td>$\ll\lambda$</td><td>$\ll\Lambda_{scattering}$</td></tr>
<tr><td rowspan="2">Point scanning (PMT acquisition)</td><td>Confocal</td><td rowspan="2">$\sim\lambda$</td><td>~10 Hz</td><td>$\sim\Lambda_{scattering}$</td></tr>
<tr><td>Multiphoton excitation</td><td>>100 Hz (resonant or AOM line scan)</td><td>$\gg\Lambda_{scattering}$</td></tr>
</table>

diffraction limit. When this grating is modulated in time, only the part of the signal that emanates from the object plane is concurrently modulated. Thus, information about the intensity distribution only in the object plane may be extracted.[32,33] A final set of methods are applicable only to physically thin objects and rely on an interferometric cavity to set up an optical field that varies in the axial direction.[34,35] This may be used to achieve axial resolution on length-scales of less than one wavelength.

3.2.1.2 Confocal Laser Scanning Microscopy

Laser scanning microscopy provides an improved means to achieve axial resolution compared with direct imaging techniques. An incoming laser beam is deflected by a pair of rotating mirrors, mounted at orthogonal angles, and is subsequently focused by a microscope objective onto the preparation. The angular deflection of the collimated laser beam causes the focused spot to move across the preparation in a specified raster pattern. The fluorescent light that is generated at the focal volume is collected by a detector as a function of time; this time series is transformed back into a position in the preparation along the trajectory of the spot. The drawback to laser scanning methods of microscopy is a reduced speed of acquisition compared with direct imaging techniques. The maximum acquisition rate is limited by the speed at which the scan mirrors can raster across the region of interest.

In confocal microscopy, axial resolution is obtained by preferentially collecting fluorescent light that is generated near the focal plane, while rejecting light that is generated outside the focal plane. Specifically, the fluorescent light is refocused along the same beam path as the excitation light. After passing back through the scan mirrors—a process referred to as "descanning"—the light is spatially filtered by a small aperture, or pinhole, whose radius is chosen so that its image in the preparation matches or exceeds the diffraction-limited radius of the illuminated region.

The pinhole preferentially passes fluorescent light generated near the focal plane. Although confocal microscopy works well for optically thin preparations, it is inefficient for preparations in which substantial scattering occurs. As the fluorescent light becomes more heavily scattered, a greater portion of the photons from the focal plane are deflected from their original trajectories and are thus blocked by the pinhole, remaining undetected. Conversely, an increasing portion of the light from outside the focal plane will be deflected to pass through the pinhole. The result is a drop in the signal-to-noise ratio.

3.2.2 Two-Photon Laser Scanning Microscopy for Strongly Scattering Preparations

Two-photon laser scanning microscopy utilizes laser light at approximately twice the wavelength of that used in traditional, one-photon microscopy. These photons carry half of the energy required to promote a molecule to its excited state, and thus the excitation process requires the simultaneous absorption of two photons from the excitation beam. The probability of two-photon absorption thus scales with the square of the incident laser intensity. This nonlinear absorption process provides an intrinsic mechanism for optical sectioning solely by the incident beam.

To gain a semi-quantitative understanding into the localized nature of two-photon excitation, consider the total fluorescence generated from a given cross-sectional plane, F_{total}. This is linearly proportional to the illuminated area, A, and a quadratic function of the incident intensity, I, defined as the incident laser power, P, divided by the area (i.e., $I = P/A$ [Figure 3.1]). The illuminated area in a given plane is proportional to the square of the axial distance of that plane from the focal plane (i.e., $A \propto z^2$) for z larger than the focal depth of the beam. Thus,

$$F_{\text{total}} \propto I^2 A \propto \left(\frac{P}{z}\right)^2.$$

The amount of fluorescence that is generated from a given axial plane is seen to decrement as a quadratic function of the distance from the focal plane (Figure 3.2). The integrated fluorescence is thus finite, that is,

$$\int_{-\infty}^{+\infty} dz\ F_{\text{total}} \propto 2 \int_{z_{\text{confocal}}}^{+\infty} dz F_{\text{total}} \propto 2 \int_{z_{\text{confocal}}}^{+\infty} dz \left(\frac{P}{z}\right)^2 \to \text{Constant}$$

where z_{confocal}, defined later (Equation 3.2), is proportional to the focus depth. Only light within the focal region contributes significantly to the signal. Thus, with TPLSM all of the fluorescent light can be collected as signal, regardless of scattering. In contrast, the above integral diverges for single photon absorption.

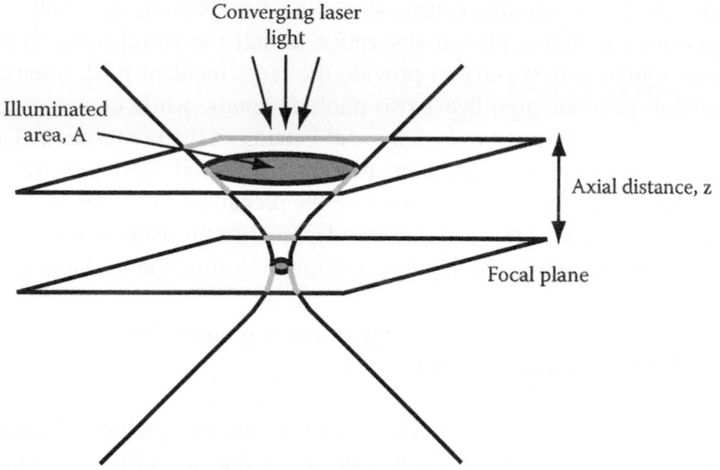

FIGURE 3.1 Cartoon of laser light near the focus of an objective. The shaded regions indicated the cross-sectional area in a given plane, normal to the direction of propagation, that is illuminated by the converging laser light. The illuminated cross-sectional area is seen to decrease as the distance to the focal plane decreases.

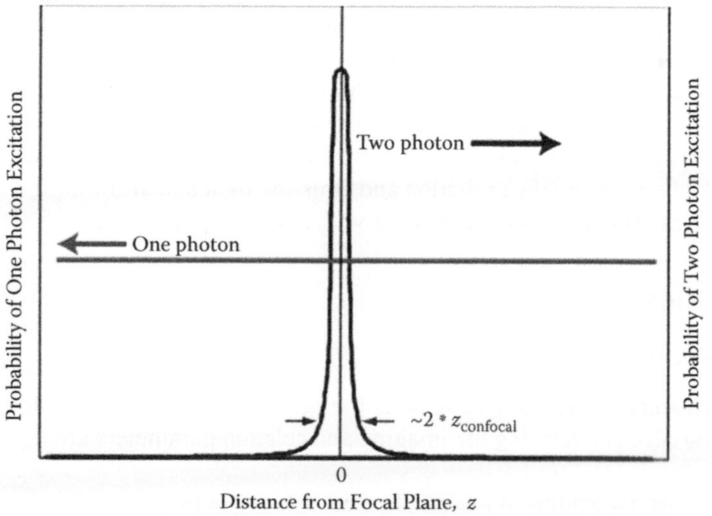

FIGURE 3.2 The integrated excitation of fluorphores within an axial plane as a function of the axial distance from the focus, denoted z. Single-photon excitation results in an excitation profile that is constant with axial distance (light line). Two-photon excitation results in an excitation profile that falls off quadratically with axial distance from the focal plane (dark line).

The probability of simultaneously absorbing two photons is small compared to the probability of single photon absorption at half the wavelength. Therefore, a pulsed laser source is necessary to provide the large incident peak intensities that are required to generate significant two-photon signals, while concurrently maintaining the average power at levels that avoid heating of the preparation. The use of ultrashort laser pulses allows for the greatest efficiency and depth penetration while reducing the potential for thermal damage to the preparation (but see Reference 36). In particular, the extent of thermal damage depends on the average power delivered by the incident beam, while the depth penetration is a function of the instantaneous peak power.

Two-photon laser scanning microscopy is uniquely suited for optically sectioning through thick preparations.[37] In brief:

1. The fluorescent signal is not descanned (i.e., the fluorescent light does not require spatial filtering with a pinhole, as for the case of confocal microscopy). This yields a higher collection efficiency for scattering preparations. It further permits a much simpler alignment of collection optics compared with confocal microscopy.
2. The use of near infrared light for two-photon excitation, as opposed to blue/UV light for single-photon excitation, is preferable with biological preparations. Wavelengths in the near-infrared are less heavily scattered, as well as less likely to be absorbed by biological molecules. In particular, the flavins and other molecules involved in bioenergetics are far less excitable by near infrared light than by UV/blue light.
3. The two-photon spectra of fluorophores are generally broader than their one photon counterparts and are spectrally separated further from the fluorescence emission peak. Both of these characteristics simplify the technical task of separating the excitation and emission light and allow for the simultaneous excitation of multiple dyes with a single laser source.

3.3 BASICS

3.3.1 Focal Volumes

The fundamental optical length scale is the radius of the beam at the focus.[38] The relevant variables in deriving the imaging and ablation parameters are

$\omega(z) \equiv$ Radius of beam a distance z from focus

$\omega(0) \equiv$ Radius of beam at the focus

$z_{\text{confocal}} \equiv$ Confocal depth (i.e., axial distance for beam to double width)

$V_{\text{excitation}} \equiv$ Optical excitation volume

$V_{\text{ablation}} \equiv$ Plasma-mediated ablation volume

$\lambda_o \equiv$ Center wavelength of laser

$\text{NA} \equiv$ Numerical aperture of objective

$n \equiv$ Optical index ($n = 1.33$ for water)

$P^{\text{incident}} \equiv$ Incident laser power
$I^{\text{incident}}(r,z,t) \equiv$ Incident laser intensity
$\text{E} \equiv$ Energy of laser pulse
$\varepsilon \equiv$ Normalized energy for plasma-mediated ablation
$F^{\text{incident}}(r,z) \equiv$ Incident laser pulse fluence

The radius of the beam at the focus, $\omega(0)$, is

$$\omega(0) = \frac{1}{\pi}\frac{\lambda_o}{NA}\sqrt{1-\left(\frac{NA}{n}\right)^2} \xrightarrow[NA \ll n]{} \frac{1}{\pi}\frac{\lambda_o}{NA}. \tag{3.1}$$

3.3.1.1 Excitation

The above expression for the radius of the beam (Equation 3.1) should be compared with the minimum lateral length-scale that can be resolved with planar illumination,[39] that is, $\lambda_o/(2 \bullet NA)$.[27] The confocal length is the distance for the beam to double in area, that is,

$$z_{\text{confocal}} = \frac{\pi\, n\, \omega^2(0)}{\lambda_o} = \frac{n}{\pi}\frac{\lambda_o}{(NA)^2}\left[1-\left(\frac{NA}{n}\right)^2\right] \xrightarrow[NA \ll n]{} \frac{n}{\pi}\frac{\lambda_o}{(NA)^2}; \tag{3.2}$$

this expression should be compared with the minimum axial length-scale that can be resolved with planar illumination,[39] that is, $(\lambda_o/n)\left[1-\sqrt{1-(NA/n)^2}\,\right]^{-1} \xrightarrow[NA \ll n]{} (2n\lambda_o)/(NA)^2$.[27] The effective excitation volume is

$$V_{\text{excitation}} = \pi\, \omega^2(0)\, 2\, z_{\text{confocal}} = \frac{2n}{\pi^2}\frac{\lambda_o^3}{(NA)^4}\left[1-\left(\frac{NA}{n}\right)^2\right]^2 \xrightarrow[NA \ll n]{} \frac{2n}{\pi^2}\frac{\lambda_o^3}{(NA)^4}. \tag{3.3}$$

For the practical case of a 40-times magnification water dipping lens with NA = 0.80 and a laser with 800 nm excitation light, the Gaussian-profile beam parameters are

$$\omega(0) = 0.25\ \mu\text{m},\ z_{\text{confocal}} \cong 0.35\ \mu\text{m},\ \text{and}\ V_{\text{excitation}} \cong 0.15\ \mu\text{m}^3.$$

3.3.1.2 Ablation

The volume of material removed by plasma-mediated optical ablation may be estimated in terms of the previously given length scales (Equations 3.1 to 3.3), the energy of the incident pulses, and a phenomenological value for the threshold fluence of ablation.[40] The intensity for a Gaussian beam that propagates along the z-axis is[38]

$$I^{\text{incident}}(r,z,t) = 2\left[\frac{P^{\text{incident}}(t)}{\pi\omega^2(z)}\right] e^{-2r^2/\omega^2(z)} \tag{3.4}$$

The relevant parameterization for cutting is the fluence,

$$F^{\text{incident}}(r,z) \equiv \int_{-\infty}^{\infty} \mathrm{d}t\; I^{\text{incident}}(r,z,t) = 2\left[\frac{E}{\pi\omega^2(z)}\right] e^{-2r^2/\omega^2(z)}, \tag{3.5}$$

where the energy per pulse is

$$E = \int_{-\infty}^{\infty} \mathrm{d}t \int_{0}^{\infty} r\mathrm{d}r \int_{-\pi}^{\pi} \mathrm{d}\theta\, I(r,0,t). \tag{3.6}$$

The dependence of the fluence on the axial distance is simplified for the paraxial approximation, valid for NA << n, that is,

$$\omega(z) = \omega(0)\sqrt{1 + \left(\frac{NA}{n}\frac{z}{\omega(0)}\right)^2}. \tag{3.7}$$

We denote the threshold value of the fluence at the focus as F_T. The extent of the ablated region along the z-axis is found from $F_T = 2E/[\pi\omega^2(z_T)]$, for which $z_T = \omega(0)\left(\frac{n}{\mathrm{NA}}\right)\sqrt{\varepsilon - 1}$ where

$$\varepsilon = \frac{2E}{\pi\, \omega^2(0)\, F_T} \tag{3.8}$$

is the normalized energy per pulse and ablation requires $\varepsilon > 1$. The ablation volume with the focal plane below the surface of the tissue is given by

$$V_{ablation} = \int_{-z_T}^{+z_T} dz\; \pi r^2(z) \;=\; \pi\, \omega^3(0)\left(\frac{n}{NA}\right)\int_{0}^{x_T} dx\; (1+x^2)\, \ln\left(\frac{1+x_T^2}{1+x^2}\right), \tag{3.9}$$

where $r(z)$ is the radius for which the fluence of the beam equals F_T, and we inverted the expression for $F(r,z)$ and normalized the integration variable, with $x_T = \sqrt{\varepsilon - 1}$. Although the integral can be done exactly, extrapolation beyond the threshold fluence is appropriate only for small values of x_T or, equivalently, for ε close to 1. We thus find our essential result,

$$V_{\text{ablation}}^{\text{minimum}} \xrightarrow[\varepsilon\approx 1]{} \left(\frac{2\pi}{9}\right)\omega^3(0)\left(\frac{n}{\mathrm{NA}}\right)(\varepsilon-1)^{3/2} \approx V_{excitation}\left(\frac{\varepsilon-1}{4.33}\right)^{3/2}, \tag{3.10}$$

which corresponds to an ellipsoidal crater of height $2z_T$ and radius $r_T = \omega(0)\sqrt{(\varepsilon-1)/6}$. At high energies per pulse, that is, $\varepsilon >> 1$, or for focal planes deep to the surface of the tissue, one needs to take into account the decrement in energy as the pulse propagates into the sample and creates a plasma.

3.3.2 Scattering by Brain Tissue

The beam will maintain a focus so long as the elastic scattering length, denoted $\Lambda_{scatter}$, is much greater than the depth of focus, that is,

$$\Lambda_{scatter} >> z_{confocal}.$$

The scattering length corresponds to the length over which the incident light is scattered away. For neocortex in living rat, $\Lambda_{scatter} \sim 200$ µm.[41–43] However, for preparations such as the granule cell layer in mouse cerebellum, where the neurons are small and densely packed, the interpretation of preliminary data (Denk and Kleinfeld, unpublished) suggests that $\Lambda_{scatter} \sim z_{confocal}$, and TPLSM is no longer effective.

The argument above suggest that the depth penetration of TPLSM is limited only by the available power. However, the situation is different if structures on the surface of the tissue are labeled in addition to structures at depth. Under this condition, the incident intensity that is required to excite an object at depth can be sufficiently strong to drive two-photon excitation of the same or different molecules near the surface. This can include endogenous fluorophores as well as exogenous labels. Nominally, the maximum distance is found by equating the exponential fall-of in power that occurs with increasing depth against the quadratic increase in intensity as light forms a focus. Further, the numerical aperture decreases with an increase in imaging depth since marginal rays propagate a greater distance than axial rays. These effects result in a practical limit of ~800 µm for the imaging depth with standard 0.8 NA dipping lenses.[43]

3.3.3 Photobleaching and Photodamage by Two-Photon Excitation

The nonlinear excitation profile of TPLSM results in localization of any photodamage, related to bleaching of the fluorophore or otherwise, that may occur. In contrast, single-photon excitation results in photodamage throughout the entire axial extent of the beam. Further, for both single- and two-photon imaging techniques the incident beam can induce heating of the nervous tissue, although this effect may be substantial only when the incident power exceeds ~0.1 W for the near-infrared wavelengths involved in TPLSM.[44]

Increased photobleaching at the focus, possibly driven by third-order or higher nonlinear processes, is a possibility with two-photon excitation. Some experimenters have observed photobleaching rates that scale linearly with absorption[36,45] or scale supralinearly with absorption only at high intensities,[46] while other experimenters have observed the supralinear relation at all intensities.[47,48] Of course, at extremely high peak powers, the laser pulse can cause ionization at the focus that permanently damages the preparation, an effect that we exploit for optical-assisted plasma-mediated ablation.

For imaging through optically thick preparations, TPLSM offers fundamental advantages over single-photon excitation microscopy. For thin preparations, the possibility of increased, but localized photodamage leaves the choice between the single- and two-photon methods unresolved.

3.3.4 Light Source

Pulsed laser light is not monochromatic. A temporally short pulse is composed of a correspondingly broad range of wavelengths. While the exact width will depend on the precise shape of the pulse, the spectral width, $\Delta\lambda$, must scale with the temporal width. The relevant parameters are

$\Delta\lambda \equiv$ Spectral width of pulsed laser light
$\tau_{pulse} \equiv$ Duration of laser pulse
$c \equiv$ Speed of light in air (3.0×10^8 m/s)

The spectra width is thus

$$\Delta\lambda \sim \frac{\lambda_o^2}{c\,\tau_{pulse}}. \quad (3.11)$$

For a sufficiently short pulse, the spectral width of the pulse can extend beyond that of the two-photon excitation spectrum for a given fluorescent molecule. In this limit, the excitation efficiency for an excessively short pulse can be lower than for a longer pulse.

A laser with an appropriate pulse width as well as center wavelength, typically between 730 and 1000 nm, must be used in a TPLSM system.[49] Ti:Sapphire-based laser systems have been reported that produce pulses with widths shorter than 10 fs, while commercially available Ti:Sapphire systems that are capable of generating roughly 100 fs duration pulses are readily available and used for our system (see Section 3.4.2). The corresponding full spectral bandwidth is ~10 nm. This width is less than the width of the two-photon absorption band of many relevant dyes (e.g., 80 nm for fluorescein). The pulse is broadened to 200 fs or more at the exit of the objective (Figure 3.27).

The fluorescent light generated by TPLSM is sensitive to fluctuations in the laser power, the center wavelength of the pulse, the pulse width, and the pulse chirp. Recall that this chirp is a shift in frequency between the start and end of a pulse or, equivalently, an evolving phase throughout the pulse. Pulses with different pulse shapes can be characterized by the same width, yet provide different peak powers and thus different excitation strengths, as demonstrated by the examples in Figure 3.3. Additionally, a temporally compact laser pulse incorporates a correspondingly disperse range of laser wavelengths, and these components can be nonuniformly distributed through the pulse envelope, as occurs during chirp (Figure 3.3c).

Ideally, the chirp of the laser pulse should be zero at the focus of the objective so that the spectral composition of the pulse is uniform across the pulse. This results in a minimum pulse width and is usually achieved with external dispersion compensation.[50] While small amounts of chirp are acceptable, heavily chirped pulses can significantly affect two-photon fluorescence efficiencies. For example, the leading edge of a positively chirped pulse can match the two-photon excitation spectrum of a fluorescent dye, while the trailing edge of the pulse corresponds to the slightly red- or Stokes-shifted emission line. Therefore, a molecule can be excited by the leading edge of a pulse, but undergo stimulated two-photon emission with the trailing

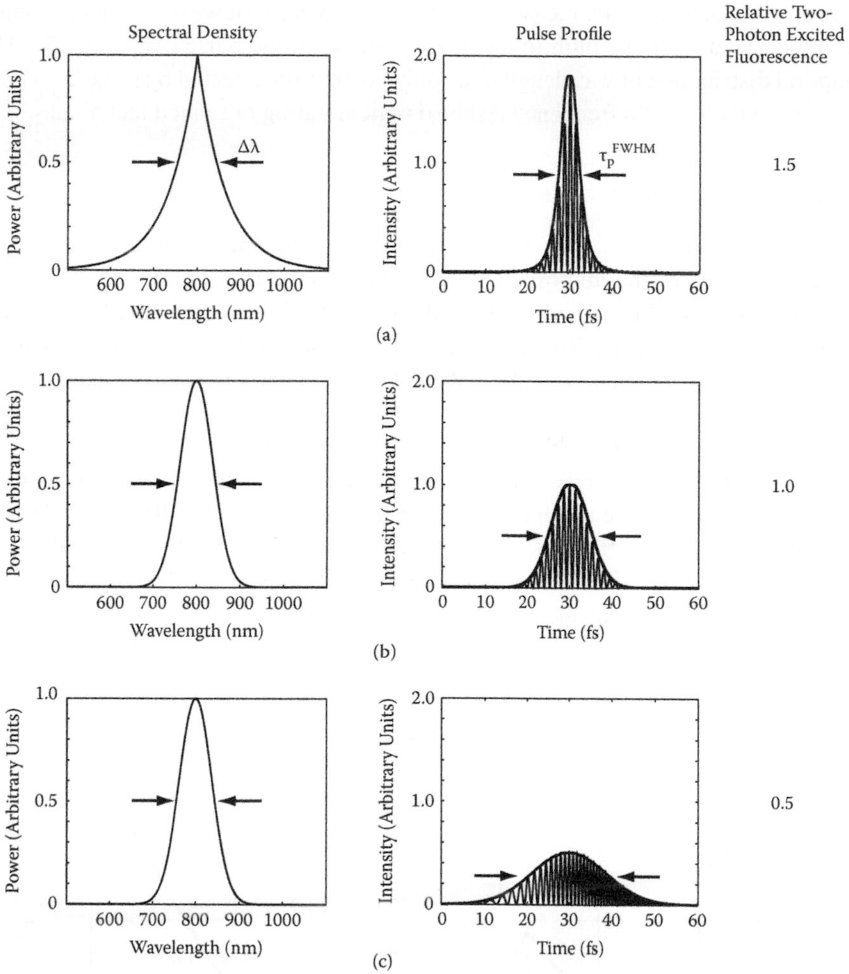

FIGURE 3.3 Pulses with different characteristics can have the same full-width half-maximal spectral width, indicated by $\Delta\lambda$. For illustrative purposes, the pulse widths are 10-times shorter than for those typically used in TPLSM experiments. (a) A pulse with an exponential spectral envelope and the corresponding Lorentzian temporal pulse shape with a full-width half-maximal pulse width of τ_p^{FWHM}. (b) A pulse with a Gaussian spectral envelope and the corresponding Gaussian pulse shape. (c) A pulse in which the constituent frequency components are non-uniformly distributed, i.e., chirped. The extent of the chirp has been grossly exaggerated from typical conditions.

edge of the pulse before single-photon fluorescence can occur.[51,52] This effect may be responsible for the effect of pulse shape on the rate of photobleaching.[53] With regard to ablation, a pulse that is broadened can lead to the production of heat.

The average output power of the laser can be monitored with a simple photodiode. The center wavelength and the spectral width of the pulse of the laser can be determined with a basic spectrometer. Note that while a fast oscilloscope can be used to display the repetition rate of the laser, even the fastest available oscilloscopes are currently unable to directly measure the laser pulse width. However, an interferometric autocorrelator can be built to monitor the shape of the pulse (Appendix B). The temporal distribution of wavelengths can only be directly detected by phase measurements with the use of a frequency resolved optical gating or related techniques.[54]

3.3.5 Delivery of Laser Light

The output parameters of a commercial laser system are unlikely to perfectly match the requirements of the microscope. The output beam may not be optimally collimated or may be collimated with a beam diameter that is either too small or too large for use in the microscope. To compensate, the beam profile is calculated along the direction of propagation and can be reshaped using either lenses or curved mirrors.

3.3.5.1 Laser Beam Profile

The laser beam diameter and divergence can be calculated using the paraxial approximation, that is, under the assumption that the intensity along the direction of propagation varies slowly on the scale of the wavelength, and the further assumption that the profile of the intensity follows a Gaussian distribution along the direction that is transverse to that of propagation,[38] with width ω(0) (Equation 3.1). This is depicted in Figure 3.4 for a divergent laser output that is refocused with an $f = +500$ mm lens; the

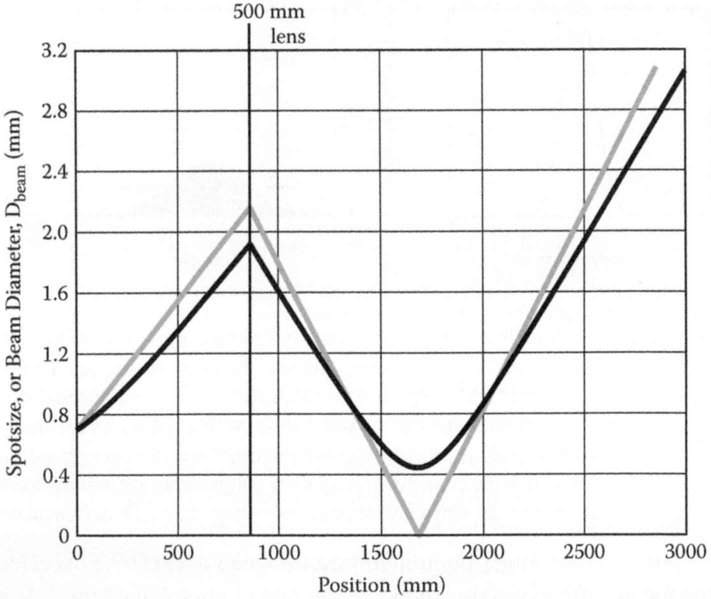

FIGURE 3.4 Calculations of the laser beam diameter, D_{beam}, as a function of distance for a beam with a starting diameter of $D_{beam} = 0.70$ mm and a divergence of 1.7 mrad at a wavelength. These values are typical for the output of a Ti:Sapphire laser. A $f = +500$ mm lens is placed 860 mm from the laser output aperture. The calculation for a Gaussian beam is shown as the dark line, while the geometric optics limit is depicted as the light line.

beam diameter is denoted by D_{beam} and equals the diameter across the points where the intensity falls by a factor of e^{-2}. Superimposed on the Gaussian beam calculation is the calculation for the same system in the geometric optics limit. The result demonstrates that, with the exception of regions close to a focus, the Gaussian beam calculation can be approximated by a geometric optics calculation with modest accuracy.

3.3.5.2 Reshaping the Laser Beam

Several different methods are available to change the divergence and diameter of a laser beam (Figure 3.5). Lenses are simple to align, and the glass will introduce only a small amount dispersion with a 100 fs pulse. Curved metal mirrors do not introduce dispersion under any conditions, and are invariably used with 10-fs pulsed sources. However, metal mirrors must be used off-axis and can thus introduce astigmatism into the laser beam.

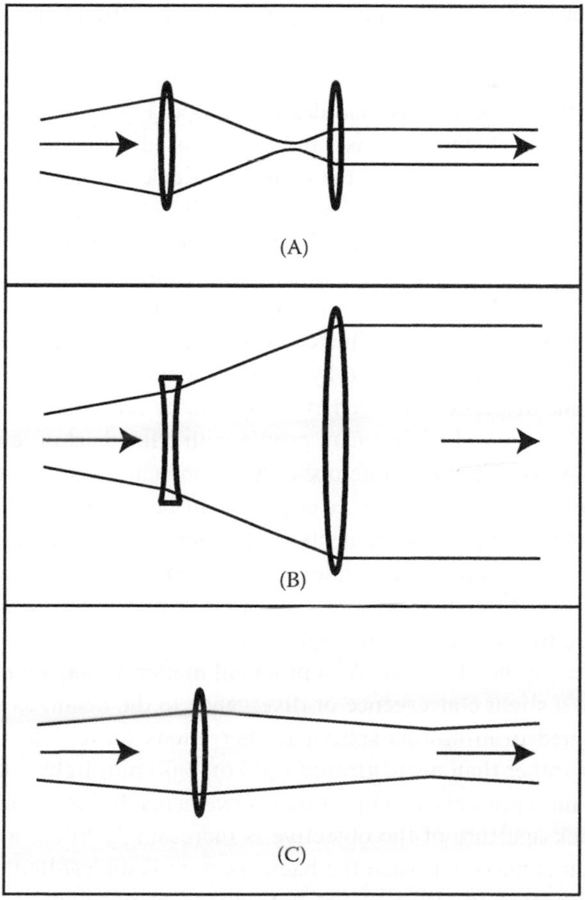

FIGURE 3.5 Four possible methods for reshaping a laser beam. (A) A Keplerian telescope. (B,D) A Galilean telescope. (C) A single, long focal length lens, as used for the calculation of Figure 3.4.

The technical characteristics involved with several lens-based telescope designs refer to the parameters:

D_{BA} ≡ Diameter of the back aperture of the objective
D_{beam} ≡ Diameter the beam at the back aperture

The choices for telescopes includes:

1. A pair of achromatic doublets can be used as a Keplerian telescope (Figure 3.5A). The lens separation should be approximately the sum of the two focal lengths, but the actual location of the telescope along the beam path is flexible. The telescope can easily be adapted to provide different magnifications by choosing different combinations of lenses at the appropriate separation. The disadvantages of this configuration include the need for rails or precision translation stages for precise alignment, and the dispersion produced by the lenses.
2. A pair of lenses, one with positive focal length and the other a negative focal length, can be used as a Galilean telescope (Figure 3.5B). This is critical for the case of amplified laser light that is used with optical ablation, as breakdown of air can occur at the focus of a Keplerian telescope.
3. A single focal length lens can be used to reshape the beam (Figure 3.5C). An achromatic doublet is placed approximately one focal length from the original focus of the laser, which is typically inside the laser cavity. Although the alignment of a single lens is simplified compared to multielement configurations, the collimated beam diameter and placement of the lens are completely determined by the focal length of the lens. As with all lenses, some dispersion is added to the system by this configuration.
4. For complete control of the beam profile with a fixed set of lenses, a four-element double-telescope system should be implemented (Figure 3.6). The laser beam diameter and the divergence can be altered by adjusting the location of two of the four elements. The deficit of this system is additional dispersion. Note that, in principle, a three-element system provides sufficient degrees of freedom but is somewhat more difficult to adjust.
5. In general, the laser beam should be collimated to minimize the spherical aberrations of the objective. As a practical matter, it may be necessary to introduce a slight convergence or divergence to the beam, as most lenses are designed to minimize spherical aberrations for visible (400 to 600 nm) light rather than near-infrared (700 to 1000 nm) light. The final size of the beam represents a compromise between resolution, which increases as the back aperture of the objective is increasingly filled, and throughput, which is maximal when the back aperture is underfilled. This trade-off is illustrated in terms of the normalized beam diameter, D_{beam}/D_{BA} (Figure 3.7); $D_{BA} \approx 6.6$ mm with the back-aperture located 35 mm from the back shoulder for the 40-times magnification, 0.8 NA dipping objective (no. 440095, Zeiss).

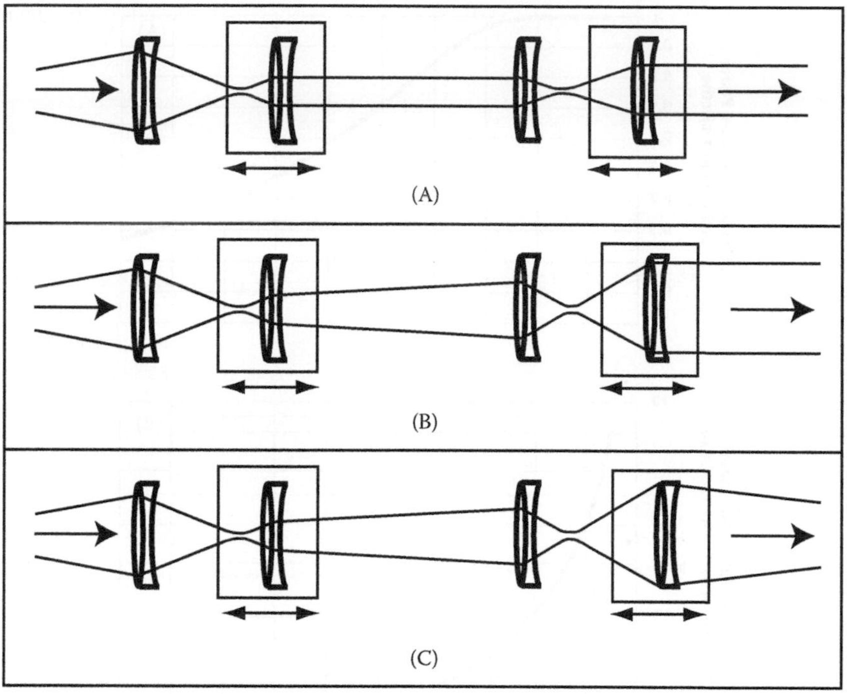

FIGURE 3.6 Ray diagrams for a dual telescope system with four lenses. The output beam diameter and divergence can be completely controlled by translating two of the four lenses. For illustration, we take the input beam to be divergent. (A) A collimated, smaller diameter output. (B) A collimated, larger diameter output. (C) A converging, larger diameter output.

3.3.6 Design of a Scan System

We consider in detail the manipulation of the path of the laser beam as a means to scan a focused spot across the preparation. The essential idea is that the objective converts the incident angle of a collimated beam into the position of a diffraction limited spot on the focal plane (Figure 3.8A). The focused laser spot can be moved across a stationary preparation by changing the angle of the incident beam. In order to change the angle without having to physically move the source, one uses a system of mirrors and auxiliary lenses according to one of two schemes.[34] The first method involves a standard 160 mm tube length objective, that is, a lens optimized for an image plane at 160 mm from the rear focal plane of the objective, in conjunction with a rotating mirror and a scan lens to deflect the rotated light back into the beam path (Figure 3.8B). A disadvantage to this configuration for TPLSM is that the incoming light to the objective is convergent, rather than parallel. This can lead to astigmatism in the focused beam if a dichroic mirror is placed directly behind the objective, as is optimal for collection of emitted light in TPLSM.[7]

An alternative scheme for the scan system is to use an infinity-corrected objective (i.e., one that is optimized for an image in the conjugate plane), together with a

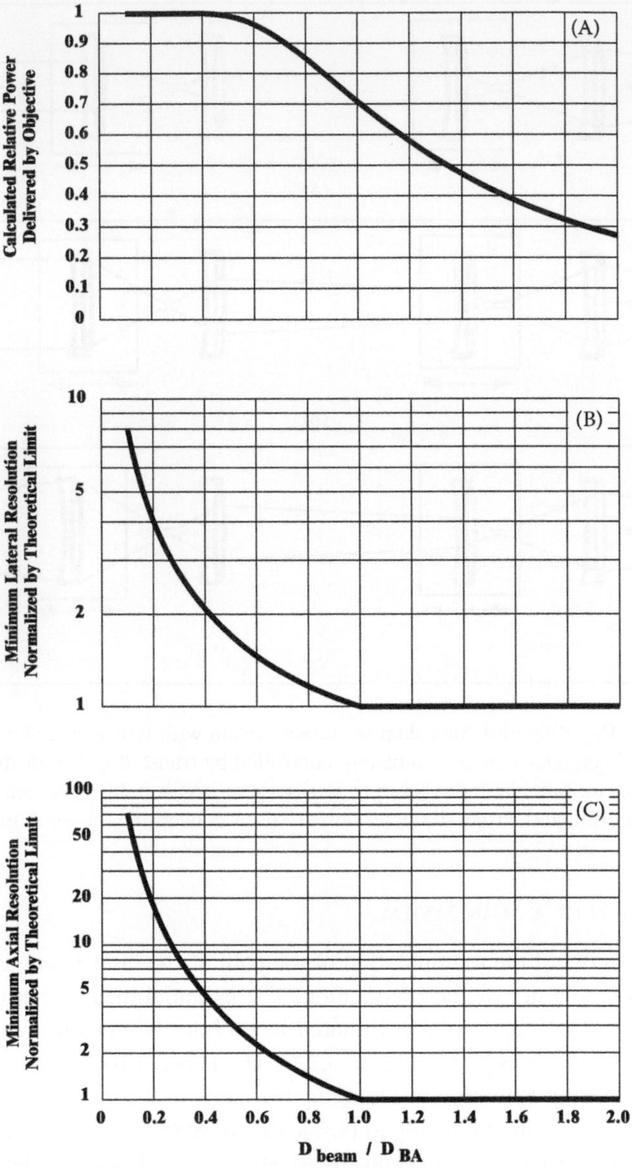

FIGURE 3.7 The trade-off between power throughput and optical resolution as a function of the size of an incident Gaussian beam, normalized as in units of Dbeam over DBA. (A) elative power versus normalized radius. (B) Lateral resolution versus normalized radius. (C) Axial resolution versus normalized radius.

rotating mirror and a tube lens as well as a scan lens (Figure 3.8C). In this configuration, the two auxiliary lenses serve to translate the angular deflection of the mirror into an incident angle at the objective, while allowing both the input to the mirror, and the input to the objective to be collimated. This is the configuration used in our

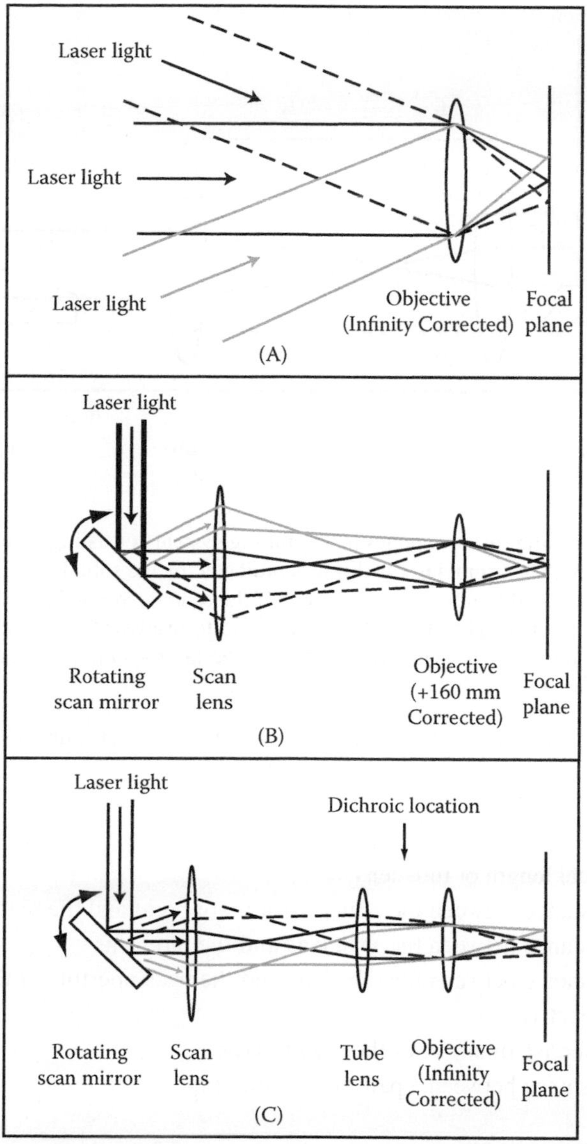

FIGURE 3.8 Illustration of different configurations for laser scanning. (A) A scanning system with only an objective. The collimated laser source must be physically moved and rotated behind an infinity-corrected objective. (B) A scanning system with a single intermediate lens, that is, the scan lens, in addition to the objective. The objective is taken as optimized for a 160-mm image distance. The rotating mirror is imaged onto the back aperture of the objective. Note that the light is convergent behind the objective. (C) A scanning system with two intermediate lenses, that is, the scan lens and the tube lens, in addition to the objective. The objective is taken as optimized for an infinite image distance. A scan lens and tube lens are placed between a rotating scan mirror and an infinity-corrected objective. The mirror surface is imaged onto the back aperture of the objective and the light between the objective and the tube lens is collimated and thus suitable for the placement of a dichroic mirror.

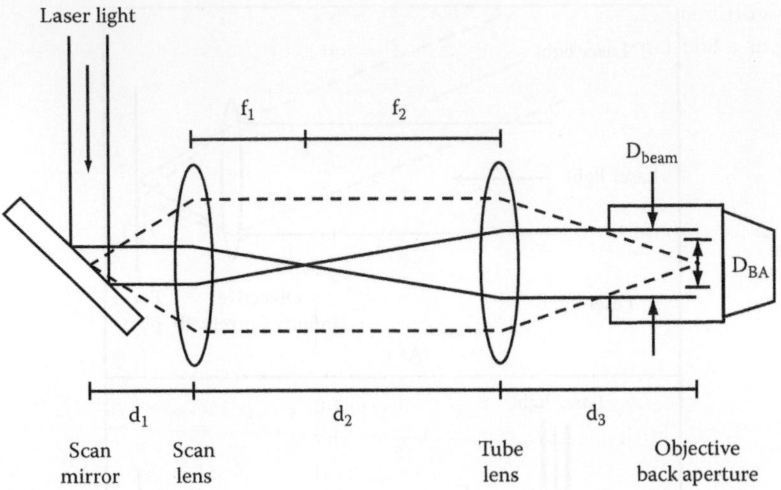

FIGURE 3.9 The alignment of optics in a scanning system consisting of an objective, tube lens, and scan lens. The actual laser pathway is indicated by the solid lines. The dashed lines depict the imaging condition between the center of the scan mirror and the center of the back aperture of the objective. The laser path is magnified to just overfill the back aperture, and the imaging condition (see text) prevents motion of the laser spot at the back aperture as the laser beam is scanned.

system, and which will now be described in greater detail (Figure 3.9). The relevant additional variables are

$f_1 \equiv$ Focal length of scan lens
$f_2 \equiv$ Focal length of tube lens
$d_1 \equiv$ Distance between the scan mirror pivot point and the scan lens
$d_2 \equiv$ Distance between the scan lens and the tube lens
$d_3 \equiv$ Distance between the tube lens and the back aperture of the objective
$D_{SM} \equiv$ Effective diameter of the scan mirror
$d_{SM} \equiv$ Distance between a pair of scan mirrors

The issue of 1D versus 2D scanning can be separated from the issue of constraining the optics in the beam path. Thus, for clarity, we focus first on a 1D scan system.

3.3.6.1 Constraints on Axial Distances and Optical Apertures

We consider the optimal alignment and layout of optics, with reference to Figure 3.9, for TPLSM. The constraints are:

1. Infinity-corrected objectives require the incident light to be collimated at the back aperture of the objective, which is a plane specified by the manufacturer that may lie deep within the lens. If the input to the scan system is a

collimated laser beam, then the scan lens and tube lens must be configured as a telescope. Therefore, their separation is given by

$$d_2 = f_1 + f_2. \tag{3.12}$$

2. The pivot point of the scan mirror should be imaged to the center of the back aperture of the objective to minimize motion (spatial deviations) of the laser path at the back aperture plane. This imaging condition is satisfied by

$$d_1 = \frac{(f_1)^2}{f_2} + f_1 - d_3\left(\frac{f_1}{f_2}\right)^2. \tag{3.13}$$

   This condition serves to maintain constant laser power through the back aperture of the objective as the scan mirror is rotated. It ensures that the intensity at the focus does not change as the mirror rotates. A graph of the required separation, d_1, as a function of the separation, d_3, for six different choices of scan lens focal lengths and a tube lens with a focal length of $f_2 = +160$ mm is shown in Figure 3.10.
3. The incident laser beam should slightly overfill the back aperture of the objective in order to utilize its full NA, that is,

$$D_{\text{beam}} > D_{\text{BA}}. \tag{3.14}$$

   This leads to the highest resolution for a given objective.

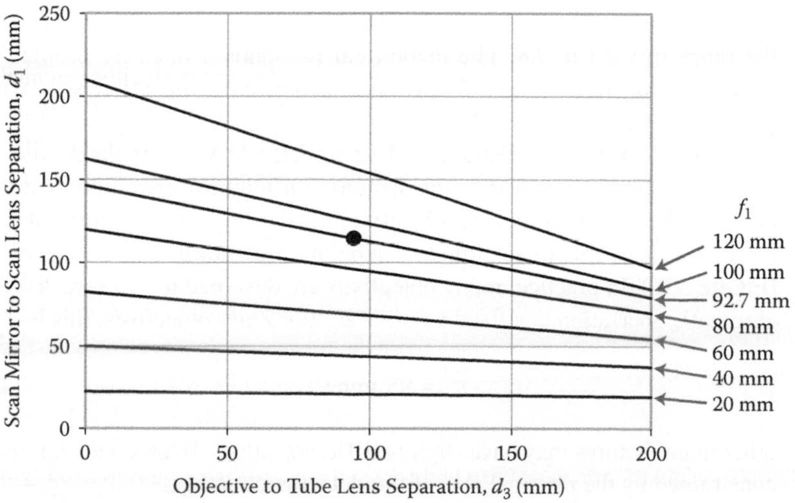

FIGURE 3.10 The calculated scan mirror to scan lens separation distance, d_1, as a function of the objective to tube lens separation distance, d_3. The calculation is shown for several choices of scan lens focal lengths, f_1. The focal length of the tube lens is assumed to be $f_2 = +160$ mm. The dot marks the actual values, $d_3 = 95$ mm and $f_1 = +92.7$ mm, used in our system.

4. Typical fast scan mirrors are only 3 mm in dimension. The simultaneous requirements that the beam should underfill the scan mirrors as a means to avoid losses due to diffraction, yet slightly overfill the back aperture of the objective as a means to maximize the spatial resolution at the focus, constitute a magnification constraint on the scan system. The required magnification, denoted m, is given by

$$\left|m\right| \equiv \frac{D_{BA}}{D_{SM}} \simeq 2\tfrac{1}{2} \text{ to } 7 \tag{3.15}$$

where the numerical values are span the range of values appropriate for the back aperture of a standard 40-times magnification dipping lens ($D_{BA} \approx 7$ mm) through a wide-field 20-times magnification dipping lens ($D_{BA} \approx 21$ mm). The scan lens should be a lens with a small *f*-number—that is, a ratio of focal length to clear aperture—that is well corrected for spherical and off-axis aberrations. These conditions are satisfied with an achromatic doublet with an *f*-number of 2. Precision scan lenses, as well as stereoscope objectives, are well corrected for off-axis aberrations and thus beneficial for applications involving large scan angles.
5. For the case of a system in which a trinocular with a standard, $f_2 = +160$ mm tube lens is used in the scan system, the required focal length of the scan lens is determined by a second constraint on the magnification, that is, $f_1 \equiv f_2/|m|$, for which the standard focal lengths of 63.5, 50.8, 38.1, 31.8, or 25.4 mm are appropriate. For the case in which a custom tube lens is used, a convenient choice in terms of availability is $f_2 = 200$ mm in conjunction with scan lenses of $f_1 = 76.2$, 63.5, 50.8, 38.1, 31.8, or 25.4 mm; these span the range m = 2.7 to 7.8. The theoretical two-photon intensity profile for different filling rations D_{beam}/D_{BA} is estimated by modeling this system with geometric optics (Figure 3.11).
6. The above constraints (Equations 3.12 to 3.15) leave a single degree of freedom in the design of the scan system. For a given separation distance between the tube lens and the objective, d_3, the imaging constraint determines the scan-lens-to-scan-mirror separation distance, d_1, or vice versa (Figure 3.10). In practice, many objectives are designed to be corrected for chromatic aberration at a fixed value of d_3. For Zeiss objectives, this is

$$d_3 \cong 95 \text{ mm}; \tag{3.16}$$

other manufactures may have slightly different values. In practice, d_3 is also constrained by the physical size of the detector assembly.

3.3.6.2 Two-Dimensional Scanning

A distortion-free, two-dimensional scan is obtained with two scan mirrors and the use of a unity magnification telescope between the scan mirrors to ensure that the centers of both mirrors and the objective back aperture are all conjugate planes of

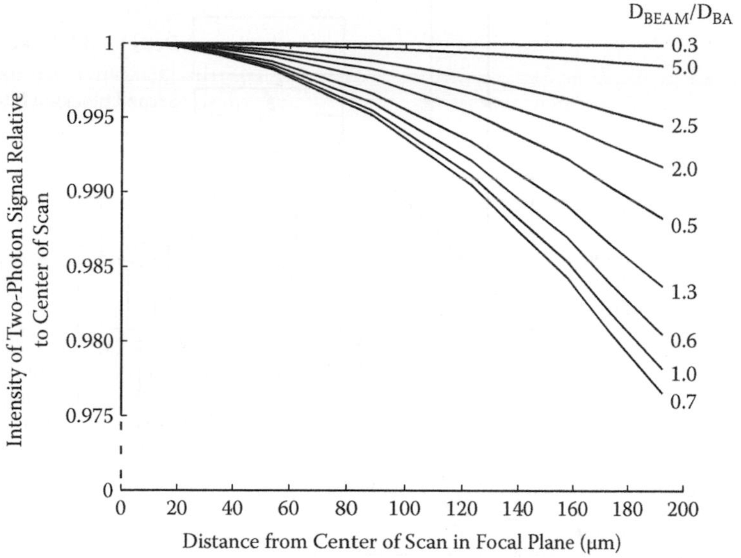

FIGURE 3.11 Theoretical calculation of the spatial inhomogeneity of the squared power, which corresponds to the two-photon excitation rate, due to the lateral motion of the scanned laser spot at the back aperture of the objective. The calculation was carried out in the geometric optics limit with the assumption of a Gaussian transverse beam profile.

one another. The addition of the intermediary optics is conceptually straightforward and was employed in commercial instruments, that is, the Biorad MRC series.

A simpler 2D scan configuration can be used that utilizes no intermediary optics if the separation between the scan mirrors can be made small compared to focal length of the scan lens, that is,

$$f_1 >> d_{SM}. \tag{3.17}$$

The conjugate plane of the back aperture is placed halfway between the two scan mirrors. The small deviation from the exact imaging conditions results principally in a slight lateral motion of the beam at the back aperture of the objective while the beam is scanned. This motion results in increased clipping (i.e., vignetting) of the beam at larger scan angles. The excitation intensity will therefore be lower at the edges of the scan than at its center (Figure 3.11).

3.3.7 Fluorescent Light Detection

The fluorescent signal generated in the illuminated region of the preparation is incoherent and is emitted over the entire 4π solid angle. A portion of the signal is collected in the backward direction by the objective, deflected by a dichroic beam-splitting mirror, and focused onto a photomultiplier tube (PMT) (Figure 3.12). Filters and additional dichroic mirrors and detectors can be used to further isolate

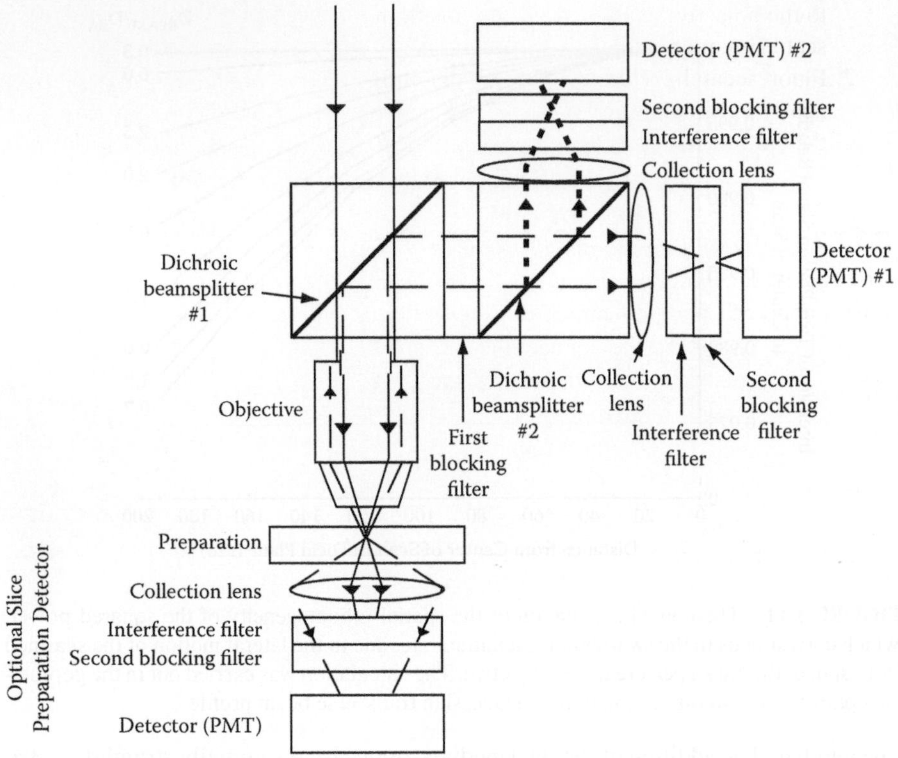

FIGURE 3.12 Schematic of the detection configuration for two-photon excited fluorescence. The excitation laser light is depicted as a solid line, while the fluorescent signals are shown as dashed and dotted lines. The collection system shown below the sample stage can be added for samples that are not too optically thick, such as brain slice.

the fluorescent signal and, if required, separate different fluorescent signals according to wavelength. For preparations that are not too optically thick (e.g., slice or cell culture), an additional detector can be placed beneath the preparation to collect the signal emitted in the forward direction[55] (Figure 3.12).

3.3.7.1 Placement of Detector Assembly

The detector assembly should be placed directly above the objective. The two main considerations in this choice of detector position are:

1. Collection efficiency is increased by placing the collection assembly directly above the objective. Although light that is emitted from the focal plane of the infinity-corrected objective should ideally be collimated as it returns through the back aperture, the actual fluorescence will diverge due to scattering and diffraction. In particular, it is mainly the scattered light that will miss the clear aperture of the tube lens. The unscattered light collected by the objective will almost exactly retrace the beam path of the excitation laser. Therefore, the placement of the detector assembly as close

to the objective as possible should result in improvements for more strongly scattering preparations.

2. Fluorescent light collected from an infinity-corrected objective will be nearly collimated just above the objective. Placing the dichroic mirror in the path at this point, rather than at a location where the beam is diverging, reduces the sensitivity of the signal to local defects on the dichroic mirror.

3.3.7.2 Constraints for Detection Elements

The backward collection pathway is shown in Figure 3.13 with the second dichroic mirror and PMT pathway eliminated for clarity. We note the following definitions:

$l_1 \equiv$ Distance between collection lens and back aperture of the objective
$l_2 \equiv$ Distance between active surface of the detector and collection lens
$D_{PMT} \equiv$ Effective diameter of the detector active surface
$f_{CL} \equiv$ Focal length of the collection lens
$D_{CL} \equiv$ Clear aperture of the collection lens

The following conditions should be maintained for the choice and placement of collection optics:

1. The back aperture of the objective should be imaged onto the active surface of the PMT to minimize motion of the fluorescent signal on the detection surface. This imaging condition is given by

$$l_2 = \frac{l_1 f_{CL}}{l_1 - f_{CL}} \tag{3.18}$$

and is graphed in Figure 3.14 for five choices of collection lens focal lengths, f_{CL}.

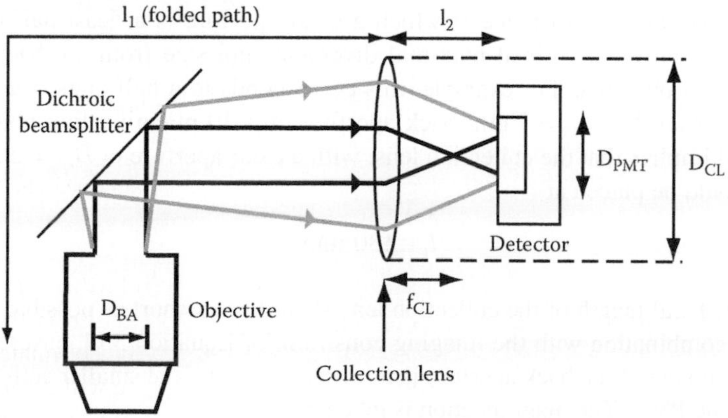

FIGURE 3.13 Placement and separations of detector elements for two-photon excited fluorescence. Unscattered fluorescent light is depicted as a dark line, while scattered fluorescent light is depicted as a gray line.

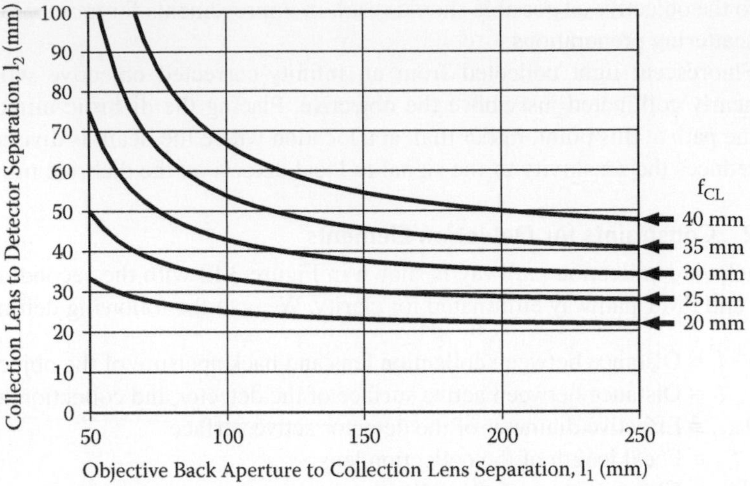

FIGURE 3.14 Graph of the imaging constraint for the distance between the detector and the back aperture of the objective for five choices of focal length for the collection lens.

2. The collection lens should have a large clear aperture to maximize collection of scattered light, some of which diverges as it exits the back aperture of the objective. In practice, physical constraints of the detector assembly (described in detail later), limit the clear aperture to

$$D_{CL} \approx 30 \text{ mm}. \tag{3.19}$$

3. The separation between the collection lens and the objective should be minimized to within physical constraints. This also serves to increase the collection efficiency of the scattered light. As a maximal limit, we arbitrarily choose a distance at which a clear aperture is at least half of the empirically determined maximal divergent spot size from the back of a typical objective. For example, this corresponds to a half-angle spread of 11° from the $D_{BA} = 7$ mm back aperture of a 40 magnification objective and implies that the collection lens, with a clear aperture of $D_{CL} = 30$ mm, should be placed at

$$l_1 \cong 130 \text{ mm}. \tag{3.20}$$

4. The focal length of the collection lens should be as short as possible. This, in combination with the imaging constraint of Equation 3.18, ensures that the image of the back aperture is minified to fit onto the smaller active area of the PMT. The magnification is given by

$$|m| = \frac{f_{CL}}{l_1 - f_{CL}}. \tag{3.21}$$

The active area of the PMT described in this chapter is 4 mm. The objective back apertures range up to 12 mm in diameter. Therefore, the magnification requirement for the described system is $|m| = 1/3$. Together with $l_1 = 130$ mm (Equation 3.21), we have

$$f_{CL} = 32.5 \text{ mm}. \tag{3.22}$$

In many imaging applications, aberrations and misalignment of optical elements distorts the spatial distribution of the collected light. This directly affects the resolution of the final image. In laser scanning microscopy, the final image is produced as a time series of collected light intensities. All of the light collected during a given time interval contributes equally to the intensity value of the corresponding spatial pixel, at least insofar as the efficiency of the detector is spatially uniform. Thus, it is more important that a consistently large portion of the fluorescent light is collected at the detector than to minimize aberrations in the collection optics.

3.3.7.3 Emitted Photon Yield per Laser Pulse

The constraints on the choice of detection electronics depends on the number of fluorescent photons emitted per incident laser pulse and on the efficiency of converting those photons into photoelectrons at the detector. For this analysis, the additional relevant parameters are

$n_{pulse}^{\text{incident}}$ ≡ Number of incident photons per laser pulse
$n_{\text{pulse}}^{\text{collect}}$ ≡ Number of emitted photons collected per pulse
$P_{pulse}^{\text{incident}}$ ≡ Peak power of incident laser pulses
$\Phi_{pulse}^{incident}$ ≡ Flux of incident photons
Γ ≡ Repetition rate of laser pulses
τ_{pulse} ≡ Width of laser pulse
β^2 ≡ Relative variance of laser pulse intensity, $\left\langle (\Phi_{pulse}^{incident})^2 \right\rangle / \left\langle \Phi_{pulse}^{incident} \right\rangle^2$
$N_{\text{molecules}}$ ≡ Number of molecules in focal excitation volume
ξ_{obj} ≡ Collection efficiency of objective
h ≡ Planck's constant (6.6×10^{-34} Js)
N_a ≡ Avogadro's number (6.0×10^{23} molecules/mole)

The number of photons per laser pulse is

$$n_{\text{pulse}}^{\text{incident}} = \frac{\lambda_o}{hc} \frac{P_{\text{average}}^{\text{incident}}}{\Gamma}. \tag{3.23}$$

For $P_{\text{average}}^{\text{incident}} = 10$ mW, below the ~ 30 mW saturation of organic fluorophores,[56] with $\lambda_0 = 800$ nm and $\Gamma = 76$ MHz, we find

$$n_{\text{pulse}}^{\text{incident}} = 5 \times 10^8 \text{ photons/pulse}.$$

The incident flux is

$$\Phi_{pulse}^{incident} = \frac{\lambda_o}{hc} \frac{P_{average}^{incident}}{\pi\, \omega^2(0)\, \tau_{pulse}\Gamma} \quad (3.24)$$

which, for NA = 0.8 so that $\omega(0) \approx 0.25$ μm (Equation 3.1) and the typical value of $\tau_{pulse} = 250$ fs for width measured at the objective (Section 3.5.2), yields $\Phi_{pulse}^{incident} \approx 1.0 \times 10^{34}$ photons/m^{2}s. The number of emitted fluorescent photons captured by the objective is

$$n_{pulse}^{collect} = \xi_{obj}\, \tfrac{1}{2}\sigma_{dye}\, \beta^2 \left\langle \Phi_{pulse}^{incident} \right\rangle^2 \tau_{pulse} N_{molecules} \quad (3.25)$$

where the $\frac{1}{2}$ represents the conversion of two incident photons to one fluorescent photon,

$$N_{molecules} \simeq [dye]\, N_a\, V_{excitation} \quad (3.26)$$

and the epifluorescent collection efficiency is

$$\xi_{obj} = \frac{1}{4\pi} \int_0^{\sin^{-1}\frac{NA}{n}} \int_0^{2\pi} \sin\theta\, d\theta\, d\phi = \frac{1}{2}\left[1 - \sqrt{1 - \left(\frac{NA}{n}\right)^2}\,\right] \xrightarrow[NA \ll n]{} \frac{1}{4}\left(\frac{NA}{n}\right)^2. \quad (3.27)$$

The typical dye concentrations varies from 10 to 50 μM (average of 20 μM) for the intracellular calcium indicator Calcium Green 1,[57] from 10 to 90 μM (average of 40 μM) for the intracellular calcium indicator Fura-2,[58] and up to 150 μm for blood serum labeled with fluorescein conjugated to dextran.[3] For the choice [dye] = 20 μM and the previous choice of NA = 0.80, for which $V_{excitation} \cong 1.5 \times 10^{-16}$ L (Equation 3.3) and $\xi_{obj} = 0.10$ (Equation 3.27), we estimate $N_{molecules} \approx 1800$. The normalized variance for a transform limited pulse is $\beta^2 = 0.6$.[7] The cross-section for Calcium Green at $\lambda_o = 800$ nm is $\sigma_{dye} = 20$ GM $= 20 \times 10^{-58}$ m^{4}s/photon, for which we estimate

$$n_{pulse}^{collect} \approx 3 \text{ photons/pulse.}$$

As discussed under "Realization" (Section 3.4.6.2), losses from the dichroic and band-pass filters used in the separation of the incident and emitted light and inefficiencies of the photodetector lead to another order of magnitude decrement in the transformation from collected photons to integrated photoelectrons. A final point is that while the excitation rate $\frac{1}{2}\sigma_{dye}\, \beta^2 \left\langle \Phi_{pulse}^{incident} \right\rangle^2 \tau_{pulse} N_{molecules}$ is independent of the NA of the objective,[7] the number of collected photons increases with increasing numerical aperture (Equation 3.25).

3.4 REALIZATION

3.4.1 OPTICAL AND MECHANICAL

Our system is laid out on a 4′ × 8′ × 1′ air table, which provides room for the lasers, ancillary optics, an interferometric autocorrelator, and a repetition rate doubler (Figure 3.15).

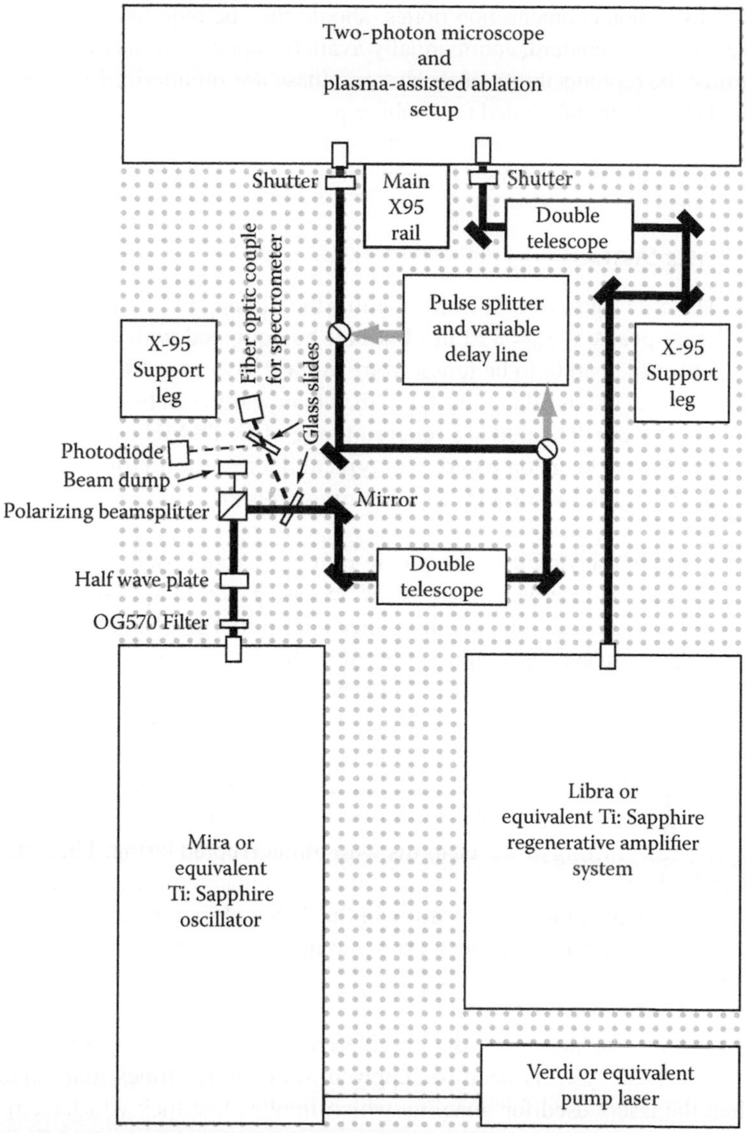

FIGURE 3.15 Schematic diagram of the layout on an optical table for delivery of laser light to our two-photon laser scanning microscope with amplified ultrashort pulses for photon-assisted plasma-mediated ablation. The main beam path is depicted by the thick dark line. The beam path through the pulse splitter and delay line, used for the optical autocorrelator, is depicted by the thick gray lines. The beam path to the additional experimental setup is depicted by the thin dark line. The beam path to the two-photon spectrometer is depicted by the thin gray line. The dashed line behind the first polarizing beamsplitter represents the path for the horizontally polarized component of the beam that is used to monitor the output power and wavelength of the laser, as well as dissipate excess power. Circles with diagonal lines represent protected silver mirrors in flipper mounts. Rectangles at 45° represent protected silver mirrors in standard mirror mounts. The dimensions of the table are 4′ by 8′.

Room for dispersion compensation optics, should they be required, is also available. Wherever possible, standard, commercially available optics and electronics were used to maximize the reproducibility of the system; these are summarized in Appendix A. Various adapters were fabricated to combine parts of noncompatible design. Machine drawings of these adapters are available by contacting the senior author. Lastly, ancillary electronics are supported by a hanging shelf above the air table.

3.4.2 Laser Systems

The most versatile source for TPLSM remains the Ti:Sapphire oscillator. The output of these lasers peaks at $\lambda_o = 800$ nm but can be equipped with broadband mirror sets that permit the output to be tuned from below 750 to over 1000 μm. The pulses that emanate from the oscillator are τ_{pulse} ~100 fs in width, can be compensated to be free of chirp, and are delivered at a repetition rate of Γ ~80 MHz. The time between pulses, Γ^{-1} ~13 ns, is long compared with typical fluorescent lifetimes of 2 to 5 ns.

There are basically two options for obtaining a system:

1. A "two-box" version uses a separate pump laser and oscillator. The diode pump laser can deliver up to 20 W of pump power at a wavelength of $\lambda = 532$ nm, close to the maximum thermal limit of the Ti:Sapphire crystal. The maximum average output power of such a laser is approximately 3 W at $\lambda_o = 800$ nm, or upward of 100 times the saturation power for typical dyes.[7] This allows for over two orders of magnitude of loss due to scattering as one probes deep into brain tissue. These laser are relatively robust.
2. A "one-box" contains in integrated pump laser and oscillator, together with automated tuning protocols. These system are simple to use and permit relatively fast shifting of wavelengths with uninterrupted lasing. The potential drawback of "one-box" systems is that the automated tuning circuitry holds the beam path at the same location in the Ti:Sapphire crystal, which results in laser-induced damage and a drop in output power over time compared to "two-box" systems.

The laser source for plasma-mediated ablation only needs to operate at a single wavelength. These systems need to supply upward of 10^3 times more energy per pulse than the lasers used for imaging, which implies that they will have relatively low repetition rates. Commercial systems use pulses from a standard Ti:Sapphire oscillator as a seed, and amplify these pulses in a regenerative scheme that involves pulse stretching and compression. While our initial system was based on the design of Murnane,[59] we have had excellent success with the "one box" Libra from Coherent Inc., which has a repetition rate of 5 kHz. Fiber-based systems, and long-baseline or cavity-dumped Ti:Sapphire oscillators, may be viable alternative sources.

3.4.3 Path from Lasers to Microscope

The delivery pathway of the laser light from the source to the scanners is depicted in Figure 3.16. A colored glass filter, type OG590, is placed immediately after the

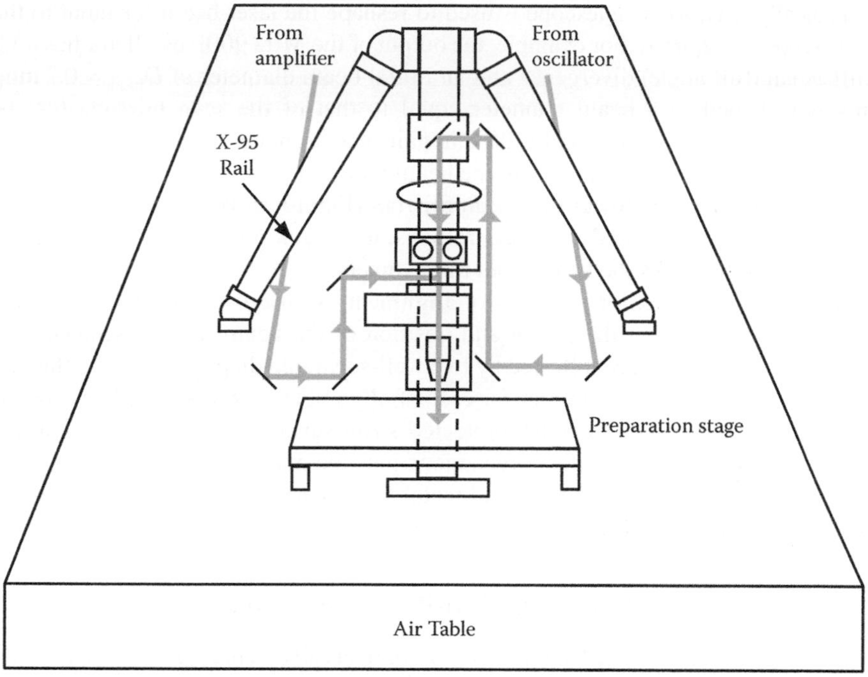

FIGURE 3.16 Illustration of the physical superstructure of the two-photon laser scanning microscope. The tripod is constructed of X-95 rail; the front leg forms the optical rail of the microscope and the side legs act as supports. The incident laser paths are shown in gray.

laser output port as a means to attenuate any residual $\lambda = 532$ nm pump light along the beam path. The output of the Ti:Sapphire laser is horizontally polarized, and thus the intensity of the laser output may be controlled by manipulating the polarization state. The polarization angle is rotated by a zero-order half-wave plate and a polarizing beamsplitter or a Glan-Thompson polarizer, is used to separate the polarized light into its horizontal, or P, and vertical, or S, components. The vertical component, which has a slightly higher reflectance by metal mirror surfaces that are mounted on the table top and positioned at an angle of 45° relative to the beam path, is deflected toward the microscope. Rotation of the output polarization from horizontal to vertical, via rotation of the half-wave plate, increases the deflection to the microscope according to the cosine of the angle. Intensity control on the millisecond time-scale can be achieved using a Pockels cell to rotate the angle of the beam, rather than a mounted half-wave plate.

For beam diagnostics, a glass slide that is positioned at 45° relative to the beam is used to pick off roughly 8% of the beam power and deflects it to a PC-based spectrometer via a fiber-optic cable (Figure 3.15). An additional glass slide is used to pick off an additional roughly 8% of the beam power and deflects it to a photodiode as a relative measure of the intensity of the laser output. Independent of which polarization is monitored by the photodiode, the pick off should be located downstream of the intensity control mechanism(s) but upstream of the shutter.

Typically, a two-lens telescope is used to reshape the laser beam for input to the microscope as required. For example, the output of the Mira 900F oscillator has a 1.7 milliradian (full angle) divergence and an initial beam diameter of $D_{beam} = 0.7$ mm. This is reshaped to a beam diameter equal to that of the scan mirrors, that is, $D_{SM} = 3$ mm, with a $f = +25$ and +100 mm pair of achromats.

A mechanical shutter is placed in the beam path just before the beam is deflected upward by a periscope toward the scan mirrors (Figure 3.15). The shutter position is computer controlled, and for safety reasons, its default position in the absence of electrical power blocks the beam from the scanners.

As a practical issue, for purposes of alignment, we use iris diaphragms as variable apertures to define the height and position of the beam. In this scheme, two well-separated mirrors are followed by two well-separated diaphragms, with the first diaphragm close to the last mirror for ease of alignment of the beam. Unless otherwise stated, all mirrors have a flat, protected-silver surface and are held in standard kinematic mounts. Lastly, safety precautions must be maintained in the use of laser light for TPLSM. Thus black anodized metal shields are used to surround the scanners and some optics that are located near normal eye level.

3.4.4 Path along Microscope: Two-Photon Excitation

The microscope is built as a tripod with X-95 optical rail (Figure 3.16). One leg serves as an optical bench, and the other two serve as supports. A customized adapter atop the tripod serves to connect the legs. A pair of mirrors mounted at 45° is used as a periscope to steer the beam to the height of the scan mirrors. The top periscope mirror mount is connected horizontally to the main X-95 rail and angled at 45°. The laser light deflected from the top periscope mirror travels horizontally to the alignment mirror, and then to the two scan mirrors. The laser then travels downward through the scan lens, the tube lens, the polarizing beamsplitter, and the dichroic beamsplitter, before being focused by the objective onto the preparation (Figure 3.17).

There are five main attachments to the main X-95 rail (Figure 3.17). At the top, two carriages are attached; one in front, and one in back. The front carriage holds the scanners and an alignment mirror, and the back carriage holds the top half of the periscope. The next attachment is the scan lens mount. A 100%/0% switching trinocular headstage, which provides the eyepieces for visual inspection of the preparation and holds the tube lens, is mounted on a separate X-95 carriage below the scan lens. The underside of the attachment for the trinocular tube also holds a polarizing beamsplitter for bringing in the amplified pulse laser source. The bottom attachment is a modular focusing unit, to which both the detector assembly and the objective holder are mounted. Lastly, the sample stage allows for "x" and "y" translation of the preparation and sits directly on the air table.

Optical scanner. The back X-95 carriage holds the top mirror of the periscope for the oscillator. An "L" bracket attaches to the mirror mount and holds it in the appropriate orientation (Figure 3.17). The front X-95 carriage is connected to a rod assembly that holds both the scanners and an alignment mirror (Figure 3.18). The alignment mirror is located at the orthogonal intersection point of the planes defined by the periscope mirrors and the first scan mirror. The scanners are mounted in a

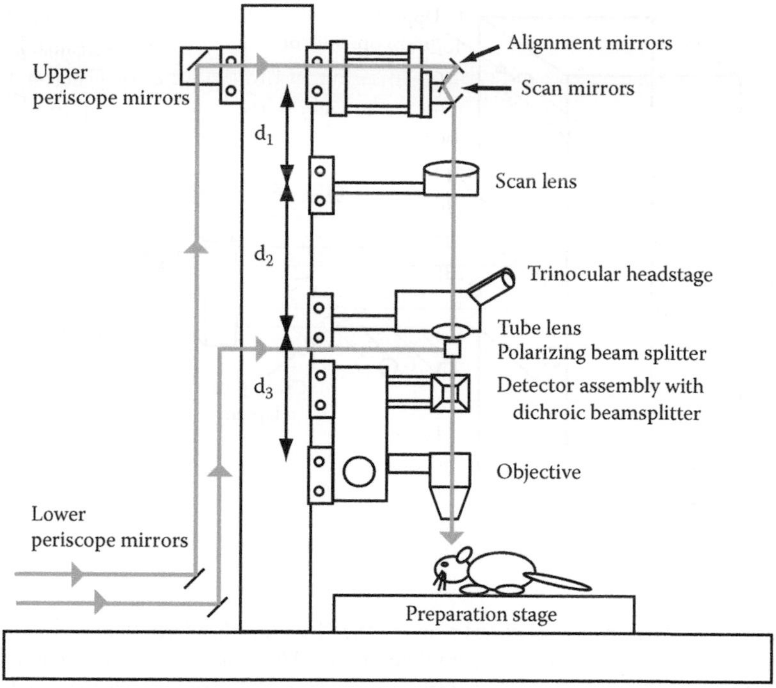

FIGURE 3.17 The main attachments to the main X-95 rail in the two-photon laser scanning microscope. The laser pathway from one of the periscopes to the objective along the optical rail that forms the microscope includes the alignment mirror, the scan mirrors, the scan lens, the tube lens, the polarizing beamsplitter, and the detector assembly. The pathway from the other periscope to the objective along the optical rail that forms the microscope includes only the polarizing beamsplitter. The collection optics for the fluorescent light are not shown. The values of the separation distances for our system are $d_1 = 114.5$ mm, $d_2 = 252.7$ mm, and $d_3 = 95$ mm.

small, orthogonally positioned mirror assembly that is a manufacturer's accessory to the scanners. The mirror assembly is attached to a custom-designed, water-jacketed base that in turn is attached to the rod assembly; water cooling is only necessary for large scan angles at the highest speeds, an atypical situation. The rod assembly itself consists of seven posts (i.e., 2 of 8", 4 of 6", and 1 of 4") and seven cross-post adapters.

Scan lens. The scan lens is a 1.0-times magnification plan dissection lens (Leica). It is held and centered on a custom-designed mount and attached to the main X-95 carriage (Figure 3.17). A major consideration for the design of the scan lens holder is mechanical stability, and thus our mount is reinforced with diagonal cross-bracing.

Headstage. The headstage is an upright trinocular (Zeiss) in which the phototube pathway is used as the laser beam pathway (Figure 3.17). The headstage contains a 100%/0% beam switcher that is used to change between visual inspection with the binocular and laser scanning microscopy. The trinocular contains a $f_2 = +160$ mm tube lens that is used as the second lens of the scan system telescope. The trinocular is

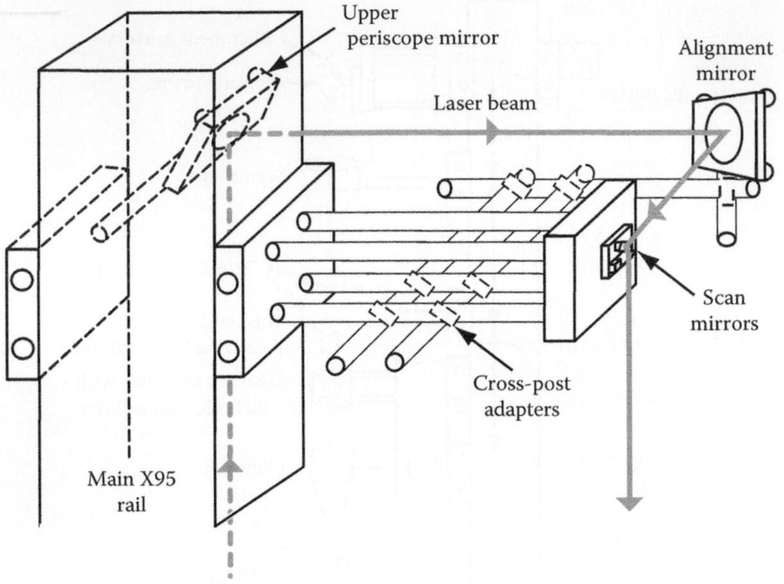

FIGURE 3.18 The uppermost attachments to the X-95 rail that form the microscope. The attachment at the rear holds the top periscope mirror. The attachment at the front holds the rod assembly, constructed of stainless steel posts, right angle cross-post adapters, and a X-95 carriage, that connects to the scan mirrors and the alignment mirror.

directly mounted to the custom-designed headstage adapter, reinforced with diagonal cross-bracing, that is in turn mounted on a X-95 carriage.

Focusing stage. Axial alignment of the objective is achieved with a commercial modular focusing unit (Nikon) (Figure 3.19). This unit allows 30 mm of vertical motion, which facilitates changing preparations and objectives. A custom-made adapter plate is attached to the focusing unit, and the objective holder is mounted to the adapter plate. The objective holder consists of a custom-made groove plate and five Microbench™ (Linos Photonics) parts in the configuration shown in Figure 3.19.

It is often desirable to step along the focus axis either as a means to automate the choice of focal plane or to systematically acquire a set of successive images. The modular focus unit is easily coupled to a stepping motor by capturing the fine focus knob in a friction-held cap. For the case of relatively fast movements along the focus axis, one can use long-range piezo-drivers such as the no. P-725 from Physik Instrumente, Karlsruhe/Palmbach, Germany.

3.4.5 Path along Microscope: Plasma-Mediated Ablation

The amplified laser beam used for ablation is either derived from the same oscillator that is used to image or from a different Ti:Sapphire oscillator, each of which will operate at a center wavelength near 800 nm. We thus combine the imaging and ablation beams with a polarizing beamsplitter that is located directly below the

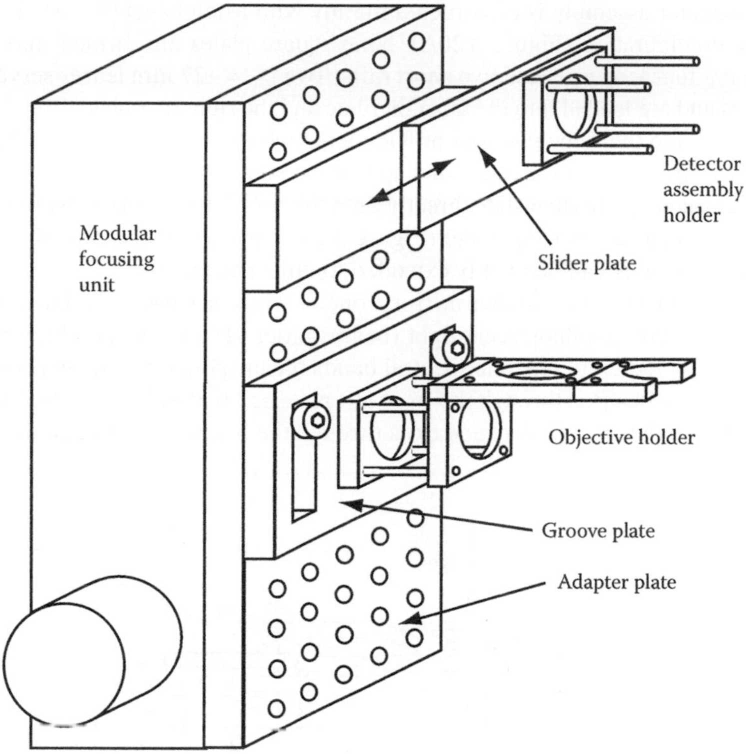

FIGURE 3.19 The focusing unit that supports the objective and detector assembly on the main X-95 rail. The objective holder consists of Microbench™ parts (Linos Photonics) connected to a Modular Focusing Unit (Nikon) by two custom-made plates; an adapter plate with 20 quarter-inch holes spaced on 1″ centers that is driven by the Modular Focusing Unit, and a groove plate that supports the objective holder. The detector assembly holder consist of a slider plate and Microbench™ parts, and is mounted on the adapter plate.

tube lens (Figures 3.16 and 3.17). We use a 25-mm beamsplitter that is glued to a mount along the back edge. The divergence of the amplified beam must be adjusted to ensure that both the focus of the imaging beam and that of the ablation beam lie in the same plane.

3.4.6 Path along Microscope: Detection

3.4.6.1 Detector Assembly

The detector assembly is attached to the focusing unit via the adapter plate. The entire detector assembly is placed on a slider plate so that the dichroic beamsplitter can be removed from the visual path, and then replaced without changing its alignment relative to the other collection optics. The back panel of a slider plate is mounted to the adapter plate, and a Microbench™ mounting plate is attached to the front panel of the slider plate. Four rods inserted into the mounting plate are used to attach the detector assembly by its rear left face.

The detector assembly is constructed mainly with Microbench™ parts in a double-cube configuration (Figure 3.20A). Nine square plates are formed into a rectangle using four long rails and two short rails. Two f_{CL} = +27 mm lenses serve as the collectors and are loaded into the far right plate and the right rear plate. Eight 2″ rods are inserted into the same plates to provide a mounting area for the glass filters and detectors. Infrared blocking filters and appropriate band-pass filters are mounted in additional square plates that slide directly onto the rods. The compact PMT detectors are mounted in a customized holder (Figure 3.20B) that also slides directly onto the rods. This assembly can be readily extended to three detectors.

The two dichroic beamsplitter mirrors, one of which separates the incident laser light from the emitted fluorescent light (beamsplitter #1) and one of which resolves the fluorescent light into separate spectral bands (beamsplitter #2), are held on prism mounts that are inserted through the front left plate and bottom right plate. The laser enters through the top left plate and exits through the bottom left plate of the double

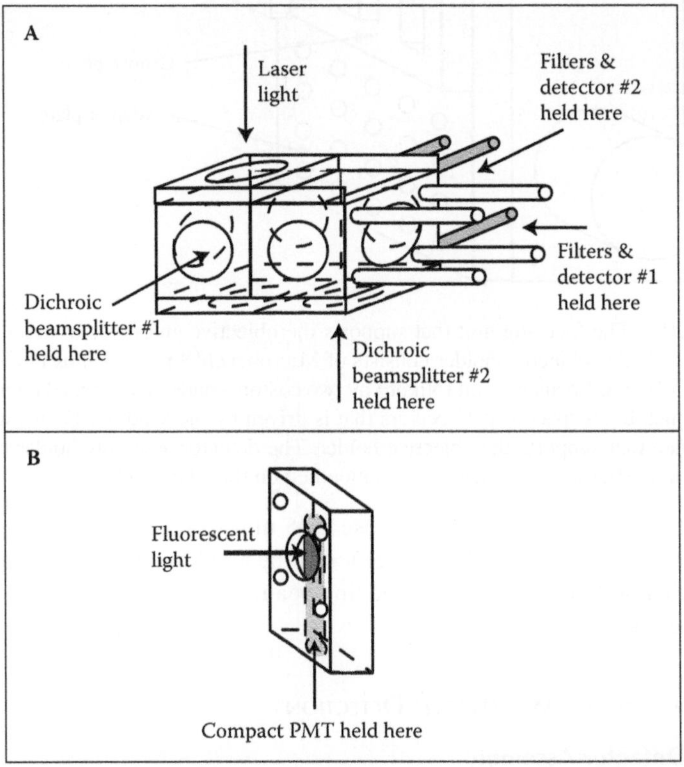

FIGURE 3.20 Illustration of the detector assembly. (A) The main assembly is fabricated from Microbench™ parts. Dichroic beamsplitters are mounted on adjustable prism mounts and placed in the labeled openings. Filter blocks and the compact photomultiplier tubes (PMTs) are held by the four rods protruding along each path. (B) The PMT case is a custom-made aluminum cage that is designed to be compatible with Microbench™ parts and the compact Hamamatsu PMTs.

cube. The edges between the plates are sealed to be light-tight with black silicone sealant (RTV 103; GE); this sealant is easily removed from anodized surfaces. Lastly, small black-anodized sleeves are inserted as baffles between the square plates that hold optics and filters to prevent stray light from entering the detector assembly.

3.4.6.2 Collection Filters and Optics

The first element in the collection pathway (Figure 3.21) is the long pass dichroic beamsplitter mirror, that is, λ_{cut} = 700 nm (no. 700DCXRU; Figure 3.22), that transmits light at the laser excitation wavelength while deflecting fluorescent light toward the detectors. This is followed by the first of two high-efficiency laser-blocking filters (E680SP-2P; Figure 3.22) to reduce the background level and associated noise that is caused by scattered incident laser light. A second dichroic is used, if necessary, to separate the fluorescent signals from two different fluorescent indicators. For example, we use a λ_{cut} = 505 nm DRLP to resolve light from yellow fluorescent protein (YFP) versus cyan fluorescent protein (CFP). The signal in either path is then imaged onto the active area of a PMT by a short focal length lens. A second laser-blocking filter (E680SP-2P) is placed in front of the detector to further reduce the background level (Figure 3.21). An alternate solution is to use 2 mm of Corning no. BG39 glass. Finally, additional interference filters, that is, λ_{pass} = 525 ± 5 nm to isolate YFP and λ_{pass} = 485 ± 11 nm to isolate CFP, are placed in the detection pathway as needed.

In the collection assembly, a f_{CL} = +27 mm lens of D_{CL} = 30 mm diameter is placed at l_1 = 120 mm from the back aperture of the objective and at l_2 = 35 mm from the detector (Figure 3.21). In this configuration, the back aperture is imaged onto the detector surface with a magnification of 0.29 times. Therefore, a D_{BA} = 7 mm back aperture is imaged to a 2.0 mm circle at the PMT active surface.

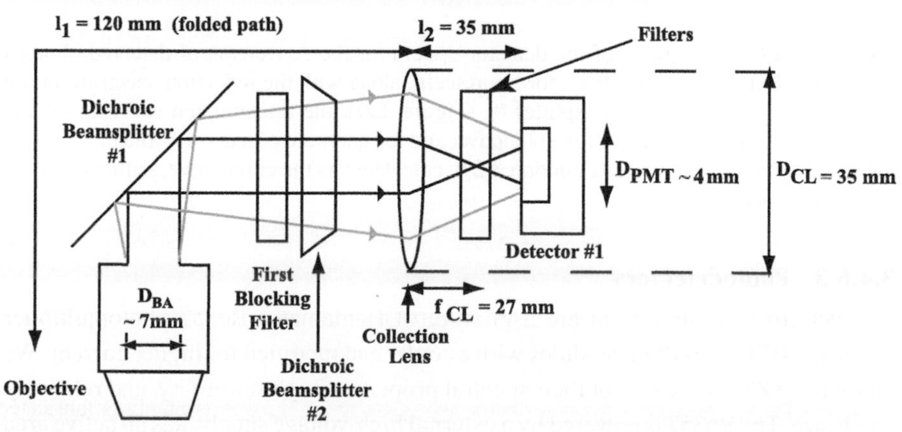

FIGURE 3.21 Detailed schematic of the detector assembly. The positions of both dichroic beamsplitters are shown; however, for clarity, we omit the optics and detector for the second, otherwise equivalent, pathway. Unscattered fluorescent light is depicted as dark lines, while scattered fluorescent light is depicted as gray lines.

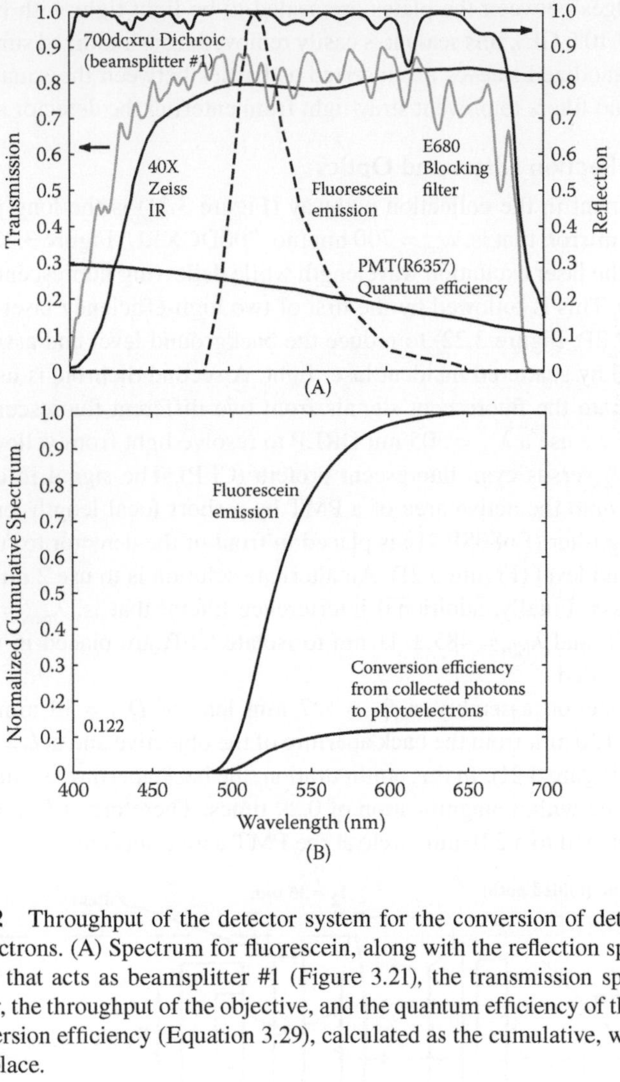

FIGURE 3.22 Throughput of the detector system for the conversion of detected photons into photo electrons. (A) Spectrum for fluorescein, along with the reflection spectrum of the dichotic filter that acts as beamsplitter #1 (Figure 3.21), the transmission spectrum of the blocking filter, the throughput of the objective, and the quantum efficiency of the photodetector. (B) Conversion efficiency (Equation 3.29), calculated as the cumulative, with two blocking filters in place.

3.4.6.3 Photodetectors

The detectors in our system are a mixture of Hamamatsu R6357 photomultiplier tubes and H7422-40 PMT modules with a cooler and modified for higher current. We chose these PMTs because of their spectral properties, high sensitivity, and relatively small size. The R6357 is powered by a external high-voltage supply, has an active area that accepts a 4-mm-diameter spot, a quantum efficiency of 0.30 at $\lambda = 550$ nm, and a maximum output current of 10 μA. This tube can be used for high photocurrents, that is, $n_{pulse}^{photoelectrons} \geq 1$ (see Section 3.3.7.1). In contrast, the H7422-40 is powered by an internal supply, has an active area that accepts a 5-mm-diameter spot, a quantum efficiency of 0.40 at $\lambda = 550$ nm, and a maximum output current of 2 μA. This tube saturates for high anode currents, and the H7422-40 MOD with a 50 μA shut-down circuit, 25-times higher than the unmodified tube, is recommended.. Independent of the detector, the output from the PMT is a pulse of current whose width is 1 to 2 ns.

3.4.6.4 Detection Electronics

The scheme to detect the emitted photons depends on the number of photoelectrons/pulse that are generated at the detector. When the number of photoelectrons/pulse is near 1, or greater, the total current is measured. When the number of photoelectrons/pulse is much less than one, photon-counting is advantageous as it eliminates extra variability from differences in the amplitude of the amplified photoelectrons, that is, pulse-height variability, as well as some thermal contributions to the noise.

We previously estimated the number of emitted photons (Section 3.3.7.1). The number of photoelectrons is found by multiplying the number of collected photon by the integrated throughput of all of the filters and the photocathode. For this analysis, we note the additional definitions:

$$n_{pulse}^{photoelectrons} \equiv \text{Number of electrons produced at PMT photocathode per pulse}$$
$$QE(\lambda) \equiv \text{Quantum efficiency of photocathode}$$
$$O(\lambda) \equiv \text{Transmission of objective}$$
$$R_{D1}(\lambda) \equiv \text{Reflection coefficient of beamsplitter \#1}$$
$$T_k(\lambda) \equiv \text{Transmission coefficient of } k\text{-th filters (or beamsplitter)}$$
$$F_{dye}(\lambda) \equiv \text{Fluorescent emission spectrum of dye}$$

Then

$$n_{pulse}^{photoelectrons} = n_{conversion}\, n_{pulse}^{collect} \tag{3.28}$$

where

$$n_{conversion} = \frac{\int_0^\infty d\lambda\; R_{D1}(\lambda)\, T_1(\lambda) \cdots T_k(\lambda)\, O(\lambda)\, QE(\lambda)\, F_{dye}(\lambda)}{\int_0^\infty d\lambda F_{dye}(\lambda)} \tag{3.29}$$

The choice of a R6357 PMT and the filters discussed above (Section 3.4.6.2) leads to a conversion of 0.12 photoelectrons per photon collected by the objective (Figure 3.22) for fluorescein dye; this number should change little with filters optimized for a given dye. Note that additional losses will occur from vignetting and reflections by the detection beam optics.

For the case of neurons labeled with Calcium Green, we estimated $n_{pulse}^{collect} \approx 3$ photons/pulse (Section 3.3.7.1), which leads to

$$n_{pulse}^{photoelectrons} \approx 0.3 \text{ photoelectrons/pulse.}$$

The probability of two or more photons arriving at the detector on a given pulse is small but non-negligable, i.e., 4%. This photocurrent may be effectively measured with either analog electronics, which is a necessity for $n_{pulse}^{photoelectrons} \sim 1$ or more, or digital electronics, which is best for $n_{pulse}^{photoelectrons} \ll 1$.

With regard to analog detection, the expected anode current from a R6357 PMT, with a gain of 4×10^6, is 15 μA for $n_{pulse}^{photoelectrons} \approx 0.3$; this current can be up to 2-times less as a result of spatiotemporal aberrations caused by the dipping objective (Section 2.5.2). While this current exceeds the 10 μA stated maximum anode current, the average current generated from a typical raster scan pattern is well within bounds. The output current of the PMT is converted to a voltage through a 10 kΩ resistor, followed an impedance buffer with a gain of 10 and a 50 Ω line driver (no. OPA637 operational amplifier followed by no. OPA633 unity gain buffer; Burr Brown, Tuscon, Arizona), for a measured time-constant of $\tau = 0.4$ μs. This is followed by either an adjustable low-pass filter (D01L8L, Frequency Devices, Ottawa, Illinois) or an integrator of local design.

Signals from weak fluorescence, such as the intrinsic in vivo levels of the flavin molecules, whose cross section is two-orders of magnitude less than that of fluorescein, or from low expression levels of green fluorescent protein (GFP) and its derivatives, are best recorded by photon counting. Here, the output of the PMT is converted to a voltage through a 50 Ω resistor, amplified by a broadband voltage-to-voltage converter, then compared with a threshold value to estimate if a photo-current pulse has occurred. The threshold operation removes noise caused by pulse-height uncertainty, and this improves the signal-to-noise ratio by a factor of $\sqrt{(1-g^{-M})/1-g^{-1}}$,[60] where g is the gain per stage and M is the number of stages of the PMT. Typically, $g = 5$ and $M = 9$, so that the increase is about 12%. Additional improvement comes from rejection of thermionic current pulses from the dynodes. A photon-counting system for TPLSM that can acquire continuously is under testing at the time of this write-up (Driscoll, Nguyen, White, Tsai, Squier, Cauwenberghs, and Kleinfeld, unpublished).

3.4.6.5 Scan Mirrors and Drivers

We use state-of-the-art galvanometer optical scanners from Cambridge Technology, Inc. These scanners are capable of rotations up to ± 20° and rates of up to 3 kHz at reduced deflection angles. We typically use angles of ± 5° or less; for a 40-times magnification, deflections of ± 1° correspond to distances of ± 40 μm in the focal plane. The scanners are designed to hold $D_{SM} = 3$ mm aperture mirrors that are provided by the manufacturer.

3.4.7 Scan and Acquisition Electronics

The scanners and associated electronics and computer boards must be chosen and configured to provide horizontal and vertical scans that are synchronized to each

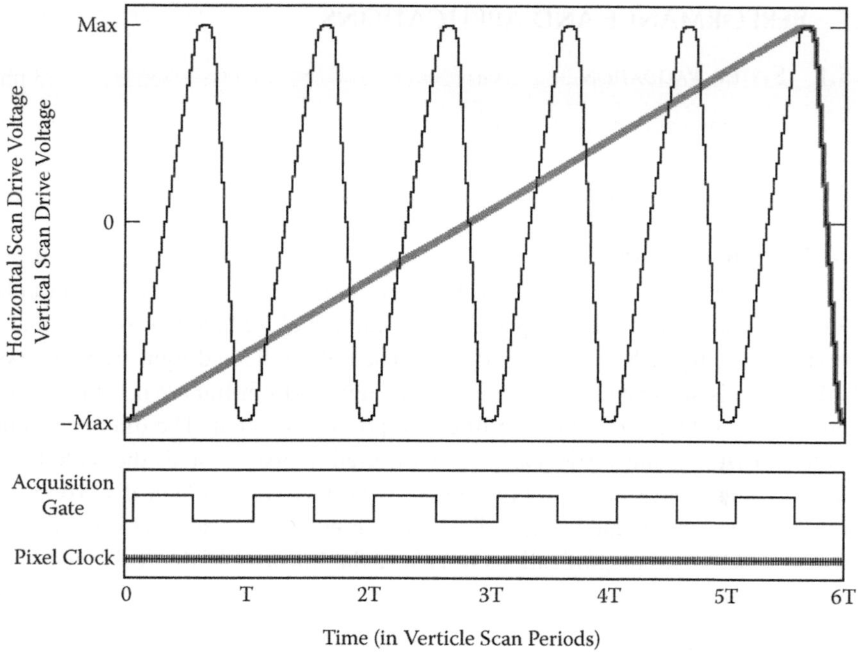

FIGURE 3.23 Cartoon illustration of the drive signals for the scan mirrors and data acquisition for a standard rater pattern. All functions are synchronized to the continuously running pixel clock. The data acquisition gate is active for a portion of the fast scan. During this time window, data is acquired at the pixel clock rate. The fast scan signal is updated which each pixel clock tick, while the slow scan signal is updated once per fast scan cycle.

other, so that they define a raster, as well as synchronized to the data acquisition. This is illustrated for a standard raster in Figure 3.23. We use the model PCI-6110 from National Instruments (Austin, Texas), which has two 16-bit digital-to-analog (DAC) channels and four analog-to-digital (ADC) channels that acquire as rapidly as 0.2 μs/sample; DACs support the horizontal and vertical scans. An additional DAC board, such as the model PCI-6711 from National Instruments, is used to control the piezoelectric *z*-focus and intensity controls. Lastly, for the case of photon counting where high-speed digital input from an external counter is necessary, we use model PCI-6534 from National Instruments, which acquires 32 bits of parallel input as rapidly as 0.05 μs/sample. All interface boards communicate with each other via the Real-Time System Integration (RTSI) bus; a 34-pin cable connects all of the RTSI connectors to form the bus.

3.4.8 Software

Software to control TPLSM has been described and, for the large part, can be adapted to different systems independent of the particulars of the hardware.[61–64] Our current system, MPScope,[62] is further described in an accompanying article and is fully integrated with the X-Y stage motors for serial ablation across regions.

3.5 PERFORMANCE AND APPLICATIONS

3.5.1 System Validation

After our system was constructed and tested, various calibrations were performed to determine the exact values of scale factors, true lateral resolution, axial resolution, and uniformity of the image.

3.5.1.1 Spatial Scales

The horizontal and vertical scale factors are determined by imaging a circular opening of known diameter—typically, a 50-μm metal pinhole that is illuminated with a 800-nm light. No fluorescent dye is used; the reflected light from the polished surface is sufficiently strong so that enough 800-nm light leaks through the dichroic beamsplitter and blocking filters to provide a signal. The diameter along both the horizontal and vertical axes, as measured in pixels, yields the scale factor at a particular zoom setting. As the image is best viewed without distortion, the gain for the horizontal and vertical directions should be identical. This calibration is repeated at several magnifications to ensure the linearity of the scale factor as a function of magnification.

3.5.1.2 Resolution

The resolving power of the microscope can be determined by imaging fluorescent beads whose diameter is below the optical resolution limit. The beads must be dilute, so that an individual bead can be identified, and they should be suspended in agarose to reduce motion during imaging. Figure 3.24 contains an image of 0.2-μm yellow-green fluorescent microspheres (Polysciences, Inc., Pennsylvania), suspended in 2% agarose, that was obtained with our microscope using a 40-times magnification water immersion objective (Zeiss). The full-width at half-maximal intensity from individual beads in the lateral dimension was calculated from the acquired images (Matlab™; Mathworks, Massachusetts). The axial resolution is obtained by compiling a series of images taken at different axial positions and calculating the intensity profile of a single bead as a function of axial position. We determined that the spatial resolution (FWHM) for our system was

$$d_{\text{lateral}} \cong 0.5\ \mu\text{m and } d_{\text{axial}} \cong 2\ \mu\text{m}.$$

The value for d_{lateral} is consistent with $2\omega_0 = 0.5\ \mu\text{m}$, while that for d_{axial} is significantly larger than $2z_{\text{confocal}} \cong 0.7\ \mu\text{m}$ but consistent the axial resolution for planar illumination.[27]

3.5.1.3 Ablation Threshold

To establish the relation between the energy of the pulses and the spatial extent of point ablations, we systematically varied both the energy per pulse and the number of pulses to generate an array of ablation sites in fixed neocortex from rat. The ablations took the form of small craters of graded sizes with the largest holes made by 510 consecutively

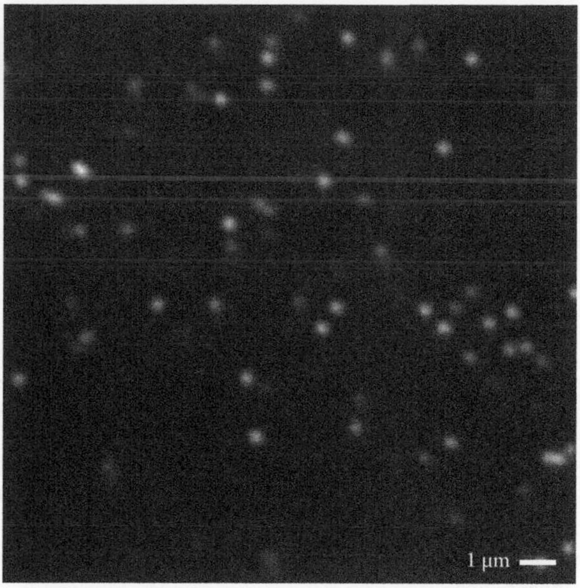

FIGURE 3.24 Image of fluorescent microspheres that are used to establish the resolution of our two-photon laser scanning microscope. The microspheres are 0.2 μm latex balls (no. 17151; Polysciences, Inc., Warrington, Pennsylvania) that are coated with fluorescein.

delivered 5 μJ pulses (Figure 3.25A). The bright border surrounding each hole stems from an increase in autofluorescence that accompanies the laser cutting.

We observed the formation of a crater at a minimum energy of 0.63 μJ for a single pulse application with this array (Figure 3.25B), which corresponds to a fluence of $F_T \approx 3$ J/cm^2 for the parameters of our beam. This is in agreement with the value of the threshold for ablation of bovine brain tissue that was freshly dissected[65] as well as ablation of various glasses.[66] The ablation volume has a greater depth along the z axis than width along the X and Y axes (Figure 3.25B), consistent with a simple model (Equation 3.9). Further, an increased number of laser pulses only weakly compensates for lower pulse energies, that is, by approximately twofold for 130 pulses Figure 3.25A). Thus, ablation is most efficient, in terms of total energy expenditure, with one or few pulses whose fluence lies above the threshold value.

3.5.1.4 Spatial Uniformity

The homogeneity of the scan was determined by imaging a spatially uniform field of fluorescence (i.e., a Petri dish filled with a 100 μM solution of fluorescein). This method finds inhomogeneities in the detection pathway as well as the in the scan pathway. The profile obtained in this manner, shown in Figure 3.25, is observed to be stable for a given configuration of the microscope, albeit slightly anisotropic (Figure 3.11). The measured profile can be used as a calibrated filter for images taken with this system; postprocessing with this filter yields a fluorescent image that is free of spatial intensity distortions.

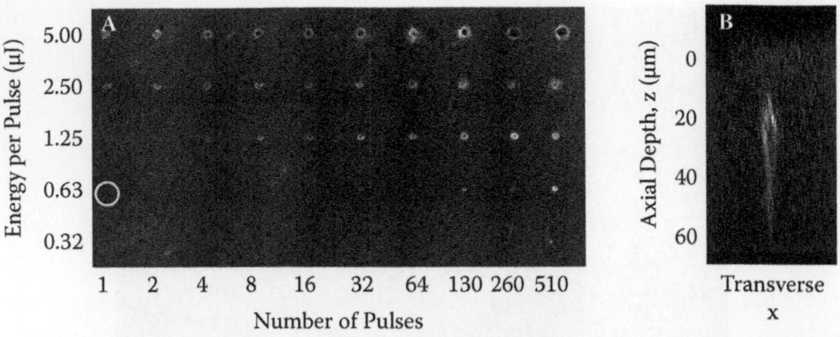

FIGURE 3.25 Threshold for plasma-mediated ablation in fixed neocortex. (A) Array of ablation craters, in fixed neocortical tissue from rat, as a function of the energy per pulse (vertical axis) and number of laser pulses (horizontal axis). Ablation was performed with a 10-times magnification, 0.2 NA air objective. The distance between craters is 100 μm. The image is a single section obtained with TPLSM at $\lambda = 760$ nm to highlight intrinsic fluorescence. The yellow circle highlights the lowest energy at this NA for clearly observed ablation in a single pulse. (B) Transverse maximal projection through the ablation volume created by a single pulse at an energy of 0.63 μJ per pulse. Note the intrinsic fluorescence that surrounds the ablated dark, inner region.

3.5.2 Pulse Width

The shape and duration of the incident laser pulse critically affects the efficiency of nonlinear optical processes (Figure 3.3). The ultrashort pulses used for multiphoton microscopy are typically one hundred to a few hundred femtoseconds in duration. Since no electronic photodetector has the necessary time resolution to measure such time-scales, optical pulse measurement techniques must be used. An autocorrelation measurement is an effective and relatively simple method of monitoring the pulse length and is simple to implement.

3.5.2.1 Autocorrelation

In an autocorrelation measurement, a laser pulse is split into two identical copies. One pulse is delayed relative to the other by a time, τ, which results in two pulses with electric fields $E(t)$ and $E(t + \tau)$. The two pulses are recombined and used to excite a nonlinear process such as two-photon fluorescence. The interferometric autocorrelation function, denoted $I_{AC}(\tau)$, is given by the sum of $E(t)$ and $E(t + \tau)$, raised to the fourth power, and averaged over time. This appears as an oscillatory signal with an envelope that depends on the delay time τ (Appendix B). A convenient feature of this autocorrelation is an 8:1 ratio of the amplitude of the maximum from zero to the amplitude of the constant background that is produced by widely separated pulses. Achieving this ratio indicates that the apparatus is well aligned. In the case that the laser pulse has essentially no chirp, the FWHM of the autocorrelation is equal to $\sqrt{2/\ln 2}$ times (1.7 times) the FWHM of the intensity of the original pulse.

Our laser system and microscope was set up without external pulse compression. We constructed an autocorrelator[67,68] that could be switched into the beam path (Figure 3.15). We found that although the commercially purchased Ti:Sapphire laser should output pulses around 120 fs, the pulses at the microscope objective are typically closer to $\tau_{FWHM} = 250$ fs, as shown by a fit of a model autocorrelation (Appendix B) to the measured autocorrelation function (Figure 3.27A). This pulse is weakly chirped (1% or ~8 nm), as evidenced by the curvature in the wings of the autocorrelation function (∗ in Figure 3.27A). The increase in pulse width represents the cumulative effect of optics throughout the entire TPLSM system (Figure 3.15). Note that, by judicious tuning of the laser, we could achieve pulses with minimal chirp and with widths of $\tau_{FWHM} = 180$ fs (Figure 3.27B and Equation 3.A.5).

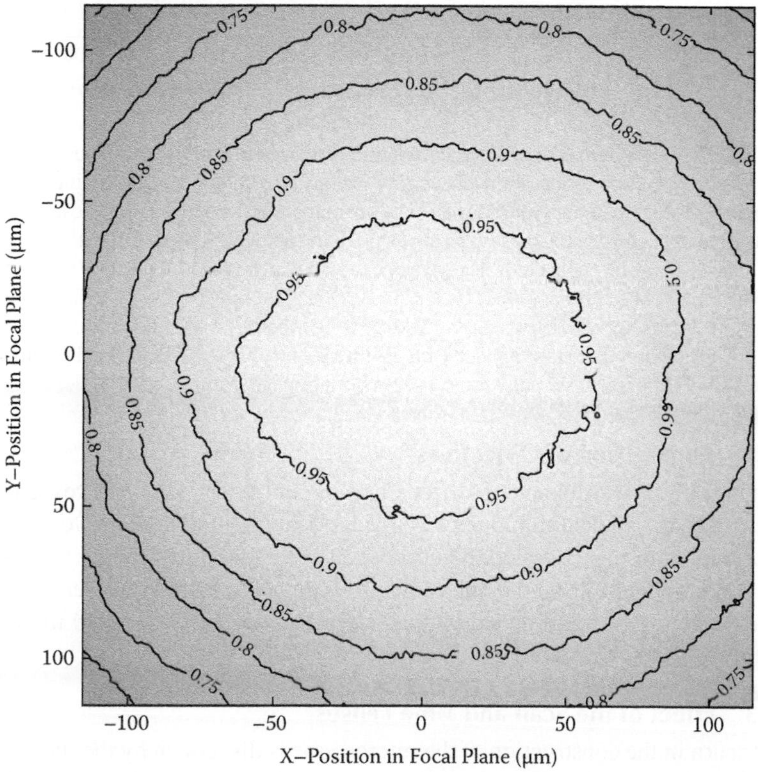

FIGURE 3.26 Contour plot of spatial uniformity of the two-photon generated and detected fluorescence measured with a 40-times magnification lens. Note that the decrement across the entire field exceeds that calculated for the generation of the fluorescence alone (Figure 3.11), but that the decrement across the central 100 μm of the field is less than 10%.

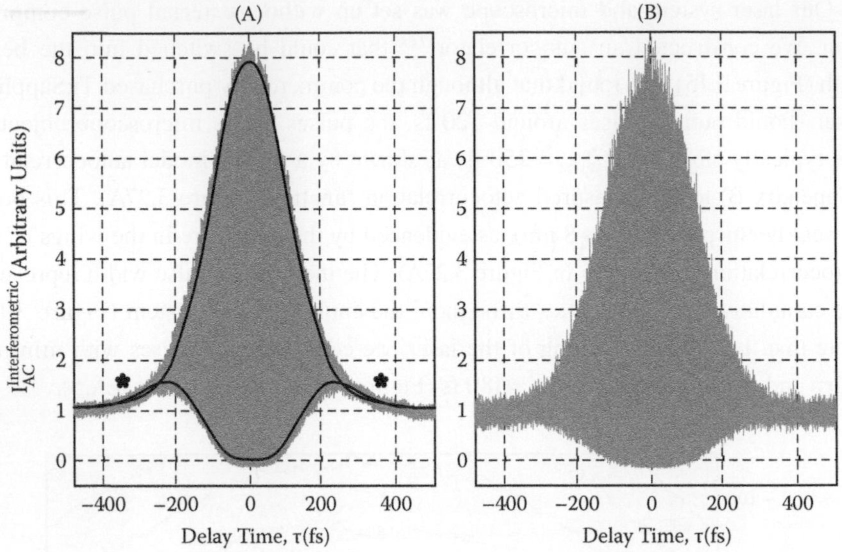

FIGURE 3.27 The interferometric autocorrelation function (Equation 3.A.4) measured for laser pulses measured at the position of the objective with our two-photon laser scanning microscope. (A) The autocorrelation under typical levels of performance. The asterisks indicate additional curvature as a result of chirp in the pulse. The envelope of the autocorrelation is fit with the calculated autocorrelation for a chirped pulse (thick envelope), from which we estimate a FWHM of the laser pulse to be 250 fs. The chirp parameter is $b = 5 \times 10^{-5}$ fs^{-2}; this corresponds to a factional shift in frequency of $2 \cdot b \cdot \tau_{FWHM} \cdot \omega = 2 \cdot b \cdot \tau_{FWHM} \cdot \lambda_0 / 2 \cdot \pi \cdot c = 0.01$, where $\lambda_0 = 800$ nm and c is the speed of light. (B) The autocorrelation for a pulse with essentially no chirp, that is, $b = 0$. For this case we estimate a FWHM of the laser pulse to be 180 fs (Equation 3.A.5).

3.5.2.2 Comparison of Objectives

To compare the performance of various brands and types of objectives, each with different glasses and detailed construction, identical laser parameters were measured using a number of objectives on the microscope. We found that the variations of the measured pulse width from the objectives were not significant at a given wavelength nor was there a significant change in pulse-width over the 780 to 880 nm range of wavelengths for the incident light (Figure 3.28).

3.5.2.3 Effect of the Scan and Tube Lenses

One concern in the construction of the microscope is dispersion by the tube and scan lenses. With the standard tube lens in the trinocular and an achromat as the scan lens, the pulse width was $\tau_{FWHM} = 230$ fs as compared with $\tau_{FWHM} = 180$ fs with only the objective in the microscope pathway.

3.5.3 Alignment of the Imaging and Ablation Foci

To achieve co-planar ablation at the center of the imaging field, the ablation and imaging beams must be collimated and co-linear when they enter the back aperture

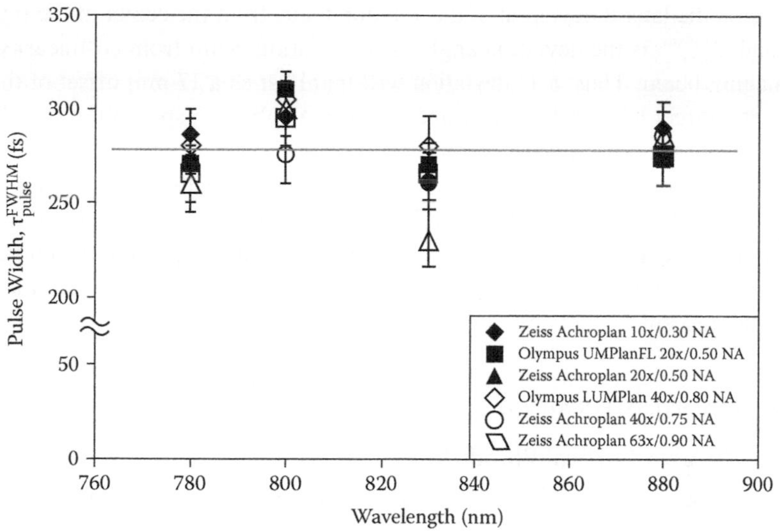

FIGURE 3.28 Plot of the FWHM of the laser pulse width measured for six different common objectives as a function of excitation wavelength. There is no significant difference in pulse width between lenses.

of the objective. This is achieved by temporarily replacing the objective with a mirror mounted at 45°, which diverts the beam to a distant target screen. The imaging beam scanners must be powered and set to a minimal scan for this alignment. Collimation of each beam is verified individually by comparing the beam profile near the 45° mirror and at the target screen. Co-linearity is achieved by overlapping the two beams at both locations. Adjustments to beam collimation are achieved by manipulation of the appropriate beam telescope (Section 3.4.3). Adjustment of co-linearity is achieved by adjustment to the two mirrors in the ablation beam path directly upstream of the polarizing beamsplitter.

The required precision of alignment can be calculated in a geometric optics limit by replacing the objective with a simple positive lens. For example, we can replace a 40-times magnification objective with a 4-mm focal length lens and calculate the beam divergence as:

$$\Delta f = f_{obj} - \left(\frac{1}{f_{obj}} - \frac{1}{f_{beam}} \right)^{-1} \tag{3.30}$$

where Δf is the axial focus shift in the sample, f_{obj} is the equivalent focal length of the objective, and f_{beam} is a measure of the beam divergence that is equal to the distance between the tube lens and the focus of the ablation beam. Thus, an ablation beam divergence corresponding to an 8-m focus will manifest as a 12% change in beam diameter at a target placed 1 m away, and result in an axial focal shift of 4 μm in the sample. The lateral deviation from co-linearity is given by:

$$\Delta r \simeq -f_{obj}\ \theta_{input}^{radians} \tag{3.31}$$

where Δr is the lateral deviation of the ablation focus from the center of the imaging field, and $\theta_{input}^{radians}$ is the deviation angle of the ablation beam from co-linearity with the imaging beam. Thus, a 1° deviation will manifest as a 17-mm offset of the two beams at a target located 1 m away and result in an ablation spot in the sample that is 70 μm from the center of the imaging field.

Experimental verification of beam alignment can be achieved by imaging and ablating the interface of a glass slide immersed in a fluorescent solution, such as aqueous fluorescein. The glass surface will appear in dark contrast to the fluorescent solution, while ablation craters formed at the interface will fill with solution and appear fluorescent.

3.5.4 Example Applications

3.5.4.1 In Vivo Cortical Blood Flow

A unique aspect of TPLSM is the ability to observe anatomical and functional signals in the brains of living animals deep to their pial surface.[4] As an example that used the microscope described above, we consider the flow of red blood cells (RBCs) in individual capillaries in rat cerebral cortex[3] (Figure 3.29). A cranial window is prepared over whisker barrel sensory cortex, and the dura is removed; a metal frame and closed surgical window are attached to the skull for stable mounting of the preparation to the microscope table.[41] The blood stream is labeled with fluorescein-conjugated-dextran (2 MDa). A single optical section obtained 100 μm below the pial surface is shown in Figure 3.29A, and the maximal projection reconstructed from 41 individual planar images, ranging in depth from 80 to 120 μm below the pial surface of the brain, is shown in Figure 3.29B. Notice that individual RBCs appear in the capillaries as dark stripes that correspond to locations where the fluorescent dye was excluded from the lumen of the vessel during the scan.

Line scans may be taken repeatedly along an individual capillary to produce a continuous measure of blood flow. The spatial coordinate is taken to be along the direction of the scan, while the perpendicular direction represents the temporal coordinate. As the RBC moves along the capillary, its position along the line scan changes from scan-to-scan, thereby tracing out a dark line, as shown for three examples (Figures 3.29C–E). The data of Figure 3.29C shows continuous flow, and the analysis of this data provides information regarding three parameters of capillary blood flow: blood cell speed, blood cell flux, and blood cell linear density, where flux = linear density × speed. The slope of the line is the inverse of the instantaneous speed, that is, speed = $|dt/dx|^{-1}$. The spacing between the lines along the temporal coordinate direction is a measure of the cell flux, while the spacing between the lines along the spatial coordinate direction is a measure of the cell density. The line scan data of Figure 3.29D contains a short vertical region, indicative of a momentary stall of the RBCs, while that of Figure 3.29E illustrates some of the high variability in flow that may be observed.

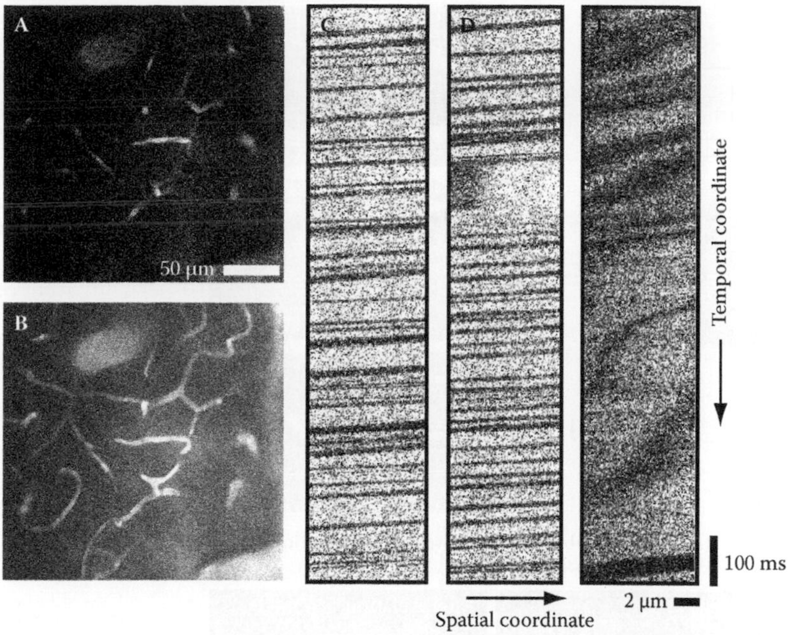

FIGURE 3.29 In vivo TPLSM used to image the capillary bed in rat vibrissa primary sensory cortex. An optical window was prepared over this region and the bloodstream was labeled with 2MDa fluorescein conjugated-dextran. (A) A single planar image from a depth of 100 μm below the pial surface of cortex. Notice that individual RBCs, which occlude fluorescent dye, appear in the capillaries as dark stripes. (B) A maximal projection reconstructed from 41 planar images, as in part A, taken between 80 and 120 μm below the pia. (C–E) Line scan data taken along single capillaries. Panel (C) shoes relatively fast, uniform flow. The thickness of the dark line may reflect the orientation of the red blood cell. Panel (D) shows flow that is interrupted by a brief stall while panel (E) shoes flow that is highly irregular.

3.5.4.2 Perturbation of Vasculature and Neuronal Processes

The integration of photo-induced plasma-mediated ablation, through the use of high-energy laser pulses, combined with imaging of fluorescently labeled tissue, through the use of TPLSM with low-energy pulses, allows the targeting and real-time monitoring of photodisruption of biological microstructures. We show examples for the cases of individual labeled blood vessels and individual labeled neuronal somata.

We first consider the use of plasma-mediated ablation to perturb the flow of blood in targeted subsurface microvessels in cortex without affecting the area surrounding the vessel.[13,69] Rodents were prepared as described above. The region of interest was imaged with TPLSM using a 40-times magnification dipping objective, and then a vessel was selected for targeting with the amplified laser source. Vessel rupture and hemorrhages could always be attained at high laser energies (Figure 3.30a), typically 1 to 10 pulses at ~0.8 μJ/pulse. Disruption with low pulse energy, ~0.3 μJ/pulse, resulted in extravasation of the fluorescein dye in the blood plasma but uninterrupted

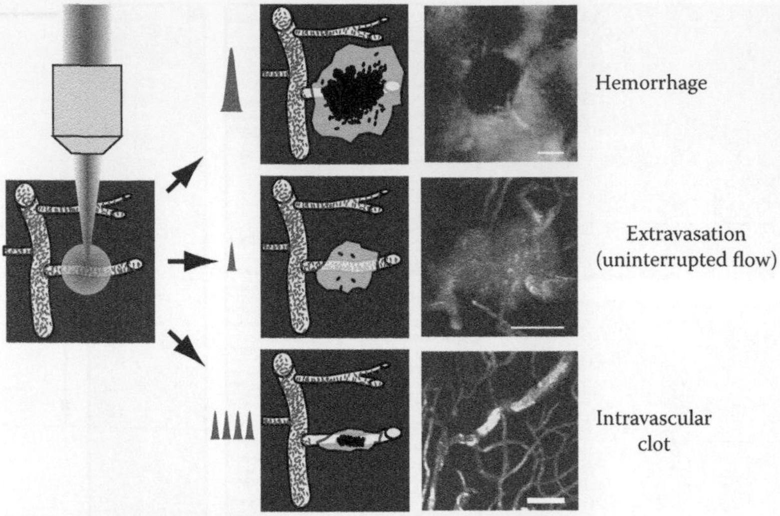

(A) Targeted subsurface microvascular disruption

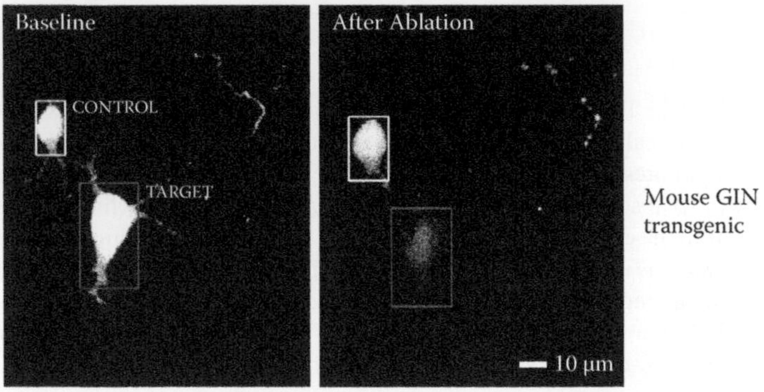

(B) Targeted somatic ablation

FIGURE 3.30 **(See color insert following 224)** In vivo TPLSM and plasma-mediated ablation in rodent cortex. (A) Schematic and examples of three different vascular lesions that are produced by varying the energy and number of laser ablation pulses. At high energies, photodisruption produces hemorrhages, in which the target vessel is ruptured, blood invades the brain tissue, and a mass of RBCs form a hemorrhagic core. At low energies, the target vessel remains intact, but transiently leaks blood plasma and red blood cells forming an extravasation. Multiple pulses at low energy leads to thrombosis that can completely occlude the target vessel, forming an intravascular clot. Scale bars, 50 μm. (B) Example of cell degeneration that is produced by a single pulse to the center of a GFP-labeled interneuron. (From Migliori, Atallah, Tsai, Scanziania, Kleinfeld, unpublished)

flow. Repeated pulses led to a clot that completely stopped RBC motion. Real-time monitoring of the vasculature with TPLSM allows one to fully characterize the actual insult and the functional response in the neighboring tissue.

We next consider the use of plasma-mediated ablation to destroy individual neurons in hippocampal slice that are highlighted through the expression of GFP.[70]

Targeting of a single pulse to a labeled interneuron quenches the fluorescence in that cell without affecting that of a neighboring interneuron (Figure 3.30B). Similar studies have used plasma-mediated ablation to cut the axon of a neuron rather than damage to soma of the neuron.[12,71]

APPENDIX A: PARTS AND VENDORS FOR THE TPLSM

MECHANICAL

The 8′ × 4′ × 1′ antivibration air table was purchased from Newport (Irvine, California). For mirror mounts, we used the Ultima® series from Newport and the Flipper Mounts from New Focus (Santa Clara, California). All mirror mounts were attached to standard 1/2″ posts and post holders from Newport that were held to the air table with clamping plates and forks from New Focus. Microbench™ parts were obtained from Linos Photonics (Milford, Massachusetts). The large X-95 rails and assorted clamps and brackets were purchased from a variety of vendors. The slider for the detector assembly was from New England Affiliated Technologies (SD Series; Lawrence, Massachusetts), and the Modular Focusing Unit was obtained from Nikon (part no. 883773; Garden City, New York). The motorized stages for X and Y translation of the preparation were obtained from Danaher Precision Systems (Salem, New Hampshire). Control for the stages was obtained with a model DMC4040 Motion Controller from Galil Motion Control (Rocklin, California). The power supplies for the controller were obtained from Acopian (Easton, Pennsylvania).

OPTICAL

All mirrors were λ/10 flatness, protected silver mirrors obtained from Newport. Water immersion objectives were purchased from Zeiss (Thornwood, NY) and Olympus. Dichroic beamsplitter #1 was a 700DCXRU (Chroma Technology; Rockingham VT); the beam blocking filters were E680-SP-2P from Chroma Technology, additional interference filters were from Omega Optical Inc. (Brattleboro, VT) and Chroma Technology (Rockingham). The wave plate, polarizing beamsplitter and intermediary optics were obtained from Newport and Linos Photonics. The scan lens is a Plan 1.0-times magnification dissection stereoscope objective with a focal length of 92.7 mm (Leica, Wetzlar, Germany).

SCANNERS

The scanning galvanometers and their drivers were obtained from Cambridge Technologies (Boston, Massachusetts). The scanners are Model 6210H Galvanometer Optical Scanners capable of ±20° maximum deflection. They have a 3-mm aperture and a 100-μs set time. We used their MicroMax™ 673XX Dual-Axis Servo Driver, which take a ±10-V analog input. The power supplies for the scan drivers were obtained from Acopian.

PHOTOMULTIPLIER TUBES

The model R6357 was specified for dark currents below 2 nA (Hamamatsu Corporation, Bridgewater, New Jersey) and powered by a model EW850-13 base (Hamamatsu Corporation) that is energized with a 0 to 2000 V DC adjustable power supply (Model PMT-20A/N; Bertan Associates, Hicksville, New York) configured to stay below the maximum rating of 1250 V that is required for this PMT. The model H7222 PMT module was specified with a modification for high threshold shut-off (Hamamatsu Corporation).

ELECTRONICS

The data acquisition board is a model PCI-6111E four channel analog input board from National Instruments (Austin, TX) that further contains two digital-to-analog converters that serve as the waveform generator for the scanning galvanometers. A model PCI-6710 Arbitrary Waveform Boards from National Instruments serves to control the Z-piezoelectric focus control. The stepping motor that controls the axial translation of the focus is a Vexta model PH268M-M1.5 (Herbach and Rademan Company, Moorestown, New Jersey). It has 400 steps per revolution, which results in 0.25 μm axial steps in combination with the Modular Focusing Unit.

LASER AND ANCILLARY EQUIPMENT

We have had good results with equipment from both Coherent, Inc. (Santa Clara, California), including a Verdi 10 W laser pumping a Mira 900F Ti:Sapphire oscillator, and Spectra Physics (a subsidiary of Newport, Inc.), including a Millennia 15 W laser pumping a Tsunami HP Ti:Sapphire oscillator. Additional equipment required for laser operation include:

1. Chillers, typically supplied with the lasers.
2. A dry nitrogen gas purge to facilitate laser operation for wavelengths longer than 900 nm.
3. An optical spectrometer to track the center frequency (e.g., USB2000+ [Ocean Optics, Inc., Dunedin, Florida])
4. An ~400-MHz bandwidth oscilloscope to aid in mode locking (e.g., Tektronix model TDS430A oscilloscope [Newark Electronics])
5. An infrared viewer (e.g., Find-R-Scope [FJW Optical Systems, Inc., Palatine, Illinois]) to aid in alignment of the microscope optics and major tuning of the oscillator.
6. A Si-based PIN diode photodiode to monitor the average laser power (e.g., Thor Labs [Newton, New Jersey] model DET10A). A calibrated energy meter (e.g., Newport model 818T-10, Newport) is used to calibrate the photodiode.
7. Safety glasses, to be worn when working near an unprotected beam line.

APPENDIX B: BASICS OF INTERFEROMETRIC AUTOCORRELATION

In the autocorrelation process, the laser pulse is split into two identical copies, and one is delayed relative to the other by a time τ. This results in two pulses with electric fields $E(t)$ and $E(t+\tau)$. These are subsequently multiplied together through a nonlinear process, such as two-photon excitation. For such a process, the instantaneous two-photon excited fluorescence for a given delay time is

$$I(t,\tau) = \left|\left[E(t) + E(t+\tau)\right]^2\right|^2. \tag{3.A.1}$$

The time resolution of the photodetector is much slower than the fluorescence duration, thus the output of the photodiode is the time average of $I(t, \tau)$ (Equation 3.A.1). This signal corresponds to the autocorrelation function, $I_{AC}(\tau)$, that is,

$$I_{AC}(\tau) = \int_{-\infty}^{\infty} dt \left|\left[E(t) + E(t+\tau)\right]^2\right|^2. \tag{3.A.2}$$

To deduce the shape of the original pulses from the autocorrelation signal $I_{AC}(\tau)$, we assume a known original pulse shape. The autocorrelation signal can be calculated in terms of parameters describing the original pulse. The measured autocorrelation signal shape can then be used to deduce the original pulse parameters. Ultrashort laser pulses usually have Gaussian or secant-squared time envelopes. Since the Gaussian is a good estimate of the actual pulse shape and facilitates calculations, we assume a Gaussian profile for our beam. The electric field of the laser pulses can be described as

$$E(t) = E_0\, e^{-\frac{t^2}{a^2} + i2\pi\, \nu_0 t} \tag{3.A.3}$$

where the full-width-half-maximum (FWHM) of the intensity of the envelope of the pulse is $a\sqrt{2\ln 2}$, ν_0 is the center frequency of the pulse, and the origin of time is arbitrary as $E(t)$ and $E(t+\tau)$ appear only in integrals over all of time (e.g., Equation 3.A.2).

The autocorrelation measurement can be set up in two ways. In an interferometric autocorrelation, fringes caused by the interference of the two pulses are observed. The two pulses are incident on the fluorescent material so that they are exactly colinear and are polarized in the same direction. In a noninterferometric autocorrelation, the fringes are smeared out either by introducing an angle between the beams or a difference in the polarization direction of the two pulses. While both autocorrelation methods give a measure of pulse length, the interferometric autocorrelation will also indicate the presence of chirp in the pulse.

The interferometric autocorrelation signal with $E(t)$ given by Equation 3.A3 is

$$I_{AC}^{\text{interferometric}}(\tau) \propto 1 + 2e^{-\frac{\tau^2}{a^2}} + 4e^{-\frac{3\tau^2}{4a^2}}\cos(2\pi\nu_0\tau) + e^{-\frac{\tau^2}{a^2}}\cos(4\pi\nu_0\tau). \quad (3.A.4)$$

The upper envelope and lower envelope can be found by setting $2\pi\nu_0\tau = \pi$ or 2π. A convenient feature of interferometric autocorrelation is the 8:1 ratio of the amplitude of the maximum from zero to the amplitude of the constant background. Achieving this ratio indicates that the apparatus is well aligned. In the case with unchirped pulses, the FWHM of $I_{AC}^{\text{interferometric}}(\tau)$ is related to the FWHM of the original pulse by

$$\tau_{FWHM} \approx \sqrt{\frac{ln2}{2}} \cdot FWHM_{AC}. \quad (3.A.5)$$

In the presence of chirp, the interferometric autocorrelation function develops curvature in the wings of the signal (* in Figure 3.26). When this feature is present, the FWHM of the autocorrelation and of the original pulse no longer have the simple relation given by Equation 3.A.5. The central peak of the autocorrelation signal will become narrower because the red-shifted, leading portion of the pulse will only incompletely interfere with the blue-shifted tail. The electric field of a linearly chirped pulse can be described as

$$E(t) = E_0\, e^{-\frac{t^2}{a^2} + i2\pi\nu_0 t - ibt^2} \quad (3.A.6)$$

where b is a measure of the chirp. The experimentally obtained autocorrelation must be fit to the autocorrelation that is calculated (Equations 3.A.2 and 3.A.6) with the two parameters, a and b, taken as variables in order to determine the pulse length. In practice, for two-photon microscopy, an absolute measurement of the chirp is generally not necessary. One typically only adjusts the laser to minimize the curvature in the wings of the autocorrelation signal.

ACKNOWLEDGMENTS

We thank Winfried Denk and Jeffrey A. Squier for their initial and continued tutelage in nonlinear microscopy. We thank Nozomi Nishimura for obtaining the unpublished spectrum in Figure 3.28, and Benjamin Migliori and Bassam Atallah for obtaining the unpublished data in Figure 3.30b. We are grateful to the David and Lucile Packard Foundation, the National Institutes of Health (NCRR, NIBIB, and NINDS), and the National Science Foundation for their cumulative support of our in vivo research program.

REFERENCES

1. Denk, W. and Svoboda, K., Photon upmanship: Why multiphoton imaging is more than a gimmick, *Neuron* 18, 351–357, 1997.

2. Svoboda, K. and Yasuda, R., Principles of two-photon excitation microscopy and its applications to neuroscience, *Neuron* 50, 823–839, 2006.
3. Kleinfeld, D. et al., Fluctuations and stimulus-induced changes in blood flow observed in individual capillaries in layers 2 through 4 of rat neocortex, *Proceedings of the National Academy of Sciences USA* 95, 15741–15746, 1998.
4. Denk, W. et al., Anatomical and functional imaging of neurons and circuits using two photon laser scanning microscopy, *Journal of Neuroscience Methods* 54, 151–162, 1994.
5. Helmchen, F. and Denk, W., Deep tissue two-photon microscopy, *Nature Methods* 2, 932–940, 2005.
6. Denk, W. et al., Two-photon laser scanning fluorescence microscopy, *Science* 248, 73–76, 1990.
7. Denk, W. et al., Two-photon molecular excitation in laser-scanning microscopy, in *Handbook of Biological Confocal Microscopy*, Pawley, J. W., Plenum Press, New York, 1995, pp. 445–458.
8. Williams, R. M. et al., Two-photon molecular excitation provides intrinsic 3-dimensional resolution for laser-based microscopy and microphotochemistry, *FASEB Journal* 8, 804–813, 1994.
9. Zipfel, W. R. et al., Nonlinear magic: Multiphoton microscopy in the biosciences, *Nature Biotechnology* 21, 1369–1377, 2003.
10. Webb, W. W., Two photon excitation in laser scanning microscopy, in *Transactions of the Royal Microscopical Society*, Elder, H. Y. Institute of Physics, Bristol, 1990.
11. Helmchen, F. and Denk, W., New developments in multiphoton microscopy, *Current Opinion in Neurobiology* 12, 593–601, 2002.
12. Yanik, M. F. et al., Functional regeneration after laser axotomy, *Nature* 432, 822, 2004.
13. Nishimura, N. et al., Targeted insult to individual subsurface cortical blood vessels using ultrashort laser pulses: Three models of stroke, *Nature Methods* 3, 99–108, 2006.
14. Tsai, P. S. et al., Ultrashort pulsed laser light: A cool tool for ultraprecise cutting of tissue and cells, *Optics and Photonic News* 14, 24–29, 2004.
15. Loesel, F. H. et al., Laser-induced optical breakdown on hard and soft tissues and its dependence on the pulse duration: Experiment and model, *IEEE Journal of Quantum Electronics* 32, 1717–1722, 1996.
16. Vogel, A. and Venugopalan, V., Mechanisms of pulsed laser ablation of biological tissues, *Chemical Reviews* 103, 577–644, 2003.
17. Joglekar, A. P. et al., Optics at critical intensity: Applications to nanomorphing, *Proceedings of the National Academy of Sciences USA* 101, 5856–5861, 2004.
18. Tsai, P. S. et al., Principles, design, and construction of a two photon laser scanning microscope for *in vitro* and *in vivo* brain imaging, in *In Vivo Optical Imaging of Brain Function*, Frostig, R. D. CRC Press, Boca Raton, 2002, pp. 113–171.
19. Kim, K. H. et al., High-speed, two-photon scanning microscope, *Applied Optics* 38, 6004–6009, 1999.
20. Duemani-Reddy, G. et al., Three-dimensional random access multiphoton microscopy for functional imaging of neuronal activity, *Nature Neuroscience* 11, 713–720, 2008.
21. Vucinić, D. and Sejnowski, T. J., A compact multiphoton 3D imaging system for recording fast neuronal activity, *PLoS ONE* 2, e699, 2007.
22. Denk, W., Two-photon scanning photochemical microscopy: Mapping ligand-gated ion channel distributions, *Proceedings of the National Academy of Sciences USA* 91, 6629–6633, 1994.
23. Potter, S. M., Two-photon microscopy for 4D imaging of living neurons, in *Imaging Neurons: A Laboratory Manual*, Yuste, R., Lanni, F., and Konnerth, A. Cold Spring Harbor Laboratory Press, Cold Spring Harbor, 2000.
24. Majewska, A. et al., A custom-made two-photon microscope and deconvolution system, *Pflugers Archives* 441, 398–408, 2000.

25. Fan, G. Y. et al., Video-rate scanning two-photon excitation fluorescence microscopy and ratio imaging with cameleons, *Biophysical Journal* 76, 2412–2420, 1999.
26. Nikolenko, V. et al., A two-photon and second-harmonic microscope, *Methods* 30, 3–15, 2003.
27. Lanni, F. and Keller, H. E., Microscopy and microscopic optical systems, in *Imaging Neurons: A Laboratory Manual*, Yuste, R., Lanni, F., and Konnerth, A. Cold Spring Harbor Laboratory Press, Cold Spring Harbor, 2000.
28. Hiraoka, Y. et al., Determination of three-dimensional imaging properties of a light microscope system: Partial confocal behavior in epifluorescence microscopy, *Biophysical Journal* 57, 325–333, 1990.
29. Agard, D. A. et al., Fluorescence microscopy in three dimensions, *Methods in Cell Biology*, 353–377, 1989.
30. Holmes, T. J. et al., Light microscopic images reconstructed by maximum likelihood deconvolution, in *Handbook of Confocal Microscopy*, 2nd edition, Pawley, J. B. Plenum Press, New York, pp. 389–402.
31. Kam, Z. et al., Computational adaptive optics for live three-dimensional biological imaging, *Proceedings of the National Academy of Sciences USA* 98, 3790–3795, 2001.
32. Neil, M. A. A. et al., Method of obtaining optical sectioning by using structured light in a conventional microscope, *Optics Letters* 22, 1905–1907, 1997.
33. Lanni, F. and Wilson, T., Grating image systems for optical sectioning fluorescence microscopy of cells, tissues, and small organisms, in *Imaging Neurons: A Laboratory Manual*, Yuste, R., Lanni, F., and Konnerth, A., Cold Spring Harbor Laboratory Press, Cold Spring Harbor, 2000, pp. 8.1–8.9.
34. Bailey, B. et al., Enhancement of axial resolution in fluorescence microscopy by standing-wave excitation, *Nature* 366, 44–48, 1993.
35. Gustafsson, et al., I5M: 3D widefield light microscopy with better than 100 nm axial resolution, *Microscopy* 195, 10–16, 1999.
36. Konig, K. et al., Pulse-length dependence of cellular response to intense near-infrared laser pulses in multiphoton microscopes, *Optics Letters* 24, 113–115, 1999.
37. Centonze, V. E. and White, J. G., Multiphoton excitation provides optical sections from deeper within scattering specimens than confocal imaging, *Biophysical Journal* 75, 2015–2024, 1998.
38. Yariv, A., *Optical Electronics: Third Edition*, Holt, Rinehart and Wilson, New York, 1985.
39. Born, M. and Wolf, E., *Principles of Optics: Electromagnetic Theory of Propagation Interference and Diffraction of Light,* 6th edition, Pergamon Press, Oxford, 1980.
40. Tsai, P. S. et al., All-optical histology using ultrashort laser pulses, *Neuron* 39, 27–41, 2003.
41. Kleinfeld, D. and Denk, W., Two-photon imaging of neocortical microcirculation, in *Imaging Neurons: A Laboratory Manual*, Yuste, R., Lanni, F., and Konnerth, A. Cold Spring Harbor Laboratory Press, Cold Spring Harbor, 2000, pp. 23.1–23.15.
42. Oheim, M. et al., Two-photon microscopy in brain tissue: Parameters influencing the imaging depth, *Journal of Neuroscience Methods* 111, 29–37, 2001.
43. Theer, P. and Denk, W., On the fundamental imaging-depth limit in two-photon microscopy, *Journal of the American Optical Society A* 23, 3139–3150, 2006.
44. Schonle, A. and Hell, S. W., Heating by absorption in the focus of an objective lens, *Optics Letters* 23, 325–327, 1998.
45. Aquirrell, J. M. et al., Long-term two-photon fluorescence imaging of mammalian embryos without compromising viability, *Nature Biotechnology* 17, 763–767, 1999.
46. Koester, H. J. et al., Ca^{2+} fluorescence imaging with pico- and femtosecond two-photon excitation: Signal and photodamage, *Biophysical Journal* 77, 2226–2236, 1999.
47. Patterson, G. H. and Piston, D. W., Photobleaching in two-photon excitation microscopy, *Biophysical Journal* 78, 2159–2162, 2000.

48. Neher, E. and Hopt, A., Highly nonlinear photodamage in the two-photon fluorescence microscopy, *Biophysical Journal* 80, 2029–2036, 2001.
49. Siegman, A. E., *Lasers*, University Science Books, Sausalito, 1986.
50. Fork, R. L. et al., Negative dispersion using pairs of prisms, *Optics Letters* 9, 150–152, 1984.
51. Squier, J. A. et al., Effect of pulse phase and shape on the efficiency of multiphoton processes: Implications for fluorescence microscopy, in *Conference on Lasers and Electro-Optics,* Optical Society of America, Baltimore, 1999, pp. 80.
52. Cerullo, G. et al., High-power femtosecond chirped pulse excitation of molecules in solution, *Chemical Physics Letters* 262, 362–368, 1996.
53. Kawano, H. et al., Attenuation of photobleaching in two-photon excitation fluorescence from green fluorescent protein with shaped excitation pulses, *Biochemical and Biophysical Research Communications* 311, 592–596, 2003.
54. Trebino, R. et al., Measuring ultrashort laser pulses in the time-frequency domain using frequency-resolved optical gating, *Review of Scientific Instrumentation* 68, 3277–3295, 1997.
55. Mainen, Z. F. et al., Two-photon imaging in living brain slices, *Methods* 18, 231–239, 1999.
56. Xu, C. et al., Multiphoton fluorescence excitation: New spectral windows for biological nonlinear microscopy, *Proceedings of the National Academy of Sciences USA* 93, 10763–10768, 1996.
57. Stosiek, C. et al., *In vivo* two-photon calcium imaging of neuronal networks, *Proceedings of the National Academy of Sciences USA* 100, 7319–7324, 2003.
58. Fink, C. et al., Intracellular fluorescent probe concentrations by confocal microscopy, *Biophysical Journal* 75, 1648–1658, 1998.
59. Backus, S. et al., 0.2-TW laser system at 1 kHz, *Optics Letters* 22(16), 1256–8, 1997.
60. Shockley, W. and Pierce, J. R., A theory of noise for electron multipliers, *Proceedings of the Institute of Radio Engineers* 26, 321–332, 1938.
61. Pologruto, T. A. et al., ScanImage: Flexible software for operating laser scanning microscopes, *Biomedical Engineering Online* 17, 13, 2003.
62. Nguyen, Q.-T. et al., MPScope: A versatile software suite for multiphoton microscopy, *Journal of Neuroscience Methods* 156, 351–359, 2006.
63. Gobel, W. and Helmchen, F., New angles on neuronal dendrites in vivo, *Journal of Neurophysiology* 98, 3770–3779, 2007.
64. Göbel, W. et al., Imaging cellular network dynamics in three dimensions using fast 3D laser scanning, *Nature Methods* 4, 73–79, 2007.
65. Loesel, F. H. et al., Non-thermal ablation of neural tissue with femtosecond laser pulses, *Applied Physics B* 66, 121–128, 1998.
66. Stuart, B. C. et al., Nanosecond-to-femtosecond laser-induced breakdown in dielectrics, *Physical Review B* 53, 1749–1761, 1996.
67. Muller, M. et al., Measurement of femtosecond pulses in the focal point of a high-numerical-aperture lens by two-photon absorption, *Optics Letters* 20, 1038–1040, 1995.
68. Diels, J.-C. M. et al., Control and measurement of ultrashort pulse shapes (in amplitude and phase) with femtosecond accuracy, *Applied Optics* 24, 1270–1282, 1985.
69. Kleinfeld, D. et al., Targeted occlusion to surface and deep vessels in neocortex via linear and nonlinear optical absorption, in *Animal Models of Acute Neurological Injuries*, Chen, J., Xu, Z., Xu , X.-M., and Zhang, J. The Humana Press, Inc., Totowa, 2008.
70. Oliva, A. A. Jr. et al., Novel hippocampal interneuronal subtypes identified using transgenic mice that express green fluorescent protein in GABAergic interneurons, *Journal of Neuroscience* 20, 3354–3368, 2000.
71. Zeng, F. et al., Sub-cellular precision on-chip small-animal immobilization, multi-photon imaging and femtosecond-laser manipulation, *Lab on a Chip* 8, 653–656, 2008.

4 MPScope 2.0

A Computer System for Two-Photon Laser Scanning Microscopy with Concurrent Plasma-Mediated Ablation and Electrophysiology

Quoc-Thang Nguyen, Jonathan Driscoll, Earl M. Dolnick, and David Kleinfeld

CONTENTS

4.1 INTRODUCTION

The MPScope[1] freeware package is a comprehensive suite of programs that run under the Microsoft Windows operating system and are designed as control and analysis elements for two-photon laser scanning microscopy (TPLSM) and various extensions. MPScope consists of MPScan, the acquisition software, MPView, the analysis program, and MPFile, a utility to open MPScope data files in user-written applications. MPScope was designed from the outset as a turn-key, stand-alone, easy-to-use software system with flexible hardware options aimed at concurrent functional imaging, (e.g., calcium fluorescence), and electrophysiology experiments. However, the versatility of MPScope is such that it can be used in many different types of experiments including in vivo cortical blood flow imaging and two-photon plasma-mediated ablation automated histology.[2] The usefulness of MPScope stems mostly from leveraging advanced software technologies available on the Windows platform.

Over 50 laboratories have registered for either using MPScope or for downloading its source code as of summer 2008. Many users run MPScope on custom-made two-photon laser scanning microscopes designed for in vivo imaging, such as the one described by Tsai and Kleinfeld[3] in Chapter 3, while others have successfully adapted MPScope to control multiphoton systems based on commercial microscopes. The widening use of MPScope had, unfortunately, the unforeseen side effect of spurring a multitude of disparate hardware configurations, some very specific to a given laboratory, with the concomitant burden of supporting them in successive releases of MPScope. To alleviate this situation for future users of MPScope, we have overhauled MPScope and formed a presumably stable baseline configuration.

4.1.1 An Overview of Software for Two-Photon Laser Scanning Microscopy

In every laser scanning microscope, luminance signals from discrete locations (i.e., pixels) are combined to form an image registered with the spatial excitation pattern of the laser beam. Synchronization of excitation signals, digitization of emitted light, and subsequent frame reconstruction and processing is best done by software. Computer programs that control two-photon laser scanning microscopes are by no means different from those used in conventional laser confocal microscopes. In both types of imaging, the frame acquisition programs are tightly related to the hardware design of the microscope, while analysis applications are much more flexible, thanks to the existence of widely used standard formats for imaging data. The challenge in developing a useful software package for in vivo TPLSM is to cover a wide range of possible experiments with a large choice of auxiliary hardware without unduly complicating the user interface and limiting the performance of the microscope. To avoid the time-consuming task of writing a two-photon microscope program *de novo*, users of two-photon microscopes based on modified Nikon Fluoview confocal platforms can take advantage of the original commercial software by extending it with custom add-ons.[4] For custom-made microscopes that use resonant scanner mirrors (e.g., the CRS series by GSI Lumonics), programs like e-Maging[5] are capable of displaying and streaming to disk up to 30 frames per second with each frame comprising 500 by 450 pixels.

4.1.2 Earlier Matlab™- and LabVIEW™-Based Two-Photon Microscope Software

Most current two-photon microscope systems, such as the one described in this book,[3] use a pair of galvanometric mirrors that are arranged orthogonally to generate the laser beam scanning pattern in the X-Y plane. Although slower than microscopes that use resonant scanners, these systems are more flexible as they provide finer control of the pixel dwell time, allow rotating full frames or line scans, and give the possibility to scan arbitrary regions of interest or to park the laser beam at a particular spot.[6] These microscopes, for example, can be controlled by ScanImage,[7] a software developed by the Svoboda Laboratory and written in the Matlab™ programming language (MathWorks, Natick, Massachusetts). Prior to MPScope, our laboratory and that of several affiliates ran a program developed in-house that was written in the graphical language LabVIEW™ (National Instruments, Austin, Texas).[8]

4.1.3 Drawbacks of Matlab™ and LabVIEW™ Development

The popular programming environments of Matlab™ and LabVIEW™ rely on some form of interpretation of the source code using a proprietary run-time engine. This approach provides, at first, a shorter development time for small- to medium-size projects, especially since many two-photon microscope users are already proficient with Matlab™ or LabView™. However, the interpreted nature of these programs imparts a significant penalty in performance and computer resource footprint. The dependence on version-bound run-time libraries makes the code particularly fragile to upgrades of the development environment. Severe backward compatibility issues occurred when Matlab™ went from versions 6.5 to 2007 and when LabVIEW™ v.6 was superseded by v.7. Perhaps the biggest drawback of developing in LabView™ or Matlab™ is that they are not designed for large and complex applications, unlike general-purpose languages such as Microsoft Visual Basic or C/C++. Although Matlab™ is essentially aimed at scientific calculations using floating-point arithmetic, a laser scanning microscope program spends most of its time dealing with integer data. Conversely, LabView™ is obviously optimized to construct data acquisition programs but is limited in its algorithmic capabilities. For both programming environments, some of these drawbacks can be alleviated by writing critical tasks in a compiled language such as C/C++, packaging these functions into Dynamic Link Libraries and calling them from the main program. This technique has the drawback of having to maintain two different code bases, with the concomitant problem of deploying a Dynamic Link Library always in synchrony with the current version of the main program.

4.1.4 MPScope 1.0: Technical Choices

The incentive to write MPScope was based on two considerations:

1. Although computer programs already exist to control two-photon microscopes, there is a considerable interest in increasing the frame rate and the on-line image processing. Both ScanImage and our past LabVIEW™ program use an asymmetrical, sawtooth-like scan pattern for the fast X mirror. A more efficient scanning scheme is to use a symmetrical waveform that allows acquisition of another scan line on the return path of the mirror.[5] Such a scan pattern increases the overall frame rate, potentially reduces photobleaching by using the retrace time to collect pixels, and has the added advantage of driving the mirrors with a waveform that contains less high frequencies than an asymmetrical waveform. From a user standpoint, it is highly desirable that the program display images with lines in the correct order while acquisition takes place. This requirement implies that the software should invert each alternate line, sampled in reverse order, in near real-time; misalignment results in comb-like features around objects viewed in full frames. Because other computer-intensive tasks have to be performed simultaneously on the incoming data, e.g., calculating and plotting intensities of region of interest, on-line fluorescent-energy resonant transfer calculations, etc., we felt that neither LabVIEW™ nor Matlab™ interpreted programs would have the required speed and future growth capabilities for fast full-frame two-photon imaging.
2. Two-photon imaging experiments are becoming more complex. In many experiments, basic functions of two-photon microscopy programs need to be successively executed automatically. A simple task, such as the acquisition of a 3D volume by repeatedly taking a stack and moving the field of view over a large preparation, is an example where customization and automation of microscope function can reduce the user workload. Inclusion of a feature to easily customize and automate tasks was a priority when developing MPScope.

4.1.4.1 Hardware Principles

The basic task of a laser scanning microscope is to output the coordinates of the laser beam position while acquiring the emitted fluorescence of the corresponding pixel during a dwell time that lasts typically from 1 to 10 μs. At these rates, programmed input/output operations and interrupt-driven input/output routines are too slow to achieve the necessary throughput; for a benchmark, see Reference 11. On the Windows platform, dispatching an interrupt takes at least a hundred instructions between the activation of the Interrupt Request Line and the servicing of the interrupt.[12] Moreover, since these techniques rely on the processor for their execution, the more they are called the less time remains for the processor to display and store the incoming pixels.

The above input/output issues are largely circumvented through the use of Application Specific Integrated Circuits located on modern data acquisition boards. For instance, the Mini-MITE circuit on the National Instrument PCI-6110 S-series board used by MPScan controls Direct Memory Access (DMA).[13] Here, the Mini-MITE takes ownership of the Peripheral Connect Interface (PCI) bus and thus becomes a "busmaster" to transfer data from/to main memory to/from the periphery without intervention of the processor. The Mini-MITE allows also several input–output-independent DMA operations (e.g., analog/digital, digital to analog, and digital input/output) to be performed simultaneously via separate paths called channels. Moreover, the Mini-MITE circuit can manage data transfer from/to buffers that appear continuous in the program memory space but are fragmented in random access memory due to virtual memory paging. This capability, called "scatter-gather DMA," allows arbitrarily large buffers in random access memory to be transferred from/to periphery.

The data transit to digital-to-analog converters or from analog-to-digital converters occurs via small first in/first out (FIFO) memory buffers that are located on the data acquisition board. The FIFOs are filled or emptied in small bursts of bus activity, usually when the FIFOs reach a half-empty state. All these complex operations are initiated, controlled, and terminated by data acquisition device drivers. In addition, the drivers are also responsible for the critical page-locking of scatter-gather DMA buffers, which prevents their swapping from main memory to disk during data transfer. In the case of National Instruments hardware, data acquisition device drivers are managed by the NI-DAQ™ or NI-DAQmx™ software layers that hide the complexity of register-level hardware programming by providing a uniform, device-independent programming interface to applications.

MPScan relies on large memory buffers to store several most recently acquired images and the entire scan pattern for a single frame (Figure 4.1). The amount of memory allocated at the application level is doubled at the device driver level. This memory, allocated by NI-DAQ™, is organized as a circular buffer to perform double-buffering DMA operations in auto-initialize mode.[14] During frame acquisition, when half of the DMA buffer zone has been filled, NI-DAQ™ copies the recently acquired frames from the DMA buffers to the MPScope buffers while acquisition takes place in the other half of the DMA buffer. NI-DAQ™ then alerts MPScan of the availability of new frames. The size of the buffers is optimized to provide enough time for MPScope to process up to one second of incoming frame data, an interval with a large safety factor, without incurring an overrun of data in the DMA buffers. A similar scheme is implemented in reverse to insure that the laser beam X and Y position data for the digital to analog converter are placed in a large buffer that contains all of the pixel coordinates of a frame. This arrangement also allows immediate updating of the X-Y pattern when the frame needs to be rotated.

Timing synchronization between the digital-analog and analog-digital conversions is ensured through the use of the same pixel clock for the two concurrent operations and by triggering on common internal signals. The routing of the clock and trigger signals is performed by the digital acquisition system timing controller, an application specific integrated circuit that is located on the data acquisition board.[15] By using the full capability offered by such application-specific integrated circuits, only one multifunction board is required to output the scan pattern while simultaneously sampling the inputs.

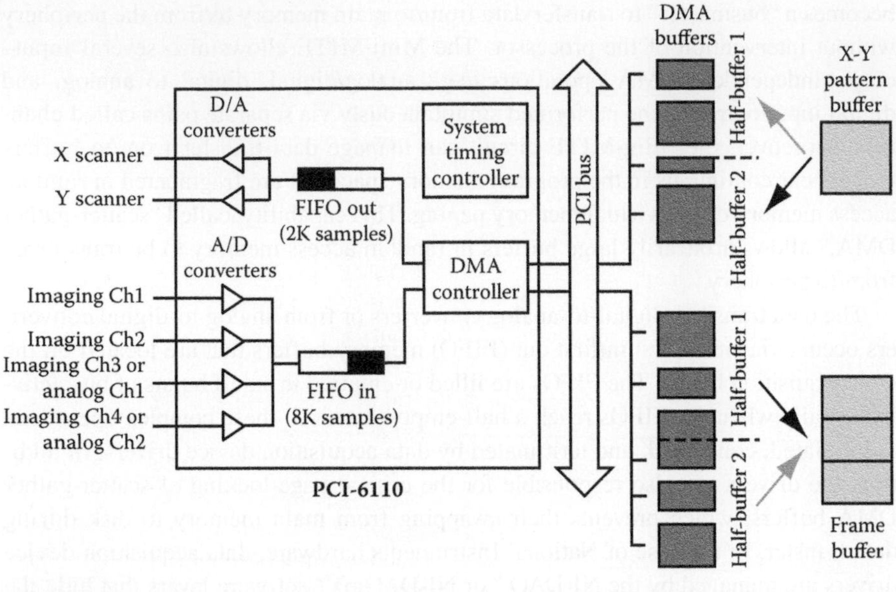

FIGURE 4.1 Data flow for MPScan acquisition and scanning. With Intel processors, DMA buffers can be fragmented into smaller, discontinuous memory zones composed of 4-kB pages of physical memory. Prior to a DMA transfer, the NI-DAQ™ driver page-locks the DMA buffers and programs the DMA controller (Mini-MITE), an application specific integrated circuit, with a linked list of memory zone descriptors. During scatter-gather operations the DMA controller goes through the list to determine the physical address of the memory locations to perform the DMA. The system timing controller is an application specific integrated circuit that synchronizes transfer operations.

4.1.4.2 Extensive Multithreading

For continuous, gap-free acquisition, the NI-DAQ™ and NI-DAQmx™ libraries enforce a cyclical data processing scheme whereby the application alternates between fetching frames from large buffers and processing the frame data. However, to enhance the responsiveness of the acquisition program, it is preferable to uncouple these two activities by splitting them into separate threads. The core of MPScan consists of a data pumping thread organized as an infinite loop that waits for the device driver to fill its frame buffer. Once this operation is performed, the data pumping thread reformats the data by splitting the sample stream into separate channels, reorders pixels in alternate lines, discards pixels at line turnarounds, reorders lines in alternate frames, saves frames to disk, and dispatches the frame data to client threads. These client threads include the frame display, the real-time frame averaging window, the real-time pixel distribution graph, the window that displays average intensity from regions of interest, and the analog data oscilloscope screen.

Care was taken to ensure proper load balancing between threads and to avoid producer/consumer deadlock problems using Windows synchronization primitives. Client threads have their own set of frame buffers that can be written by the data-pumping thread when authorized by a guarding semaphore or event. This allows client threads to process their data without interference from the data pumping thread, or to have them go to sleep in a very efficient way for the central processor unit, while waiting for the data pump thread to signal the availability of new frames. Other threads created by MPScan include the one that communicates with particular X-Y-Z stage controllers. The communication thread method allows background processing of acknowledgment replies sent by the controller, which can occur, especially in the case of a long displacement, several seconds after MPScan sends a command. The creation of a communication thread thus prevents the entire program from hanging up after commands to the X-Y-Z stage are issued.

The multithreaded architecture of MPScan is ideally matched for current multiprocessor and multicore computers. This computing power allows several tasks to be performed entirely in software that, in single processor computers, were performed by specialized hardware with embedded central processor units or digital signal processors. This includes rotation of the scan directions and various timer/stimulators. A problem that multithreading does not solve, however, is the generation of highly reproducible timing intervals. The most accurate timer under Windows is the multimedia timer that allows, at best, only a 1 ms resolution with possible unpredictable delays. The time resolution and accuracy of the multimedia timer are insufficient to provide sub-millisecond, high-precision intervals that are critical to stimulate live preparations. For these and related tasks, additional PCI hardware featuring onboard timers is necessary. Finally, the large memory capacity associated with current computers, i.e., at least 2 Gbytes of random access memory, is adequate enough to store vast numbers of successive frames simultaneously in memory.

4.1.4.3 Object-Oriented, Native-Compiled Code

The choice of an object-oriented native code compiler to develop MPScope, as opposed to compilers outputting interpreted byte codes such as .Net compilers or Java, offers several advantages over Matlab™ and LabVIEW™. Speed is the primary issue, since repetitive processing has to be performed in the data pumping thread in order to rearrange the incoming pixel stream into frames, especially since MPScan outputs symmetric line scans and thus has limited fly-back time.

We developed MPScope in the Object Pascal language with the Delphi™ Integrated Development Environment (CodeGear, Scotts Valley, California). Delphi™, a modern descendent of Turbo Pascal, has a particularly short compilation time. This feature was especially appreciated during the development of MPScan, which required frequent trial-and-error runs to adjust NI-DAQ™ function call parameters in order to obtain the desired data acquisition behavior. Delphi™ comes with an object-oriented framework of software components called the Visual Component Library that permits easy development of Windows applications, while still allowing complete access to the Windows Application Programming Interface. One advantage of using a framework that relies on native Windows controls, unlike LabVIEW™, for instance, is the lack of constraints on the "look and feel" of the application. For concreteness, we show the MPScan user interface during a two-photon automated histology session in Figure 4.2.

The use of the native Windows Application Programming Interface is also critical as a means to optimize bus-intensive graphical operations that handle device-

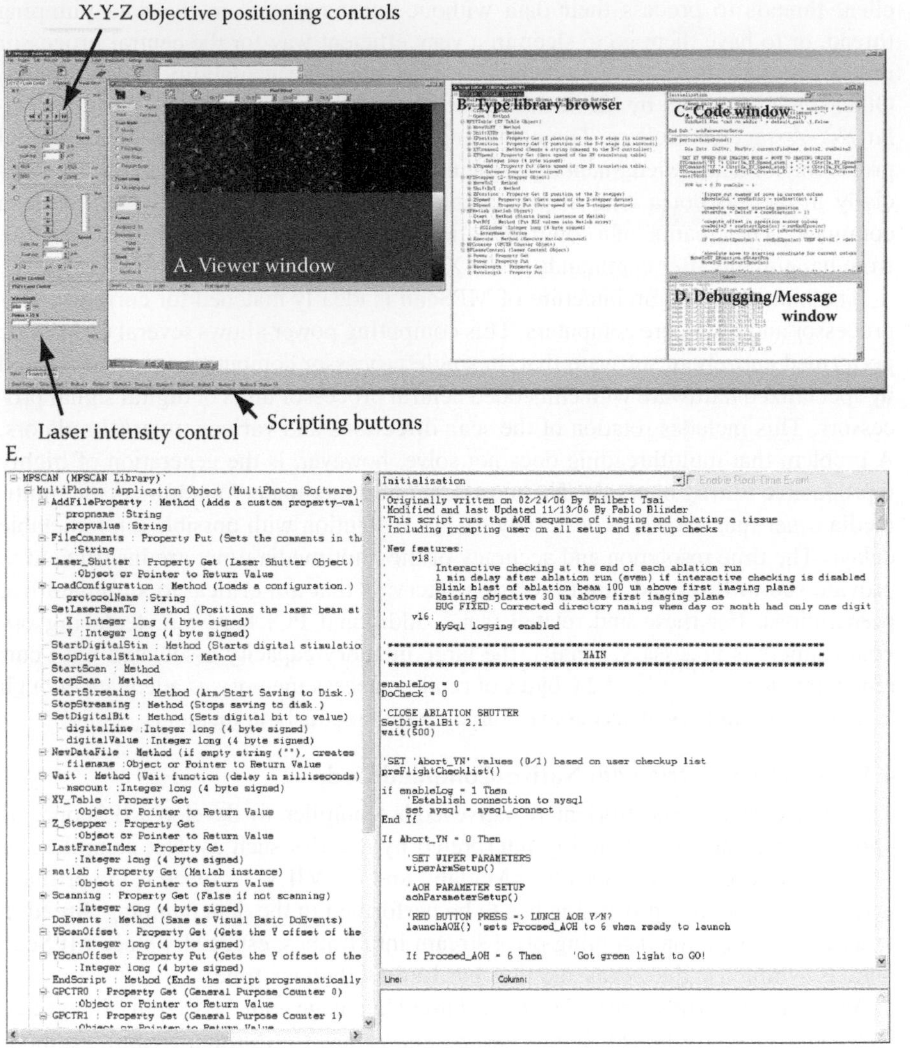

FIGURE 4.2 MPScan user interface and integrated scripting environment. (A) The main window for the display of data; three channels are shown here. Saturation levels are indicated by assigning red to the maximum A/D level. (B) Left panel displays, by default, the Type Library of MPScan; which describes objects and methods available in MPscope to other programs. This allows one to find the objects created in MPScan and to identify their properties and associated remarks. Any other type library can be displayed as well. (C) Right panel display shows the main coding window for control and scripting. The editor shows either the initialization code of the script (equivalent to the "main" function in a program written in the language C). Events include run-time imaging events and clicks on user-customizable buttons. (D) The Debugging/Messages window. This window is used by the script engine to provide scripting status and debugging notifications. Scripts can also write messages to this window during their execution. (E) Expanded view of the left and right display in panels B and C.

independent bitmaps. For instance, MPScan uses the Video for Windows Application Programming Interface to display frames from an off-screen composing bitmap with a function designed for hardware accelerated streaming video. In principle, the DirectX library, which is geared toward writing video games and is used in e-Maging,[5] should provide even better graphical speed thanks to the ability to perform hardware flips of video surfaces from the main memory to the video card. Although the performance gain is clearly warranted in the case of video-rate two-photon imaging, i.e., of order 30 frames per second, it is not clear if the speed gain is worthwhile when imaging is done at our typical rate of 3 frames per second, given the complexity of programming a windowed DirectX application.

As a general design issue, choosing an object-oriented language has clear implications for the architecture of a two-photon microscope program. The cardinal concepts of object orientation, namely, encapsulation, inheritance, and polymorphism, were systematically used in MPScan. This ensures that all end-users would operate the microscope in a consistent manner regardless of the details of the auxiliary hardware that is incorporated into the microscope system, i.e., X-Y translation table, Z-focus stepper, Z-focus piezoelectric objective driver, laser shutter, micromanipulators, etc. To deal with the multiplicity of these devices, MPScan conceptualizes the auxiliary instruments as software objects that are grouped into categories defining their role and common behavior. All device objects are instantiated from a base class called TMPScanDevice; T stands for type, a naming convention to differentiate classes from instances of objects. As shown in Figure 4.3, in the case of X-Y translation table all actual devices descend from the TXYTable class, which implements a common behavior for all translation stages as a set of virtual methods. For instance, when the user clicks on a screen control that moves the objective by a given distance in the X direction, the XYTable object changes the color of a virtual light-emitting diode to red on the front panel, indicating an impending motion, sends the command to perform a relative move, waits for the move to be completed, toggles the light-emitting diode color to green, reads the actual position of the stage, and updates the X position value on screen. To implement a new translation stage, one only needs to create a descendent of TXYTable that will automatically inherit the behavior of the base class. The new class can either add its own methods or override preexisting ones. Encapsulation of properties and methods in each derived classes prevents the unwanted interaction of the code that controls device objects.

Instantiation of objects representing actual devices is done when MPScan starts. MPScan reads the hardware configuration of the microscope, which is stored in the Windows registry. Depending on the settings of the configuration, MPScan creates the device objects from their respective class, e.g., a TXYDMC4040 for a DMC-4040 controller, and attempts to connect them to the hardware they support via their preferred communication interface, e.g., RS-232, USB, TCP-IP, etc., by dynamically loading their manufacturer-provided library if needed. Following their creation and connection, these objects are treated by the rest of the program as instances of their ancestor class using polymorphism; for the case of the DMC-4040 object, the ancestor class is the TXYTable.

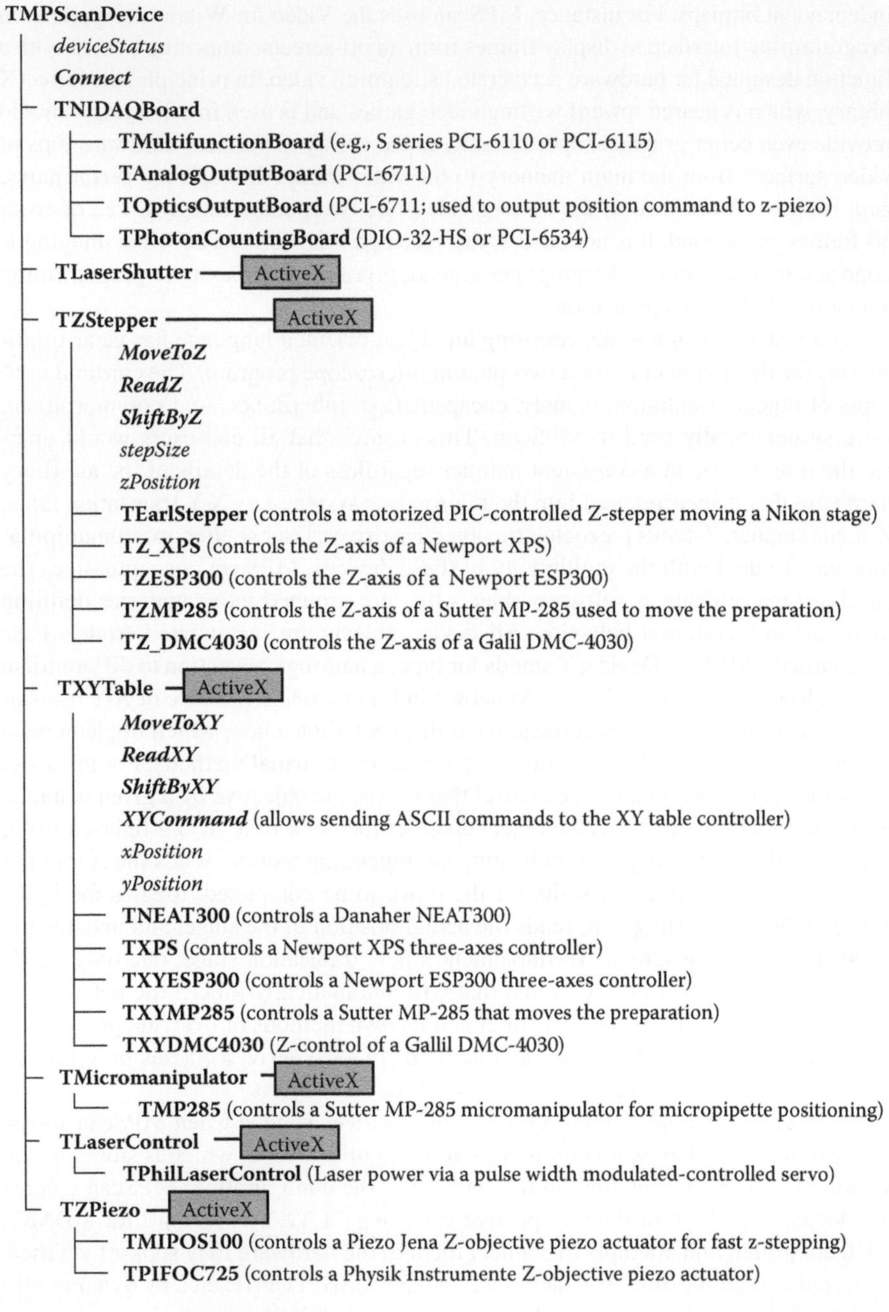

FIGURE 4.3 Class hierarchy of MPScan physical devices. Bold: base classes and descending classes of objects controlling physical devices. Single-inheritance model. *Italic*: properties. ***Bold Italic***: virtual methods. For the sake of clarity, many properties and methods have been omitted from this diagram. ActiveX: ActiveX Automation object created by the class for scripting purposes.

4.1.4.4 Componentization

The cornerstone of MPScope interoperability with other programs is the widespread use of Component Object Model (COM) technologies; the reader is referred to References 16 and 17 for background on COM. Although the Microsoft .Net application framework is replacing parts of COM on Windows machines, the need to deploy the .Net 2.0 run-time libraries with any .Net program, and the fact that .Net applications rely on interpretation of bytecodes instead of compiled native code, ruled out the use of .Net as the component foundation of the entire MPScope suite.

The object-oriented design of MPScan provides a coherent architecture to control MPScan programmatically using the COM-based ActiveX Automation standard. Viewed from an outside client, MPScan does not appear as a monolithic program but as an application server consisting of a hierarchy of COM components, some of which represent common operations performed on the microscope and others that abstract a real device connected to the microscope. These components expose an ActiveX Automation interface that can be invoked by COM clients either at run-time, via a late-binding IDispatch interface, or more effectively via direct early-binding v-table function calls that are discovered at compile-time by consulting the Type Library; this library describes objects and methods available in MPscope to other programs and is embedded in the MPScan executable file. An added advantage of using well-proven COM technologies is the possibility to control MPScan over a network from a remote computer, or to output frame data in real-time from MPScan to a remote client through the use of Distributed COM (see Reference 10) providing that there is no firewall between the computers. Furthermore, MPScan can source specific COM events and thus allows a scripting client to be called automatically when scanning starts, scanning ends, or when a frame has been acquired.

The COM-based ActiveX Automation standard is also used extensively in MPScope data files. These files have a fairly unique layout, thanks to the adoption of the Structured Storage technology available on the Windows platform. The organization of MPScope files does not rely on a flat structure but is akin to a directory storing subdirectories and data streams, with each stream containing demultiplexed imaging or analog channel. Gap-free electrophysiology data simultaneously acquired with imaging, and subsequently down-sampled, can be efficiently stored alongside and independently of imaging frames. The use of Structured Storage also considerably eases the writing of analysis programs since they do not have to deal with version-dependent file offsets and pointer calculations to reach a particular piece of data. In addition, imaging metadata are stored in a stream that acts as an extensible header that can be queried and browsed from the Windows Explorer Shell without having to open the file (Figure 4.4). Custom tags can be added to MPScope files in MPScan, e.g., by scripting, without interfering with the basic structure of the file. Overall, using Structured Storage in MPScope facilitates data management as it allows consolidation of all experimental data of a given run into a single file.

To ease interoperability with other programs, MPScope files can be exported as Windows bitmaps, TIFF, or AVI files. In addition, MPView can automatically transfer frames to Matlab™. Another way of opening MPScope files in programs other than MPView is to use the MPFILE.OCX ActiveX component. This add-on, part of

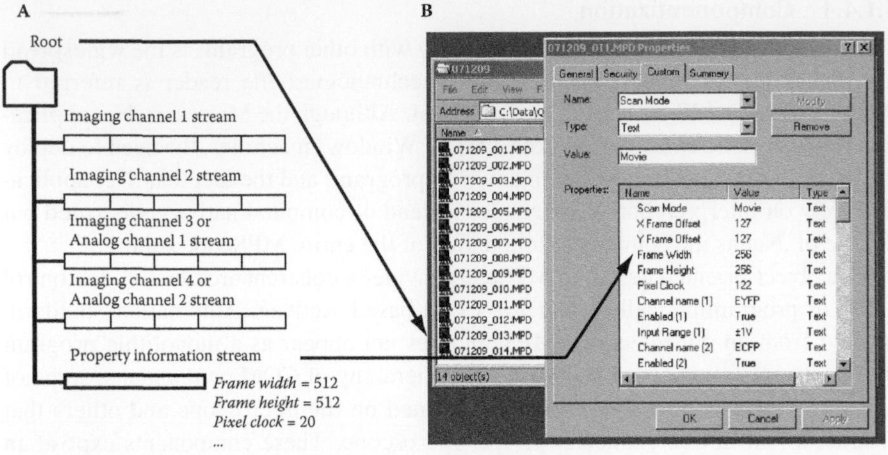

FIGURE 4.4 MPScope file format. (A) Internal layout of a MPScope file. (B) Screen shot of the File Property window, displayed by left-clicking the file name in the Windows graphical shell.

the MPScope suite, provides a simple way of opening MPScope files to obtain frame pixel values or metadata in programs written in languages such as C/C++, Matlab™, Visual Basic, or Delphi™. Until recently, however, MPScope files could only be opened by programs running under Windows. Fortunately, Microsoft has just made the internal format of Structured Storage public,[18] which will allow the development of platform-independent code to deal with MPScope data files. Moreover, several software libraries that are able to read Structure Storage files are available for other operating systems, such POIFS,[19] which is written in Java. The existence of these libraries opens the possibility for analyzing native MPScope data under MacOS X or Linux.

4.1.4.5 ActiveX Scripting

A general problem that is often encountered in microscopy, and a defining feature of optically assisted, plasma-mediated ablation experiments,[3,9] is the repetition of the imaging/ablation cycles and the timely execution of various operations. In principle, each experiment could be automated by modifying the source code of the two-photon acquisition program. Overall, the entire MPScope package represents over 63,000 lines of code. Rebuilding of source code for each application would be inordinately time-consuming and inefficient. The best solution is to avoid source code modification and enable end-users to control the two-photon imaging software with small, external programs written by each user. This approach, termed "scripting," is extensively used in large and complex microscope applications. The prime example of embedded scripting is the inclusion in the Zeiss™ two-photon software of a full-featured programming environment based on the Microsoft Visual Basic for Applications language.

Application scripting requires two software features. The first is a programming language that allows programming scripts to be easily written by end-users. Scripting languages can be mastered in a few hours by nonprogrammers, thanks to a simple grammar that eschews strongly typed constants, explicit memory allocation, and pointers. To allow users to test their code immediately, scripting languages are usually interpreted. Although interpretation of code with untyped variables leads inevitably to slow execution, speed increase can be achieved by tokenization of the source code before interpretation of bytecodes.

The second feature of scripting is a common protocol between the client scripts and the two-photon server software to pass and retrieve variables and execute commands across process boundaries. On a single Windows computer, client-server communications can be implemented by sending simulated keystrokes or by passing private messages between windows; the latter method is used, for instance, by fluorescence imaging software to control the MDS pClamp electrophysiology program. These communication methods are, unfortunately, extremely rudimentary, do not allow two-way dialogues, and are ill-suited for passing large sets of data. However, adoption of the COM ActiveX Automation standard endows MPScope with well-understood programming entry points with the added possibility of performing Remote Procedural Calls between clients and the two-photon software across a local network via Distributed COM. An added advantage of using ActiveX Automation in scripting is the possibility to use any prewritten ActiveX component that can be either developed with a general-purpose Windows programming language, e.g., C/C++, Visual Basic, Delphi™, etc., or provided by a vendor. For instance, scripts can control an ActiveX component developed in house that interfaces with a 1608-FS USB-based data acquisition board (Measurement Computing, Malboro, Massachusetts). This component adds to MPScan up to eight analog input lines with 16-bit resolution that can be read by scripts.

While Visual Basic for Applications has commendable features, it is too expensive to be used as a scripting language in an academic freeware. However, Microsoft's COM-based ActiveX Scripting technology is a suitable alternative to develop, run, and debug scripts in a language independent manner. Scripting interpreters only need to conform to the ActiveX Scripting specification; in principle, an ActiveX Scripting-enabled application can choose from a wide variety of scripting languages at run-time to execute scripts in a given language. This allows developers to write their scripts either in Basic or Java-derivative languages such as Microsoft Visual Basic for Scripting Edition (VBScript) and Java Script (JScript), Python, or Haskell. The advantage of VBScript is that the VBScript interpreter is free and is available in every copy of Windows as the default scripting engine of Microsoft Explorer Web browser.

MPScan tightly integrates VBScript in its scripting development environment to provide a simple yet powerful way to customize MPScan without having to modify the MPScan source code. Although the scripting development environment features simple editing capabilities with basic debugging capabilities, ten user-customizable buttons provide script writers with a limited Graphical User Interface programming capability. Scripts can be highly complex, especially those automating all optical histology.[3] Figure 4.2 shows such an intricate script that repeatedly performs 256 by 256 μm^2 fast stacks down to 700 μm in depth at greater than 50 different tiled locations in a fixed brain specimen using automatic laser intensity control to ensure consistent brightness of sections located at different depths. The script subsequently controls plasma-mediated ablation of the previously imaged region by opening the shutter of the ablation laser line and moving the stage in a raster. Following ablation, the script raises the objective, cleans it using a wipe mounted on a servo, and dips it back into the imaging medium for another pass of imaging and ablation. The script runs overnight and generates 45 Gbyte of imaging data. MPScan scripts can, in addition to controlling MPScan, send commands to other applications. For instance, the scripts can communicate with the MySQL database to store or retrieve sets of record via ActiveX Automation.

4.2 MPSCOPE 2.0

4.2.1 General Features

MPScope 2.0 is based on the same design choices of the earlier version, which emphasized fast scanning using symmetrical line scans, intuitive turnkey operation, independence from run-time environment, and a highly scriptable architecture. The analysis program MPView provides common frame operations such as averaging, maximal projection on three axes, several options to export frames, and the possibility to automatically find regions of interest and plot their intensity across time. Most of the new features in MPScope 2.0 are in the acquisition program MPScan. These include:

1. The maximum frame size is extended to 4096 by 4096 pixels from 512 by 512 pixels in the original code, which was adequate for physiology but not for the larger brain areas processed with all optical histology using lenses that have a 4 mm field of view.
2. The recommended option to control the objective and preparation positions uses a model DMC-4040 industrial motion controller from Galil Corporation, which replaces a now obsolete Danaher NEAT300 controller and a custom-built motorized Z-stepper.
3. Support for analog integration and fully digital photon counting on up to four imaging channels to increase the signal/noise ratio under conditions of low emission.
4. Fine control of excitation power through a power attenuation kit from Newport Corporation, which can monitor true root-mean-square laser power via a USB-controlled model 1918-C power meter and implement computer-controlled power feedback by moving a half-wave plate with a model PR50 compact rotation stage via a model SMC-100 controller communicating with the computer using a RS-232 link.[20]
5. Synchronous triggering of the laser pulses within the scan cycle for plasma-mediated ablation.
6. Enhanced support for Z-piezoelectric objective actuators. These devices allow faster Z-motion of the objective position compared with conventional motorized Z-focus controllers but have relatively limited range; thus, a Z-piezoelectric actuator is used together with a Z-focus stepper. At present, MPScope 2.0 supports piezoelectric actuators having a range of either 100 or 400 μm. These include the PiezoJena model MIPOS 100 and Physik Instrumente PIFOC P-725.4C series, which are configured in close-loop

mode to ensure accuracy and repeatability of the motions at the expense of speed. In our laboratory, these piezoelectric actuators are used to generate 3-dimensional X-Z-T movies of 256 by 256 pixels at a rate of 2 frames per second, with 100 μm spans of Z-travel, or 4-dimensional X-Y-Z-T in vivo movies of mouse cerebral vasculature. In the Helmchen Laboratory, these are used to make 4D line-scans that spiral through a depth of cells.[21] Control of these devices from MPScan is straightforward by using a PCI-6711 digital/analog board that provides an analog control voltage proportional to the desired displacement. MPScan can be configured to move the objective either after every line scan or after every frame and to repeat this pattern indefinitely. This allows acquiring tilted frames in addition to the X-Z-T and X-Y-Z-T modes mentioned above. Repeated Z-scanning is achieved by outputting a sawtooth-like command waveform.

4.2.2 Hardware Requirements

The minimal configuration to run MPScope 2.0 consists of:

1. A computer running Microsoft Windows 2000 or XP loaded with the NI-DAQ™ library, the latest version of which can be downloaded from the National Instruments Web site, and at least 1 Gbyte of random access memory. At the time of writing, MPScope 2.0 has not been tested under Microsoft Vista.
2. One National Instruments PCI-6110 data acquisition board.
3. A breakout box with BNC connectors in the front, e.g., National Instruments NI BNC-2090, is highly recommended to interface with the 68-pin connector of the PCI-6110.

For additional hardware control, up to two PCI-6711 analog output boards can be used to send positional commands to a Z-piezoelectric objective actuator and to generate voltage commands for electrophysiological instrumentation. All auxiliary boards have to be connected to the main PCI-6110 board using a National Instrument Real-Time System Integration bus.

4.2.3 Motion Control Support

We use an industrial, off-the-shelf three-axes motion solution based on stepper motors controlled by a commercial 4-channel controller (DMC-4040, Galil Corp., Rocklin, California). Each stage includes a brake, encoders, and limit switches; the wiring for these is given in Figure 4.5 and Tables 4.1 and 4.2. Minimal displacement along the Z axis is 0.3125 μm for our stages, which is smaller than the axial resolution of the microscope. Motion and position reporting commands issued by MPScan are sent to the DMC-4040 controller via Ethernet. Inside MPScan, the DMC-4030 is controlled by its own thread that communicates to the hardware layer via a Dynamic Link Library provided by Galil. Adding DMC-4040 support to MPScan did not require any major software change, since its software object was developed by inheriting its behavior from a virtual, idealized X-Y translation stage and Z-focus stepper.

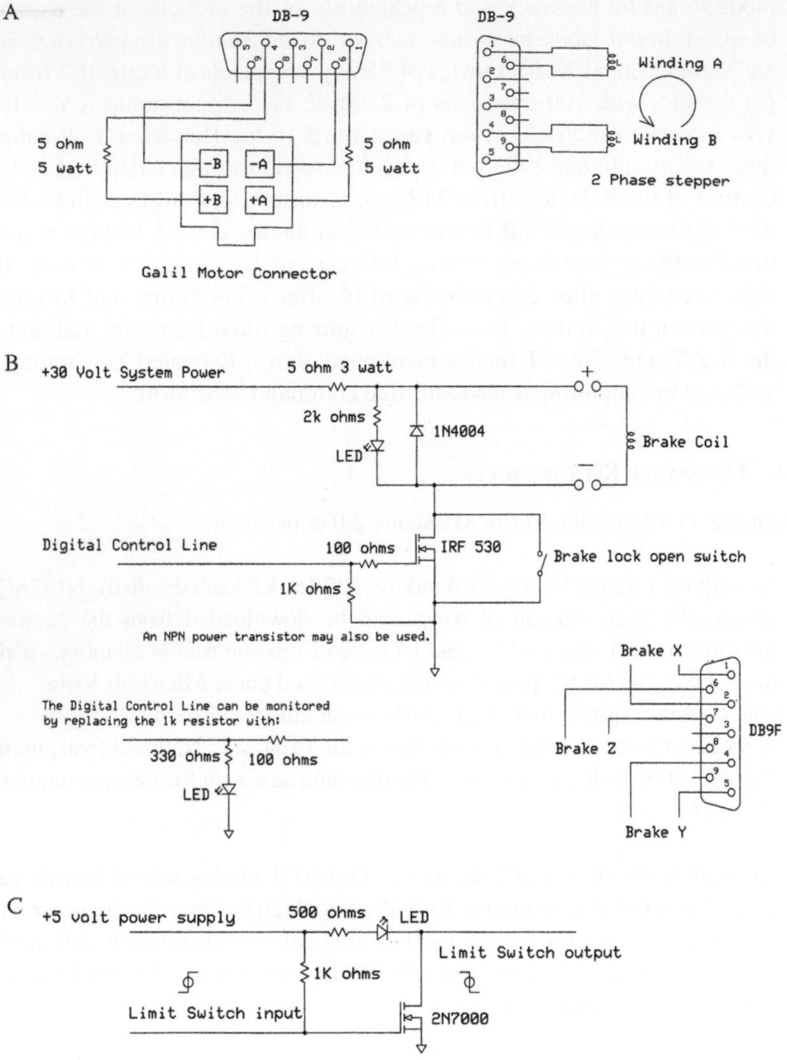

FIGURE 4.5 Schematic of interface circuits for control of X, Y, and Z axes by the Galil DMC-4040. (A) Connections to each of the three stepping motors. The Galil unit provides current pulses to the motor windings, and the voltage can be as high as 30 V. The resistors serve to protect the outputs and also reduce the time-constant of the winding inductances. The DB-connectors serve only as an interface. (B) Connections to each of the three electromagnetic brakes using digital control lines (Table 4.2). The DB-connectors serve only as an interface. (C) Connections to each of the six limit switches (Table 4.2).

As a practical matter, it is useful to maintain manual control of the stages at set-up time. We made use of a three-axis joy stick controller (Feteris S30-JHK-ZT-30R3G-000) with a fourth channel used to read a single-turn potentiometer that

TABLE 4.1
Pin-outs for a Galil DMC-4040 quadrature encoder I/O (15 connector)

Pin		Pin		Pin	
1		6		11	Aux A+
2	B+	7	B−	12	Aux B−
3	A+	8	A−	13	
4	Aux B+	9		14	
5	Ground	10		15	+15 V

Note: Pin-out is for each of up to four separate channels and connectors. Main quadrature encoder is A-B and auxiliary encoder is Aux A–Aux B.

TABLE 4.2
Pin-outs for Galil DMC-4040 external driver digital I/O (44-pin connector)

Pin		Pin	
1		23	Reverse limit switch (Y)
2		24	Reverse limit switch (Z)
3		25	
4		26	
5		27	DO-2 (brake Y)
6	LS-Comm (+5 V)	28	
7		29	
8		30	+5 V
9		31	Ground
10		32	
11	O-PWR (+5 V)	33	
12	DO-3 (brake Z)	34	
13		35	Ground
14	O-return (ground)	36	Forward limit switch (X)
15	+5 V	37	Forward limit switch (Y)
16		38	Forward limit switch (Z)
17		39	
18		40	Ground
19		41	DO-1 (brake X)
20		42	
21		43	
22	Reverse limit switch (X)	44	

Note: Pin-out is for all four channels on a single connector.

sets the scale for the speed of movement. The DMC-4040 contains an 8-channel analog to digital converter port that is convenient for this task; the wiring is shown in Figure 4.6. Firmware code for joystick resides in the DMC-4040; see Appendix. The manual controller is overridden by MPScope to prevent conflicts while the program is running.

4.2.4 Analog Integration

Photomutiplier Tube (PMT) currents are converted into voltage signals by preamplifiers, which can be either purchased commercially or, in our case, built in the laboratory. These preamplifiers are either of the transimpedance type or consist of a voltage preamplifier measuring the voltage drop caused by the PMT current output across a grounded resistor. Numerous companies produce preamplifiers suitable for PMT operations, i.e., Stanford Research Systems in the United States and Femto in Germany.

Simple signal conditioning of the preamplifier output can be performed either by digital integration or by low-pass filtering. The first method consists of sampling the preamplifier analog signal several times during the pixel duration and adding these values. Drawbacks of this method include an increase in samples collected and more importantly the inability to deal with fast preamplifiers. For instance, the transimpedance preamplifiers in our laboratory output a voltage pulse lasting 50 to 100 ns in response to a single photon. These responses are too fast to be adequately digitized by the analog/digital converters of the PCI-6110, which have a maximal sampling frequency of 5 MHz. The alternative of low-pass filtering the PMT signal can be done either in the preamplifier module if this option is built in, or by an external low-pass filter unit. In either case, the cutoff-frequency f_c of the filter should be set to $f_c = 2/(\pi\tau_p)$ where τ_p is the pixel integration time.[22]

An efficient analog signal conditioning solution is to use a true pixel integrator. This device offers several advantages over low-pass filtering:

1. Suppression of correlation between adjacent pixels
2. Simplicity of operation, since there is no need for cut-off frequency adjustment whenever the pixel dwell time changes
3. Better image quality thanks to an overall variance of the image as much as twice as small as that obtained with low-pass filtering[20]

A further refinement of true pixel integration consists of using alternatively switched integrators to prevent blanking-off time when one integrator needs to be reset. This scheme, known as multiphase pixel integration, is used in commercial confocal microscopes such as the Nikon C1 with the name of Dual Integration Signal Processing.

A challenge in designing a stand-alone multiphase pixel integrator is to generate the timing signals that switch alternate integrators and reset them in synchrony with the pixel clock. Fortunately, the PCI-6110 board, under MPScan control, can output several signals that can be combined to control a multiphase integrator. The schematics of a simple multiphase pixel integrator in use in our laboratory can be found in Figure 4.7. In brief, each channel of this circuit consists of a Linear Technologies

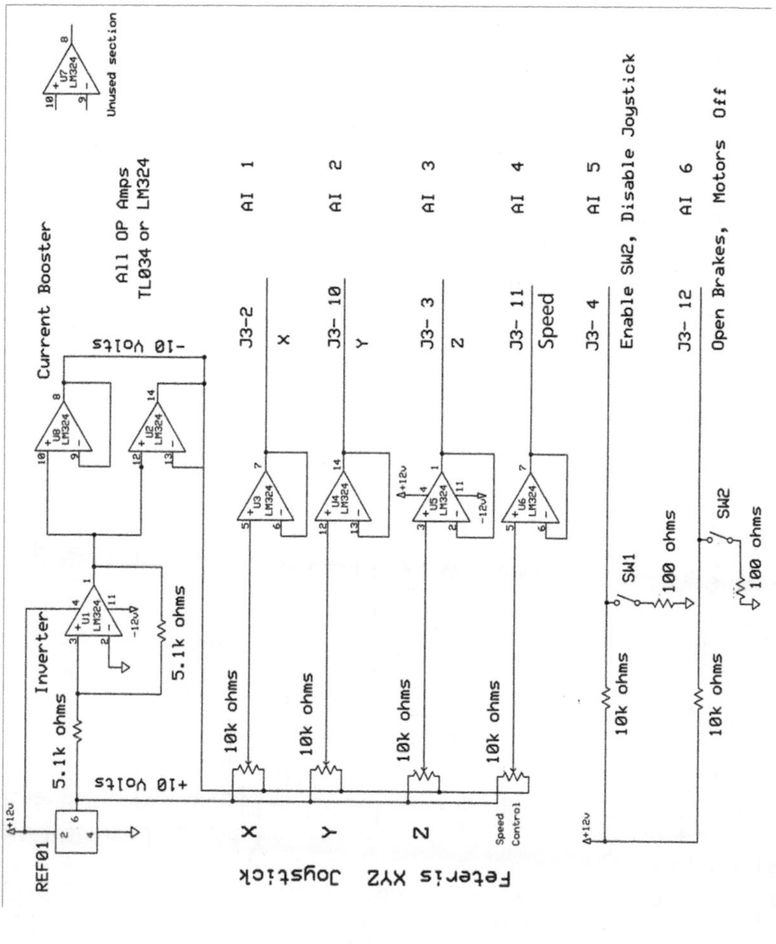

FIGURE 4.6 Schematic of the circuit for a resistive 3-channel joystick to control X, Y, and Z axes. The joystick is energized by power from the Galil DMC-4040 and the output of each potentiometer is read by analog-to-digital converters in the Galil unit and used to control the position of the stages; see Appendix for code. A fourth potentiometer is read to set an overall speed level and a fifth and sixth are utilized as binary switches since there are no digital inputs on the analog input connector to the Galil unit.

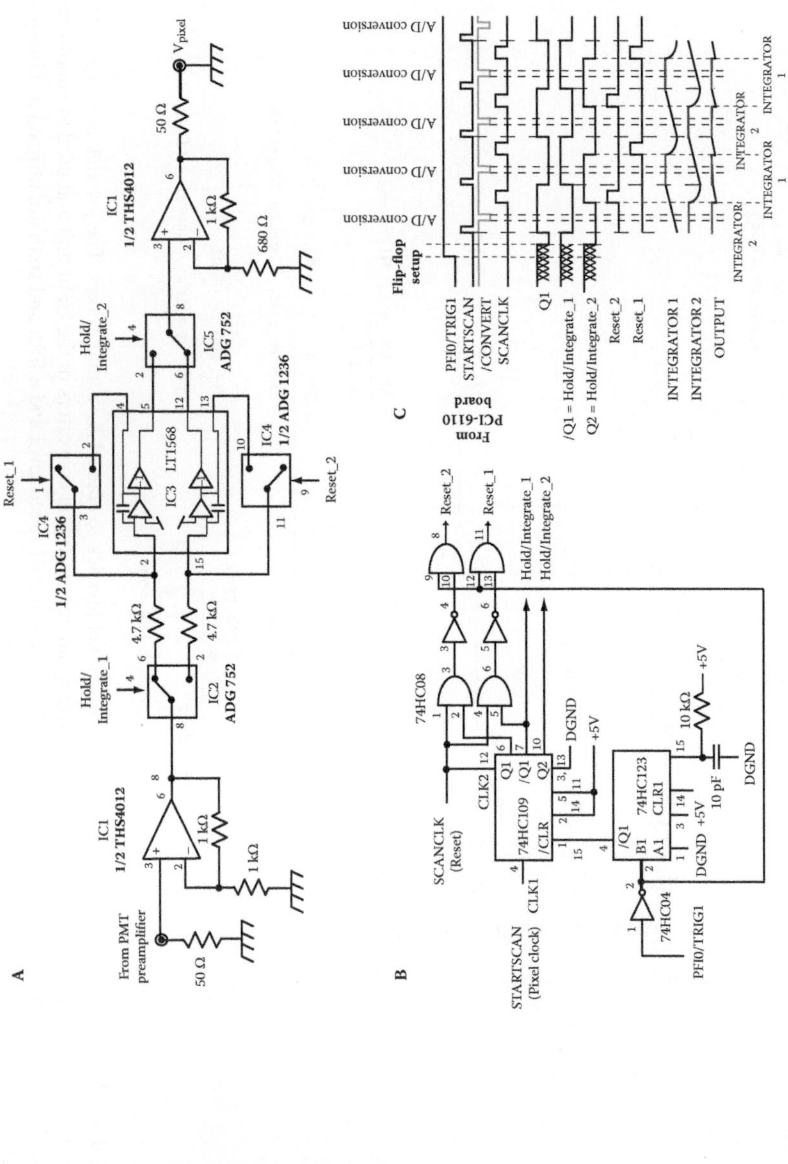

FIGURE 4.7 Multiphase pixel integrator schematics. (A) Multiphase pixel integrator for one imaging channel. (B) Control logic. (C) Timing diagram of the control logic. The /CONVERT signal is not used by the control logic but is added to show that the actual digitization window occurs between the STARTSCAN and SCANCLK signals.

LTC1568 filter chip. This integrated circuit features two operational amplifiers with bandwidth of 50 MHz each coupled to its own capacitor. The two capacitors, with a nominal value of C_f = 105 pF, are conveniently matched to within 0.5 %. In addition, each op amp can be followed by an inverter contained in the same package. A dual voltage integrator with an output of $V_{out} = (R_f C_f)^{-1} \int_{T_p} V_{in}\, dt$, where T_p is the pixel duration and V_{in} is the input voltage, is created by adding a resistor R_f = 4.7 kΩ before each of the LTC1568 inputs. Each integrator is reset by shortcircuiting its capacitor using an Analog Device ADG1236 dual switch, which has a low charge injection. Switching between integration and reset is performed by two ADG752 ultra-fast switches with t_{on} = 8 ns and t_{off} = 3 ns. Input and output signal conditioning is performed by a Texas Instrument THS4012 dual op-amp. TTL circuits that control the switches receive their inputs from the PFI0/TRIG1, STARTSCAN, and SCANCLK signals issued by the PCI-6110 board (Figure 4.7A). PFI0/TRIG1 is used to reset the 74HC109 flip-flops at the beginning of an imaging session via a 74HC123 one shot, while STARTSCAN and SCANCLK create the Integrate/Hold and Reset pulses (Figure 4.7B). The control logic can issue the timing signals simultaneously to several banks of integrators, each assigned to an imaging channel.

A refinement of integration that is appropriate for low light levels is the hybrid photon counting technique.[22,23] In this method, the input of the integrator consists of short TTL pulses triggered by single photon pulses from the PMT outputs, which are discriminated by amplitude thresholding to both remove PMT dark noise and regularize the amplitude of the photoelectric pulse.

4.2.5 Dissemination and Support

As was previously the case with MPScope 1.0, MPScope 2.0 is available free of charge to academic users by contacting the authors. MPScope users have also the possibility to become members of the MPScope Google discussion group, which deals with scripting, recompilation, optoelectronics, and hardware support issues. In addition, MPScope 2.0 can be found at the NeuroImaging Tools and Resources Clearinghouse (NITRC) Web site (http://www.nitrc.org/) administered by the National Institutes of Health. Continuous development of MPScope by the authors and the MPScope community will ensure that MPScope remains compatible with the latest technological changes, including the possible replacement of galvanometric mirrors with faster scanning devices, such as acousto-optic modulators, and the incorporation of galvanometric mirrors or acousto-optic modulators for additional laser beams to be used for ablation or uncaging.

ACKNOWLEDGMENTS

We thank T. Baldacchini of the Technology and Applications Center, Newport Corp., Irvine, California, for the loan of a variable laser attenuator kit and P. Blinder and P. S. Tsai for the example script of Figure 4.2. This work was supported by the NIMH (MH071566), NIBIB (EB003832), NCRR (RR021907), and NSF (DBI 0455027).

APPENDIX

Firmware for Galil DMC-4040 controller to operate three-axis stage through an analog joystick

```
#AUTO                          ;'allows the program to execute
                                 when power is applied
#A                             ;'this section initializes the
                                 stepper motor parameters
MT -2.0,-2.0,-2.0
YA 16,16,16                    ;'16X stepper resolution
YB 400,400,400                 ;'basic pulses per revolution
SP 3000,3000,3000              ;'speed in rpm
LD 0,0,0                       ;'enable limit switches
CN -1,-1,-1,0,0                ;'limit switch active low
DP 0,0,0                       ;'set initial stage position to
                                 zero
VX=0                           ;'zero all variables
VY=0
VZ=0
VXA=0
VYA=0
VZA=0
OB 1,0                         ;'brakes on X axis
OB 2,0                         ;'brakes on Y axis
OB 3,0                         ;'brakes on Z axis
#B                             ;'start of main loop
  VMIN = .5                    ;'minimum acceptable value from any
                                 joystick axis
  V4 = @AN[4]                  ;'read the "Rate", i.e., multiplier
                                 potentiometer value
  VMLT = (V4+10)*(V4+10)*2     ;'velocity multiplier for X and Y
                                 axes
  VMZ = (V4+10)*(V4+10)*0.5    ;'velocity multiplier for Z axis
  VS1 = @AN[5]                 ;'read limit switch 1
  VS2 = @AN[6]                 ;'read limit switch 2
  JP #C, (VS2>VMIN)            ;'if VS2 high, continue normally,
                                 otherwise check VS1
  IF (VS1<VMIN)
    OB 1,1                     ;'if VS1 low, then release all
                                 brakes
    OB 2,1
    OB 3,1
    MO                         ;'turn off motors
  ELSE                         ;'if VS1 high, apply all brakes and
                                 enable motors
    OB 1,0
    OB 2,0
```

```
  OB 3,0
  SH                              ;'turn on motors (servo here)
 ENDIF
 JP#B                             ;'continue checking switches until
                                    VS1 is high
 #C                               ;'if reached, apply all axis brakes
 OB 1,0
 OB 2,0
 OB 3,0
 SH                               ;'turn on motors just in case
 #XAXIS
  V1=@AN[1]                       ;'read X joystick value
  V1A = @ABS[V1]                  ;'calculate absolute value
 JP #XEND, V1A<VMIN               ;'if below VIN then clear value,
                                    set brake, stop motion
 JP #XAXIS1,(_LFX=0)&(V1<0)       ;'OK to move in X if positive
                                    switch set and X negative
 JP #XAXIS1,(_LRX=0)&(V1>0)       ;'OK to move in X the negative
                                    switch set and X positive
 JP #XEND,(_LFX=0)|(_LRX=0)       ;'stop X motion for any other case
 #XAXIS1
  VX=V1*VMLT                      ;'if X more than VMIN then multiply
                                    value by VMLT
  OB 1,1                          ;'release X brake
  JG VX,0,0                       ;'set Jog mode for the X axis with
                                    the VX value
  BGX                             ;'begin X motion
 JP #XAXIS                        ;'cycle until joystick is released
 #XEND                            ;'end of X axis operation
  VX = 0                          ;'clear X value
  OB 1,0                          ;'apply X brake
 STX                              ;'stop X motion
 #YAXIS                           ;'begin to process Y and Z axes
                                    same as X axis
  V2 = @AN[2]
  V2A = @ABS[V2]
 JP #YEND, V2A<VMIN
 JP #YAXIS1,(_LFY=0)&(V2<0)
 JP #YAXIS1,(_LRY=0)&(V2>0)
 JP #YEND,(_LFY=0)|(_LRY=0)
 #YAXIS1
  VY = V2*VMLT
  OB 2,1
  JG 0,VY,0
  BGY
 JP #YAXIS
 #YEND
  VY = 0
```

```
    OB 2,0
    STY
  #ZAXIS
    V3 = @AN[3]
    V3A = @ABS[V3]
  JP #ZEND,V3A<VMIN
  JP #ZAXIS1,(_LFZ=0)&(V3<0)
  JP #ZAXIS1,(_LRZ=0)&(V3>0)
  JP #ZEND,(_LFZ=0)|(_LRZ=0)
  #ZAXIS1
    VZ = V3*VMZ
    OB 3,1
    JG 0,0,VZ
    BGZ
  JP #ZAXIS
  #ZEND
  VZ = 0
  OB 3,0
  STZ
JP #B                          ;'loop through X, Y, and Z axes
#LIMSWI                        ;'limit switch interrupt processing;
                                 all 6 switches are checked
IF (_LFX=0)                    ;'check if forward X switch is
                                 asserted
   #TSTFX
   JS #WAITX                   ;'wait for joystick X channel to be
                                 released
ENDIF
IF (_LRX=0)                    ;'check if reverse X switch is
                                 asserted
   #TSTRX
   JS #WAITX
ENDIF
IF (_LFY=0)                    ;'check if forward Y switch is
                                 asserted
   #TSTFY
   JS #WAITY
ENDIF
IF (_LRY=0)                    ;'check if reverse Y switch is
                                 asserted
   #TSTRY
   JS #WAITY
ENDIF
IF (_LFZ=0)                    ;'check if forward Z switch is
                                 asserted
   #TSTFZ
   JS #WAITZ
ENDIF
```

```
IF (_LRZ=0)                    ;'check if reverse Z switch is
                                 asserted
#TSTRZ
   JS #WAITZ
ENDIF
ZS 1                           ;'reset stack pointer, cancel
                                 interrupt by reducing stack by
                                 one
JP #B                          ;'return to main loop
#WAITX                         ;'subroutine to watch X chan.
                                 joystick; returns when X below
                                 VMIN
   VX1 = @AN[1]
   VXAB= @ABS[VX1]
JP #WAITX,(VXAB > VMIN)
EN
#WAITY                         ;'subroutine to watch Y channel
                                 joystick
   VY1 = @AN[2]
   VYAB = @ABS[VY1]
JP #WAITY, (VYAB > VMIN)
EN
#WAITZ                         ;'subroutine to watch Z channel
                                 joystick
   VZ1 = @AN[3]
   VZAB = @ABS[VZ1]
JP #WAITZ,(VZAB > VMIN)
EN
```

REFERENCES

1. Nguyen, Q.-T. et al., MPScope: A versatile software suite for multiphoton microscopy, *Journal of Neuroscience Methods*, 156, 351–359, 2006.
2. Tsai, P.S. and Kleinfeld, D. In vivo two-photon laser scanning microscopy with concurrent plasma-mediated ablation: Principles and hardware realization, in *Methods for* In vivo *Optical Imaging,* 2nd edition (Frostig, R. D., Ed.), CRC Press, Boca Raton, 2009, pp. 59–114.
3. Tsai, P.S. et al., All-optical histology using ultrashort laser pulses, *Neuron*, 39, 27–41, 2003.
4. Nikolenko, V. et al., A two-photon and second harmonic microscope, *Methods*, 30, 3–15, 2003.
5. Nguyen, Q.-T. et al., Construction of a 2-photon microscope for real-time Ca^{2+} imaging, *Cell Calcium*, 30, 383–393, 2001.
6. Rietdord, J. and Stelzer, E.H.K., Special optical elements, in *Handbook of Biological Confocal Microscopy,* 3rd ed. (James B. Pawley, Ed.), Springer, New York, 2006, pp. 43–58.
7. Pologruto, T.A. et al., ScanImage: Flexible software for operating laser scanning microscopes, *Biomedical Engineering Online,* http://www.biomedical-engineering-online.com/content/2/1/13, 2003.

8. Tsai, P.S. et al., Principles, design, and construction of a two-photon laser-scanning microscope for in vitro and in vivo brain imaging, in In vivo *Optical Imaging of Brain Function* (Frostig, R. D., Ed.), CRC Press, Boca Raton, 2002, pp. 113–171.
9. Nishimura, N. et al., Targeted insult to individual subsurface cortical blood vessels using ultrashort laser pulses: Three models of stroke, *Nature Methods,* 3, 99–108, 2006.
10. Nguyen, Q.-T. and Miledi, R., e-Phys: A suite of electrophysiology programs integrating COM (Component Object Model) technologies, *Journal of Neuroscience Methods*, 128, 21–31, 2003.
11. Dempster, J., *The Laboratory Computer: A Practical Guide for Physiologists and Neuroscientists* (Biological Techniques Series), Academic Press, San Diego, 2001.
12. Oney, W., *Programming the Microsoft Windows Driver Model*, Microsoft Press, Redmond, 1999.
13. PCI E Series Register-Level Programmer Manual, National Instruments, 1998.
14. NI-DAQ User Manual for PC Compatibles, National Instruments, 2000.
15. DAQ-STC Technical Reference Manual, National Instruments, 1999.
16. Brockschmidt, K., *Inside OLE*, Microsoft Press, Redmond, 1995.
17. Box, D., *Essential COM* (The DevelopMentor Series), Addison-Wesley, Reading, 1998.
18. Windows Compound Binary File Format Specification, Microsoft Open Specification Promise, Microsoft Corporation, 2007.
19. Apache POI-POIFS-Java implementation of the OLE 2 Compound Document format. The Apache POI Project. http://poi.apache.org/poifs. 2007.
20. Computer Controlled Variable Attenuator for Lasers, Application Note 31, Newport Corporation, 2007.
21. Göbel, W. et al., Imaging cellular network dynamics in three dimensions using fast 3D laser scanning, *Nature Methods,* 4, 73–76, 2007.
22. Art, J., Photon detectors for confocal microscopy, in *Handbook of Biological Confocal Microscopy,* 3rd ed., (James B. Pawley, Ed.), Springer, New York, 2006, pp. 251–264.
23. Driscoll, J. et al., Gigahertz photon counting and gating for improved detection in two-photon microscopy (in preparation).

5 In Vivo Observations of Rapid Scattered Light Changes Associated with Neurophysiological Activity

David M. Rector, Xincheng Yao, Ronald M. Harper, and John S. George

CONTENTS

5.1 INTRODUCTION

Cognitive processes and other advanced neural functions rely on spatial and temporal interplay within linked neural networks. To study this interplay and test network level models, imaging techniques are required to capture dynamics of neural interaction. Of the many available imaging techniques, including PET, MRI, MEG, and EEG, recent developments in optical techniques offer significant advantages and a unique complement to other methods. Much effort has been invested in techniques that use light to acquire images of brain activity. Sensitive optical techniques have demonstrated spatial organization of visual cortex columnar structures in a fashion that complements electrophysiological recording.[1–3] Spatial patterns of sensory activation in human temporal cortex[4] and rodent sensory cortex[5,6] have also been visualized. Thus, hemodynamic and other metabolic indicators have successfully mapped the dynamics of neural activity in vivo using spectroscopic and oximetry techniques.[7,8] Light absorbance changes associated with metabolic and hemodynamic processes are robust and relatively easy to obtain noninvasively, but spatial and temporal resolutions are limited by the anatomy and physiological regulation of cerebral perfusion. Fundamentally, the spatial resolution is limited by microvasculature organization, and the temporal resolution is limited by the rate of vessel diameter fluctuations and hemoglobin deoxygenation, which can occur as rapidly as 150 to 250 ms.[9]

Fast optical signals that correspond more closely to electrical activation have remained difficult to detect above noise for in vivo and noninvasive measurements, because fast signals are small relative to other physiological events (including hemodynamics) and electrical noise. However, no other existing method offers the resolution, specificity, and field of view required for such work, making fast optical techniques the holy grail for in vivo neurophysiology.

Several laboratories have recorded relatively fast optical changes noninvasively using fiber optics and modulated light[10] or continuous illumination,[11] but with relatively low spatial and temporal resolution. Detailed investigations of the coupling between neurovascular signals and electrophysiological patterns are under way,[12] and several investigators have started to combine optical and magnetic resonance imaging modalities to investigate the sources of signals from both methods.[13,14] Diffusion tensor MRI has recently gained popularity as a functional imaging methodology, because at least part of these signals appear to originate from similar cellular swelling mechanisms as may underlie fast optical signals.[15,16] Our results suggest that optical signals can track neurophysiological dynamics at high speed. Thus, this chapter will focus on methods and results for using optical signals for recording fast neural events.

In the past decade, it has become increasingly important to record simultaneously from large neural populations to assess their interactions to perform complex tasks. Theoretical arguments and experimental observations suggest that correlated firing across many individual neurons may encode relationships within the data stream,[17] and recent studies have found significant information in the synchrony of neural populations.[18,19] Such work has demonstrated that it is important to know not only when a neuron fires but also how such discharge occurs in relation to activity in other cells. Most procedures for assessing activity of many neurons involve the use of multiple electrode arrays. The density of such arrays has grown rapidly from a few electrodes to a hundred or more electrodes in close proximity. Although electrode arrays provide excellent temporal resolution of neural activity, spatial resolution and sampling density are limited, and invasive electrodes have the potential to damage tissue. Moreover, single unit recordings by microelectrode arrays can be biased by preferential sampling of large neurons.

Although optical measurements can temporally resolve the submillisecond dynamics of action potentials,[20,21] such measurements have typically employed single channel detectors for speed and sensitivity. Even when fast-changing signals are enhanced through the use of voltage sensitive dyes, most investigators have used limited photodiode arrays consisting of a few dozen detectors (i.e., comparable to the resolution of typical electrode arrays). We have demonstrated the feasibility of imaging fast optical signals associated with neural activity using solid state imagers with a large number of detectors, such as CCDs. To date, our in vivo measurements have mostly been limited to averaged evoked activity of neural populations acting in synchrony. However, we recently demonstrated dynamic visualization of stimulus-evoked neural activity in isolated retina, with subcellular spatial resolution. Further, in some cases we can record functional images from large collections of individual cells in single passes.

Our current studies involve tissue illumination with light of specific wavelengths, while scattered light is typically collected through microscope optics or by a coherent fiber optic image conduit, and conveyed to a charged coupled device (CCD) camera. The present technology allows continuous long-term measurements from a 2D tissue surface, with image capture rates up to 2000 Hz. Some versions of our imager allow recording from deep brain structures in freely behaving animals without disrupting normal behavior. Such techniques are essential to assess the role of the brain in spontaneous state-related and motor behaviors.

We performed a number of physiological experiments to study the nature of light-scattering changes in vivo, and to investigate brain functioning in acute preparations, isolated retina, and in freely behaving animals. Images of optical changes from the dorsal hippocampus, ventral medulla, and whisker barrels showed clear regional patterns in response to physiologic manipulations or state alterations, which corresponded to neural activation of these structures. Fast components of the optical responses demonstrated improved spatial specificity and temporal signatures over the slower metabolic signals, and exhibited consistent changes after repeated stimulation or state changes. The procedure allowed assessment of activity components that were difficult to measure or were inaccessible with standard microelectrode techniques in freely behaving animals. Relationships between light scattering changes and neural activation were established by analyzing reflectance changes during synchronous "spontaneous" and evoked electrical activity, pharmacologically induced activity, and spontaneous state changes.[22–24] Synchronous oscillatory activation produced detectable light-scattering changes at similar frequencies to those observed in concurrent electroencephalographic recording.

We believe that a large portion of the signals obtained during these studies result from changes in light scattering with some absorbance component. Since we typically use 660 nm or longer illumination wavelengths, absorbance by hemoglobin is low.

Indeed, through the use of spectral component modeling, Malonek and Grinvald[25] claim the contribution of light scattering to the mapping components was larger than 70% at longer wavelengths. Additionally, the use of dark field illumination around the perimeter of the region forces the light to enter the tissue and to be scattered before returning to the detector. Thus, scattering events play a more prominent role in changes that we see, especially since the vasculature (the locus of the dominant absorbance changes) is located primarily on the surface of the tissue.

Earlier reports of fast optical signals associated with neural activity described scattering changes and polarization (birefringence) changes.[20,21] While polarized-light illumination is required for cross-polarized measurements, nonpolarized light is typically used for assessment of scattering changes accompanying neural activation. The best evidence for polarization signals would be direct measurement of a change in the angular distribution. Demonstration of a flat spectral dependence across a polarity change in the Hgb oxygenation difference spectrum (e.g., 780–820 nm) would also be a strong indication.

Many questions in neural interaction require temporal resolution in excess of the resolution provided by electroencephalographic measurements (which reflect integrated signals over large neural populations), and our recent efforts have been directed toward assessment of faster neural changes by optical means. We found that Schaeffer's collateral stimulation, in vivo, activated hippocampal cell populations and produced light-scattering changes concomitant with evoked electrical responses. These fast optical changes have been imaged and further characterized in recent studies of the rat dorsal medulla and whisker barrels.

Recordings from isolated lobster nerve have been useful in optimizing scattered light changes resulting from volleys of action potentials. We have also used isolated retinas for better characterization of fast intrinsic optical signals associated with neural activation. Such signals can be found in single trials, and show optical changes that occur on the submillisecond time scale, comparable with ionic flux across the neural membrane. Studies with isolated nerve are ongoing in our laboratories to investigate the biophysical mechanisms of fast-scattered light changes associated with membrane potentials.

5.2 MECHANISMS OF OPTICAL RESPONSES

A number of classes of optical changes have been observed during neural activation. Absorption, refraction, birefringence, transmission, and reflectance processes all can affect the measurement of scattered light changes during neural discharge. The majority of studies examine changes in blood volume and oxygenation associated with neural activation. Oxygenated hemoglobin molecules primarily absorb green light. Thus, hemoglobin concentration will influence green-light reflectance; red light, which is absorbed less efficiently, will be modified to a lesser extent (500 to 1000 times lower as determined from standard hemoglobin absorbance curves[26]) by this mechanism. Hemoglobin also mediates absorption changes in the near infrared, in principle allowing deeper measurements with diffuse light techniques.

5.2.1 Scattering and Polarization

Light is transmitted through neural tissue with varying degrees of scattering. A process that increases light transmittance will result in decreased reflected light. Increased transmission (for example, by reduced back-scattering) occurs when light-scattering particles become more dilute in a medium, as may be the case with cell swelling. Another potential mechanism involves reflectance changes. A reflective surface that alters its orientation or curvature will change the angle at which it reflects light, resulting in more or less light return. However, a systematic characterization of mechanisms that influence light scattering in response to neural activity remains to be accomplished.

Polarization signals may result from dynamic scattering and/or birefringence changes of neural tissue during neural activation. Birefringence involves a change in the rotation of polarized light, a process often associated with conformational changes in protein structure, and changes in the spacing of closely packed membranes. Many optical changes during tissue activation result from changes in polarized light.[20,21] Large protein molecules change conformation in response to membrane potential differences. For example, the voltage-activated sodium channel undergoes a dramatic twist in response to depolarization; the degree of this twist has been indexed by polarized light rotation.[27,28] Strong birefringence is also observed in multilayer lipid vesicles. It is likely that changes in the microscopic architecture or packing of cellular membranes or intracellular organelles might also contribute to a birefringence signal. Our recent experimental and theoretical investigations suggest that dynamic physical changes (i.e., swelling or shrinking) of the activated neurons can contribute to both scattering and polarization signals.[29]

5.2.2 Absorption

Important molecules for tissue light absorption include hemoglobin, water, cytochromes, amino acids, lipids, nucleic acids, and sugars. Absorption measurements have been used to assess quantitative concentrations of oxygenated and deoxygenated hemoglobin. Additionally, the local blood volume of neural tissue increases with activation.[3,25,30–34] Typically, red light is used to visualize hemodynamic changes because tissue is more transparent to red light. Green light (515 to 530 nm) absorption can also be applied for determining local activation and is particularly useful for visualizing vasculature. A disadvantage of measuring hemoglobin concentration changes is that the spatial resolution is limited by microvasculature, and the temporal resolution is limited by vessel diameter fluctuation rates and hemoglobin deoxygenation. Cytochromes capture blue-to-ultraviolet light in conjunction with increased metabolism, and rapidly alter their optical properties with cellular activation. Such absorption changes can be measured with single-cell resolution. However, the relatively low efficiency of blue light detection requires high, and potentially damaging, illumination levels.

5.2.3 Cellular Swelling

Hill, over 50 years ago,[35,36] reported decreased opacity of squid giant axon with action potential generation. After offset adjustment to background illumination transmitted through a nerve fiber, relative changes as large as 10% of residual background were observed with repeated stimulation. Increased axon volume in conjunction with opacity changes suggested that increased transmission resulted from dilution of intracellular contents. MacVicar and Hochman[37] examined white light transmission through a blood-free, in vitro hippocampal slice following electrical stimulation of Schaeffer's collaterals, and found increased local broad spectrum light transmission through the tissue. When anion channels were blocked with furosemide, post-synaptic potentials were elicited, but optical changes were not observed, suggesting that transient changes in optical properties of the slice may result from volume changes associated with cellular swelling. Within in vivo preparations, component analysis of spectral data reveals a wavelength-independent process, which suggests light-scattering changes with neural activation.[25] Intracellular and secondary extracellular volume changes provide a basis for relating increased neural discharge to diminished reflectance via the cellular swelling hypothesis.[38] Volume changes modify tissue density, thereby changing the refractive index and reflective properties of tissue. Changes in neural activity in the CA1 region of the hippocampal slice inversely relate to changes in the extracellular volume fraction.[39] Additionally, increased synchronous firing accompanies shrinkage of extracellular space, even when synaptic transmission is blocked.[40–43]

Lipton provided evidence that activation of a blood-free cortical slice produced decreased light reflected from the tissue.[44] Grinvald and colleagues[1] observed 0.1% absolute light reflectance decreases in 665–750 nm reflectance of the cat visual cortex in response to visual stimulation. In these experiments, functional maps of the visual cortex were obtained by illuminating the cortical surface and imaging reflectance with a high-sensitivity CCD camera.

5.2.4 Temporal Resolution of Light-Scattering Changes

Early studies using reflected light to map visual cortex activation patterns often required stimulation for many seconds to several minutes before changes could be measured.[1,45] The optical-response time course depends largely on illumination wavelength, and sensitivity to various physical mechanisms for optical changes. Optical changes occurring rapidly enough to follow the action potential typically required voltage sensitive dyes (VSD) or polarized light with high sensitivity photodiodes,[20] high-speed video technology,[46] or two-photon microscopy.[47] Indeed, VSDs have been getting much better over the past decade with lower toxicity, improved tolerance to bleaching, and easier loading into the tissue. However, VSDs remain invasive to the tissue, and our ultimate goal is to perform neural imaging using scattered light with noninvasive procedures. While there is still much work to do to achieve this goal, we believe optical technology is rapidly evolving to enable this possibility.

Our studies suggest that with sufficient sensitivity, light-scattering changes associated with neural activation can be measured and imaged.[48–51] With sufficient temporal resolution, it may be possible to make sophisticated measurements of time-correlated network behaviors within neural populations. However, achieving sufficient temporal resolution for resolving neurophysiological processes remains a major challenge for imaging studies. Other physiological processes, such as vascular and tissue motion, cause light-scattering changes that are often 10 times larger than the fast neural optical signals, making fast signals difficult to resolve. Thus, more sensitive techniques are needed to detect these small changes.

5.2.5 Cell Types Contributing to Optical Responses

The cellular origin of light-scattering changes has generated much controversy. Hill[35,36] showed that increased giant squid axon volume corresponds to increased transmitted light. However, neural tissue contains glial cells, which constitute a major proportion of tissue volume. Glial cells also swell or shrink with changes in extracellular ion concentration. Dendrites swell in response to activation.[52] Thus, at least three possible volume-related contributors to light-scattering changes exist: increases in neuronal cell body, dendrite size, and glial size. Cohen[20] suggested that compact clusters of small axons, such as in the crab nerve bundle, produce larger changes in reflectance than a single large axon such as the squid giant axon. Our own experience suggests that densely packed tissues, such as gray matter, provide larger fractional signals than isolated nerves.

Since proteins and lipids compose a large percentage of dry cell weight, these components probably contribute most of the scattering changes. A structure that possesses more scattering elements (such as dense dendrite/axon clusters, mitochondria, or other intracellular organelles) would be expected to exhibit a greater change in scattering. When measuring light scattering from brain tissue in vivo, both small processes and large-cell body diameters contribute to local changes observed with the probe, but high density components may contribute more to changes than large-cell bodies. Indeed, MacVicar and Hochman[37] found a larger degree of light-scattering changes in the dendritic layers of the hippocampus, which also contain many small interneurons. Glial cells may also contribute to the spatially organized changes associated with neuronal discharge, because they typically respond to osmotic changes in the local micro-environment. However, the time course for glial swelling is probably significantly slower than neural swelling rates.

5.2.6 Peripheral versus Direct Illumination

Many reflectance studies using lens-coupled systems illuminate the tissue surface and detect light at a distance. Relatively high light levels are required, and most of the illumination detected by the camera consists of specular reflection from the tissue surface or scattering from superficial tissue. Only a small proportion of the light detected in this configuration interacts with tissue beneath the surface of the cortex. Most investigators employ high-sensitivity, cooled camera equipment with slow scan rates (1–10 fps) for detection of such small changes. Recently, we overcame some of these problems in studies of isolated retina by using transmitted light or oblique (a form of dark field) illumination, together with faster cameras.

We developed another method of detecting scattered light using fiber optic image conduit. The image conduit is in direct contact with both the tissue and the camera. Light is delivered to the tissue around the perimeter of the probe as in "dark field" illumination. Thus, before entering the image conduit fibers, light must be scattered by objects below the tissue surface. The use of fiber optics in reflectance imaging greatly increases the detection sensitivity for light-scattering changes over lens-coupled devices. Because of improved collection efficiency, this technique does not require high light levels, allowing use of miniature LEDs for illumination. The geometry of light paths from illumination fibers to detection fibers is such that a large portion of reflected light is perturbed by cells that change their scattering properties with activation. Such procedures greatly enhance the contrast of tissue activity changes.

Imaging systems incorporating a gradient index (GRIN) lens can also be placed in contact with the tissue surface, which eliminates a number of optical artifacts that may arise from mechanical motion of the tissue surface. These systems may incorporate optics that allow focusing into tissue beyond the end of the imaging probe. This feature, along with optical geometry, facilitates advanced imaging strategies, such as confocal microscopy. However, the GRIN lens is much less efficient at capturing and transmitting light than fiber optic image conduit because of smaller acceptance angles, thus much higher illumination levels are required.

5.3 METHODS

Our imaging and data acquisition system is optimized for spatial and temporal resolution and sensitivity for detecting small changes in scattered light. The optical system utilizes perimeter illumination to mimic dark-field microscopy to optimize contrast for scattered light changes. Components directly in contact with the tissue reduce mechanical artifacts from blood pressure and respiration cycles. The optical system also provides the ability to form confocal images for better depth discrimination. The imaging system consists of one of several high speed CCD sensors with good signal to noise (66 dB or better), coupled with custom scanning and digitizing circuitry.

5.3.1 Imager Design

Fiber optic perimeter illumination with image conduit or a GRIN lens provides an efficient arrangement for illumination. The device uses a straight piece of image conduit with illumination fibers arranged around the image conduit perimeter[53] (Figure 5.1). The arrangement precludes direct illumination; however, our studies showed that dark-field illumination doubled the field depth-of-view (to 600 mm) over direct illumination methods[54] due to increased modulation contrast, and enhanced the magnitude of light scattering signals.[49,55]

Optical techniques using coherent fiber optic image conduit[54,56,57] formed excellent images of the tissue surface. Also, since illumination surrounded the imaged area, dark-field methods eliminated specular reflectance and provided scattering information from deeper tissue. However, light from deeper structures was out of focus, blurring the resultant images. Because cells of interest are frequently located several hundred microns below the surface, focusing at deeper levels is desirable to appropriately measure such structures. We developed an advanced imaging system using a gradient index lens which serves as a relay lens for tissue, and provides firm tissue contact to minimize movement and specular reflectance. A microscope objective projects an image onto a high-speed, miniature

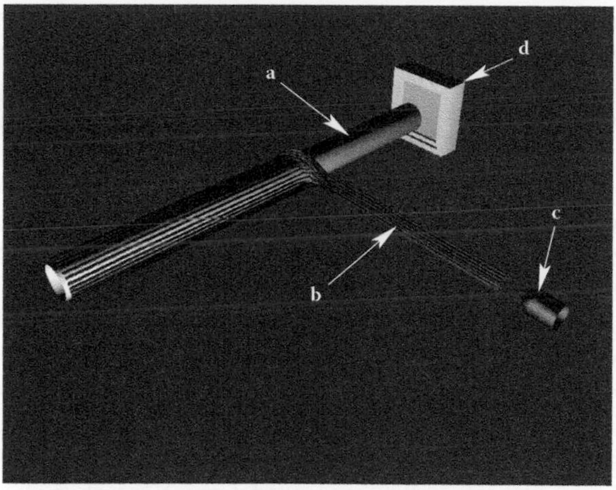

FIGURE 5.1 A 3D rendering of the image conduit probe illustrates a coherent fiber image conduit (a) with plastic illumination fibers (b) around the perimeter of the image conduit, which provides dark-field illumination from a light source (c), typically a light-emitting diode. Back-scattered light from the neural tissue is collected by the image conduit, and an image from the neural surface is transmitted to a CCD camera (d).

CCD camera.[49] The microscope objective and camera assembly is adjusted independently of the GRIN lens, allowing focal depth changes without disturbing the probe/tissue interface. Fiber optics around the probe perimeter provide diffuse illumination from a 780-nm laser, or 660- and 560-nm light-emitting diodes. The device has the potential for miniaturization to be chronically implanted in freely behaving animals. Related prototype devices we developed can generate confocal and spectral images.

5.3.2 Digitizing Hardware

We built a custom device that digitizes video and multiple analog signals simultaneously at 12-bit resolution.[57,58] A computer system displays intensity histograms, as well as raw or averaged images and dynamic average pixel intensity over the image or specified subregion along with a strip-chart-style display of physiological data. In our experience, continuous, real-time feedback of physiological and image data is extremely useful for consistent experimental success.

To increase sensitivity for detection of very small changes in light intensity, we employ techniques for increasing the effective dynamic range of the video digitizing system. One method involves building video amplifiers that allow judicious selection of black level and gain. With such a system, even an 8-bit system can achieve adequate dynamic range. The black level is first set so that a digital value of 16 represents the dimmest pixel in the image, and a value of 240 represents the brightest pixel in the image. This procedure increases the ability to detect intensity changes by a factor of ~7. One disadvantage of this procedure is that it requires an extra calibration to recover absolute intensity values. A second and more difficult method for increasing dynamic range involves digitizing a reference video frame, storing that frame in memory, and using a differential video amplifier to subtract the reference frame from incoming frames. Such a procedure introduces many potential noise sources, although commercial systems based on this strategy are available.

The current digitizing system is built on a PCI interface card for high-speed data transfers. It contains 8 million bytes of double-buffered memory for synchronous acquisition and download capability, and uses 7 ns programmable logic devices (PLDs) to execute the various control functions for digitizing and CCD control. The acquisition hardware gains flexibility through field programmable arrays (FPGAs) and embedded digital signal processors (DSPs). The PCI interface allows 132 MB/s burst and 66 MB/s continuous data streaming to the host computer for archiving, on-line display, and on-line analysis.[58]

This system has performed well for rapid and flexible video and electrophysiological acquisition. Data are stored in the on-board memory in a standard file format IFFPHYS, an internationally recognized format for physiological data. The PCI card plugs into any standard Intel motherboard. Simple drivers were written for a UNIX operating system to control the card and download buffered data. We recently undertook a redesign of this system to employ newer and higher performance technology.

In addition to real-time display during acquisition, our system performs analyses that recall a series of image frames in sequence, define image groups of interest, and produce averaged group images. To create animated sequences, the software performs sequential, time-triggered subtraction, ratio, and calibration calculations (Figure 5.2). Analyses includes frame-by-frame standard deviation and FFT computations.

5.3.3 Retinal Imaging

Isolated retinas provide a simple model for understanding the transient intrinsic optical signals associated with neural activity from a tissue more structurally analogous to brain tissue than the isolated lobster nerve. Since much of the invertebrate signal is observed because all the fibers are in parallel, the retinal preparation enables the development of optical techniques for fast neural imaging that are closer to in vivo imaging, without significant deep-tissue scattering and motion artifacts. We validated fast near-infrared (NIR) light microscopy of transmitted light changes associated with retinal neural activity using subcellular spatial resolution.[59–61]

In order to spatially resolve the optical changes across different tissue depths, new techniques are required to remove scattering contributions from above and below the depth of interest. Optical coherence tomography (OCT) is an optical technique where collimated light is projected onto the tissue, and then reflected light is mixed with light from a reference beam. The resulting image results from interference patterns such that any light that is not at the same depth relative to the reference beam is rejected. By moving the position of the reference beam, depths up to 1 mm can be probed. By observing changes in the OCT light across time, small changes in light scattered from specific tissue depths can be imaged. Thus, OCT is an attractive method as a function probe for depth-resolved light-scattering changes; however, it is limited in depth by the small number of photons that

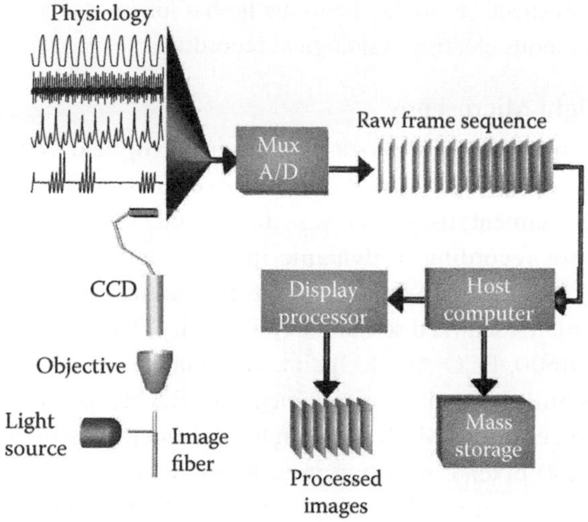

FIGURE 5.2 A diagram of system components shows a light source projecting light into flexible fiber optics mounted around the perimeter of a coherent fiber optic image conduit or gradient index lens. The imaging device placed in contact with the neural structure of interest collected back-scattered light and formed an image through a microscope objective onto a CCD camera. A change in the focal plane was achieved by moving the CCD camera and objective relative to the image fiber, which remained fixed in position. Video frames from the CCD camera and 16 channels of electrophysiology were multiplexed and digitized continuously during the experiment. A LINUX/PC-based host computer collected the raw frame sequences, and streamed them continuously onto a mass storage device. The host computer can search image sequences for stimulus trigger events, average frames, and transfer those averages to a display processor via a fast ethernet connection for on-line analysis and visualization. (Modified from Rector et al., *NeuroImage*, 14, 977–994, 2001. With permission.)

successfully back-scatter from depths more than 1 mm. We constructed a rapid functional OCT system to demonstrate depth-resolved recording of reflected light changes in the activated retina.[62]

5.3.3.1 Preparation of Isolated Retinas

Isolated frog (*Rana pipiens*) retinas were obtained by rapid euthanizing and decapitation, and by double pithing before the eyes were removed. The procedure was conducted in a dark room with dim red illumination. The dissection of the retina was performed in Ringer's solution containing (in mM):[63] 110 NaCl, 2.5 KCl, 1.6 $MgCl_2$, 1.0 $CaCl_2$, 22 $NaHCO_3$, and 10 D-glucose. After removing the intact eye, we hemisected the globe below the equator with fine scissors to remove the lens and anterior structures before removing the retina. The retina was sliced radially, or a small wedge of tissue was used to allow the retina to lie flat in the recording chamber. Finally, the retina was transferred into the recording chamber (which typically contained an array of recording electrodes) for intrinsic optical imaging. During an experiment, the retina was immersed in a chamber filled with Ringer's solution. The retina was pressed to the bottom of the sample chamber with moderate pressure using a micromesh sheet[62] to make the tissue lie flat for imaging and to improve the quality of simultaneous electrophysiological recordings.

5.3.3.2 NIR Light Microscopy

Figure 5.3 presents a simplified block diagram of the optical system for NIR light imaging of transient intrinsic optical changes associated with retinal neural activity. During an experiment, the retina was illuminated continuously with NIR light (800–1000 nm) for recording of dynamic intrinsic optical changes, and a white light flash was used to activate the retina. We used an iris or slit aperture with visible light to stimulate localized retinal neural activity. The system used a fast digital camera (PCO1600, PCO AG, Kelheim, Germany) with 14-bit dynamic range, which has 2-GB built-in random access memory (RAM) for fast image recording with a transfer speed of 80 MB/s. Although the camera has relatively high resolution (1600 × 1200 pixels), we often used lower resolutions for high frame rate recording. By employing pixel binning and reading out only a specified area of

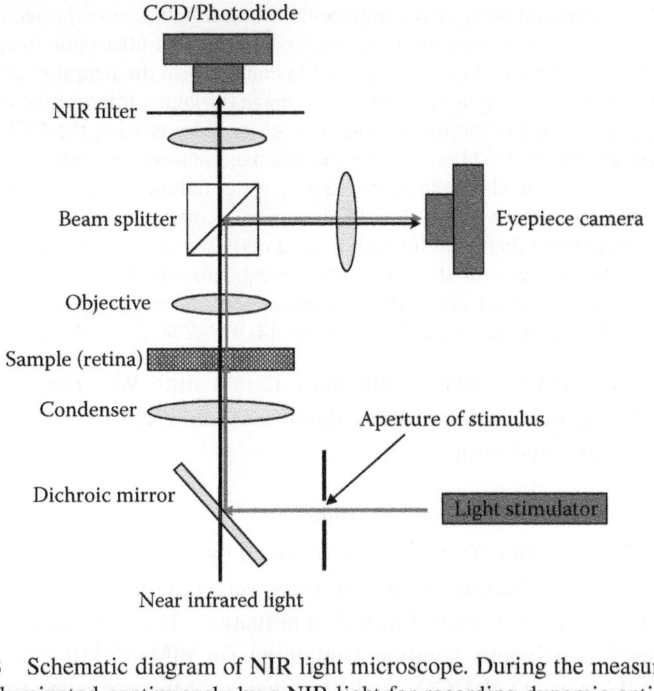

FIGURE 5.3 Schematic diagram of NIR light microscope. During the measurement, frog retina was illuminated continuously by a NIR light for recording dynamic optical changes, and a white light emitting diode (LED) was used to produce a visible light stimulus. Optical and electrophysiological responses were measured simultaneously. While the high speed CCD camera (or photodiode) was used for NIR light recording of transmitted light changes associated with retinal activation, the eyepiece camera could simultaneously monitor the visible light stimulation. At the dichroic mirror, the visible stimulus light was reflected, and the NIR light passed through.

interest, it was practical to achieve imaging speed >100 frames/second, with frame resolution better than 200 × 200 pixels.[60] A 10-s image sequence was typically recorded with 2-s prestimulus baseline recording.

5.3.3.3 NIR Light OCT

Figure 5.4 shows a simplified block diagram of our OCT system for reflected light detection of transient intrinsic optical changes associated with retinal neural activity. In a time-domain OCT system, the optical phase (effective depth) of the reference beam is periodically modulated. Rapid, vibration-free modulation is critical to ensure high speed and signal-to-noise ratios (SNR). A no-moving-part electro-optic phase modulator (EOPM) has been used for vibration-free phase modulation in conventional laser interferometers.[64] However, the introduction of a new component in the reference arm of an OCT system is undesirable, because the optical dispersion can reduce the SNR and axial (depth) resolution for OCT imaging.[65] Recently, we developed an EOPM-based (Model 370, Conoptics Inc. Danbury, California) OCT,[62] and validated the feasibility of compensating for the optical dispersion of the EOPM by introducing comparable dispersion in the sample arm. Before we compensated for the optical dispersion of the EOPM, the OCT depth resolution was 63 um. After we compensated, the optical dispersion of EOPM, the depth resolution was ~19 um, which agreed well with the theoretical estimate.[62]

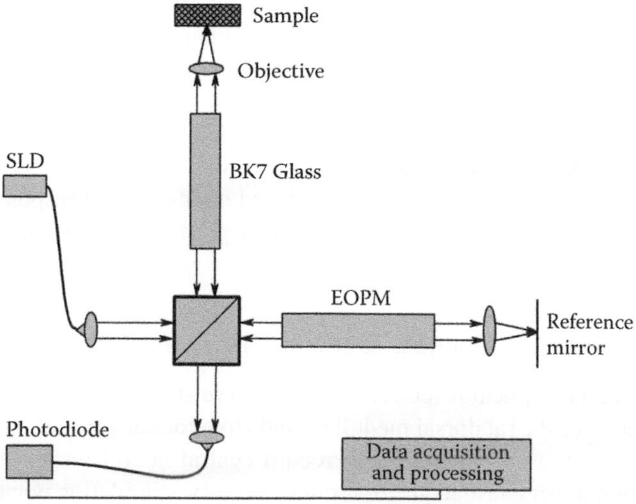

FIGURE 5.4 Schematic diagram of NIR light OCT. A superluminescent diode (SLD) with a 793-nm center wavelength and a 15nm full-width-half-maximum (FWHM) spectral width. BK7 glass block was used to compensate chromatic aberration of the electro-optical phase modulator (EOPM). A 10× (NA = 0.25) objective was used to illuminate the sample with a light power ~150 uW and to collect reflected light. (Modified from Yao et al., *Appl. Opt.*, 44, 2019–2023, 2005. With permission.)

5.4 BRAIN IMAGING IN INTACT ANIMALS

We performed numerous studies of light-scattering changes within freely behaving animals, as well as in several acute preparations. In all studies, we observed changes in back-scattered light corresponding to increased neural activity. Our procedures employ the contact image probe and data acquisition system described earlier to acquire data continuously to mass storage and to provide real-time display image processing.

5.4.1 Correlations with Evoked Potentials from Cat Hippocampus

Acquisition of scattered light at 2.4 kHz from a single point on the dorsal hippocampal surface with a high sensitivity photodiode revealed a strong inverse correlation between evoked potentials and light scattering changes in the hippocampus.[48] Stimulation produced a complex electrical response with at least two components: an early population spike within 10 ms of the stimulus artifact, and a population postsynaptic potential occurring within 50 ms. The light-scattering signal showed an inverse relationship to the electrical signal with at least three components: (1) a slow response which corresponded to the population postsynaptic potential, (2) a very slow response lasting several seconds, which may result from vascular processes, and (3) when the time scale of the plot is expanded at the time of stimulus, we observe a small, fast response, the onset of which preceded the early population spike by 5 ms (Figure 5.5B). The amplitude of the evoked light-scattering change was dependent on a stimulus position corresponding to a laminar spatial representation of contralateral projections of the Schaeffer's collaterals.[66] When the photodiode was replaced with a CCD camera and individual images captured at the peak of the evoked response, a topographical pattern of activation could be recorded when different regions of the dorsal hippocampus were stimulated in an 8 × 2 mm-square-region on the contralateral side. Some stimulating positions produced strong optical responses with significant spatial structure (Figure 5.5C), and other positions produced little response (Figure 5.5D).

5.4.2 Brainstem and Cortical Recording from the Rat

We investigated fast optical responses to vagal nerve stimulation within the nucleus tractus solitarius of the rat dorsal medulla[50] and from the rat whisker barrel cortex.[51] These experiments used our imager to record central neural responses to peripheral stimulation at very fast acquisition rates. Sensory stimulation elicited response patterns, detected as changes in back-scattered light. Regional responses disclosed four distinct time courses with components that paralleled fast electrical evoked responses as well as slower hemodynamic signals. Additionally, different response components could be optimally recorded at different depths.[49,50]

For cortical studies, back-scattered light images were continuously digitized (300 fps) during whisker stimulation (0.2 ms pulse, 1–2 s random interval, 400 pulses), together with blood pressure, EKG, tracheal pressure, and field potentials from a macro wire inserted under the probe. To determine changes in light scattering over

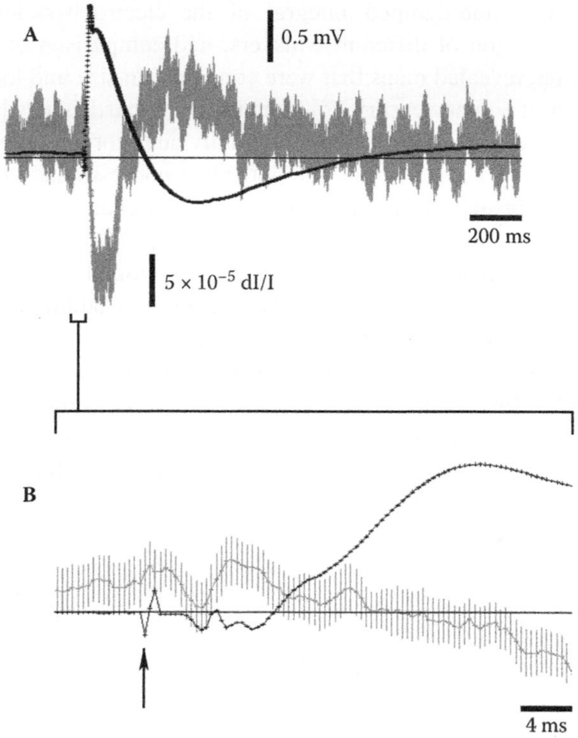

FIGURE 5.5 A fiber optic probe with perimeter illumination was placed on the surface of the dorsal hippocampus of a cat. Back-scattered light was collected with a photodiode at 2.4 kHz, or a CCD camera at 100 Hz during stimulation of the contralateral hippocampus. Stimulation elicited evoked electrical responses (dark lines, Panels A and B) and concurrent optical responses (gray traces). Panel B shows an expanded view of the electrical and optical response around the stimulus artifact (arrow). (Modified from Rector et al., *J. Neurophysiol.*, 78: 1707–1713, 1997. With permission.)

time after the stimulus, images from each sequence were averaged on a pixel-by-pixel basis and divided by the average of the pre-stimulus baseline. To prevent physiologically induced phase locking of the cardiac cycle to the stimulus, random stimuli (uniformly distributed, 1- to 2-s interstimulus interval) were presented to single whiskers on the left side using upward deflections of a 17-gauge hypodermic tubing attached to a miniature 8-ohm speaker. The speaker was driven with an 8 volt, 0.2-ms pulse, preceded by a 5-ms pre-trigger pulse. Time triggered averages (typically 500 to 1000 trials) were processed during the on-line analysis. An average baseline image (constructed from 25 images collected during the pre-stimulus period) was subtracted from the average image sequences.

Fast intrinsic scattered light signals tightly coupled to electrical responses were imaged in response to physiological stimulation of the rat whisker barrel cortex. When the whisker barrels were imaged with the CCD camera, the integrated signal across the image field showed a dynamic optical response with an early temporal transient corresponding to a time-damped integral of the electrophysiological response (Figure 5.6). Stimulation of different whiskers, and comparison of the regions of apparent activation, revealed maps that were consistent in size and location with the expected position of somatosensory columns to which particular whiskers project. Optical responses were initially localized to individual cortical columns that represent single whiskers. Later in time, the optical response spread across the entire barrel field, with a reversal in direction of the optical change about 500 ms after stimulation.

The temporal pattern of optical response waveforms found in this study is similar to that found by electrically stimulating the vagal nerve bundle and imaging the dorsal medulla, with four components.[50] The early optical response was dominated by a signal closely associated with the postsynaptic evoked response. On a time scale of a few hundred milliseconds, we saw evidence of a response with a slightly wider spatial distribution consistent with an initial dip in tissue oxygenation. At later times, we observed a reversal in the direction of the optical response that may correspond to an overcompensatory hemodynamic response flooding the region with oxygenated blood. Our results suggest that optical signals can track neurophysiological dynamics at very high frequencies. We saw that fast optical signals can also be recorded in rodent whisker barrel cortex in response to single whisker twitches.[51] In addition to evoked responses, we observed high-frequency (400–600Hz) oscillations in photodiode measurements, highly correlated with electrophysiological records. Optical techniques may provide the spatial and temporal resolution required to investigate the role of very fast dynamic responses in cortical information processing by somatosensory systems.[18]

Previous intrinsic signal imaging studies disclosed initial activation during single whisker stimulation with a broad peak centered over the simulated barrel.[67] In our studies, the initial pattern of activation disclosed by the fast-scattering signals is rather focal, perhaps limited to the extent of a single barrel. There is a suggestion in these data that the early metabolic response (the so-called optical mapping signal) may be somewhat more diffuse, but our measurements need to be optimized (perhaps employing different illumination wavelengths) to better distinguish between early light scattering signals and fast absorbance changes. In agreement with the Polley et al. study[67] and other reports, the slow absorbance changes corresponding to the hemodynamic (fMRI BOLD) response were broadly distributed across the barrel field, relatively independent of the particular whisker stimulated.

5.5 FAST-SCATTERED LIGHT CHANGES IN ISOLATED NERVES

Nerves from lobsters, crabs, squid giant, and isolated retinal preparations provide an excellent means to evaluate scattered light changes that accompany action potentials.[20,21] Crayfish claw nerves provide large signals, although the nerve is 10 times smaller than the lobster nerve, and total light intensity and scattered light changes caused by action potentials are smaller than detected in other invertebrates. Since invertebrate nerve bundles are typically well-aligned along one axis, strong polarization signals can be obtained, and are typically 10 times larger than scattering signals.[68] We have been using

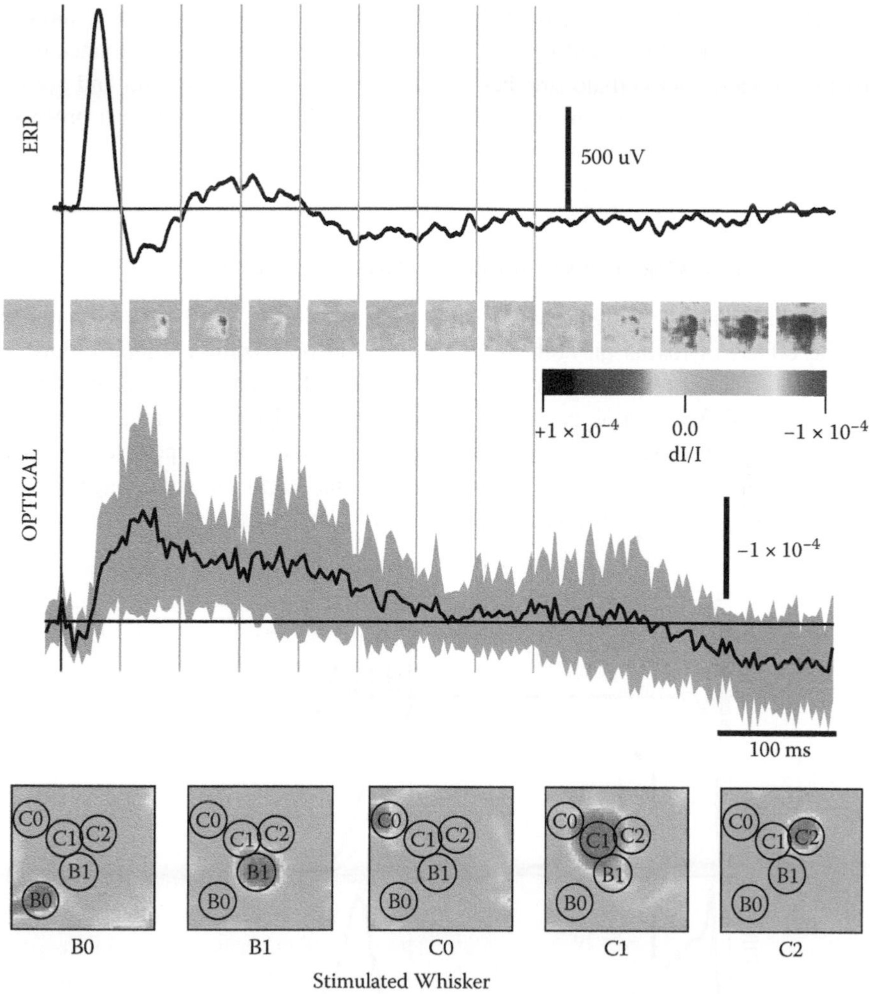

FIGURE 5.6 **(See color insert following page 224)** Evoked-response potential (ERP) and optical signal averaged across all pixels of the imager (OPTICAL) in a CCD experiment show a response to single-whisker stimulation. The optical trace has been inverted to show similarity in temporal structure. An average of 25 baseline images (before stimulation), subtracted from an average of 25 images collected during the early peak response shows a discrete region that corresponds to the cortical column associated with the stimulated whisker. Each image was collected from a 3 mm by 3 mm region of somatosensory cortex over the whisker barrels. A later epoch reveals a spreading of the optical response across space that changes polarity after 500 ms and covers a large portion of the imaged area. In pseudocolor images, warm colors represent decreases in scattered light and cool colors represent scattered light increases. In the lower 5 images, stimulating particular whiskers produced early response images, with discrete regions that localize to the expected map of cortical columns. Each of the five images (3 mm × 3 mm) was collected after stimulating the whisker indicated below the image. For spatial comparison, circles were drawn on all images, showing the activated regions. (Modified from Rector et al., *NeuroImage*, 26[2], 619–627, 2005. With permission.)

invertebrate nerve preparations to study optical properties of isolated neural tissue to optimize techniques for recording fast correlates of neural activity. The isolated retina provides a good intermediate step between the invertebrate preparation and cortical tissue, because the tissue contains multiple neuronal laminae with lateral processes randomly oriented, but such measurements are not subject to many of the physiological noise sources (such a blood flow) that in vivo cortical preparations possess.

5.5.1 Teasing Apart Polarization and Swelling Effects

We visualized scattered light changes in the isolated lobster leg nerve to characterize the effects of toxins on both the scattering and polarization signal. Figure 5.7 shows polarization responses an order of magnitude larger than scattering signals, and closely follow the time course of voltage-sensitive dye (VSD) measurements.[69] For all measurements, the scattering signal was delayed with respect to the polarization signal, and was affected differently by toxins. Earlier studies suggested that polarization signals result primarily from dynamic birefringence changes of activated

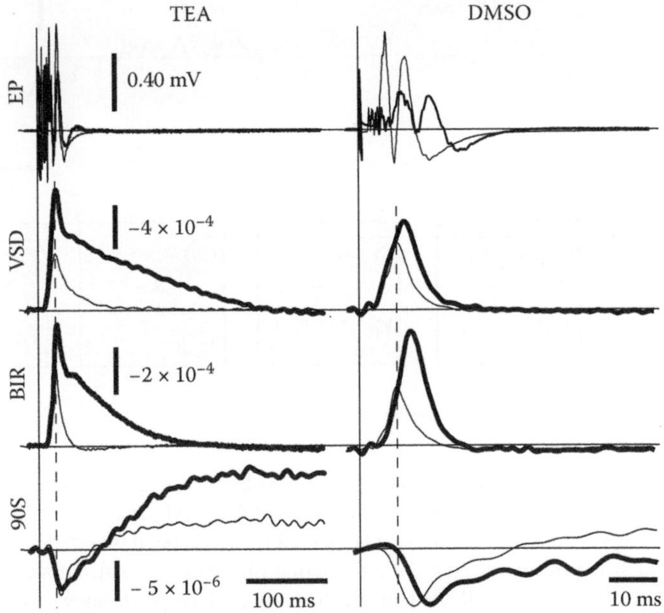

FIGURE 5.7 Field potential (EP), voltage-sensitive dye (VSD), birefringence (BIR), and 90° scattering (90S) recordings taken before (thin traces) and after (thick traces) 250 mM tetraethylammonium chloride (TEA) application and 10% dimethyl sulfoxide (DMSO) application. The solid vertical lines indicate the time of stimulus, and the dotted vertical lines mark the peak of the VSD and BIR response, which arrives approximately 2 ms earlier than the 90S signal. Each trace contains 100 averages. To emphasize the changes in time course for the DMSO recording, the time scale has been expanded for the right panel. EP1, EP2, BIR, and 90S were recorded simultaneously, while the VSD recordings represent separate experiments. (Modified from Foust A.J. and Rector D.M., *Neuroscience*, 145[3]: 887–899, 2007. With permission.)

tissues. However, we did not expect birefringence to be closely associated with VSD measurements, because birefringence changes should arise from molecular events, while membrane potential is an electrical phenomenon. Although VSD and polarization measurements are both reportedly potential-dependent,[70,71] we suspect that endogenous scattering events could confound a small portion of these signals and cause fluctuations not directly related to membrane potential. The 90° scattered (90S) light change, typically attributed to cellular swelling, developed a long-lasting reversal with a potassium channel blocker (tetraethylammonium chloride, TEA), and implies that cellular mechanisms other than swelling, refractive index, and polarization could generate optical scattering signals, which may be used to optimize in vivo recording techniques. Dimethyl sulfoxide (DMSO) disrupted membrane phospholipids, and altered both the 90° scattering and the BIR responses, but in different ways, again suggesting that different physical mechanisms contribute to the two optical signals.

Several investigators have linked possible biophysical mechanisms of optical changes to mechanical,[72–76] molecular,[27,28,77] and electrical[78] events associated with membrane excitation. Vesicle fusion,[79] cytoskeletal,[80] and axoplasmic[81] events may also participate. In combination with cellular swelling, the phospholipids in the membrane bilayer may change orientation, causing birefringence changes. While all of these mechanisms are probably operating at some level, there has been little consensus as to which element or combination of elements dominates these fast optical signals. Cellular swelling may underlie at least part of the scattering change.[29] However, since the scattering signal peak lags 1–3 ms behind the birefringence peak in simultaneous lobster nerve recordings, these optical changes must be mediated by a different combination of biophysical phenomena.[68]

There is evidence from our work and others,[49,79] that the fastest observed optical response may precede the arrival of the electrical signal. This intriguing possibility could arise through changes in the membrane channels that initiate the flow of ions. However, our findings suggest that the fast response develops earlier and propagates faster than would be expected from a purely local mechanism. We hypothesize that the earliest response could be caused by a compression wave established at the stimulus site (and at sites of subsequent activation) that travels through the nerve at the speed of sound. This might provide an alternative explanation for soliton-like propagation of neural signals[82] while remaining consistent with the model of action potential propagation proposed by Hodgkin and Huxley.[83]

5.5.2 Fast CCD Optical Imaging of Retinal Neural Activity

We initiated NIR light recording of transient intrinsic optical signals in retina with a fast photodiode system.[60] Although the total transmitted NIR light changes recorded by the fast photodiode in isolated retina were dominated by a positive optical response, high resolution CCD imaging disclosed both positive and negative optical responses with larger fractional changes (Figure 5.8). At the single-cell level, the maxima of both positive and negative responses were ~5% dI/I. Dark-field and polarization imaging could further improve fractional optical changes by lowering the background light intensity.[59] However, greater illumination intensity was required to ensure a sufficient number of photons collected by the CCD camera for high-quality imaging.

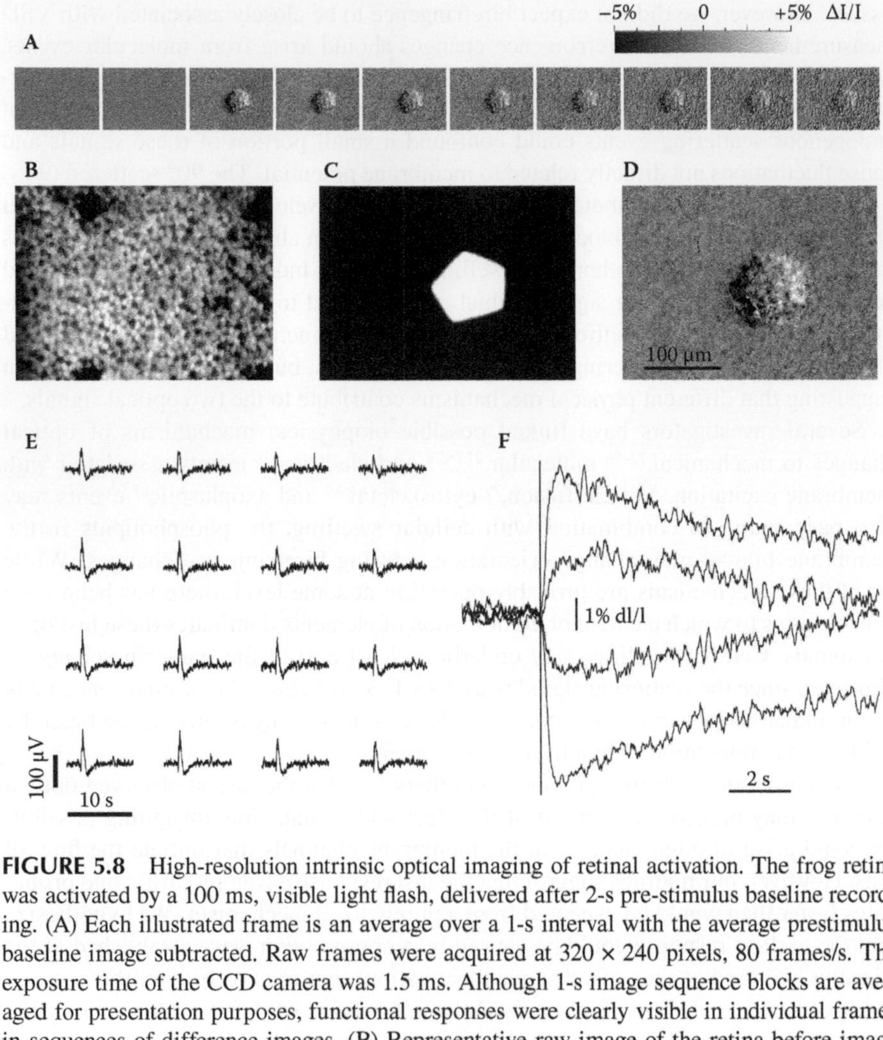

FIGURE 5.8 High-resolution intrinsic optical imaging of retinal activation. The frog retina was activated by a 100 ms, visible light flash, delivered after 2-s pre-stimulus baseline recording. (A) Each illustrated frame is an average over a 1-s interval with the average prestimulus baseline image subtracted. Raw frames were acquired at 320 × 240 pixels, 80 frames/s. The exposure time of the CCD camera was 1.5 ms. Although 1-s image sequence blocks are averaged for presentation purposes, functional responses were clearly visible in individual frames in sequences of difference images. (B) Representative raw image of the retina before image sequence processing. (C) An image of the stimulus pattern, collected by replacing the sample with a mirror. (D) Enlarged image of the fourth frame of top line. (E) Corresponding electrophysiological measurement of 4 × 4 channels. For the microelectrode array (MEA), the diameter of each electrode was 30 μm, and the electrode distance was 200 μm. Four of the electrodes can be seen in image (B) as dark blobs along the upper and lower edges of the anatomical image, two of which are aligned near the right edge of the image field. (F) Fast optical responses at 5 μm × 5 μm resolution (average of 25 pixels). Vertical lines indicate the stimulation time. These traces show different polarities and timescales of optical responses. (Modified from Yao, X.C. and George, J.S., *J. Biomed. Opt.*, 11, 064030, 2006. With permission.)

In coordination with high spatiotemporal resolution imaging, we used variable strength stimulation to characterize the fast intrinsic optical signals associated with retinal activation. For simultaneous depiction of temporal and spatial dynamics of the intrinsic optical responses, Figure 5.9 presents M-sequence images recorded

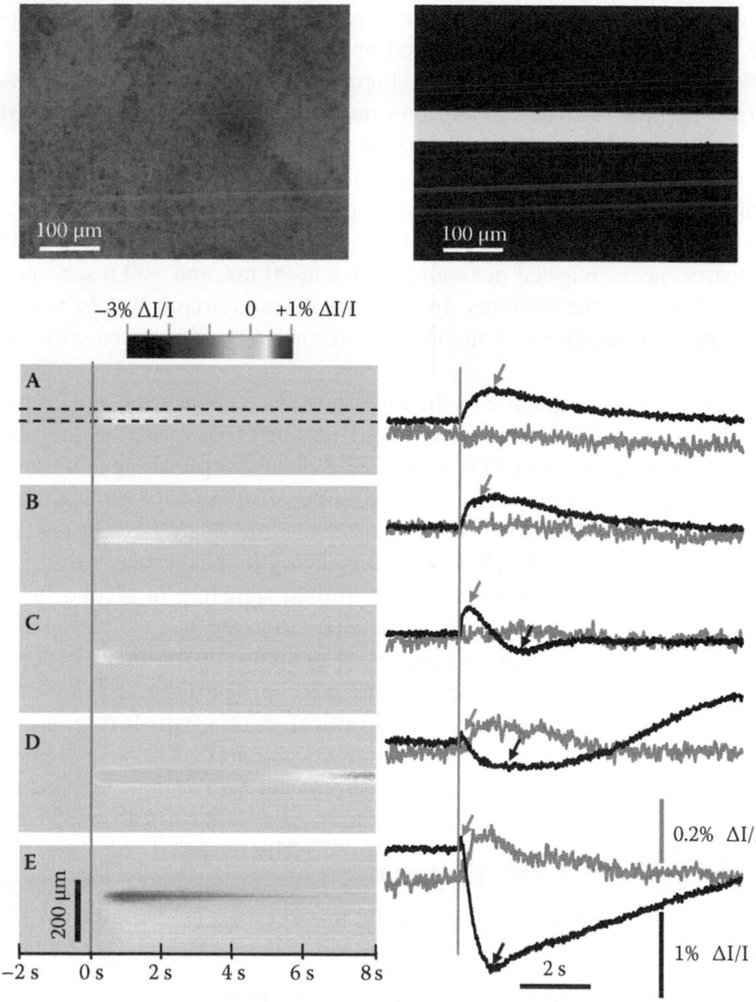

FIGURE 5.9 **(See color insert following page 224)** Intrinsic optical signals elicited by variable strength stimuli. Top left: Representative raw image of the retina before differentiating processing. Top right: An image of the slit stimulus pattern, collected by replacing the sample with a mirror. The slit was roughly 50 μm wide and 500 μm long. Bottom left: image sequences showing fast intrinsic optical signals. Bottom right: Black traces and gray traces represent temporal changes of averaged intrinsic optical signals of a stimulated retinal area (the area between the two longitudinal dashed lines indicated in the left of panel (A) and a control (unstimulated) area, respectively. Raw CCD images were acquired at 80 frames/s. The exposure time of the CCD camera was 1 ms. Irradiances of the visible light stimuli were (A) 8×10^{-6}, (B) 8×10^{-5}, (C) 8×10^{-4}, (D) 8×10^{-3}, and (E) 8×10^{-2} μJ, respectively. The top image sequence, (A) was recorded first, and the following sequences were recorded at 2-min intervals. The vertical gray lines indicate the delivery time of the stimuli. Gray and black arrows in the bottom right panels indicate amplitude peaks of positive and negative optical responses, respectively. (Modified from Zhao, Y.B. and Yao, X.C., *Opt. Lett.*, 33, 342–344, 2008. With permission.)

from a frog retina activated by variable stimulus.[61] From Figure 5.9, we observed that the intrinsic optical signals activated by a low strength light stimulus, such as 8×10^{-6} μJ (Figure 5.9A) or 8×10^{-5} μJ (Figure 5.9B), consist of a fast positive peak (gray arrow pointed) with a relatively slow negative-going recovery phase. When the stimulus was strengthened to 8×10^{-4} μJ, an additional negative peak (black arrow) could be observed following the positive peak (Figure 5.9C). With stronger stimuli, the latencies, relative to the delivery time of the stimuli, of the positive and negative peaks were shortened (Figure 5.9D,E). In Figure 5.9e, the rapid positive and subsequent negative peaks reached peak amplitudes at 90 ms, and 900 ms, respectively, after the delivery of the stimulus. In the retinal area covered by the visible light stimulus, the peak amplitude of the negative response was increased with strengthening stimulus.

However, the peak amplitude of the rapid positive response was not correspondingly increased, but slightly reduced. Based on our experimental observation and previous electroretinogram (ERG) studies, the positive optical signals most likely result from post-photoreceptor responses, and negative optical signals may be primarily associated with the photoreceptor transduction. The relatively low sensitivity of the positive optical signals to stimulus strength change may result from the nonlinear amplification mechanisms of the retinal signaling in second- and third-order neurons.[84–86] With such an automatic adaptation mechanism, the photoreceptor signal can be amplified or moderated, depending on the lighting conditions, allowing the visual system to be less sensitive in a bright light and more sensitive in a dim light environment. The negative optical signals, which responded sensitively to the stimulus strength change, may reflect significant contributions of activated photoreceptors. Previous optical measurements with a single channel detector (i.e., a fast photodiode) have disclosed negative-going optical responses from photoreceptors activated by visible light flashes.[87,88] The observed phenomenon that a stronger stimulus shortened the latency of the negative-going response to reach the amplitude peak agrees with the transduction model proposed by Lamb et al.[89]

5.5.3 Fast OCT Recording of Retinal Activation

Using the in-lab-built, rapid, vibration-free OCT (Figure 5.4), we demonstrated depth-resolved recording of intrinsic optical signals in stimulus-activated frog retina. By focusing the OCT instrument at different depths of the retina, we recorded and differentiated light-scattering changes from the photoreceptor and ganglion layers.[62] In Figure 5.10C, the scattered light from the photoreceptor layer was reduced during retinal activation. In contrast, scattered light from the ganglion layer was increased, as shown in Figure 5.10D.

Although we successfully demonstrated depth-resolved recording of retinal neural activity, it was not practical for us to achieve fast 3D imaging of intrinsic optical signals because of the limited SNR and requirement of time-consuming scanning of the single OCT probe. We anticipate that further development of fast 3D optical imaging systems, with improved imaging speed and SNR, will allow dynamic visualization of fast neural activity in three dimensions. Advanced 3D intrinsic optical imaging may advance the study of dynamic visual processing in the retina, and can

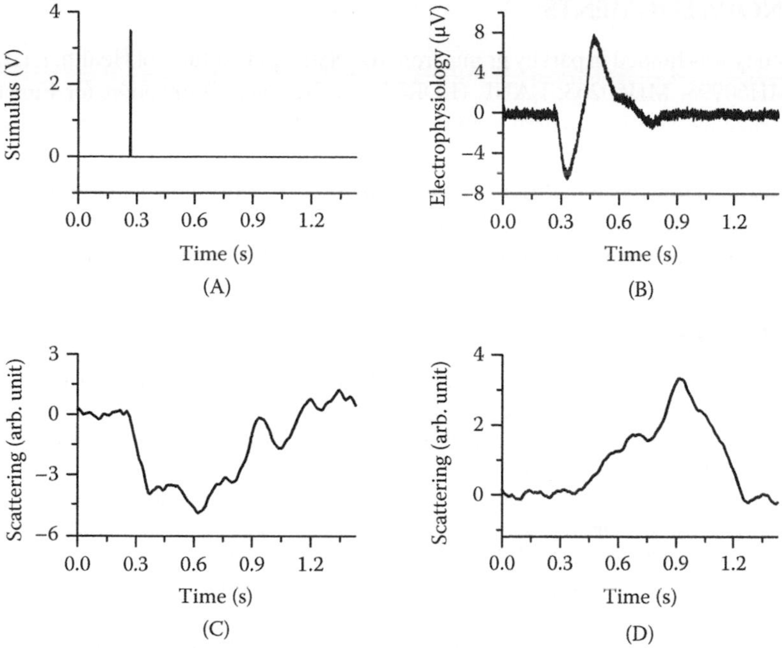

FIGURE 5.10 Rapid OCT recording of retinal neural activity. (A) Voltage pulse of 10 ms was used to drive a white LED, and the light flash stimulated the frog retina. (B) Electrophysiological response associated with the light stimulus. (C) Scattering response at the photoreceptor layer. (D) Scattering response at the ganglion layer. Each trace is an average of 100 trials, and the recording interval was 1.5 s. (Modified from Yao, et al., *Appl. Opt.*, 44, 2019–2023, 2005. With permission.)

also provide new methodology for early detection of retinal diseases and reliable assessment of treatment outputs.

5.6 SUMMARY

Optical signals indicative of different aspects of neural signaling can now be imaged with very high spatial and temporal resolution. The signals can be associated with structures that remain in focus deep below the cortical surface, and be captured at relatively high signal-to-noise ratios by judicious illumination of activated structures to minimize extraneous scattering. Optical signals from various depths of the dorsal medulla parallel evoked electrical potentials elicited by stimulation of afferent nerves. Very rapid optical correlates of activity from invertebrate axons result in smaller optical signals than from central neural sites, but have the potential to reveal insights into propagation mechanisms that may not be apparent by electrophysiologic recording. Because near-infrared light can penetrate deeply into tissue, even passing through bone, it may be possible to perform imaging or optical tomography of cortical activation noninvasively,[6] perhaps in human subjects.[10,11]

ACKNOWLEDGMENTS

This study was funded in part by grants from the National Institutes of Health: HL 60296 and MH60993, MH60263; LANL (LDRD); the National Foundation for Functional Brain Imaging/MIND Research Network; and the U.S. Department of Energy Artificial Retina Project. DR was a director's fellow at LANL and a Beckman Foundation Young Investigator Fellow. MH60993 is a Human Brain Project/Neuroinformatics research project funded by the National Center for Research Resources, National Institute of Mental Health, National Institute on Drug Abuse, and the National Science Foundation.

REFERENCES

1. Grinvald et al. Functional architecture of the cortex revealed by optical imaging of intrinsic signals. *Nature*, 324: 361–364, 1986.
2. Ts'o et al. Functional organization in the primate visual cortex as revealed by optical imaging of intrinsic signals. *Science*, 249: 417–420, 1990.
3. Frostig et al. Cortical functional architecture and local coupling between neuronal activity and the microcirculation revealed by in vivo high-resolution optical imaging of intrinsic signals. *Proc. Natl. Acad. Sci. U.S.A.*, 87: 6082–6086, 1990.
4. Haglund et al. Optical imaging of epileptiform and functional activity in human cerebral cortex. *Nature*, 358: 668–671, 1992.
5. Narayan et al. Mapping functional activity in rodent cortex using optical intrinsic signals. *Cereb. Cort.*, 4(2): 195–204, 1994.
6. Masino et al. Characterization of functional organization within rat barrel cortex using intrinsic signal optical imaging through a thinned skull. *Proc. Natl. Acad. Sci. USA*, 90(21): 9998–10002, 1993.
7. Villringer A. and Chance B. Non-invasive optical spectroscopy and imaging of human brain function. *Trends Neurosci.*, 20(10): 435–442, 1997.
8. Strangman et al. Non-invasive neuroimaging using near-infrared light. *Biol. Psychiatry,* 52(7): 679–693, 2002.
9. Sandman et al. Is there an evoked vascular response? *Science,* 224(4655): 1355–1357, 1984.
10. Gratton, et al. Noninvasive detection of fast signals from the cortex using frequency-domain optical methods. *Ann. N.Y. Acad. Sci.,* 820: 286–299, 1997.
11. Steinbrink, et al. Somatosensory evoked fast optical intensity changes detected non-invasively in the adult human head. *Neurosci. Lett.,* 291(2): 105–108, 2000.
12. Franceschini, et al. Coupling between somatosensory evoked potentials and hemodynamic response in the rat. *Neuroimage*, 41(2): 189–203, 2008.
13. Kleinschmidt, et al. Simultaneous recording of cerebral blood oxygenation changes during human brain activation by magnetic resonance imaging and near-infrared spectroscopy. *J. Cereb. Blood Flow Metab.*, 16(5): 817–826, 1996.
14. Boas, et al. A vascular anatomical network model of the spatio-temporal response to brain activation. *Neuroimage,* 40(3): 1116–1129, 2008.
15. Werring, et al. A direct demonstration of both structure and function in the visual system: Combining diffusion tensor imaging with functional magnetic resonance imaging. *Neuroimage*, 9(3): 352–361, 1999.
16. Le Bihan D. Looking into the functional architecture of the brain with diffusion MRI. *Nat. Rev. Neurosci.*, 4(6): 469–480, 2003.
17. Fries, et al. Synchronization of oscillatory responses in visual cortex correlates with perception in interocular rivalry. *Proc. Natl. Acad. Sci. USA*, 94(23): 12699–12704, 1997.

18. Jones, M.S. and Barth, D.S. Spatiotemporal organization of fast (>200 Hz) electrical oscillations in rat Vibrissa/Barrel cortex. *J. Neurophysiol.*, 82(3): 1599–1609, 1999.
19. Lindsey, et al. Repeated patterns of distributed synchrony in neuronal assemblies. *J. Neurophysiol.*, 78(3): 1714–1719, 1997.
20. Cohen, L.B. Changes in neuron structure during action potential propagation and synaptic transmission. *Physiol. Rev.*, 53(2): 373–413, 1973.
21. Tasaki, I. and Byrne, P.M. Rapid structural changes in nerve fibers evoked by electrical current pulses. *Biochem. Biophys. Res. Comm.*, 188(2): 559–564, 1992.
22. Rector, et al. Imaging the dorsal hippocampus: Light reflectance relationships to electroencephalographic patterns during sleep. *Brain Res.*, 696: 151–160, 1995.
23. Rector et al. Imaging of hippocampal and neocortical neural activity following intravenous cocaine administration in freely behaving animals. *Neuroscience*, 54(3): 633–641, 1993.
24. Rector, et al. Imaging of the goat ventral medullary surface activity during sleep-waking states. *Am. J. Physiol.*, 267: R1154–R1160, 1994.
25. Malonek, D. and Grinvald, A. Interactions between electrical activity and cortical microcirculation revealed by imaging spectroscopy: Implications for functional brain mapping. *Science*, 272(5261): 551–554, 1996.
26. Perkampus, H.H. Ed. *UV Atlas of Organic Compounds,* Vol. V. New York: Plenum Press, 1971.
27. Landowne, D. Optical activity change with nerve impulses. *Biophys. J.,* 61: A109, 1992.
28. Landowne, D. Molecular motion underlying activation and inactivation of sodium channels in quid giant axons. *J. Membrane Biol.*, 88: 173–185, 1985.
29. Yao, et al. Cross-polarized reflected light measurement of fast optical responses associated with neural activation. *Biophysical J.*, 88: 4170–4177, 2005.
30. Raichle et al. Correlation between regional cerebral blood flow and oxidative metabolism. *Arch. Neurol.,* 33: 523–526, 1976.
31. Baron et al. Noninvasive measurement of blood flow, oxygen consumption, and glucose utilization in the same brain regions in man by positron emission tomography: Concise communication. *J. Nucl. Med.*, 23(5): 391–399, 1982.
32. Belliveau et al. Functional mapping of the human visual cortex by magnetic resonance imaging. *Science*, 254: 716–719, 1991.
33. Hayaishi, O. Oxyhemoglobin increases and deoxyhemoglobin decreases in the circulation of the brain of the rhesus monkey during REMS but not during slow wave sleep. In *Slow Wave Sleep: Its Measurement and Functional Significance.* Chase, M.H. and Roth, T., Eds., Los Angeles, CA: Brain Information Service/Brain Research Institute, University of California, Los Angeles, 1990.
34. Onoe et al. REM sleep-associated hemoglobin oxygenation in the monkey forebrain studied using near-infrared spectrophotometry. *Neurosci. Lett.*, 129(2): 209–213, 1991.
35. Hill, D.K. The effect of stimulation on the opacity of a crustacean nerve trunk and its relation to fibre diameter. *J. Physiol.*, 111: 283–303, 1950.
36. Hill, D.K. The volume change resulting from stimulation of a giant nerve fibre. *J. Physiol.*, 111: 304–327, 1950.
37. MacVicar, B.A. and Hochman, D. Imaging of synaptically evoked intrinsic optical signals in hippocampal slices. *J. Neurosci.*, 11(5): 1458–1469, 1991.
38. Ballanyi, K. and Grafe, P. Cell volume regulation in the nervous system. *Renal. Physiol. Biochem.*, 3–5: 142–157, 1988.
39. McBain et al. Regional variation of extracellular space in the hippocampus. *Science*, 249: 674–677, 1990.
40. Dudek et al. Osmolality-induced changes in extracellular volume alter epileptiform bursts independent in chemical synapses in the rat: Importance of non-synaptic mechanisms in hippocampal epileptogenesis. *Neurosci. Lett.*, 120: 267–270, 1990.

41. Taylor, C.P. and Dudek, F.E. Excitation of hippocampal pyramidal cells by an electrical field effect. *J. Neurophysiol.*, 52(1): 126–142, 1984.
42. Taylor, C.P. and Dudek, F.E. Synchronization without active chemical synapses during hippocampal after discharges. J. *Neurophysiol.*, 52(1): 143–155, 1984.
43. Roper et al. Osmolality and nonsynaptic epileptiform bursts in rat CA1 and dentate gyrus. *Ann. Neurol.*, 31(1): 81–85, 1992.
44. Lipton, P. Effects of membrane depolarization on light scattering by cerebral cortical slices. *J. Physiol.,* 231: 365–383, 1973.
45. Blasdel, G.G. Visualization of neuronal activity in monkey striate cortex. *Annu. Rev. Physiol.*, 51: 561–581, 1989.
46. Zecevic, et al. Imaging nervous system activity with voltage-sensitive dyes. *Curr Protoc Neurosci.*, chap. 6: Unit 6.17, 2003.
47. Kuhn et al. In vivo two-photon voltage-sensitive dye imaging reveals top-down control of cortical layers 1 and 2 during wakefulness. *Proc Natl Acad Sci USA,* 105(21): 7588–7593, 2008.
48. Rector, et al. Light scattering changes follow evoked potentials from hippocampal schaeffer collateral stimulation. *J. Neurophysiol.*, 78: 1707–1713, 1997.
49. Rector et al. A focusing image probe for assessing neural activity in-vivo. *J. Neurosci. Meth.*, 91: 135–145, 1999.
50. Rector, et al. Scattered light imaging in vivo tracks fast and slow processes of neurophysiological activation. *NeuroImage*, 14, 977–994, 2001.
51. Rector, et al. Spatio-temporal mapping of rat whisker barrels with fast scattered light signals. *NeuroImage*, 26(2), 619–627, 2005.
52. Van Harreveld, A. Changes in the diameter of apical dendrites during spreading depression. *Am. J. Physiol.*, 192: 457–463, 1958.
53. Rector et al. Fiber optic imaging of subcortical neural tissue in freely behaving animals. In *Optical Imaging of Brain Function and Metabolism*. Dirnagl, U. et al., Eds., New York: Plenum Press, pp. 81–86, 1993.
54. Rector et al. A miniature CCD video camera for high-sensitivity light measurements in freely behaving animals. *J. Neurosci. Meth.*, 78: 85–91, 1997.
55. Holthoff, K. and Witte, O.W. Intrinsic optical signals in rat neocortical slices measured with near-infrared dark-field microscopy reveal changes in extracellular space. *J. Neurosci.*, 16(8): 2740–2749, 1996.
56. Rector, D.M. and Harper, R.M. Imaging of hippocampal neural activity in freely behaving animals. *Behav. Brain Res.*, 42: 143–149, 1991.
57. Rector et al. A data acquisition system for long-term monitoring of physiological and video signals. *Electroenceph. Clin. Neurophysiol.*, 87: 380–384, 1993.
58. Rector, D.M. and George, J.S. Continuous image and electrophysiological recording with real time processing and control. *Methods*, 25(2): 151–163, 2001.
59. Yao, X.C. and George, J.S. Near-infrared imaging of fast intrinsic optical responses in visible light-activated amphibian retina, *J. Biomed. Opt.*, 11, 064030, 2006.
60. Yao, X.C. and George, J.S. Dynamic neuroimaging of retinal light responses using fast intrinsic optical signals, *Neuroimage*, 33, 898–906, 2006.
61. Zhao, Y.B. and Yao, X.C. Intrinsic optical imaging of stimulus-modulated physiological responses in amphibian retina, *Opt. Lett.*, 33, 342–344, 2008.
62. Yao, et al. Rapid optical coherence tomography and recording functional scattering changes from activated frog retina, *Appl. Opt.*, 44, 2019–2023, 2005.
63. Sieving et al. Push-pull model of the primate photopic electroretinogram: A role for hyperpolarizing neurons in shaping the b-wave, *Vis Neurosci.*, 11, 519–532, 1994.

64. Huntley et al. Phase-shifted dynamic speckle pattern interferometry at 1 kHz, *Appl. Opt.*, 38, 6556–6563, 1999.
65. Hitzenberger et al. Dispersion induced multiple signal peak splitting in partial coherence interferometry, *Opt. Commun.*, 154, 179–185, 1998.
66. Tamamaki, N., and Nojyo, Y. Disposition of slab-like modules formed by axon branches originating from single CA1 pyamidal neurons in the rat hippocampus. *J. Comp. Neurol.* 291: 559–564, 1992.
67. Polley et al. Varying the degree of single-whisker stimulation differentially affects phases of intrinsic signals in rat barrel cortex. *J. Neurophysiol.*, 81(2): 692–701, 1999.
68. Carter et al. Simultaneous birefringence and scattered light measurements reveal anatomical features in isolated crustacean nerve. *J. Neurosci. Meth.*, 135, 9–16, 2004.
69. Foust A.J. and Rector D.M. Optically teasing apart neural swelling and depolarization. *Neuroscience*, 145(3): 887–899, 2007.
70. Cohen et al. Changes in light scattering that accompany the action potential in squid giant axons: Potential-dependent components. *J. Physiol.*, 224: 701–725, 1972.
71. Cohen, et al. Changes in axon fluorescence during activity: Molecular probes of membrane potential. *J. Membr. Biol.*, 19: 1–36, 1974.
72. Cohen et al. Light scattering and birefringence changes during nerve activation. *Nature*, 218: 438–441, 1968.
73. Tasaki, et al. Changes in fluorescence, turbidity, and birefringence associated with nerve excitation. *Proc. Natl. Acad. Sci. USA*, 61: 883–888, 1968.
74. Tasaki et al. Mechanical changes in crab nerve fibers during action potentials. *Jpn. J. Physiol.*, 30: 897–905, 1980.
75. Watanabe A. Mechanical, thermal, and optical changes of the nerve membrane associated with excitation. *Jpn. J. Physiol.*, 36: 635–643, 1986.
76. Yao et al. Optical lever recording of displacements from activated lobster nerve bundles and Nitella internodes. *Appl. Opt.*, 42: 2972–2978, 2003.
77. Landowne D. Measuring nerve excitation with polarized light. *Jpn. J. Physiol.*, 43: S7–S11, 1993.
78. Stepnoski, et al. Noninvasive detection of changes in membrane potential in cultured neurons by light scattering. *Proc. Natl. Acad. Sci. USA*, 88: 9382–9386, 1991.
79. Salzberg et al. Large and rapid changes in light scattering accompany secretion by nerve terminals in the mammalian neurohypophysis. *J. Gen. Physiol.*, 86: 395–411, 1985.
80. Oldenbourg et al. Birefringence of single and bundled microtubules. *Biophys. J.*, 7: 645–654, 1998.
81. Watanabe A. and Terakawa S. Alteration of birefringence signals from squid giant axons by intracellular perfusion with protease solution. *Biochim. Biophys. Acta*, 436: 833–842, 1976.
82. Heimburg T. and Jackson A.D. On soliton propagation in biomembranes and nerves, *Proc. Natl. Acad. Sci. USA*, 102(28): 9790–9795, 2005.
83. Hodgkin A.L. and Huxley A.F. Propagation of electrical signals along giant nerve fibers. *Proc. R. Soc. Lond. B. Biol. Sci.*, 140(899): 177–183, 1952.
84. Baylor D.A. and Hodgkin A.L. Changes in time scale and sensitivity in turtle photoreceptors, *J. Physiol.*, 242, 729–758, 1974.
85. Mizunami, M. Nonlinear signal transmission between second- and third-order neurons of cockroach ocelli, *J. Gen. Physiol.*, 95, 297–317, 1990.
86. Dong, C.J. and Hare, W.A. Contribution to the kinetics and amplitude of the electroretinogram b-wave by third-order retinal neurons in the rabbit retina, *Vision Res.*, 40, 579–589, 2000.
87. Harary et al. Rapid light-induced changes in near infrared transmission of rods in *Bufo marinus*, *Science*, 202, 1083–1085, 1978.

88. Pepperberg, et al. Photic modulation of a highly sensitive, near-infrared light-scattering signal recorded from intact retinal photoreceptors, *Proc. Natl. Acad. Sci. USA*, 85, 5531–5535, 1988.
89. Lamb, T.D. and Pugh, E.N., Jr., A quantitative account of the activation steps involved in phototransduction in amphibian photoreceptors, *J. Physiol.*, 449, 719–758, 1992.

6 Imaging the Brain in Action

Real-Time Voltage-Sensitive Dye Imaging of Sensorimotor Cortex of Awake Behaving Mice

Isabelle Ferezou, Ferenc Matyas, and Carl C.H. Petersen

CONTENTS

6.1 INTRODUCTION

Electrophysiological measurements have demonstrated that information can be processed on the time scale of milliseconds in the mammalian brain. Neuronal electrical changes are probably the fastest events occurring in the nervous system, and they are likely to orchestrate many of the subsequent slower events, such as changes in second-messenger concentration, structural alterations, and regulation of gene expression. Whereas electrophysiological recordings from individual electrodes have revealed many important aspects of brain function, it is also clear that complex neuronal processing of information does not derive from the activity of individual neurons, but rather results from the concerted actions of many neurons distributed across different brain areas. Indeed, considerable progress has been made toward increasing the number of electrodes in electrophysiological recordings in order to begin to understand the coordinated function of neuronal networks.[1–2] Despite these very important technical developments, it remains clear that the spatial organization of neuronal electrical activity will be difficult to study with electrophysiological approaches, even using arrays of more than 100 electrodes.

Optical methods allow high-resolution imaging and therefore provide a useful tool to explore the spatial distribution of neuronal activity, especially in superficial brain structures such as the cerebral cortex. Indeed, quite remarkably, there are intrinsic changes in reflected light from the cortical surface, which can be easily measured and provide high spatial resolution maps of cortical organization.[3–6] However, these intrinsic changes in light absorption and scattering of brain tissue do not relate directly to neuronal electrical activity, but instead result primarily from hemodynamic changes, in a similar way to the BOLD fMRI signal. In order to make optical measurements that relate directly to electrical activity, it has so far proven necessary to apply dyes that change their absorption or emission spectra in a manner depending upon membrane potential. Such compounds are termed *voltage-sensitive dyes.*

6.2 PRINCIPLES OF THE VOLTAGE SENSITIVE DYE (VSD) IMAGING TECHNIQUE

6.2.1 Optical Measurement of Membrane Potential

The first VSDs applied to invertebrate preparations showed that it was possible to optically record fast membrane potential changes like action potentials in single neurons.[7–8] These pioneering recordings inspired the next decades of research, much of which has been targeted toward making measurements from the intact brain during the processing of sensory information. The first *in vivo* VSD recordings of sensory processing in the vertebrate brain were obtained in the optic tectum of the frog[9] and subsequently in the rat somatosensory and visual cortex.[10] These early investigations were made difficult by small signal amplitudes, large hemodynamic signals, phototoxicity, and limitations in detector and computer technology. Many of these problems have begun to be solved and, in this chapter, we try to provide information on the current state-of-the-art techniques for *in vivo* VSD imaging with the specific goal of measuring neuronal activity in awake behaving animals.

6.2.2 Voltage-Sensitive Probes for *In Vivo* Applications

One of the largest problems for *in vivo* VSD imaging is the prominence of heartbeat-related artifacts. With each heartbeat, the recorded signals are contaminated by movement artifacts from the physical pulsation of blood vessels and, in addition, by the shift of the hemoglobin absorption spectrum, which depends strongly on its oxygenation level. The most frequently used wavelengths for fluorescence measurements are indeed strongly absorbed by hemoglobin with large differences between deoxyhemoglobin and oxyhemoglobin. Heartbeat-related signals can therefore contribute significantly to optical fluorescence measurements.

To minimize these artifacts, Grinvald and coworkers suggested the application of fluorescent dyes with absorption and emission spectra in the far red, longer than 600 nm. In these wavelengths, hemoglobin absorbs little light, and the difference between oxygenated and deoxygenated forms is not so dramatic. In order to test this idea, they generated a new series of VSDs (RH1691, RH1692, and RH1838), which have revolutionized *in vivo* VSD imaging.[11–12] With these new dyes, together with advances in camera technology, imaging cortical dynamics at high temporal and spatial resolution have become routine measurements in many laboratories.[13–38] In our own studies, we have only used RH1691, and this review will therefore focus on this VSD. Although RH1691 works well in the rodent neocortex, it does not work in the rodent olfactory bulb[16] so care is needed in choosing and validating the VSD in individual applications.

RH1691 (Figure 6.1A) is a water-soluble aromatic anionic oxonol compound based on a vinylogous carboxylate conjugate system; its conformation changes are executed via transferring electrons. Upon application to cells, it is thought to insert mostly into the lipid bilayer of plasma membranes (Figure 6.1B) where it presents the properties of a fluorescent dye: it can be excited by light at ~630 nm, and emits fluorescent photons with wavelengths >665 nm. The electrical potential difference (on the order of a hundred millivolts) across the plasma membrane (with a thickness of ~5 nm) generates a strong electric field, which affects its fluorescence properties. Changes in the membrane potential are linearly related to the measured fluorescence of RH1691 (Figure 6.1C,D), when it is applied to *Xenopus* oocytes under two-electrode voltage clamp.[27] Voltage-clamp jumps are accompanied by rapid fluorescence changes (Figure 6.1E), which follow the membrane potential with submillisecond resolution.[27] The fluorescence of RH1691 is therefore rapidly and linearly related to membrane potential, similar to other voltage-sensitive dyes,[39] making interpretation of measurements relatively straightforward.

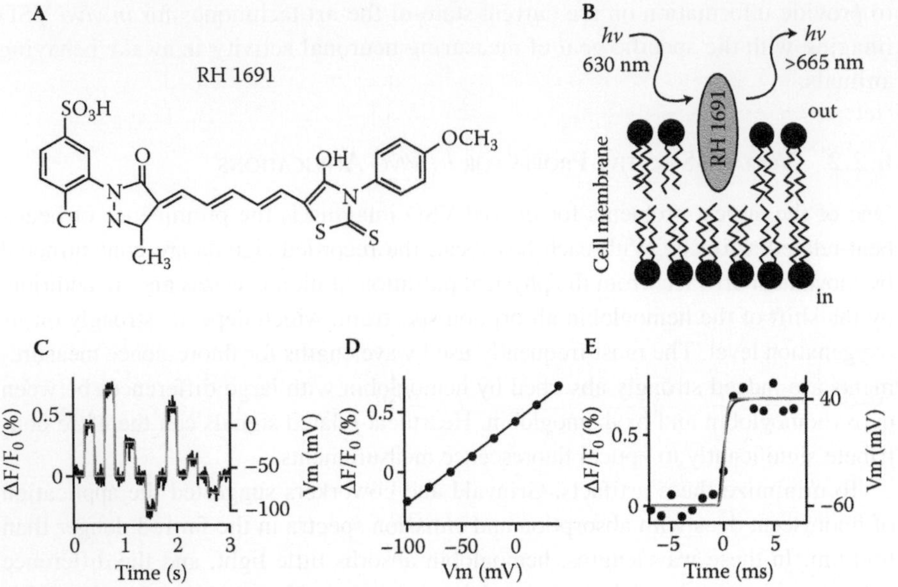

FIGURE 6.1 RH1691 is a fast and linear voltage-sensitive dye. (A) Chemical structure of the voltage sensitive dye RH1691. (B) The VSD RH1691 is a fluorescent dye excited at ~630 nm and emitting fluorescence >665 nm. The dye is thought to insert primarily into the plasma membrane lipid bilayer and its electrochemical structure is modified depending upon the transmembrane electrical field. This change in structure alters both the absorption and emission properties of the dye. (C) Biophysical measurements of RH1691 fluorescence applied to voltage-clamped *Xenopus* oocytes. Changing membrane potential (Vm, grey trace) results in well-correlated fluorescence changes ($\Delta F/F_0$, black trace). (D) RH1691 fluorescence changes linearly as a function of membrane potential (Vm) in measurements carried out in *Xenopus* oocytes. (E) Higher time resolution view of RH1691 fluorescence changes (black data points, grey trace shows a sigmoidal fit to the fluorescence data) following voltage changes (solid black line) in *Xenopus oocytes.* RH1691 follows membrane potential changes with sub-millisecond resolution. (Panels C, D, and E are modified from Berger, T., Borgdorff, A.J., Crochet, S., Neubauer, F.B., Lefort, S., Fauvet, B., Ferezou, I., Carleton, A., Luscher, H.R., and Petersen, C.C.H. 2007. *J. Neurophysiol.* 97: 3751–3762. With permission.)

6.3 BASIC EXPERIMENTAL SETUP FOR *IN VIVO* VSD EXPERIMENTS

6.3.1 Surgery and Cortical Staining with RH1691

In order to make *in vivo* measurements, the dye must be introduced into the brain. The simplest way to obtain an even staining of the cortical surface is to apply the dye dissolved in Ringer's solution directly to the cortical surface. A craniotomy must therefore be prepared over the cortical region of interest, and the dura subsequently removed (although a recent technical report from Lippert and collaborators[30] shows that RH1691 can stain through the rat dura). This must be done with extreme care not to damage the cortex, especially during removal of the dura. The VSD RH1691 is applied at a concentration of 1 mg/ml directly to the cortical surface and left to diffuse into the brain for a period of ~1 hour. Subsequently, the unbound dye must be washed away; the craniotomy can eventually be covered with agarose (~1–2%), which stabilizes the exposed cortex and reduces pulsation-related movement artifacts. A coverslip is then placed on the top of the preparation, forming a sealed chamber that, in addition, prevents the agarose from drying during the recording session. To combine VSD imaging with other experimental approaches (i.e., electrophysiological recordings, microstimulation, or drug injections), it is also possible to preserve an open access to the cortex on the side of the coverslip, but in this case, the experimenter might need to add extra Ringer's solution during the experiment to keep the preparation hydrated.

6.3.2 Acquisition of the Fluorescence Signals

In order to image VSD signals *in vivo* it is necessary to collect large amounts of light at low magnification. To obtain a high numerical aperture with a large working distance, it is useful to use tandem-lens optics.[40] The magnification is then given by the ratio of the focal length of the two lenses, which are mounted front-to-front, sandwiching the dichroic mirror. The VSD RH1691 is excited with band pass filtered (630 ± 15 nm) light, reflected using a 650 nm dichroic, and focused on the cortical surface with the lower lens. The excitation light is typically delivered by a halogen light source, which has very high stability compared to an arc lamp. The emitted fluorescent light is collected via the same optical pathway, but without the reflection of the dichroic, long-pass filter (>665 nm), and is focused onto a detector via the upper lens. Because VSD signals are typically small (on the order of 0.1%), it is critical to collect many photons to get beyond the physical limitations of shot noise, which follows the square root of the number of collected photons. In order to resolve a 0.1% change in light, it is necessary to collect ~1,000,000 photons. Specialized CMOS (complementary metal oxide semiconductor) cameras allow the acquisition of images at high time resolution with each pixel having a very large well depth. Fortunately, there are now a number of commercially available imaging systems that are well-suited for *in vivo* VSD imaging.

6.3.3 Data Analysis

The recorded VSD signals are usually affected by the bleaching of fluorescence which can be corrected offline. Subtraction of a best-fit double-exponential is well-suited when analyzing single trials. For experiments carried out with anesthetized animals, subtraction of heartbeat-synchronized and averaged blank sweeps is an alternative that has the additional advantage of removing the artifacts arising from blood pulsation–related movements. However, experiments performed with anesthetized mice revealed that the heartbeat-related signals often have a minimal impact on VSD recordings with RH1691 (Figure 6.2A,B). No modulation of the optical signals was observed with respect to the respiratory rhythm in these recordings[24] (Figure 6.2A,C).

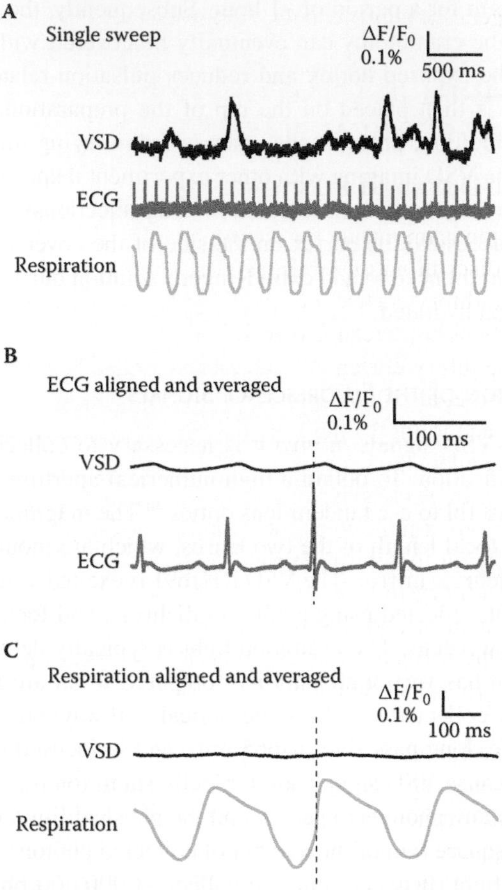

FIGURE 6.2 Heartbeat- and respiration-related artifacts have a minimal impact on the VSD recordings using RH1691 in anesthetized mice. (A) Simultaneous recording of the voltage-sensitive dye signal, respiration (from recording movement of the rib cage with an optical displacement sensor) and heartbeat (from recording the electrocardiogram [ECG]). There is no obvious correlation of heartbeat or respiratory rhythm contributing to the VSD signal. (B) The VSD signal was aligned to a fixed point on the ECG and many heart beats (n = 1249 from four mice) were averaged in order to specifically study the contribution of heartbeat-related artifacts to the VSD signal. The VSD signal was indeed modulated by heartbeat, but these artifacts were an order of magnitude smaller than the electrical signal reporting cortical membrane potential changes (see the trace in panel A). (C) The VSD signal was aligned to a fixed point in the respiratory cycle and many cycles (n = 407 from four mice) were averaged. No modulation of the VSD signal was found with respect to the respiratory rhythm. (Modified from Supplementary Figure 2 of Ferezou, I., Haiss, F., Gentet, L.J., Aronoff, R., Weber, B., and Petersen, C.C.H. 2007. *Neuron* 56: 907–923. With permission from *Neuron*, Cell Press, Elsevier.)

6.4 INTERPRETATION OF THE VSD SIGNALS

6.4.1 Origins of Recorded Optical Signals

Brain slices can be prepared in order to investigate what the dye stains under these conditions. In both rat[19,30] and mouse[24] barrel cortex, the RH1691 primarily labels the supragranular layers (Figure 6.3A). The dye is found in the neuropil, consistent with being bound to plasma membranes rather than being internalized (Figure 6.3B). The peak of the fluorescence profile with respect to cortical depth is found in layer 2/3 (Figure 6.3C). Presumably because the dye is water soluble, it diffuses out of layer 1 during the rinsing period after the staining, although other unknown structural features in layer 1 could also decrease RH1691 binding. Based on the linear relationship of fluorescence and membrane potential (Figure 6.1), in addition to the dye distribution following *in vivo* labeling (Figure 6.3), the fluorescence signals of RH1691 are therefore most likely to relate to changes of membrane potential in layer 2/3.

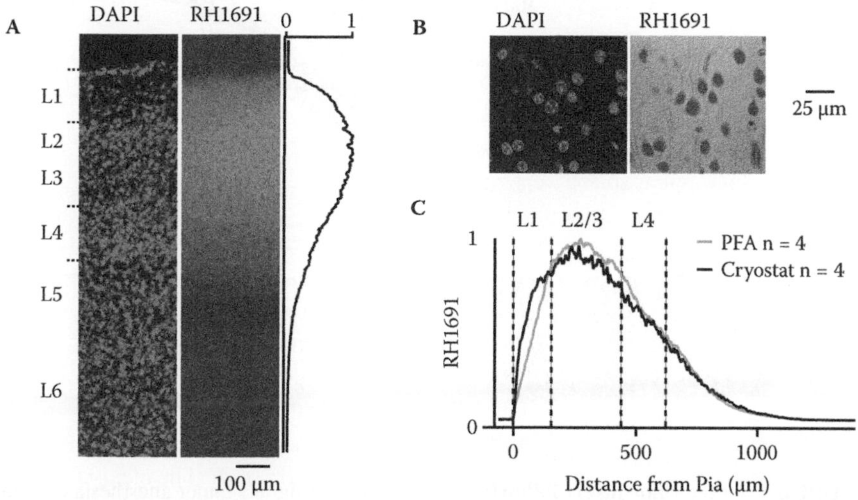

FIGURE 6.3 Applying RH1691 to the cortex stains the supragranular layers. (A) Topical application of RH1691 to the mouse barrel cortex labels mainly the supragranular layers. Parasagittal slices (100 μm thick) from stained barrel cortex, counterstained with DAPI and observed with confocal microscopy, reveal that the dye appears to wash out of layer 1, resulting in the highest dye concentrations in layer 2/3. (B) The dye does not enter inside the cells but rather labels membranes. (C) Quantification across different experiments reveals consistent labeling of supragranular layers, with a peak of RH1691 fluorescence in layer 2/3, independent of whether the brain is fixed with paraformaldehyde (PFA) or rapidly frozen and sectioned using a cryostat. (All panels modified from Ferezou, I., Bolea, S., and Petersen, C.C.H. 2006. *Neuron* 50: 617–629. With permission from *Neuron*, Cell Press, Elsevier.)

6.4.2 RH1691 Reports Supragranular Membrane Potential Changes in Anesthetized Mice

In order to directly correlate the fluorescence signals from RH1691 with membrane potential, we have combined whole-cell recordings with VSD imaging in the barrel cortex of urethane anesthetized rodents.[19,20,24,27] Remarkably close relationships were observed between the membrane potential of individual neurons and the VSD signal for both evoked and spontaneous cortical activity (Figure 6.4).

The rodent barrel cortex is a region of the primary somatosensory cortex, which is closely involved in the processing of tactile information from the mystacial vibrissae.[41] This cortical area is highly organized with an obvious anatomical mapping of

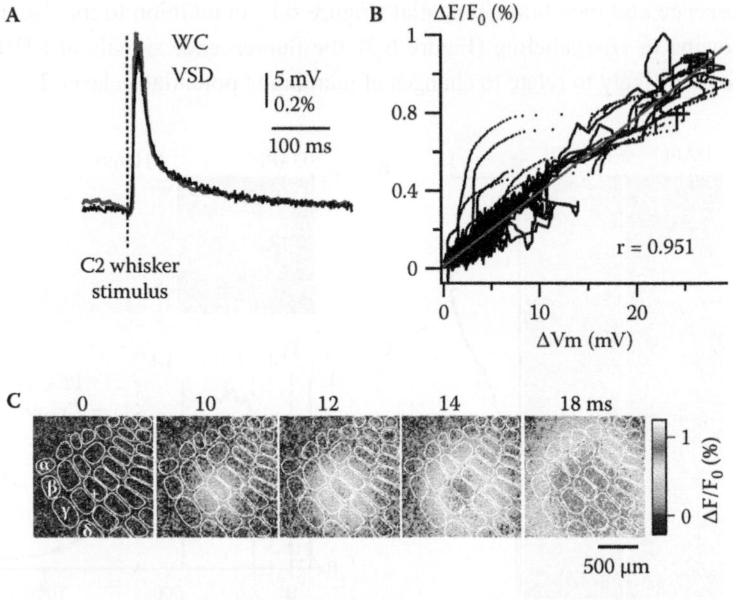

FIGURE 6.4 **(See color insert following page 224)** VSD signals under anesthesia correlate with subthreshold membrane potential changes in the supragranular layers. (A) Simultaneous whole-cell (WC) recording and VSD imaging *in vivo* under urethane anesthesia from the mouse somatosensory barrel cortex. The C2 whisker is briefly deflected for 2 ms (vertical dotted line), evoking a depolarizing sensory response. The averaged WC recording (red trace) and VSD signal measured from the C2 column location (black trace) are superimposed (average of six sweeps). The kinetics of the VSD responses and the membrane potential are almost identical. (B) The changes in VSD fluorescence are plotted as a function of change in membrane potential for each individual sweep (averages shown in panel A). The red line shows a linear fit with a Pearson correlation coefficient r = 0.95. (C) The spatiotemporal dynamics of the sensory response corresponding to the data in panels A and B. The red cross in the C2 barrel indicates the location targeted with the whole-cell recording. The C2 whisker is deflected at time 0 ms and about 10 ms later, a localized response is imaged in the C2 barrel column (the barrel map is superimposed in white). Over the next milliseconds, the sensory response spreads rapidly across the entire barrel map. (All panels modified from Ferezou, I., Bolea, S., and Petersen, C.C.H. 2006. *Neuron* 50: 617–629. With permission from *Neuron*, Cell Press, Elsevier.)

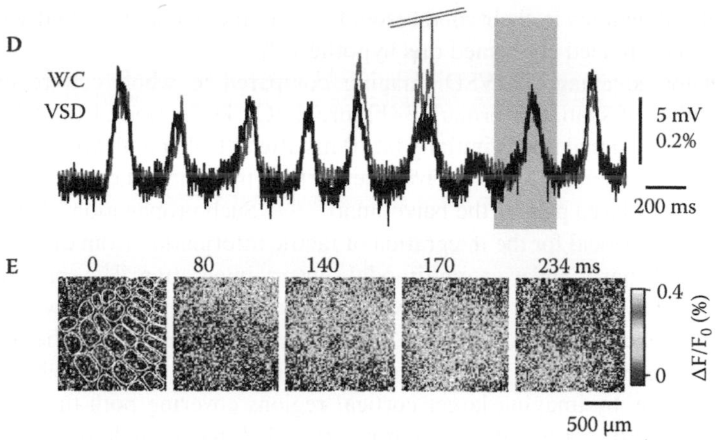

FIGURE 6.4 (continued) VSD signals under anesthesia correlate with subthreshold membrane potential changes in the supragranular layers. (D) The neocortex under urethane anesthesia is spontaneously active in the form of up and down states. Subthreshold spontaneous activity is also closely correlated between the whole-cell recording (red trace) and the VSD signal (black trace). Action potentials (truncated) in individual neurons are not accompanied by large VSD signals. (E) VSD images corresponding to a spontaneous depolarization event recorded in WC (highlighted in grey in panel D). Such images indicate that spontaneous activity often occurs as propagating waves, which can be followed by VSD. (All panels modified from Ferezou, I., Bolea, S., and Petersen, C.C.H. 2006. *Neuron* 50: 617–629. With permission from *Neuron*, Cell Press, Elsevier.)

each individual whisker. Because the whiskers and the corresponding barrel cortex are laid out in a stereotypical pattern, they have been given labels to allow identification. In our studies we have focused primarily on the so-called C2 whisker. The simplest experiment that one can perform with respect to comparing VSD signals with membrane potential is to briefly deflect a single whisker and measure the evoked cortical response. A brief deflection of the C2 whisker evokes activity in the somatosensory cortex, which is signaled via the trigeminal nerve to the brain stem, which in turn sends information to the ventrobasal thalamus from where thalamocortical neurons project to the primary somatosensory cortex. There are only two synapses between the whisker and the cortex, the first in the brain stem and the second in the thalamus. Sensory information can therefore reach the neocortex rapidly following a whisker deflection, with latencies typically around 10 ms. *In vivo* whole-cell recordings of layer 2/3 neurons show that a single brief 2 ms whisker deflection evokes a depolarizing sensory response lasting some tens of milliseconds (Figure 6.4A). Simultaneous optical measurement indicates that the kinetics of the VSD signals are remarkably close to the subthreshold membrane potential changes (Figure 6.4A): plotting VSD fluorescence as a function of membrane potential shows a linear relationship (Figure 6.4B). This correlation is particularly remarkable because the VSD signals result from changes in a large population of neurons, whereas the whole-cell recording relates to only a single neuron. From this observation, it would therefore seem likely that most neurons, within a few hundred microns of each other, have similar dynamics in their subthreshold membrane potentials. Dual whole-cell recordings have indeed confirmed this hypothesis.[19,20]

The major advantage of VSD imaging compared to whole-cell recording is the availability of spatial information (Figure 6.4C). Deflection of the C2 whisker evokes a sensory signal that is initially (at 10 ms after whisker deflection) confined to the corresponding barrel column. Over the next milliseconds, the sensory response spreads across a large part of the barrel map.[19,20,24] Such propagation of the evoked activity may be critical for the integration of tactile information from different whiskers, other body parts, and even multimodal sensory integration. The spreading VSD signals evoked by deflection of single whiskers are in good agreement with the large subthreshold receptive fields measured in layer 2/3 barrel cortex.[42–44] The spatiotemporal properties of the cortical activation induced by sensory input have been further studied, in mice, by imaging larger cortical regions covering both the somatosensory and the motor cortex. Such recordings revealed that a single deflection of the C2 whisker evokes, in addition to the initial spreading somatosensory response, a secondary localized region of activation in the motor cortex, which also spreads. Intracortical microstimulation experiments demonstrated that this secondary activated region corresponds to the region of the motor cortex that controls whisker movements. VSD imaging of cortical responses induced by the stimulation of different whiskers in the same conditions revealed a somatotopic arrangement of the activated areas within the primary motor cortex.[29] These first images of motor cortex activation upon sensory stimulation indicate that the whisker-related motor cortex is likely to integrate sensory input with motor commands.

Because of the relatively good signal-to-noise ratio of the recordings obtained with RH1691, it has also been possible to optically record spontaneous cortical activity, which by necessity demands single trial resolution. Under anesthesia, the membrane potential of cortical neurons often exhibits slow and large-amplitude fluctuations, which have been termed UP and DOWN states.[20,45–48] These spontaneous membrane potential changes correlate closely with the simultaneously measured VSD signal (Figure 6.4D). Again, since the whole-cell recording reports single neuron activity, whereas the VSD signal arises from a population, these data would seem to indicate that many neurons are behaving electrically in a very similar manner. Dual whole-cell recordings from anesthetized animals confirm that membrane potential fluctuations indeed occur synchronously in neurons within a local cortical region.[20,49,50] Although the subthreshold membrane potential changes recorded in individual neurons correlate closely with the VSD signals, the same cannot be said of the suprathreshold activity. Action potentials recorded in single neurons occur during the depolarized UP state, but whereas the whole-cell recordings show a large electrical signal change during the action potential, there is no indication of this event in the VSD record. Because the VSD is sufficiently fast to follow action potential waveforms, we must conclude that only very few nearby neurons spike at the same time. This would indicate a sparse action potential coding in layer 2/3 barrel cortex, as also suggested by electrophysiological recordings.[19,20,44,48,51]

The spatiotemporal dynamics of the spontaneous activity imaged by VSD reveal a great deal of complexity. In many cases the spontaneous depolarized active periods (UP states), appear as propagating waves of activity (Figure 6.4E). They can travel as planar waves, spirals, or even more complex patterns. In some cases, interesting correlations between spontaneously active regions have been observed. For example, in the mouse, we have observed synchronous spontaneous activity between the somatosensory and the motor cortex.[29] In the cat visual cortex, the spontaneous activity under anesthesia often takes the form of orientation maps.[13,18] In the rat parietal association area, which lies between the primary visual and the barrel cortex, imaging reveals a clear preference for propagation of the spontaneous waves within the cross-modal axis.[36] These observations from different species suggest that the spontaneous patterns of activity imaged under anesthesia are likely to relate intrinsic properties of the cortical functional connectivity, which do not depend upon the vigilance state of the brain.[52]

The close correlation *in vivo* under anesthesia between the VSD signal from RH1691 and the membrane potential in layer 2/3 neurons, for both evoked and spontaneous activity, suggests that the optical signals are open to a simple and straightforward interpretation. Because action potentials in layer 2/3 neurons occur rarely under our experimental conditions, the VSD signal is almost entirely dominated by subthreshold membrane potential changes. We conclude that VSD imaging reports the spatiotemporal dynamics of subthreshold membrane potential changes of the supragranular cortical layers.

6.5 VOLTAGE-SENSITIVE DYE IMAGING

6.5.1 Experimental Strategies

The relative simplicity and robustness of the *in vivo* VSD imaging technique and our ability to interpret the signals in a simple manner has allowed us to investigate the spatiotemporal cortical dynamics of awake mice with the aim of obtaining direct correlations between behavior, sensory input, and cortical activity. In order to perform VSD imaging of awake mice, we developed two different experimental strategies. The first approach consisted in imaging the cortex of head-restrained mice,[20,29] (Figure 6.5A) and the second method, based on the use of a flexible fiber optic bundle, allowed us to image cortical activity while the mouse was moving freely within a limited area[24] (Figure 6.5B). Both strategies yielded interesting and complementary results.

Mice accept head-restraint with surprising ease. Following implantation of a metal fixation post on the occipital bone, the mouse is trained to sit calmly in the experimental recording setup for a few days, on the basis of a couple of habituation sessions per day. At first the mouse is head-restrained for a few minutes and, as the mouse gradually adapts, the period of head-fixation can be increased up to a period of a couple of hours. On the day of the imaging experiment, the standard staining procedures are carried out using reversible gas anesthesia. After complete recovery from the anesthesia, it is then possible to image the cortical dynamics in the awake head-fixed mouse (Figure 6.5A). This approach has the great advantage that it should be possible to combine awake VSD imaging with other experimental approaches such as simultaneous electrophysiological recordings or targeted drug injections.

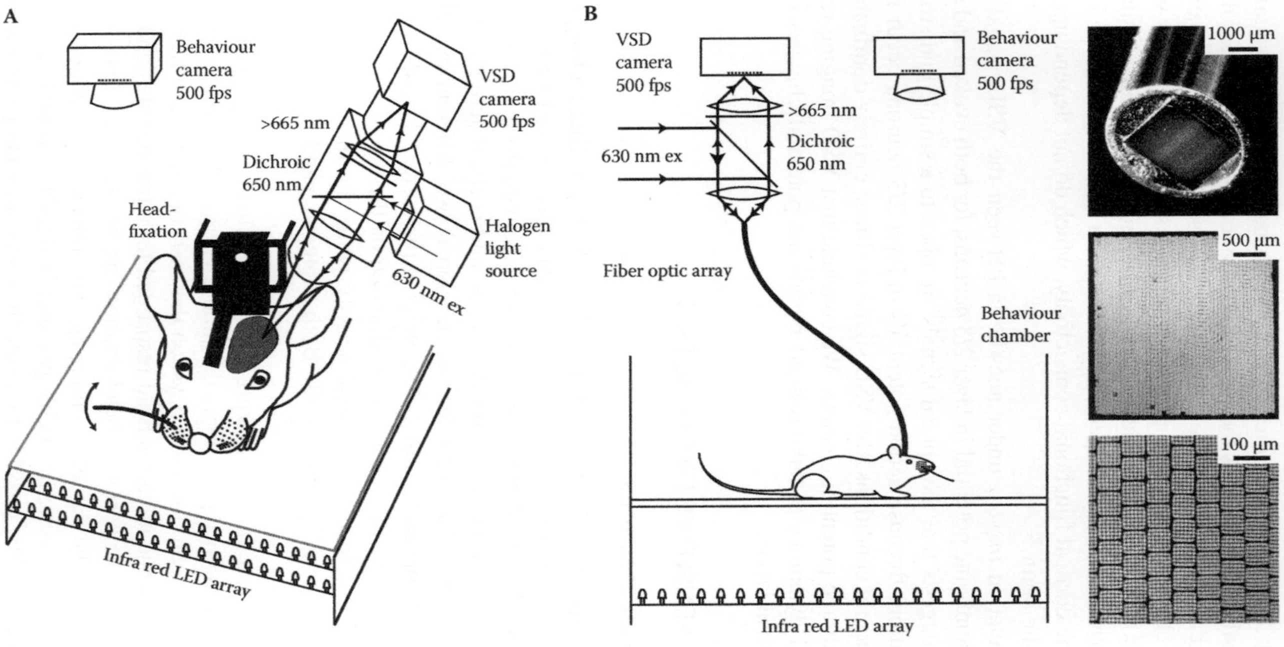

FIGURE 6.5 Fluorescence VSD imaging of awake mice with conventional epifluorescence and fiber optic imaging. (A) Mice readily accept head-fixation and conventional optics can therefore be applied for fluorescence imaging of awake mice. (B) Fiber optics can also be applied to image VSD signals in freely moving mice. Details of the fiber tip and the arrangement of the individual fibers are shown on the right. (All panels modified from Ferezou, I., Bolea, S., and Petersen, C.C.H. 2006. *Neuron* 50: 617–629. With permission from *Neuron*, Cell Press, Elsevier.)

However, head-restraint necessarily reduces the behavioral repertoire of animals, which in some cases is helpful in reducing the complexity of the experimental environment but also raises concerns relating to the physiological relevance of the data collected. It is therefore also important to attempt to record cortical activity of freely moving animals that can exhibit more natural behaviors. In this aim, we have begun to develop a relatively simple fiber optic imaging approach, which we applied to freely moving mice[24] (Figure 6.5B). The fiber we used consisted of a well-ordered array of 300 × 300 individual fibers with 8-μm cores and high numerical aperture (0.6 NA). The fibers were rigidly arranged at each end to obtain reliable image transfer but, between the ends, they were loose, allowing great flexibility and rotation. This flexibility is crucial for the ability of the mouse to move and turn around relatively freely. Compared to direct imaging, we found large light losses using the fiber optic approach, with total collected fluorescence being only about a quarter of the normal. Nonetheless, there remained sufficient VSD signal to resolve cortical dynamics in single trials with direct correlation to the ongoing behavior.

There is, therefore, growing interest now in VSD imaging as we enter a new era, where—for the first time—we can resolve the real-time spatiotemporal dynamics of cortical activity in awake-behaving animals. In the final sections of this review, we present our first glimpses of cortical activity in awake mice, and we discuss our hopes for the future.

6.5.2 State-Dependent Sensory Processing

Sensory perception does not arise purely from the ongoing inflow of sensory stimuli arriving at any moment. Perception should rather be viewed as a creative process, where these sensory signals are interpreted and the percept constructed based in large part on our previous experiences. It is therefore not surprising that many studies have found profound differences in how sensory signals are processed in the brain, depending upon the behavior of the animal.[41,53] Such so-called top-down processing of sensory information is also strongly suggested by the intricate anatomical feedback loops that are present throughout the neocortex and other brain areas.

With respect to the whisker sensorimotor system, when the animal is awake, there are two obvious states. During some periods the whiskers are still, whereas during other periods, the whiskers move rapidly and rhythmically following forward and backward movements, in a behavior that has been termed "whisking." Through VSD imaging of head-fixed awake mice,[29] we found that evoked responses in the mouse sensorimotor cortex were very different, depending upon the ongoing behavior of the mouse (Figure 6.6A). These results agreed well with independent measurements made using whole-cell recordings in awake head-fixed mice.[54] During quiet periods, when the mouse was not whisking, a small brief deflection of a whisker evoked a sensory response that began in the barrel cortex after ~10 ms. Over the next milliseconds, the response spread over the somatosensory cortex. In addition, a secondary localized region of activation was observed in the whisker-related motor cortex, which also subsequently propagates. Clearly, even a single, brief whisker deflection evokes a highly distributed sensory response. The spreading responses observed under anesthesia (Figure 6.4C) therefore also appear to be relevant to physiological

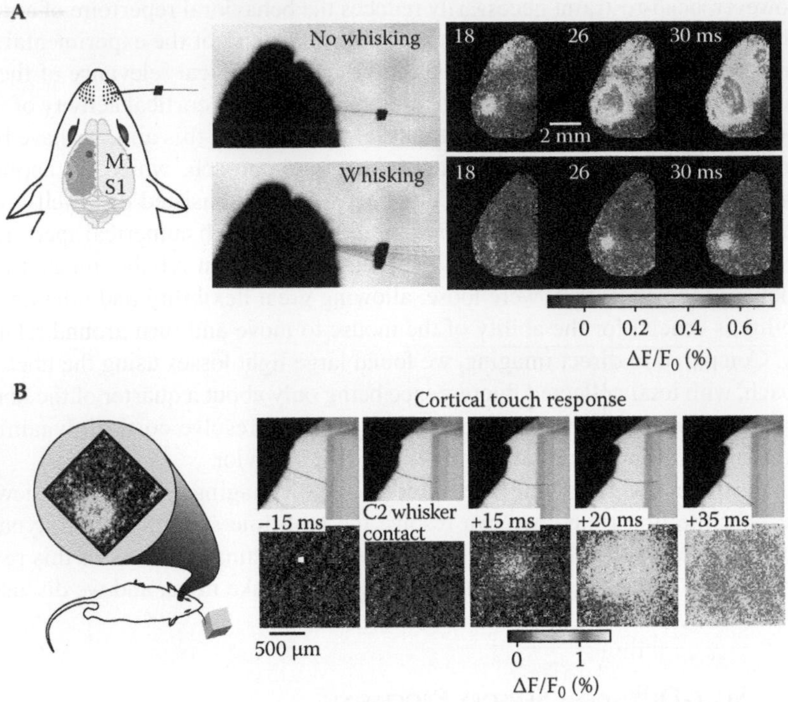

FIGURE 6.6 **(See color insert following page 224)** Imaging sensorimotor processing in awake-behaving mice. (A) VSD imaging of sensorimotor cortex in a head-restrained mouse. A small metal particle is attached to the C2 whisker, and whisker deflections are evoked by brief 2 ms magnetic pulses. A craniotomy covering a large fraction of somatosensory and motor cortex (green shaded region on the scheme) was stained with the VSD dye RH1691. The red dot indicates the position of the C2 barrel column of primary somatosensory cortex, and the blue dot is located in motor cortex. Photographs (middle) show 25 superimposed frames taken over a 100-ms period immediately before the stimulation. If the whisker stimulus is delivered when there is no spontaneous whisking (upper images), then a large amplitude sensory response is observed in the primary somatosensory barrel cortex (S1, 18 ms), followed by a secondary localized activation in the primary motor cortex (M1, 26 ms). Both S1 and M1 signals spread as propagating waves, which can propagate across a large fraction of the cortex. However, when the same stimulus is delivered during active whisking (lower images), there is only a signal in S1 on this trial. The ongoing behavior clearly makes a major impact upon the evoked sensory response. (Modified from Supplementary Figure 2 of Ferezou, I., Haiss, F., Gentet, L.J., Aronoff, R., Weber, B., and Petersen, C.C.H. 2007. *Neuron* 56: 907–923. With permission from *Neuron*, Cell Press, Elsevier.) (B) Using fiber optic imaging, we imaged the spatiotemporal dynamics of sensory processing during active touch of a whisker encountering a real object during freely moving exploration. The response began in a localized region and subsequently spread across a large part of the somatosensory cortex. (Modified and reproduced from Ferezou, I., Bolea, S., and Petersen, C.C.H. 2006. *Neuron* 50: 617–629. With permission from *Neuron*, Cell Press, Elsevier.)

brain states. These large and distributed sensory responses might correspond to a "wake up" call for driving the cortex into an active state. Sensory-evoked responses in the whisker-related motor cortex correlated with the initiation of whisker movement.

The motor cortex activity might therefore contribute to driving whisker movements, which could be directed to gather information relating to the immediate environment, and could be viewed as an attempt by the mouse to understand what caused the whisker deflection.

The same stimulus during whisking evoked a very different cortical response (Figure 6.6A). The response began in a similar manner compared to during quiet wakefulness, with a localized response in the primary somatosensory barrel cortex. However, the amplitude was smaller, and it did not spread. During the active brain state, sensory processing can therefore remain confined to highly localized cortical regions. From these observations, it is clear that the cortical response to a simple whisker deflection depends strongly on the behavioral state. Future experiments must determine the mechanisms underlying this state-dependence and the relationships to perceptual thresholds. Furthermore, it would also be interesting to combine VSD imaging in behaving animals with electrophysiological measurements to reveal how the activity of single neurons relates to the population VSD signals and how the different signals are influenced by behavioral states.

6.5.3 Imaging the Cortical Response during Active Touch

A mouse whisks as it explores a novel environment, and it is therefore thought that this behavior is an active processing mode during which the mouse tries to acquire sensory information with its whiskers. Thus, in the same way that humans would reach out to touch and manipulate an object in order to gauge its size, shape, and texture, mice actively contact objects with their whiskers to obtain tactile information. Applying the VSD fiber imaging approach to freely moving mice,[24] we obtained the first images of the cortical processing of touch as a mouse approached and contacted an object with a whisker (Figure 6.6B). We found that active touch of real objects evoked large spreading signals in the barrel cortex. They were again initiated in a small localized region, with a diameter similar to that of an individual barrel, and colocalized with the sensory-evoked response mapped under anesthesia. In the milliseconds that followed the first response, the VSD signal spread over a large cortical area. Future experiments must investigate how sensory responses to the touch of real objects (Figure 6.6B) compare quantitatively to the artificial deflections evoked by passive stimuli (Figure 6.6A). These experiments are just the beginnings of investigations into the cortex of awake animals, and there is obviously much further work to be done. We find it encouraging that the VSD imaging technique appears to provide useful information relating to cortical dynamics during behavior.

6.6 DISCUSSION

Sensory information is rapidly processed in highly distributed cortical networks. The spatiotemporal dynamics of cortical membrane potential changes can be resolved *in vivo* with millisecond temporal and columnar spatial resolution through VSD imaging. Application of this technique to awake-behaving mice reveals prominent state-dependent processing of whisker-related information within sensorimotor cortex. This

VSD imaging technique opens the door to investigation of the distributed processing underlying active acquisition of tactile sensory information.

However, it is clear that there are major limitations to the current VSD imaging approach discussed in this review: (1) VSD signal amplitudes are small, (2) membranes are nonspecifically stained, and (3) the technique is restricted to the supragranular layers of the neocortex. These limitations mean that it is not currently possible to image the membrane potential of a network of neurons *in vivo* with cellular or subcellular resolution. Equally, right now it is not possible to differentiate between membrane potential changes in different subtypes of neurons, for example, GABAergic neurons versus excitatory neurons. Furthermore, it would clearly be very interesting to investigate the spatiotemporal dynamics of membrane potential changes in deep brain structures and to correlate them with the complex dynamics observed in the neocortex (Figure 6.6).

Fortunately, over the past years, there have been a number of technical advances that offer hope for improvements to *in vivo* imaging, which may allow *in vivo* imaging of subcellular compartments, specific neuronal subtypes and deep brain structures.

6.7 FUTURE PERSPECTIVES

6.7.1 New Dyes and Nonlinear Optics

The development of RH1691, RH1692, and RH1838 operating at long wavelengths[12] has helped to increase the signal-to-noise ratio by reducing the contribution of heartbeat-related artifacts. However, the actual signal amplitude remains very small, usually well below 1% $\Delta F/F_0$. Calcium-sensitive dyes in comparison offer signals that are 100 times bigger. These large signals from calcium dyes, together with the development of methods for high-resolution two-photon imaging[55] and loading populations of cells,[56] have allowed *in vivo* calcium imaging of networks with single cell resolution.[48,57–61] However, the calcium dye signals report almost exclusively action potential activity, and the subthreshold membrane potential dynamics remain hidden from this technique.[27] Increasing the signal amplitude for VSDs could therefore result in dramatic advances in our understanding of brain function.

The most obvious approach is simply to keep synthesizing new dyes, and by persistent screening one might find dyes with better characteristics than the currently used ones.[62,63] An equally promising route has been to excite the VSD with laser light, in order to optimize the excitation wavelength and therefore to obtain larger signal amplitudes.[62] So far, these approaches have yielded moderate but significant improvements for *in vitro* imaging that may in the future also contribute to improving *in vivo* imaging.

An exciting prospect is to use nonlinear optical phenomenon to enhance VSD signals. With the increasing availability of easy-to-use pulsed laser sources, multiphoton microscopy is becoming a widely used technology for *in vivo* imaging. The most important advantage offered by two-photon microscopy is the inherent optical sectioning resulting from the need for near-simultaneous absorption of two photons per excitation event, which in practice occurs primarily in the focal plane. In contrast to water soluble dyes, VSDs insert into the plasma membranes and are therefore oriented in a specific direction. All VSD molecules on a given patch of membrane are oriented in the same direction, and this can contribute to an interesting nonlinear phenomenon termed *second harmonic generation* (SHG). In the SHG process, two photons are near simultaneously absorbed as for two-photon fluorescence. However, whereas in fluorescence the excited molecule emits a longer wavelength photon, in SHG the two photons are converted into a single photon with twice the energy. The SHG signal depends upon the hyperpolarizability of the VSD, enhanced by resonance of the excitation electromagnetic field, in addition to the electric field experienced by the VSD molecule. Recent SHG measurements show promising results for high-resolution VSD imaging using the steryl dye FM4-64.[64,65]

6.7.2 Voltage-Sensitive Fluorescent Proteins

Increasing the signal amplitudes of fluorescent dyes will be extremely helpful for improving the resolution of VSD measurements; however, if membranes are nonspecifically labeled, then we will still not know to which neurons the VSD signal relates. Neurons have very fine intermingled axonal and dendritic processes, and glial cells also have extremely intricate membranes. If all membranes are similarly labeled, then it will not be possible to distinguish nearby fluorescently labeled dendrites, axons, or glia. One interesting approach would be to label single neurons by dye intracellular injections, but this approach limits the possibilities for investigating network activity in many neurons. Labeling of specific cells types in the brain has been accomplished most elegantly by genetic methods. Through cell-type-specific promoters and advanced combinatorial genetic strategies of gene regulation, it has become possible to label different populations of neurons with fluorescent proteins. For example, in one mouse line, all GABAergic neurons express the green fluorescent protein (GFP).[66] A great deal of hope for the future development of voltage imaging has therefore been placed in the development of genetically encoded voltage-sensitive fluorescent proteins.

The first genetically encoded indicators responded to changes in calcium concentration,[67] and since then, a number of improved genetically encoded calcium-sensitive indicators (GECIs) have been developed and found to be functional in the mammalian brain.[68–70] Fluorescent proteins sensitive to changes in membrane potential have also been developed, most of which are based upon GFP attached to different voltage-gated ion channels.[71–73] So far these approaches have not been successful in the mammalian brain, mainly because the proteins are not trafficked efficiently to the plasma membrane.[74] However, renewed hope now comes from coupling a voltage-sensitive phosphatase to GFP.[75]

6.7.3 Fiber Optic VSD Imaging

The ability to image through fiber optics[24,76,77] has two major advantages over conventional imaging. Firstly, the animal is free to move around, and such measurements are therefore inherently more physiological. However, this advantage is

31. Sharon, D., Jancke, D., Chavane, F., Na'aman, S., and Grinvald, A. 2007. Cortical response field dynamics in cat visual cortex. *Cereb. Cortex* 17: 2866–2877.
32. Wallace, D.J., and Sakmann, B. 2008. Plasticity of representational maps in somatosensory cortex observed by in vivo voltage-sensitive dye imaging. *Cereb. Cortex* 18: 1361–1373.
33. Xu, W., Huang, X., Takagaki, K., and Wu, J.Y. 2007. Compression and reflection of visually evoked cortical waves. *Neuron* 55: 119–129.
34. Yang, Z., Heeger, D.J., and Seidemann, E. 2007. Rapid and precise retinotopic mapping of the visual cortex obtained by voltage-sensitive dye imaging in the behaving monkey. *J. Neurophysiol.* 98: 1002–1014.
35. Chen, Y., Geisler, W.S., and Seidemann, E. 2008. Optimal temporal decoding of neural population responses in a reaction-time visual detection task. *J. Neurophysiol.* 99: 1366–1379.
36. Takagaki, K., Zhang, C., Wu, J.Y., and Lippert, M.T. 2008. Crossmodal propagation of sensory-evoked and spontaneous activity in the rat neocortex. *Neurosci. Lett.* 431: 191–196.
37. Yoshida, T., Sakagami, M., Katura, T., Yamazaki, K., Tanaka, S., Iwamoto, M., and Tanaka, N. 2008. Anisotropic spatial coherence of ongoing and spontaneous activities in auditory cortex. *Neurosci. Res.* 61: 49–55.
38. Nauhaus, I., Benucci, A., Carandini, M., and Ringach, D.L. 2008. Neuronal selectivity and local map structure in visual cortex. *Neuron* 57: 673–679.
39. Zhang, J., Davidson, R.M., Wei, M.D., and Loew, L.M. 1998. Membrane electric properties by combined patch clamp and fluorescence ratio imaging in single neurons. *Biophys. J.* 74: 48–53.
40. Ratzlaff, E.H., and Grinvald, A. 1991. A tandem-lens epifluorescence macroscope: Hundred-fold brightness advantage for wide-field imaging. *J. Neurosci. Methods* 36: 127–137.
41. Petersen, C.C.H. 2007. The functional organization of the barrel cortex. *Neuron* 56: 339–355.
42. Moore, C.I., and Nelson, S.B. 1998. Spatio-temporal subthreshold receptive fields in the vibrissa representation of rat primary somatosensory cortex. *J. Neurophysiol.* 80: 2882–2892.
43. Zhu, J.J., and Connors, B.W. 1999. Intrinsic firing patterns and whiskers-evoked synaptic responses of neurons in the rat barrel cortex. *J. Neurophysiol.* 81: 1171–1183.
44. Brecht, M., Roth, A., and Sakmann, B. 2003. Dynamic receptive fields of reconstructed pyramidal cells in layers 3 and 2 of rat somatosensory barrel cortex. *J. Physiol.* 553: 243–265.
45. Steriade, M., Nunez, A., and Amzica, F. 1993. A novel slow (<1 Hz) oscillation of neocortical neurons in vivo: Depolarizing and hyperpolarizing components. *J. Neurosci.* 13: 3252–3265.
46. Cowan, R.L., and Wilson, C.J. 1994. Spontaneous firing patterns and axonal projections of single corticostriatal neurons in the rat medial agranular cortex. *J. Neurophysiol.* 71: 17–32.
47. Sachdev, R.N., Ebner, F.F., and Wilson, C.J. 2004. Effect of subthreshold up and down states on the whisker-evoked response in somatosensory cortex. *J. Neurophysiol.* 92, 3511–3521.
48. Kerr, J.N., Greenberg, D., and Helmchen, F. 2005. Imaging input and output of neocortical networks in vivo. *Proc. Natl. Acad. Sci. USA* 102: 14063–14068.
49. Lampl, I., Reichova, I., and Ferster, D. 1999. Synchronous membrane potential fluctuations in neurons of the cat visual cortex. *Neuron* 22: 361–374.

50. Volgushev, M., Chauvette, S., Mukovski, M., and Timofeev, I. 2006. Precise long-range synchronization of activity and silence in neocortical neurons during slow-wave oscillations. *J. Neurosci.* 26: 5665–5672.
51. de Kock, C.P., Bruno, R.M., Spors, H., and Sakmann, B. 2007. Layer- and cell-type-specific suprathreshold stimulus representation in rat primary somatosensory cortex. *J. Physiol.* 581: 139–154.
52. Vincent, J.L., Patel, G.H., Fox, M.D., Snyder, A.Z., Baker, J.T., Van Essen, D.C., Zempel, J.M., Snyder, L.H., Corbetta, M., and Raichle, M.E. 2007. Intrinsic functional architecture in the anaesthetized monkey brain. *Nature* 447: 83–86.
53. Gilbert, C.D. and Sigman, M. 2007. Brain states: Top-down influences in sensory processing. *Neuron* 54: 677–696.
54. Crochet, S., and Petersen, C.C.H. 2006. Correlating whisker behavior with membrane potential in barrel cortex of awake mice. *Nat. Neurosci.* 9: 608–610.
55. Denk, W., Strickler, J.H., and Webb, W.W. 1990. Two-photon laser scanning fluorescence microscopy. *Science* 248: 73–76.
56. Stosiek, C., Garaschuk, O., Holthoff, K., and Konnerth, A. 2003. In vivo two-photon calcium imaging of neuronal networks. *Proc. Natl. Acad. Sci. USA* 100: 7319–7324.
57. Ohki, K., Chung, S., Ch'ng, Y.H., Kara, P., and Reid, R.C. 2005. Functional imaging with cellular resolution reveals precise micro-architecture in visual cortex. *Nature* 433: 597–603.
58. Ohki, K., Chung, S., Kara, P., Hübener, M., Bonhoeffer, T., and Reid, R.C. 2006. Highly ordered arrangement of single neurons in orientation pinwheels. *Nature* 442: 925–928.
59. Kerr, J.N., de Kock, C.P., Greenberg, D.S., Bruno, R.M., Sakmann, B., and Helmchen, F. 2007. Spatial organization of neuronal population responses in layer 2/3 of rat barrel cortex. *J. Neurosci.* 27: 13316–13328.
60. Dombeck, D.A., Khabbaz, A.N., Collman, F., Adelman, T.L., and Tank, D.W. 2007. Imaging large scale neural activity with cellular resolution in awake mobile mice. *Neuron* 56: 43–57.
61. Sato, T.R., Gray, N.W., Mainen, Z.F., and Svoboda, K. 2007. The functional microarchitecture of the mouse barrel cortex. *PLoS Biol.* 5: e189.
62. Kuhn, B., Fromherz, P., and Denk, W. 2004 High sensitivity of stark-shift voltage-sensing dyes by one- or two-photon excitation near the red spectral edge, *Biophys. J.* 87: 631–639.
63. Zhou, W.L., Yan, P., Wuskell, J.P., Loew, L.M., and Antic, S.D. 2007. Intracellular long-wavelength voltage-sensitive dyes for studying the dynamics of action potentials in axons and thin dendrites. *J. Neurosci. Methods* 164: 225–239.
64. Dombeck, D.A., Sacconi, L., Blanchard-Desce, M., and Webb, W.W. 2005. Optical recording of fast neuronal membrane potential transients in acute mammalian brain slices by second-harmonic generation microscopy. *J. Neurophysiol.* 94: 3628–3636.
65. Nuriya, M., Jiang, J., Nemet, B., Eisenthal, K.B., and Yuste, R. 2006. Imaging membrane potential in dendritic spines. *Proc. Natl. Acad. Sci. USA* 103: 786–790.
66. Tamamaki, N., Yanagawa, Y., Tomioka, R., Miyazaki, J., Obata, K., and Kaneko, T. 2003. Green fluorescent protein expression and colocalization with calretinin, parvalbumin, and somatostatin in the GAD67-GFP knock-in mouse. *J. Comp. Neurol.* 467: 60–79.
67. Miyawaki, A., Llopis, J., Heim, R., McCaffery, J.M., Adams, J.A., Ikura, M., and Tsien, R.Y. 1997. Fluorescent indicators for Ca2+ based on green fluorescent proteins and calmodulin. *Nature* 388: 882–887.
68. Hasan, M.T., Friedrich, R.W., Euler, T., Larkum, M.E., Giese, G., Both, M., Duebel, J., Waters, J., Bujard, H., Griesbeck, O., Tsien, R.Y., Nagai, T., Miyawaki, A., and Denk, W. 2004. Functional fluorescent Ca2+ indicator proteins in transgenic mice under TET control. *PLoS Biol.* 2: e163.

69. Díez-García, J., Matsushita, S., Mutoh, H., Nakai, J., Ohkura, M., Yokoyama, J., Dimitrov, D., and Knöpfel, T. 2005. Activation of cerebellar parallel fibers monitored in transgenic mice expressing a fluorescent Ca2+ indicator protein. *Eur. J. Neurosci.* 22: 627–635.
70. Heim, N., Garaschuk, O., Friedrich, M.W., Mank, M., Milos, R.I., Kovalchuk, Y., Konnerth, A., and Griesbeck, O. 2007. Improved calcium imaging in transgenic mice expressing a troponin C-based biosensor. *Nat. Methods* 4: 127–129.
71. Siegel, M.S., and Isacoff, E.Y. 1997. A genetically encoded optical probe of membrane voltage. *Neuron* 19: 735–741.
72. Sakai, R., Repunte-Canonigo, V., Raj, C.D., and Knopfel, T. 2001. Design and characterization of a DNA-encoded, voltage-sensitive fluorescent protein. *Eur. J. Neurosci.* 13: 2314–2318.
73. Ataka, K., and Pieribone, V.A. 2002. A genetically targetable fluorescent probe of channel gating with rapid kinetics. *Biophys. J.* 82: 509–516.
74. Baker, B.J., Lee, H., Pieribone, V.A., Cohen, L.B., Isacoff, E.Y., Knopfel, T., and Kosmidis, E.K. 2007. Three fluorescent protein voltage sensors exhibit low plasma membrane expression in mammalian cells. *J. Neurosci. Methods* 161: 32–38.
75. Dimitrov, D., He, Y., Mutoh, H., Baker, B.J., Cohen, L., Akemann, W., and Knöpfel, T. 2007. Engineering and characterization of an enhanced fluorescent protein voltage sensor. *PLoS ONE* 2: e440.
76. Helmchen, F., Fee, M.S., Tank, D.W., and Denk, W. 2001. A miniature head-mounted two-photon microscope: High resolution brain imaging in freely moving animals. *Neuron* 31: 903–912.
77. Flusberg, B.A., Cocker, E.D., Piyawattanametha, W., Jung, J.C., Cheung, E.L., and Schnitzer, M.J. 2005. Fiber-optic fluorescence imaging. *Nat. Methods* 2: 941–950.

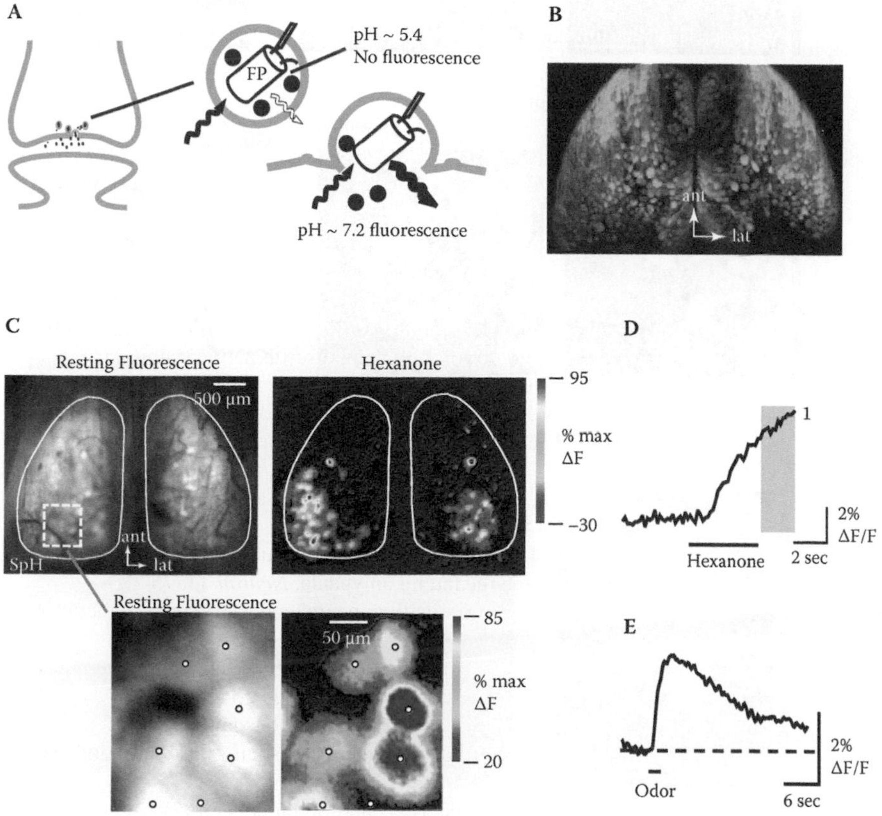

A
pH ~ 5.4
No fluorescence
FP
pH ~ 7.2 fluorescence
B
ant
lat
C
Resting Fluorescence
500 µm
ant
lat
SpH
Hexanone
95
% max
ΔF
–30
Resting Fluorescence
50 µm
85
% max
ΔF
20
D
1
2%
ΔF/F
Hexanone
2 sec
E
Odor
2%
ΔF/F
6 sec

FIGURE 1.2

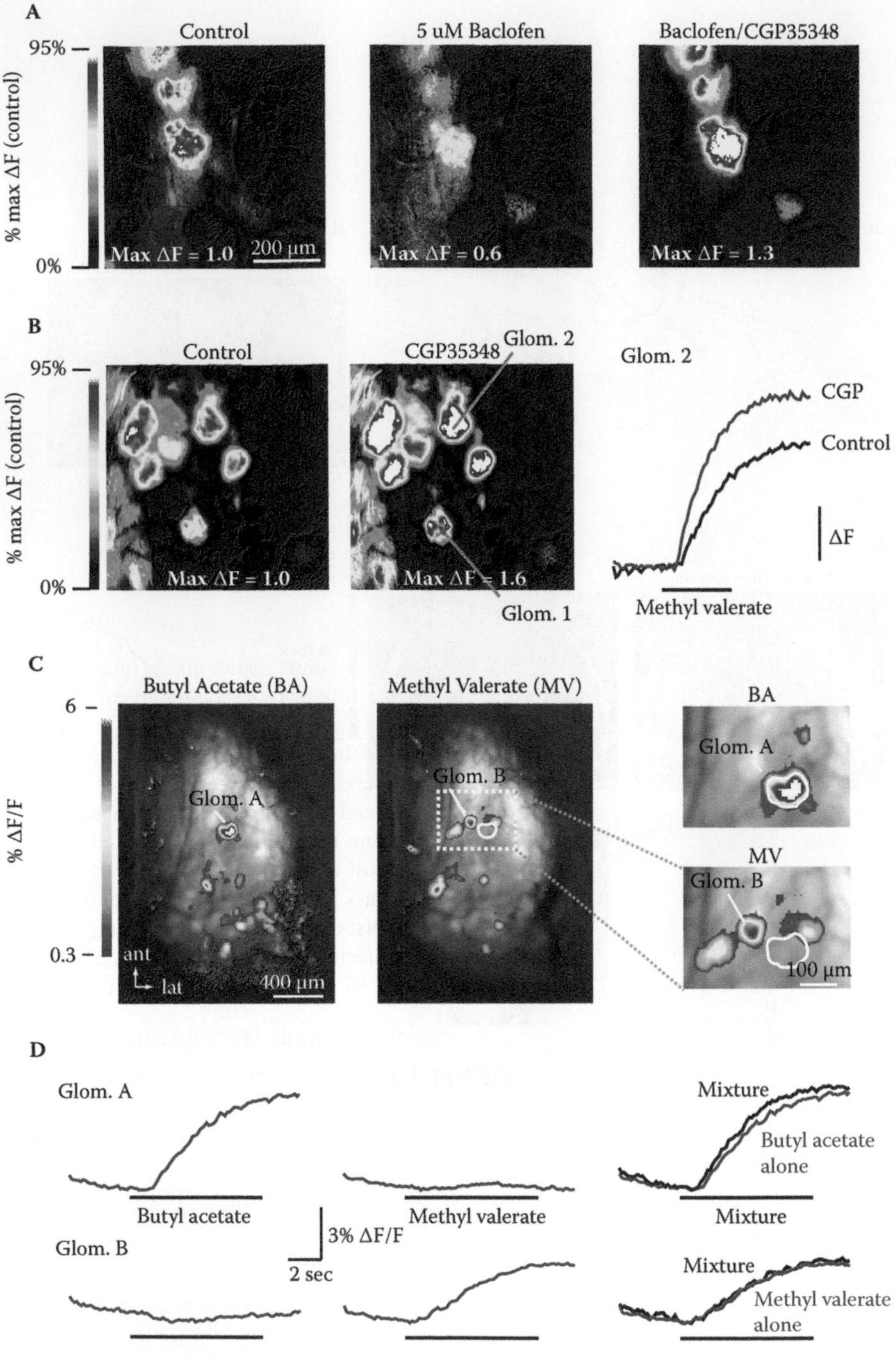
A
Control
5 uM Baclofen
Baclofen/CGP35348
95%
% max ΔF (control)
0%
Max ΔF = 1.0
200 μm
Max ΔF = 0.6
Max ΔF = 1.3
B
Control
CGP35348
Glom. 2
Glom. 2
95%
% max ΔF (control)
0%
Max ΔF = 1.0
Max ΔF = 1.6
Glom. 1
CGP
Control
ΔF
Methyl valerate
C
Butyl Acetate (BA)
Methyl Valerate (MV)
6
% ΔF/F
0.3
Glom. A
Glom. B
ant
lat
400 μm
BA
Glom. A
MV
Glom. B
100 μm
D
Glom. A
Butyl acetate
Methyl valerate
Mixture
Butyl acetate alone
Mixture
3% ΔF/F
2 sec
Glom. B
Mixture
Methyl valerate alone

FIGURE 1.11

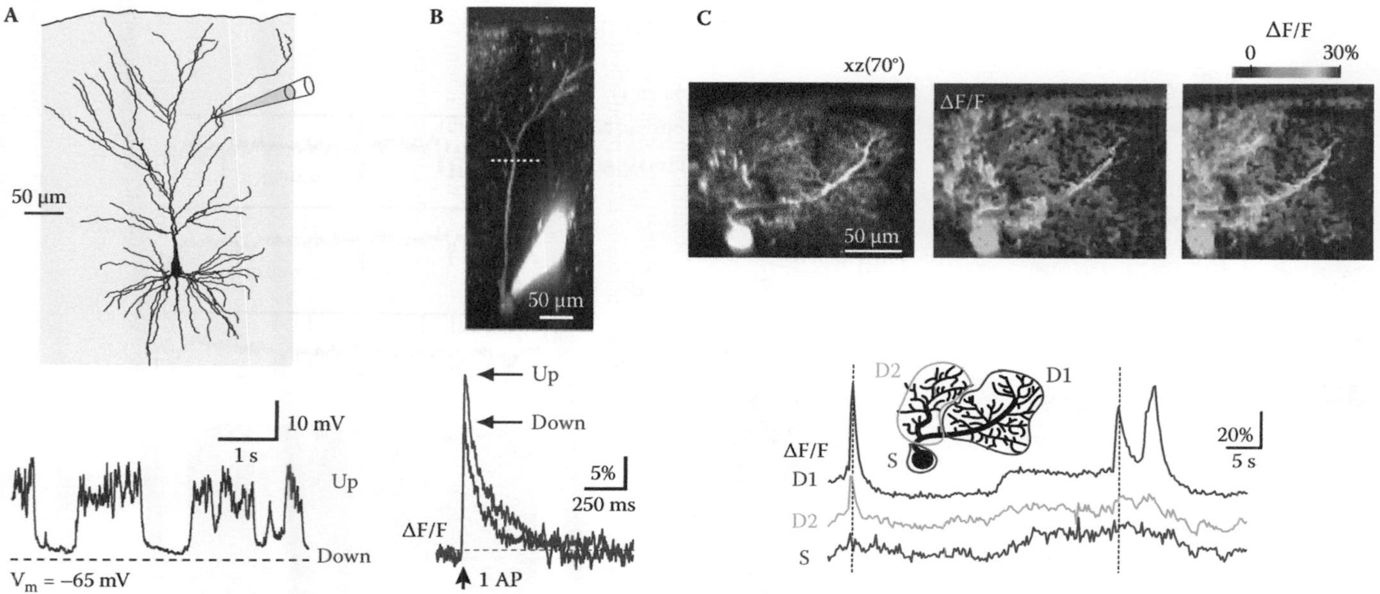

A
50 µm
10 mV
1 s
Up
Down
Vm = –65 mV
B
50 µm
Up
Down
5%
250 ms
ΔF/F
1 AP
C
xz(70°)
50 µm
ΔF/F
ΔF/F
0
30%
D2
D1
S
ΔF/F
D1
D2
S
20%
5 s

FIGURE 2.5

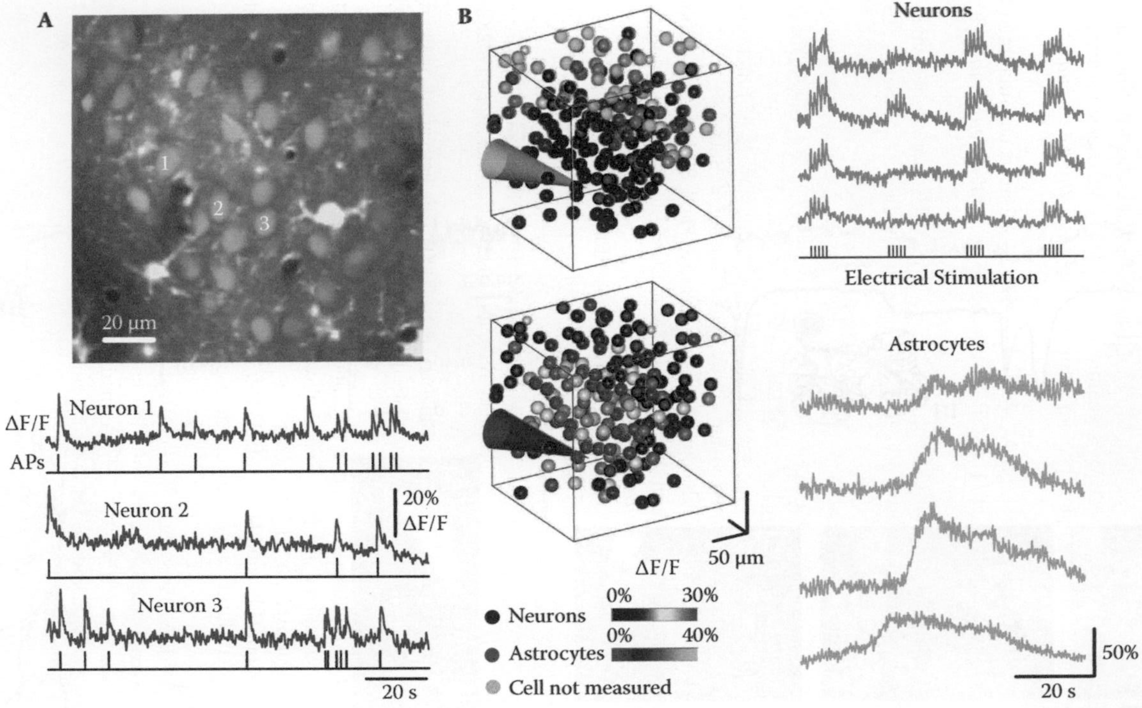
A
20 μm
1
2
3
Neuron 1
ΔF/F
APs
Neuron 2
20%
ΔF/F
Neuron 3
20 s
B
50 μm
ΔF/F
0%
30%
Neurons
0%
40%
Astrocytes
Cell not measured
Neurons
Electrical Stimulation
Astrocytes
50%
20 s

FIGURE 2.6

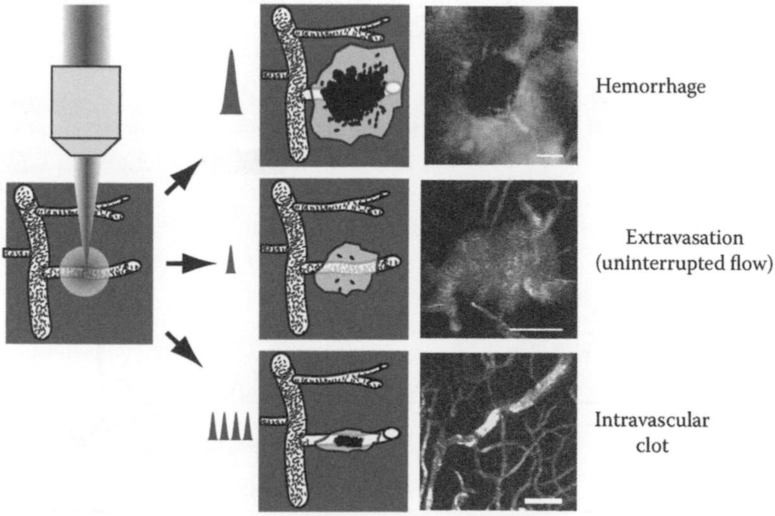

(a) Targeted subsurface microvascular disruption

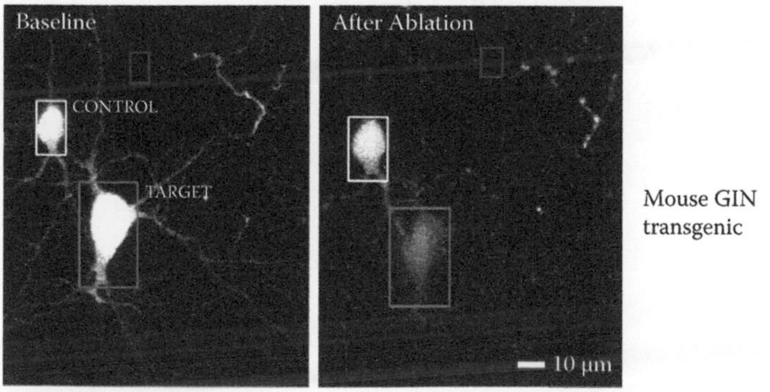

(b) Targeted somatic ablation

FIGURE 3.30

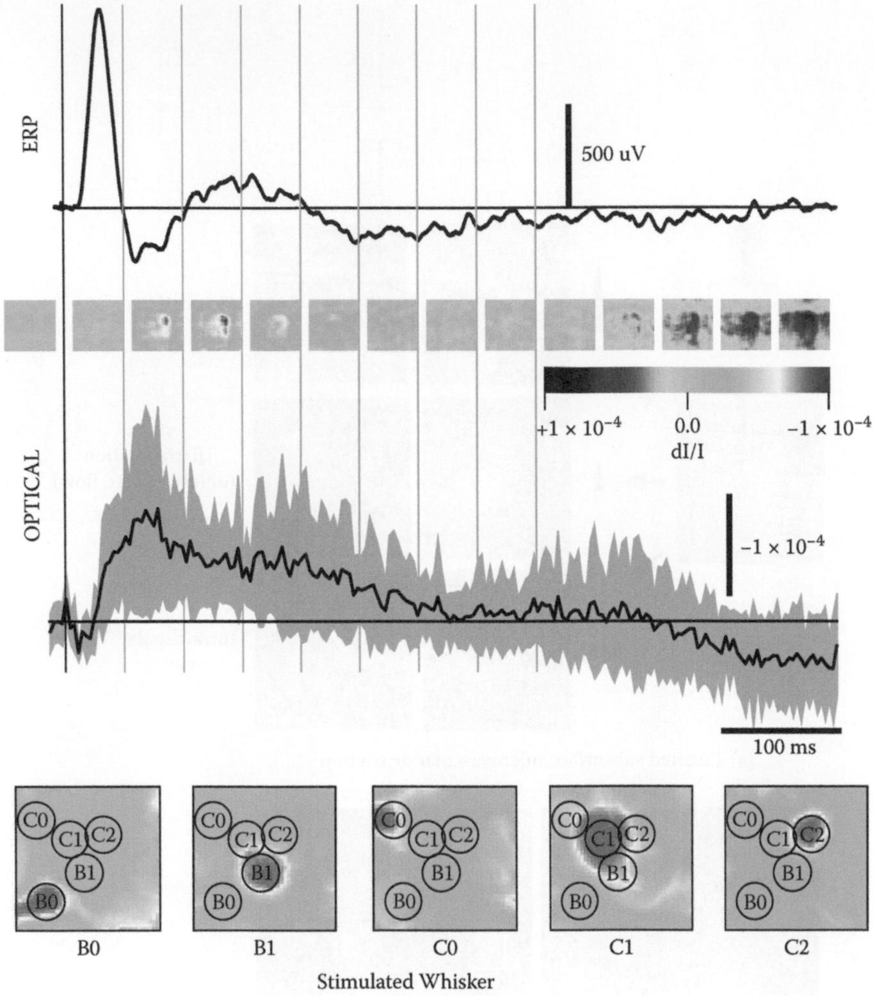

ERP
500 uV
+1 × 10−4
0.0
−1 × 10−4
dI/I
OPTICAL
−1 × 10−4
100 ms
C0
C1
C2
B1
B0
B0
B1
C0
C1
C2
Stimulated Whisker

FIGURE 5.6

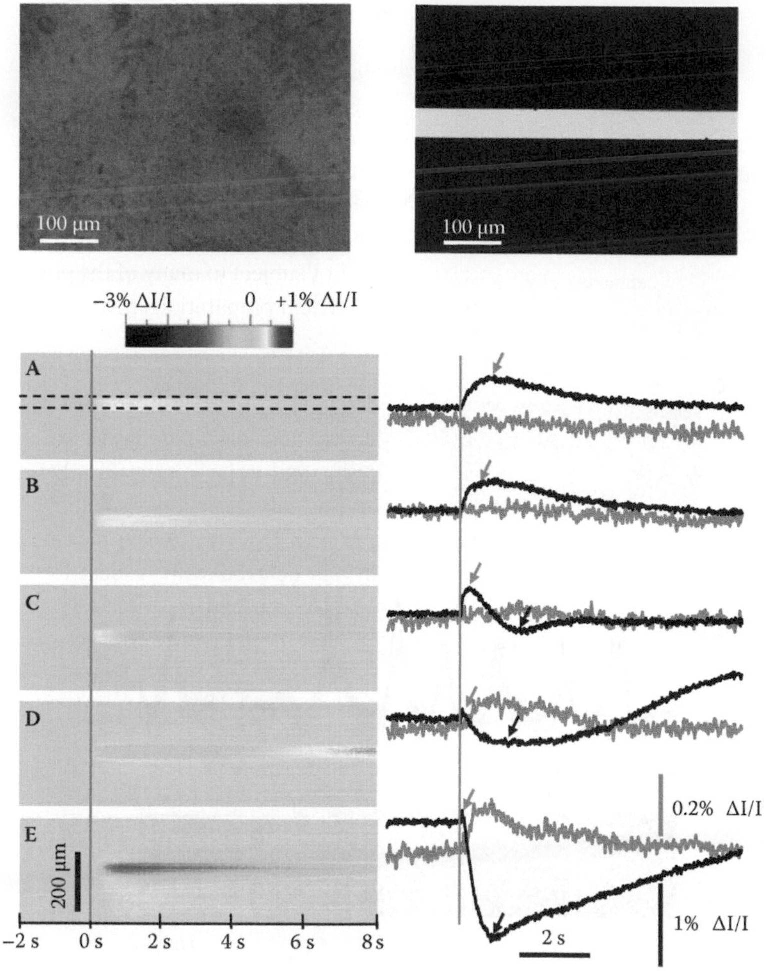
100 μm
100 μm
−3% ΔI/I
0
+1% ΔI/I
A
B
C
D
E
200 μm
−2 s
0 s
2 s
4 s
6 s
8 s
2 s
0.2% ΔI/I
1% ΔI/I

FIGURE 5.9

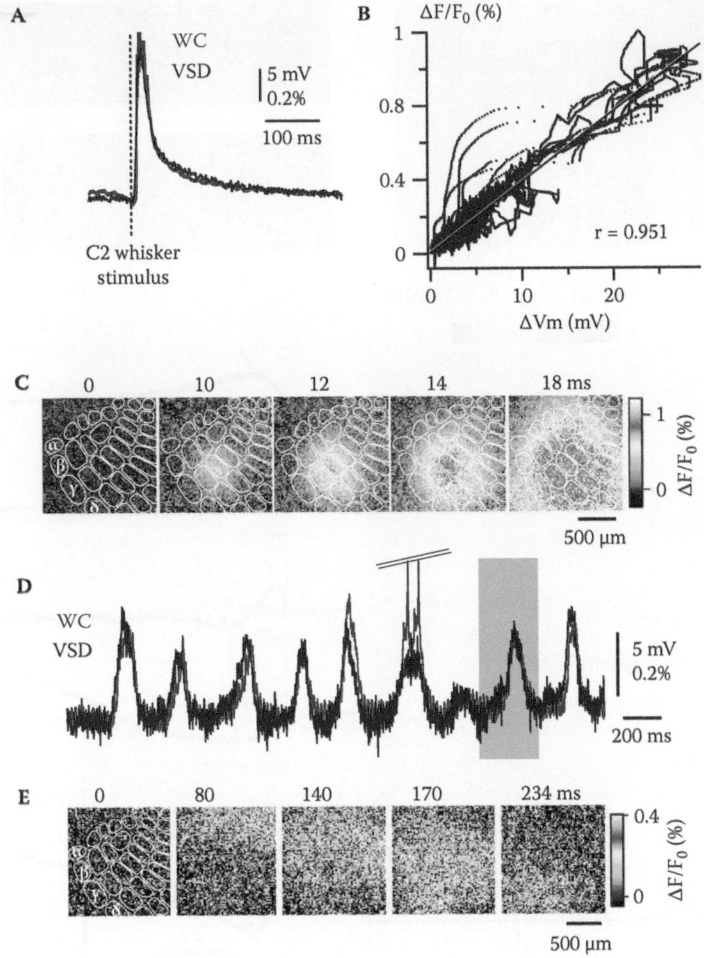
A
WC
VSD
5 mV
0.2%
100 ms
C2 whisker stimulus
B
ΔF/F0 (%)
1
0.8
0.4
0
r = 0.951
0
10
20
ΔVm (mV)
C
0
10
12
14
18 ms
1
0
ΔF/F0 (%)
500 μm
D
WC
VSD
5 mV
0.2%
200 ms
E
0
80
140
170
234 ms
0.4
0
ΔF/F0 (%)
500 μm

FIGURE 6.4

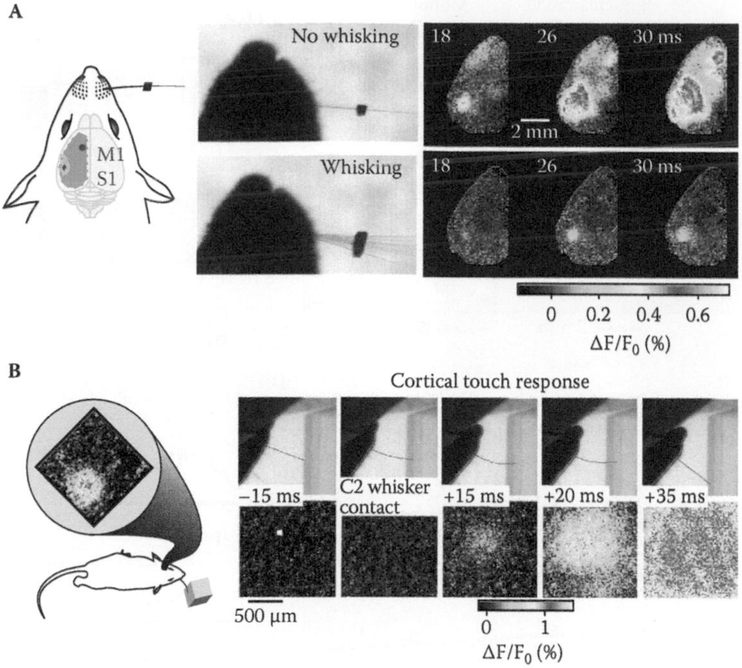
A
No whisking
Whisking
M1
S1
18
26
30 ms
2 mm
0
0.2
0.4
0.6
ΔF/F0 (%)
B
Cortical touch response
−15 ms
C2 whisker contact
+15 ms
+20 ms
+35 ms
500 μm
0
1
ΔF/F0 (%)

FIGURE 6.6

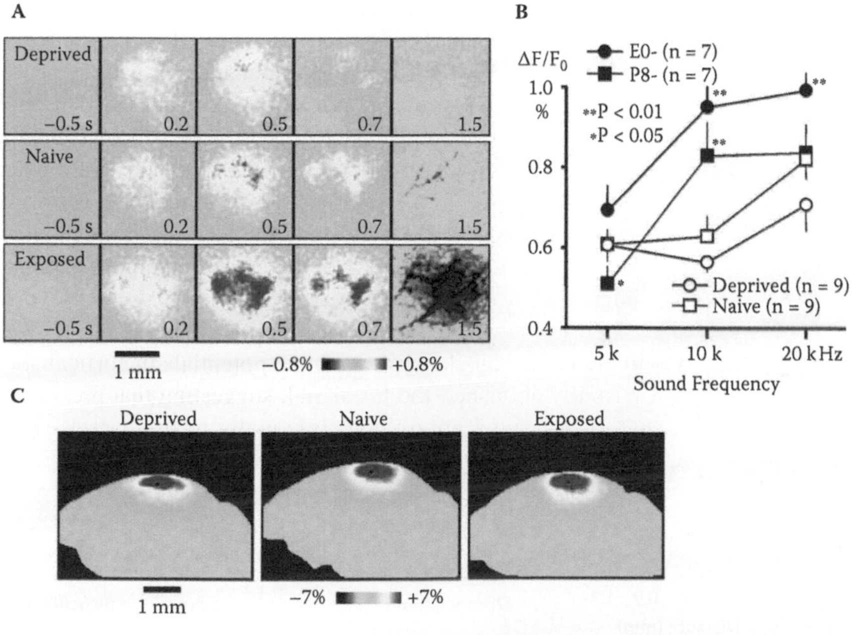
A
Deprived
Naive
Exposed
−0.5 s
0.2
0.5
0.7
1.5
1 mm
−0.8%
+0.8%
B
ΔF/F0
%
E0- (n = 7)
P8- (n = 7)
**P < 0.01
*P < 0.05
1.0
0.8
0.6
0.4
Deprived (n = 9)
Naive (n = 9)
5 k
10 k
20 kHz
Sound Frequency
C
Deprived
Naive
Exposed
1 mm
−7%
+7%

FIGURE 7.3

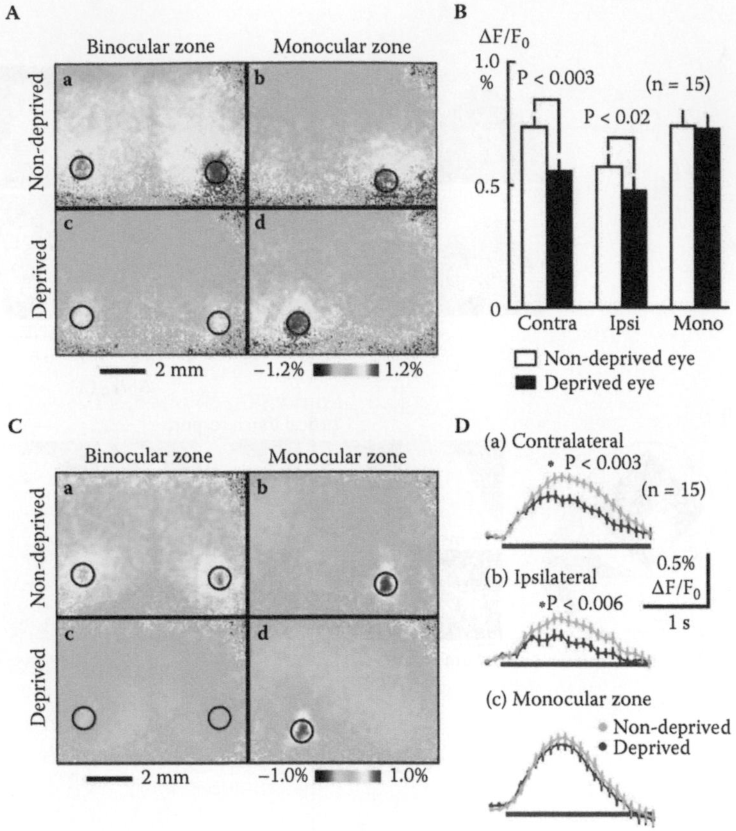

FIGURE 7.5

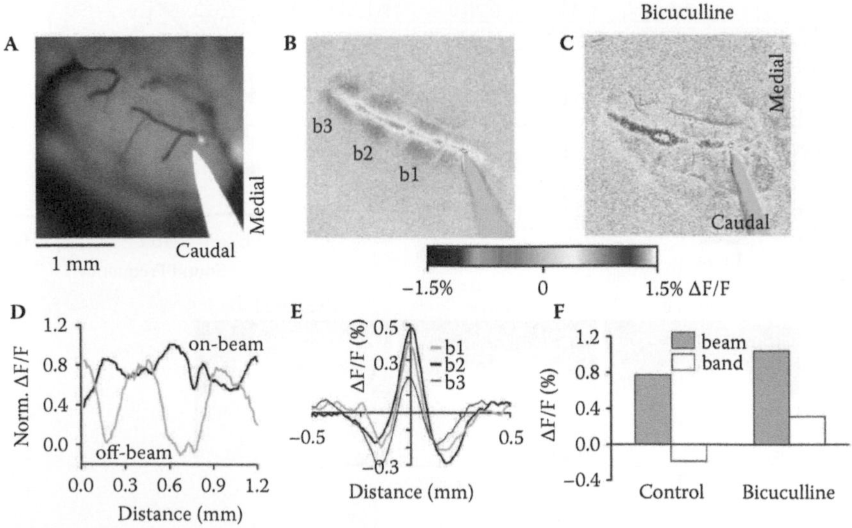

FIGURE 8.9

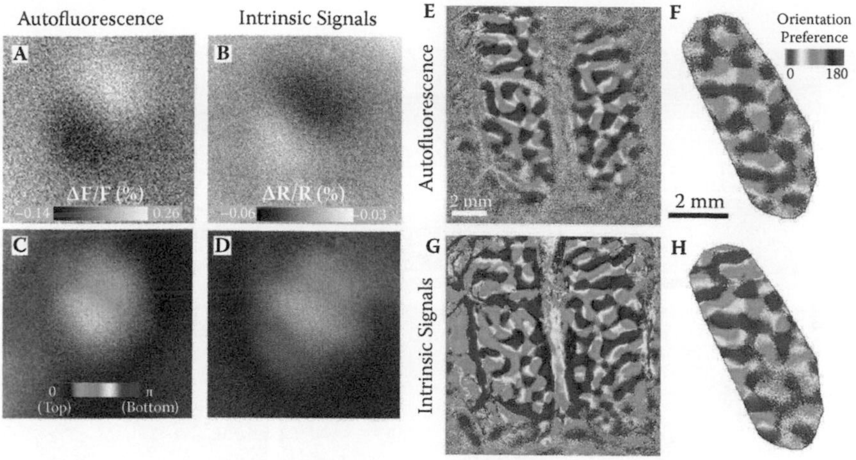

FIGURE 8.10

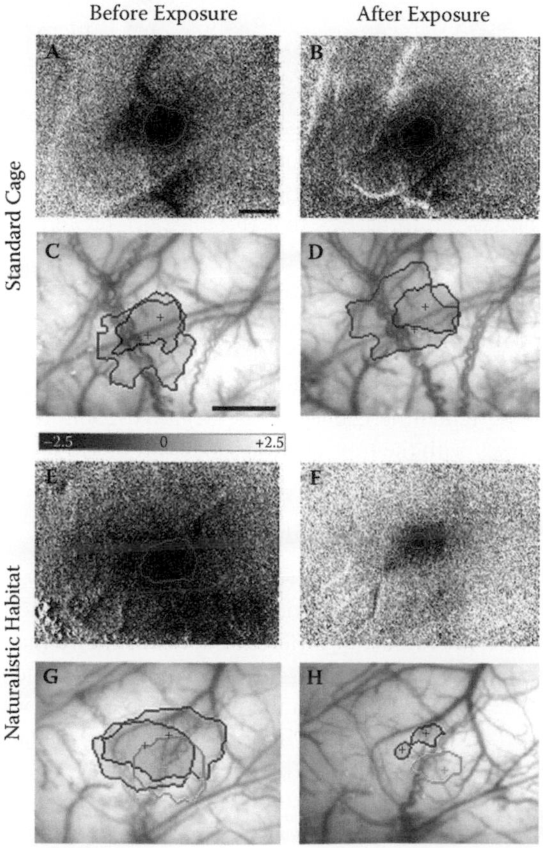

FIGURE 9.11

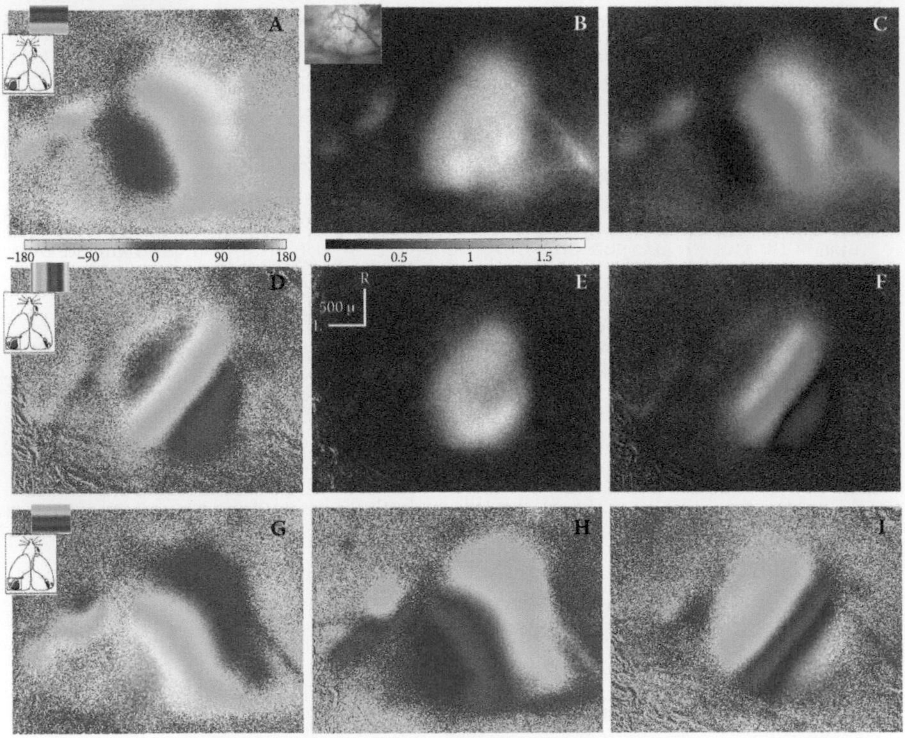

FIGURE 10.4

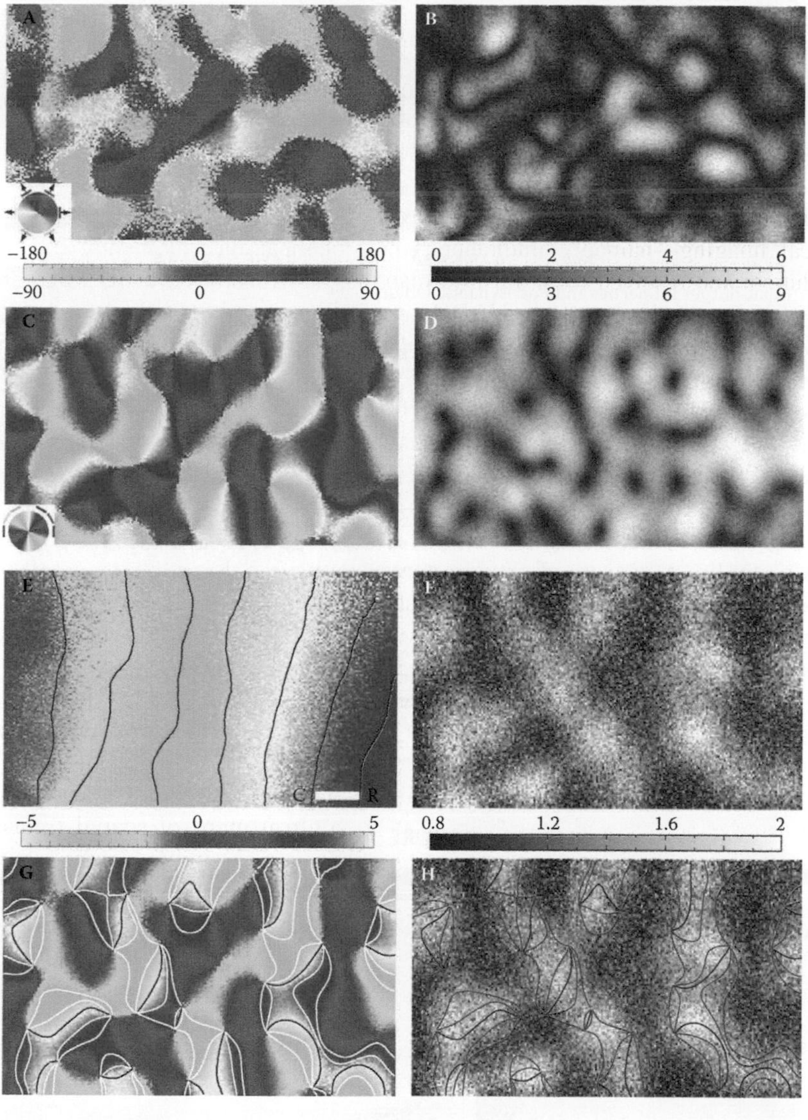

FIGURE 10.5

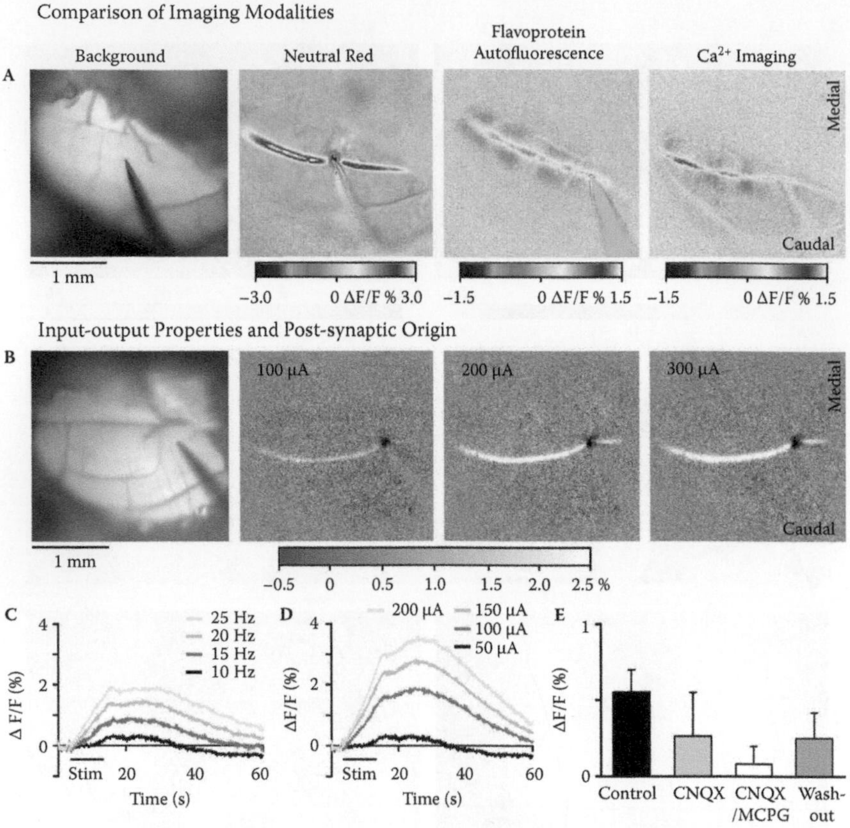

FIGURE 11.4

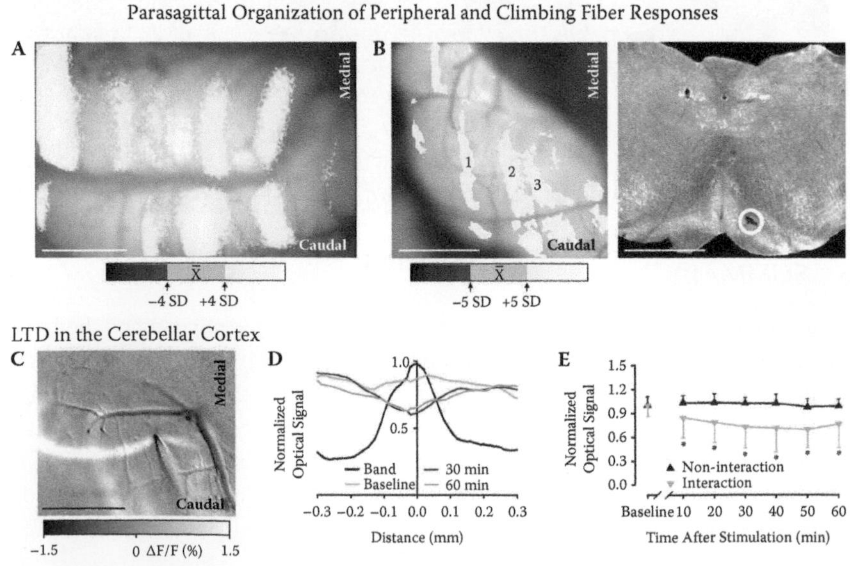

FIGURE 11.5

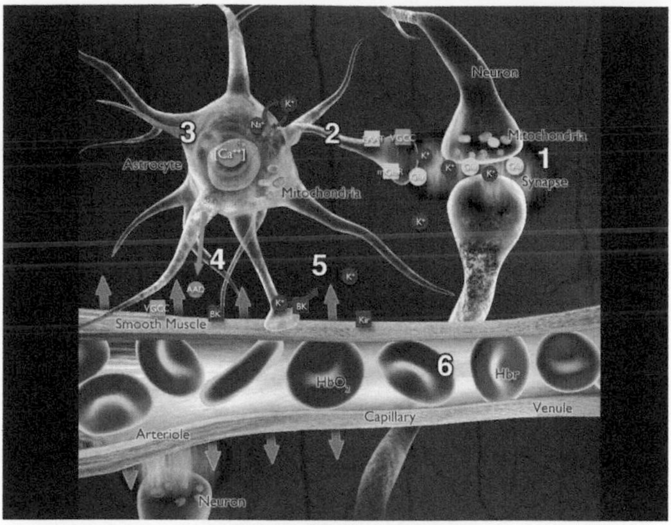

FIGURE 13.1

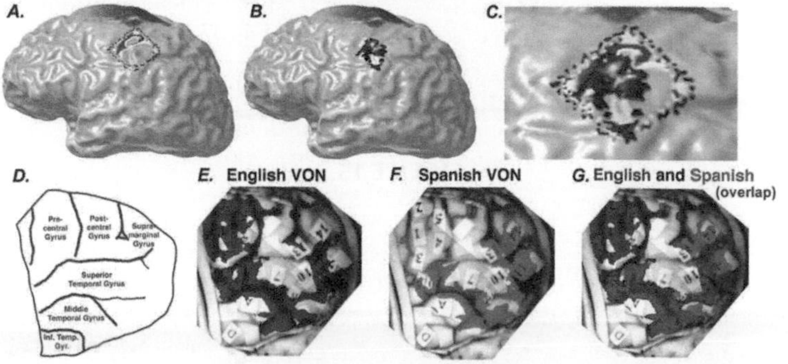

FIGURE 13.6

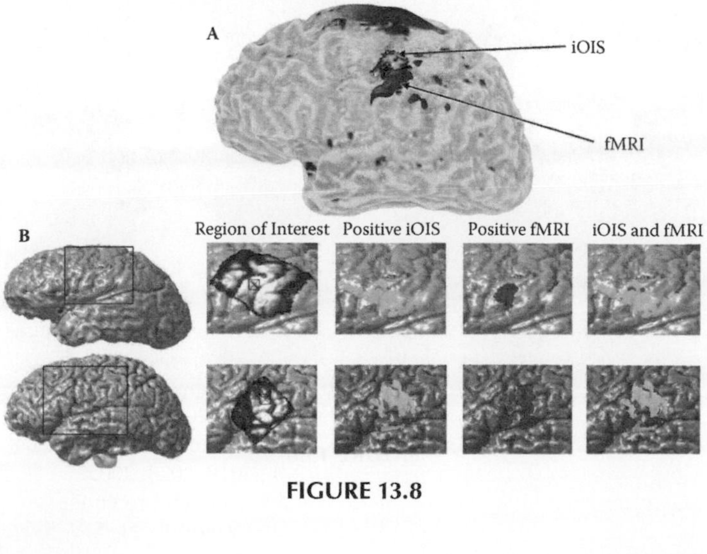

FIGURE 13.8

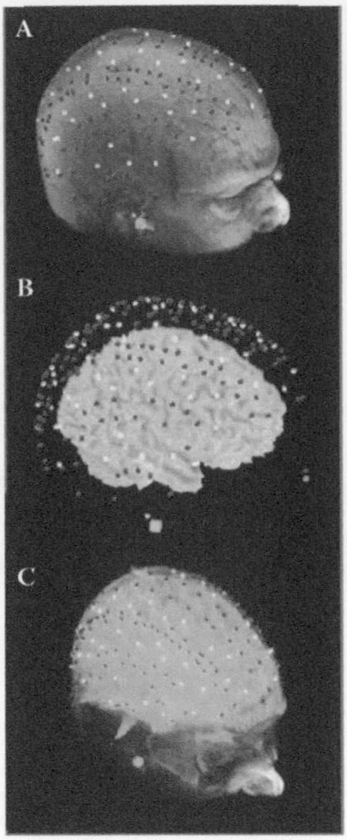

FIGURE 15.3

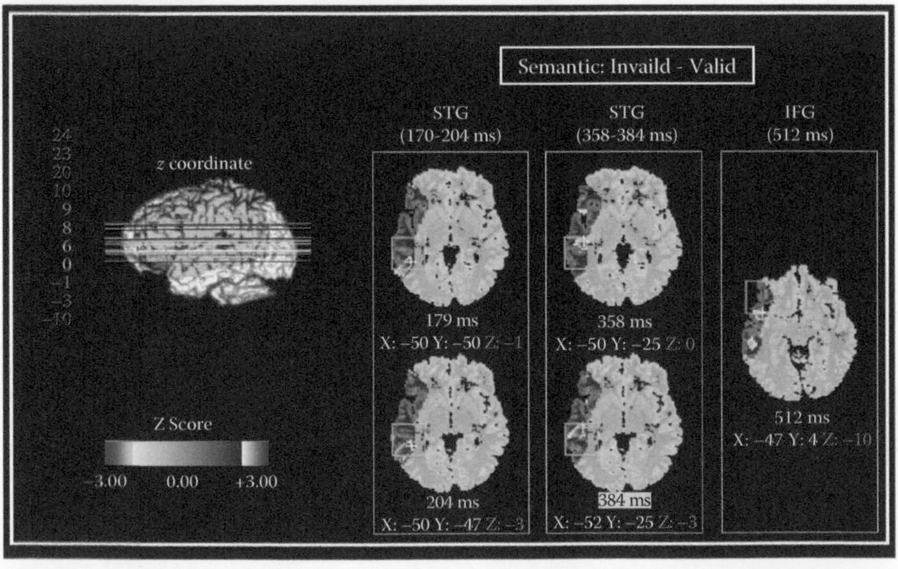

FIGURE 15.6

7 Flavoprotein Fluorescence Imaging of Experience-Dependent Cortical Plasticity in Rodents

Katsuei Shibuki, Ryuichi Hishida, Manavu Tohmi, Kuniyuki Takahashi, Hiroki Kitaura, and Yamato Kubota

CONTENTS

7.1 INTRODUCTION

In this chapter, flavoprotein fluorescence imaging is described, with a focus on transcranial imaging for investigating experience-dependent plasticity in mice. The molecular mechanisms underlying higher brain functions are important subjects of neuroscience research. The cerebral cortex, which is expected to have essential roles in higher functions, has mainly been investigated in primates. However, the analytical methods for elucidating molecular mechanisms in the primate brain are rather limited. An alternative approach is to investigate higher cortical functions in rodents. In particular, mice are very useful for investigating molecular mechanisms, because many genetically manipulated strains are available. However, the cortical functions of rodents can not be studied intensively, partly because the cortex is fragile, and electrophysiological analysis is sometimes difficult. In contrast, mice have thin skulls that are transparent enough to allow transcranial imaging of the cortical activities through the intact skull (Schuett et al. 2002, Shibuki et al. 2007). Auditory, visual, and somatosensory cortices are visible through the intact skull (Figure 7.1A). Transcranial analysis of cortical functions has a number of technical merits: the operation to remove the skull is omitted so that surgical damages on the cortex can be avoided, and the fragile cortex of mice is kept intact within the skull during the recording experiments. Optical imaging of cortical activities usually requires experimental skills, whereas the transcranial imaging of the mouse cortical activities may be performed even by beginner neuroscientists or students.

In transcranial analysis of cortical functions, application of exogenous indicators, such as voltage-sensitive dyes or calcium indicators, is difficult. Transcranial imaging may be performed in mice that express with protein sensors derived from GFP (McGann et al. 2005, Diez-Garcia et al. 2007). Alternatively, it may be performed using some intrinsic signals reflecting cortical activities. Most of intrinsic signals reflecting brain activities are coupled with activity-dependent facilitation of aerobic energy metabolism (Fein and Tsacopoulos 1988, Shibuki 1989, 1990, Vanzetta and

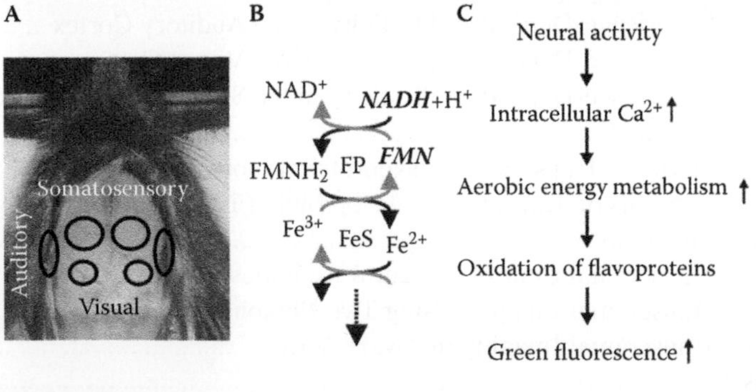

FIGURE 7.1 Transcranial imaging of cortical activities. (A) Transcranial observation of the cortical surface in an anesthetized mouse. (B) Endogenous fluorescence substances in the electron transfer system in mitochondria. The oxidized flavoproteins (FMN) and NADH are fluorescent, whereas the reduced flavoproteins ($FMNH_2$) and NAD^+ are not. (C) Coupling between neural activities and changes in endogenous green fluorescence.

Grinvald 1999). One of the results of facilitated energy metabolism is that oxyhemoglobin in the capillary is converted to deoxyhemoglobin, and that the light absorption properties of hemoglobin are changed (Frostig et al. 1990). Endogenous fluorescence changes are also produced by facilitated energy metabolism in the brain (Chance et al. 1962), because fluorescence substances such as NADH or flavoproteins are involved in aerobic energy metabolism (Figure 7.1B).

Activity-dependent changes in endogenous fluorescence derived from NADH have been used to monitor brain activities (Chance et al. 1962, Rosenthal and Jöbsis 1971, Lothman et al. 1975, Lewis and Schuette 1976). However, decreases in fluorescence signals derived from NADH might also be caused by activity-dependent increases in blood flow (Kitaura et al. 2007), because both excitation light and emitted fluorescence are easily absorbed by hemoglobin. Although activity-dependent increases in flavoprotein fluorescence signal are unlikely to be mimicked by hemodynamic responses, flavoprotein fluorescence changes had not been used to monitor brain activities in previous studies, as these signals were not clearly recorded (Aubin et al. 1979). Endogenous fluorescence derived from NADH or flavoprotein is weak compared with that derived from fluorescence dyes. Therefore, clear separation of the faint fluorescence from the strong excitation light is essential. The faint fluorescence must be detected by a sensitive camera with a low noise level. Therefore, application of endogenous fluorescence derived from flavoproteins to functional brain imaging requires recent development of advanced optical devices (Shibuki et al. 2003, Reinert et al. 2004, Coutinho et al. 2004, Weber et al. 2004, Husson et al. 2007).

7.2 THE PRINCIPLE UNDERLYING FLAVOPROTEIN FLUORESCENCE IMAGING

Although neuronal excitation does not produce clear optical signals by itself, activity-dependent facilitation of aerobic energy metabolism has fast dynamics (Fein and Tsacopoulos 1988, Shibuki 1989, 1990, Vanzetta and Grinvald 1999), and produces a number of intrinsic optical signals. It has been known that intracellular energy metabolism can be reflected in endogenous fluorescence changes, because endogenous fluorescence substances such as NADH and flavoproteins are involved in oxygen metabolism (Chance et al. 1962, Benson et al. 1979). NADH is produced from nonfluorescent NAD^+ by the Krebs-cycle dehydrogenases and subsequently oxidized by the respiratory chain in mitochondria. Therefore, activity-dependent decreases in fluorescence signals derived from NADH may be used as an intrinsic signal reflecting energy metabolism. NADH is also produced from NAD^+ at the glyceraldehyde 3-phosphate dehydrogenase step of glycolysis, and subsequently reoxidized by the conversion of pyruvate to lactate in the cytoplasm. NADPH is another important source of intrinsic fluorescence with similar optical properties to NADH. However, analytical measurements of NADH and NADPH in rodent brain have shown that both the resting fluorescence and the fluorescence changes are dominated by NADH rather than NADPH (Meryll and Guynn 1981, Klaidman et al. 1995). Furthermore, NADPH is the principal electron donor in reductive biosynthesis, whereas NADH is the principal electron donor in oxidative metabolism.

Therefore, NADH, but not NADPH, is a major fluorescence substance reflecting aerobic energy metabolism in the brain.

High levels of the flavin adenine nucleotide of a-lipoamide dehydrogenase are present in the brain: this flavin emits green fluorescence with a peak at approximately 520 nm in the presence of blue light (Kunz and Gellerich 1993). This flavoprotein is a major source of mitochondrial flavin fluorescence (Hassinen and Chance 1968, Chance et al. 1979). Enhanced mitochondrial energy metabolism converts reduced flavoproteins, emitting almost no fluorescence, to their oxidized forms, emitting green fluorescence (Hassinen and Chance 1968, Ragan and Garland 1969, Chance et al. 1979). Therefore, neural activities and the resultant enhancement in aerobic energy metabolism can be monitored as an increase in the green fluorescence derived from flavoproteins. The coupling between neural activities and aerobic energy metabolism is mediated by intracellular calcium changes (Figure 7.1C), which have important roles in neural functions such synaptic transmission or induction of neural plasticity. Mitochondria are widely distributed in various cellular compartments, including the presynaptic terminals (Billups and Forsythe 2002, Kosterin et al. 2005), and the metabolic load produced by neuronal excitation and synaptic transmission is efficiently covered by the enhanced energy metabolism in mitochondria.

Fluorescence signals derived from NADH and flavoproteins exhibit biphasic responses after electrical stimulation of neural tissue (Lipton 1973, Shuttleworth et al. 2003). It is suggested that the earlier phase corresponds to aerobic energy metabolism in neurons and that the later phase reflects glial glycolysis (Kasischke et al. 2004). Of these two phases, the earlier phase is suitable for functional imaging, because it occurs faster than the subsequent hemodynamic responses (Kitaura et al. 2007). In previous studies, NADH signals rather than flavoprotein signals were used to monitor neural activities (Rosenthal and Jöbsis 1971, Jöbsis et al. 1971, Lothman et al. 1975, Lewis and Schuette 1976), because signals derived from flavoproteins were reported to be much smaller than the NADH signal (Aubin et al. 1979). However, recent studies using advanced optical equipment have demonstrated that the flavoprotein fluorescence signal in the earlier phase is suitable for functional brain imaging in the cerebral cortex (Shibuki et al. 2003, Murakami et al. 2004, Weber et al. 2004), cerebellar cortex (Reinert et al. 2004, Coutinho et al. 2004), and hippocampal slices (Shuttleworth et al. 2003). Flavoprotein signals have some technical merits over NADH signals. Firstly, exposure to ultraviolet light, required for exciting NADH signals, can be detrimental to brain tissue, whereas blue light required for exciting flavoprotein signals penetrates deeper into tissue than ultraviolet light. Two-photon fluorescence microscopy of NADH circumvents the need for ultraviolet light (Bennett et al. 1996, Kasischke et al. 2004). Similarly, two-photon imaging of flavoprotein fluorescence has also been demonstrated and might provide an alternative means of one-photon imaging (Huang et al. 2002). Secondly, fluorescence increases in flavoprotein signals can be easily differentiated by subsequent hemodynamic signals, which produce a depression in fluorescence intensities, and it is difficult to differentiate between the initial depression of NADH fluorescence and subsequent hemodynamic responses.

Pioneering studies have utilized endogenous fluorescence changes to monitor the metabolic state of the brain during anoxia (Chance et al. 1962, Tomlinson et al. 1993) and spreading depression (Mayevsky and Chance 1974, Lothman et al. 1975, Haselgrove et al. 1990, Hashimoto et al. 2000). Static distribution of endogenous fluorescence has been measured to investigate the oxidation-reduction state in mitochondria in freeze-trapped brain samples (Chance et al. 1979). More dynamic changes in energy metabolism have been measured during direct cortical stimulation and epileptic activities (Rosenthal and Jöbsis 1971, Jöbsis et al. 1971, Lothman et al. 1975, Lewis and Schuette 1976). Measurement of endogenous fluorescence has also been conducted in cultured cells and brain slices using usual one-photon microscopy (Aubin et al. 1979, Duchen 1992, Mironov and Richter 2001, Schuchmann et al. 2001, Shuttleworth et al. 2003). Optical methods based on fluorescence changes have no limitation in spatial resolution up to the subcellular level, and fluorescence imaging has been performed using two-photon excitation microscopy (Bennet et al. 1996, Kasischke et al. 2004). Therefore, it is possible to identify the cellular species responsible for activity-dependent fluorescence changes, such as neurons, astrocytes, and other non-neural cells in neural tissue.

7.3 EXPERIMENTAL PROCEDURES

7.3.1 Surgical Procedures Required for Imaging Experiments in Anesthetized Animals

Although flavoprotein fluorescence signals are applicable for functional imaging in various preparations, transcranial imaging of mouse cortical activities is one of the most useful applications (Takahashi et al. 2006, Tohmi et al. 2006). This technique is an easy and powerful method for investigating cortical functions in mice. Because mice are relatively resistant to urethane, they were anesthetized at high doses (1.6–1.7 g/kg, i.p.). After subcutaneous injection of a local anesthetic, disinfected skin was incised, and the skull covering the cerebral cortex was exposed and cleaned. The head of a mouse could be fixed using a stereotaxic frame to record activities in the somatosensory or visual cortices. When the activities in the auditory cortex were investigated, a piece of metal was attached to the skull with dental cement, and the metal piece was fixed to a manipulator. The surface of the intact skull was covered with liquid paraffin to prevent drying and to keep the skull transparent. Surgical procedures were usually terminated within 30 min. An additional dose of urethane (0.2 g/kg, s.c.) may be administered when necessary. Some recording experiments have also been performed using other anesthetic agents, such as pentobarbital (Tohmi et al. 2006), although the cortical activities were less clearly recorded.

Rats may also be used for flavoprotein fluorescence imaging although the skull has to be removed (Shibuki et al. 2006). Rats have a merit in that they are more easily trained in behavioral learning. When rats were used for imaging experiments, they were anesthetized with urethane (1.5 g/kg, i.p.). Mannitol (2 g/kg, i.p.) was administered to prevent brain edema. Rectal temperature was monitored and kept at 38°C using a heating pad. After subcutaneous injection of lidocaine, disinfected skin covering the skull was incised. The skull was removed and the exposed cortex was covered by 2% agarose (Type I-B, Sigma) dissolved in saline.

7.3.2 Fluorescence Imaging

Fluorescence measurements were usually started about 1 h after the introduction of anesthesia. We used a cooled CCD camera system (AQUACOSMOS/Ratio with ORCA-ER camera, Hamamatsu Photonics, Hamamatsu, Japan). Other cooled CCD camera systems with low noise and high sensitivity may also be used. However, some camera systems developed for intrinsic signal imaging might be not suitable for fluorescence imaging, because such camera systems are expected to operate under bright illumination. In our experiments, the camera system was attached to an epifluorescence binocular microscope with a 75 W xenon light source (MZ FL III, Leica). In this microscope, the light path for excitation light was independent from that for emitted fluorescence. This design was important for observing endogenous fluorescence, which was sometimes very faint compared with those derived from exogenous fluorescence dyes. The use of dichroic mirrors in the common light port for both excitation light and fluorescence signals sometimes causes mixing between the two, and faint fluorescence signals are easily masked by such contamination. In our experiments, endogenous green fluorescence (λ = 500–550 nm) was excited by blue light (λ = 470–490 nm), and images of cortical areas, including the region of interest were recorded at 9 frames/s. The frame rate may be increased (Kubota et al. 2008), but the rate of 9 frames/s or so was sufficient in most experiments, because of the slow time courses of fluorescence changes. When neural activities in sensory cortices were investigated, fluorescence images were captured in a recording session, during which mice were given sensory stimuli in trials repeated at 20–60-s intervals. This interval was required because the fluorescence responses and subsequent vascular responses were completely terminated and recovered during this interval time.

7.3.3 Analytical Methods

One of the merits of transcranial imaging using flavoprotein signals is its reproducibility and little variability from animal to animal because surgical damage to the brain can be avoided using the transcranial approach. Furthermore, mitochondria are richly present in brain tissue with little variations in density. Another merit of transcranial imaging using flavoprotein signals is that there is an almost linear relationship between the amplitude of fluorescence responses and the magnitude of neural activities (Shibuki et al. 2003, Tohmi et al. 2006). For quantitative analysis of fluorescence responses, images elicited by particular stimuli are usually averaged over 20–40 trials. Spatial averaging may also be used to improve the quality of images. In some trials, neural responses may be recognized, even in a single trial (Shibuki et al. 2003, Tohmi et al. 2006). However, responses are variable from trial to trial, so that averaging is required to reduce this biological variability.

The images are normalized to represent relative fluorescence changes ($\Delta F/F_0$), in which ΔF is a fluorescence change elicited by a particular stimulus, and F_0 is the baseline fluorescence intensity before the stimulus. Response amplitudes are evaluated as values of $\Delta F/F_0$ in a particular window. The response peaks are usually found 0.5–1.0 s after stimulus onset. Temporal averaging of the frames around the response peaks may also improve the signal to noise ratio. The range of the temporal averaging is usually between 0.5 and 1.0 s after stimulus onset. This range is selected to minimize the effects of neurovascular responses following the neural activities.

7.4 TRANSCRANIAL IMAGING OF EXPERIENCE-DEPENDENT CORTICAL PLASTICITY

7.4.1 Experience-Dependent Plasticity in the Auditory Cortex

Auditory information is important for communication in mice, and mice have well-developed sound discrimination abilities (Ehret and Riecke 2002, Geissler and Ehret 2002). The auditory cortex does not have an essential role in the discrimination of sound frequencies, because discrimination of tonal pitch is performed by rats in which the auditory cortex is bilaterally lesioned (Diamond and Neff 1957, Ohl et al. 1999, Ono et al. 2006). However, the locations of cortical auditory neurons are arranged according to the tonotopic maps determined by the sound frequencies, which stimulate the neurons with the lowest intensity (Willott et al. 1993, Stiebler et al. 1997). In electrophysiological unit recordings, cortical auditory neurons respond to various stimuli when the stimulus intensities are sufficiently strong. Therefore, it is difficult to determine the function of a particular neuron from its specific stimuli. Auditory information is probably represented by the patterns of activities in many neurons, so that imaging experiments may be useful. Intrinsic signals have been performed to investigate the functions of the auditory cortex (Bakin et al. 1996, Hess and Scheich 1996, Harel et al. 2000, Versnel et al. 2002, Nelken et al. 2004).

A merit of transcranial flavoprotein fluorescence imaging is that quantitative analysis of fluorescence signals is possible because of its reproducibility. Therefore, analysis of experience-dependent plasticity is a suitable subject for this technique. In young animals, exposure to tonal stimuli at a particular frequency causes enlargement of the corresponding cortical auditory area in the tonotopic map (Zhang et al. 2001, Nakahara et al. 2004). Even in adult animals, the neural circuits in the auditory cortex are modified after injury in the cochleae (Harrison et al. 1993, Rajan et al. 1993) or after behavioral training using attended tonal stimuli at a particular frequency (Recanzone et al. 1993, Pantev et al. 1998, Gao and Suga 2000, Fritz et al. 2003).

We investigated the experience-dependent plasticity induced by sound exposure using transcranial flavoprotein fluorescence imaging in mice (Takahashi et al. 2006). When naïve mice anesthetized with urethane were stimulated by tonal stimuli, increases in endogenous fluorescence appeared in the auditory cortex 0.2 s after the stimulus onset and the peak responses were observed about 0.5 s after the stimulus onset (Figure 7.2A,B). The response amplitude was less than 1% in $\Delta F/F_0$. The amplitude of the intrinsic signal using green light reflection is about 1% in the auditory cortex (Versnel et al. 2002, Nelken et al. 2004). However, the signal has a long latency of about 1 s after stimulus onset (Versnel et al. 2002), so that it could be strongly affected by activity-dependent hemodynamic responses.

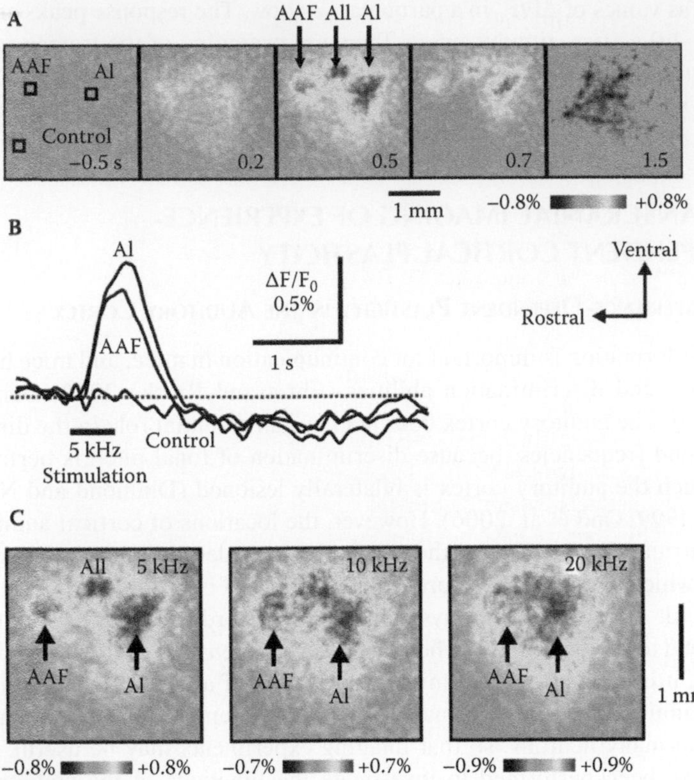

FIGURE 7.2 Transcranial imaging of cortical auditory activities. (A) Fluorescence responses elicited by a 5-kHz tonal stimulus at 40 dB SPL in a naïve mouse. Responses can be recognized in the areas corresponding to the right primary auditory cortex (AI), anterior auditory field (AAF), and the secondary auditory cortex (AII). The numbers shown in each image are the time after stimulus onset in seconds (minus value represents the time before stimulus onset). (B) Time course of endogenous fluorescence responses in the windows of 10 × 10 pixels in AI, AAF and outside the auditory cortex. (C) Endogenous fluorescence responses in the right auditory cortex elicited by 5, 10, and 20 kHz tonal stimuli at 40 dB SPL in the same mouse. The images were taken 0.5 s after stimulus onset. (Modified from Takahashi, K., Hishida, R., Kubota, Y., Kudoh, M., Takahashi, S., and Shibuki, K. 2006. *Eur. J. Neurosci.* 23: 1265–1276. With permission.)

When the stimulus frequency was changed between 5 kHz and 20 kHz, the fluorescence responses were observed according to the tonotopic maps in separate areas corresponding to the primary auditory cortex (AI) and the anterior auditory field (AAF) (Figure 7.2C). Such mirror-symmetrical maps obtained by fluorescence imaging agree with the tonotopic maps obtained by electrophysiological recordings in mice (Willott et al. 1993, Stiebler et al. 1997). Fluorescence responses were sometimes observed in another area that might correspond to the secondary auditory cortex (AII).

To induce experience-dependent plasticity in the auditory cortex, mice were exposed to a 10-kHz tonal stimulus, and the effects on the cortical responses to 5-,

10-, and 20-kHz stimuli were tested. Exposure to 10-kHz tonal stimuli potentiated the fluorescence responses to 10 kHz in comparison to those recorded in naïve mice (Figure 7.3A). The endogenous fluorescence responses in sound-deprived mice were slightly depressed in comparison to those in naïve mice. Potentiation was clearly observed with respect to the poststimulus response duration and the amplitudes of the responses. The effects of sound exposure were specific to the stimulus frequency: significant potentiation ($P < 0.01$, Mann Whitney U-test) was observed in the responses elicited by a 10-kHz sound, whereas the responses elicited by a 5- or 20-kHz sound were less clearly affected by sound exposure (Figure 7.3B). No particular critical period was observed before P42, because potentiation was observed

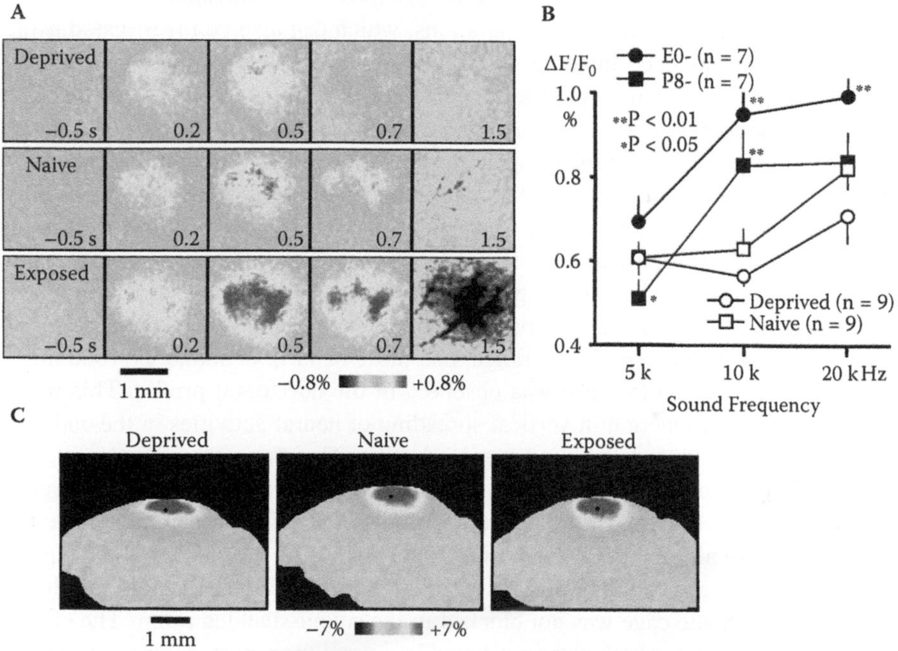

FIGURE 7.3 **(See color insert following page 224)** Experience-dependent plasticity of cortical auditory activities. (A) Examples of endogenous fluorescence responses elicited by a 10-kHz sound in sound-deprived (Deprived), naive (Naive) and sound-exposed (Exposed) mice. (B) Means and S.E.M.s of the response amplitudes in various groups of mice were plotted against the stimulus sound frequency used to elicit fluorescence responses. Sound exposure was started before pregnancy (E0-) or on postnatal day 8 (P8-). The responses to a 10-kHz sound in all of the sound-exposed groups were significantly potentiated compared with those of sound-deprived mice ($P < 0.01$, Mann Whitney U-test). The responses to a 20-kHz sound were significantly potentiated only in E0- mice ($P < 0.01$) and those to a 5-kHz sound were depressed only in P0- mice ($P < 0.05$). (C) Endogenous fluorescence images elicited by repetitive stimulations at 10 Hz for 0.5 s in slices obtained from sound-deprived, naïve and sound-exposed (E0-) mice. These images were recorded 2 s after stimulus onset. Stimulation was applied at supragranular layers (black spots). Note the difference in the vertical thickness of the response areas. (Modified from Takahashi, K., Hishida, R., Kubota, Y., Kudoh, M., Takahashi, S., and Shibuki, K. 2006. *Eur. J. Neurosci.* 23: 1265–1276. With permission.)

in all sound-exposed mice exposed to sound stimuli during this period. In contrast, tonotopic map plasticity investigated with electrophysiological recordings shows a critical period between P9 and P30 (Zhang et al. 2001, Nakahara et al. 2004). The lack of an apparent critical period in fluorescence responses suggest that intracortical circuits may have important roles in producing the fluorescence responses elicited by suprathreshold stimuli, because cortical responses elicited by supra-threshold stimuli are amplified by intracortical circuits (Douglas et al. 1995) and intracortical circuits are plastic in slices obtained from adult rats (Kudoh and Shibuki 1997, Kudoh et al. 2002).

In studies of experience-dependent plasticity, a disadvantage of transcranial imaging is that the identification of plastic synapses is not possible. However, this may be overcome by using slice preparations, which can also be investigated using flavoprotein fluorescence imaging. To identify the synapses modified by acoustic environments, we prepared auditory cortical slices from sound-exposed mice. To activate intracortical circuits, repetitive stimulation was applied to the supragranular layers. Endogenous fluorescence responses were observed in horizontally distributed areas in the supragranular layers. The amplitude of the response peak was usually observed 2 s after stimulus onset, when the experiments were performed at room temperature. The response amplitude was not significantly affected by sound exposure. However, the depth of the response areas into the vertical direction was modified (Figure 7.3C). In the vertical profile, fluorescence responses in sound-deprived mice were observed in a narrower range than those in naïve or sound-exposed mice. In contrast, no clear difference was observed in the horizontal profile. This result indicates that the reduction in vertical spreading of neural activities in the auditory cortex may be produced in sound-deprived mice. Therefore, the effects of sound exposure might appear in the vertical intracortical circuits. No apparent difference was found between the naïve mice and sound-exposed mice, probably because the naïve mice were also exposed to environmental sounds of various spectral frequencies. The sound-deprivation in this study was very mild, because the noise produced by mice in the home cage was not blocked in the sound-shielded room. The effects of sound deprivation, which were not found in a previous report using electrophysiological recordings (Zhang et al. 2001), could be seen using the fluorescence imaging, probably because this technique is not strongly affected by variations in each neuron. Exposure to noise has significant effects on the normal development of the auditory cortex in young animals (Chang and Merzenich 2003). The clear difference between the sound-deprived group and other groups (Figure 7.3C) suggests that the sound intensity of natural acoustic environments may be very important for regulating the development of the auditory cortex.

7.4.2 Experience-Dependent Plasticity in the Visual Cortex

Ocular dominance plasticity in the visual cortex is one of the most extensively investigated forms of experience-dependent plasticity (Wiesel and Hubel 1965, Hubel and

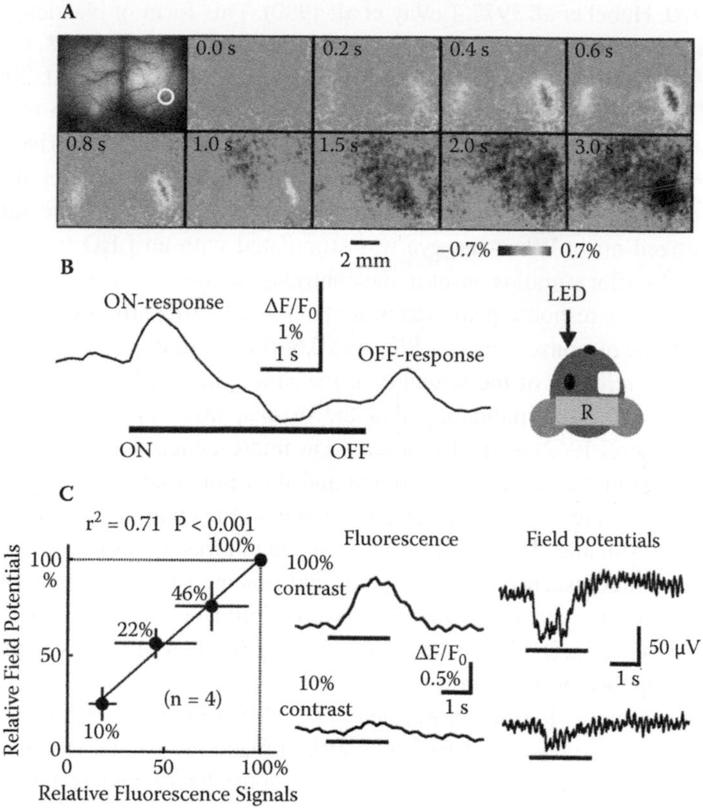

FIGURE 7.4 Transcranial imaging of cortical visual activities. (A) Original fluorescence image and serial images in $\Delta F/F_0$ of the left and right visual cortices. In this mouse, the right eye was covered, and an LED light placed in front of the mouse was turned on for 4 s. Time after stimulus onset is shown in each image. Note that on-responses in the contralateral right visual cortex are shown together with faint responses in the ipsilateral left cortex. Reduction in intensity of fluorescence signals, partly due to activity-dependent hemodynamic responses, was observed even during the presence of the light stimulus. (B) Time courses of $\Delta F/F_0$ in a circular window in (A) placed on the response center in the contralateral right visual cortex. The horizontal bar shows the timing of LED light stimulation. Schematic drawings of LED stimulation and recording (R) are also shown. (C) Relationship between amplitudes of field potentials and those of fluorescence responses elicited by grating patterns of 100, 46, 22, and 10% in contrast. Responses were normalized against maximal responses elicited by 100% contrast patterns in each mouse. Means and S.E.M.s obtained from four mice are shown. A statistically significant correlation ($R^2 = 0.71$, $P < 0.001$) was observed between the two parameters. Traces of fluorescence changes and field potentials elicited by grating patterns with 100% or 10% contrast are also shown. Fluorescence changes were slightly slower, but the signal to noise ratio was comparable or better than that of field potential recordings. (Modified from Tohmi, M., Kitaura, H., Komagata, S., Kudoh, M., and Shibuki, K. 2006. *J. Neurosci.* 26: 11775–11785. With permission.)

Wiesel 1970, Hubel et al. 1977, LeVay et al. 1980). This form of plasticity has been also studied in mice using electrophysiological recordings (Dräger 1978, Gordon and Stryker 1996), intrinsic signal imaging (Antonini et al. 1999, Cang et al. 2005, Hofer et al. 2006), and histological analysis (Antonini et al. 1999, Tagawa et al. 2005). We investigated experience-dependent plasticity after monocular deprivation (MD) using transcranial flavoprotein fluorescence imaging in mice (Tohmi et al. 2006).

Fluorescence images of the visual cortex were taken through the intact skull of anesthetized mice. When the eye was stimulated with an LED light, responses appeared 0.2 s after stimulus onset in the contralateral and ipsilateral visual cortices (Figure 7.4A). The response peak was observed 0.6–.0 s after stimulus onset, and the amplitude of the response peak in $\Delta F/F_0$ was 0.6–0.7%. Fluorescence responses were observed after removal of the stimulus in the same places where the on-responses appeared (Figure 7.4B), indicating that the off-responses in the cortical neurons (Hubel and Wiesel 1962) were also observed as fluorescence changes. We compared the magnitudes of fluorescence responses and field potential recordings elicited at the same cortical sites by grating patterns, in which the contrast of stimuli was varied between 100% and 10% (Figure 7.4C). As the contrast of stimuli decreased, the magnitudes of fluorescence responses and field potential responses simultaneously decreased, indicating that fluorescence responses could be used for quantitative measurement of neural activities in the visual cortex. Retinotopic maps could also be produced using flavoprotein fluorescence signals.

We induced ocular dominance plasticity using MD starting on P28 for 4 days, and bilateral cortical responses to LED light stimuli placed in front of the mouse were recorded. The deprived eye responses in both the contralateral and ipsilateral cortices were clearly depressed compared with the nondeprived eye responses, whereas no apparent change was observed in the monocular zone (Figure 7.5A,B). No effect of MD was found at the age of 6 weeks. The effects of MD were most clearly observed when it was started at 4 weeks of age, and were saturated after 4 days of MD. These characteristics are compatible with previous results obtained using electrophysiological recordings in mice (Dräger 1978, Gordon and Stryker 1996). A technical merit of flavoprotein fluorescence imaging is that neural activities in a wide area, including bilateral visual cortices, can be simultaneously observed, so that fluorescence responses elicited via the deprived eye in the right/left cortex, and those elicited via the nondeprived eye in the left/right cortex can be directly compared. Responses in the monocular zone can be easily recorded as a control. These findings suggest that transcranial flavoprotein fluorescence imaging is applicable for investigating experience-dependent plasticity in the visual cortex. When the surface of the skull is covered with transparent dental resin, the skull maintains its transparency for several days after surgery (Takao et al. 2006). Using this technique, we found that nondeprived eye responses increased after MD and deprived eye responses decreased after MD. From these results, we concluded that MD-induced plasticity in young mice has two components: potentiation of nondeprived eye responses and depression of

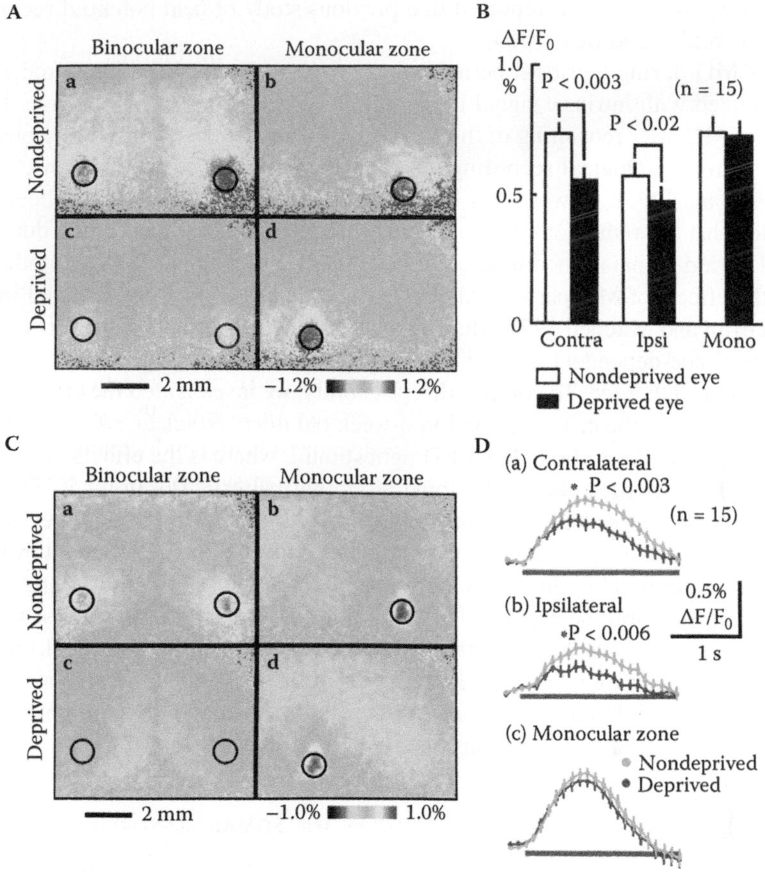

FIGURE 7.5 **(See color insert following page 224)** Experience-dependent plasticity of cortical visual activities. (A) Fluorescence responses elicited by LED light stimuli in the visual cortex after monocular deprivation (MD) of the right eye. Nondeprived eye responses in the binocular zone (a) and monocular zone (b), and deprived eye responses in the binocular zone (c) and monocular zone (d) are shown. These images were obtained from the same mouse. (B) Amplitudes of fluorescence responses ($\Delta F/F_0$) in the contralateral binocular zone, ipsilateral binocular zone and contralateral monocular zone. Means and S.E.M.s from 15 mice are shown. Statistical differences were evaluated by Mann Whitney U-test. (C) Fluorescence responses elicited by moving grating patterns 2 weeks after MD performed between P26 and P33. Nondeprived eye responses in the binocular zone (a) and the monocular zone (b), and deprived eye responses in the binocular zone (c) and the monocular zone (d) are shown. These images were obtained from the same mouse. Deprived eye responses were reduced after MD compared with nondeprived eye responses in the binocular zone, whereas those in the monocular zone were not. (D) Time course of fluorescence responses elicited by moving grating patterns in the contralateral (a) and ipsilateral (b) binocular zones, and contralateral monocular zone (c) of the visual cortex 2 weeks after MD, performed between P26 and P33. (Modified from Tohmi, M., Kitaura, H., Komagata, S., Kudoh, M., and Shibuki, K. 2006. *J. Neurosci.* 26: 11775–11785. With permission.)

deprived eye responses, as reported in a previous study of field potential recordings in mice (Frenkel and Bear 2004).

Prior MD during a critical period enhances plasticity in the adult visual cortex, as visualized with intrinsic signal imaging (Hofer et al. 2006), whereas long-lasting effects of MD after reopening of the deprived eye have not been previously observed from electrophysiological recordings in rodents (Sawtell et al. 2003, Hensch 2004, Tagawa et al. 2005). However, in behavioral studies using mice, it has been demonstrated that the reduction in visual acuity produced by MD performed during the critical period is maintained in adult mice (Prusky and Douglas 2003). In other animals, the effects of MD performed during the critical period were detected in adult mice (Giffin and Mitchell 1978, Murphy and Mitchell 1987). Therefore, it is expected that experience-dependent plasticity produced by MD during the critical period is preserved in adult mice. To confirm this possibility, we investigated the effects of MD performed during the critical period in 6-week-old mice. No clear effect of MD was found in the responses elicited by LED light stimuli, whereas the effects of MD were found in the responses elicited by moving grating patterns (Figure 7.5C,D). These findings indicate that the effects of MD performed during the critical period were preserved in visual responses elicited by complex patterned stimuli, but not by simple LED light stimuli. This difference between LED light stimuli and moving grating patterns can probably be explained by assuming that the long-lasting effects of MD might be observed only in intracortical circuits, but not in thalamocortical synapses: LED light stimuli might be appropriate for activating thalamocortical synapses, but not for activating intracortical circuits, whereas intracortical circuits might be effectively stimulated by moving grating patterns.

7.4.3 Experience-Dependent Plasticity in the Somatosensory Cortex

Learning using somatosensory cues is known to produce cortical plasticity in the somatosensory cortex (Recanzone et al. 1992a, 1992b, Wang et al. 1995). These cortical changes are thought to have important roles in sensory information processing. However, the details have not been clearly determined. Rats are suitable for behavioral learning. Therefore, we trained rats to measure the ability to discriminate flutter stimulation frequencies applied to the skin, and investigated cortical activities in the primary sensory cortex (SI) after learning using flavoprotein fluorescence imaging (Shibuki et al. 2006). A form of neural plasticity specific to temporal stimulus patterns has been reported to occur in the primary auditory cortex (Kilgard and Merzenich 1998, Bao et al. 2004). Therefore, we expected that the intracortical circuits in SI might be responsible for stimulus-specific plasticity, which could explain some of the neural mechanisms underlying the behavioral discrimination of flutter frequency.

Rats were trained to discriminate floor vibrations at rewarded (S+) and unrewarded (S–) frequencies (Figure 7.6A,B). After this discrimination learning was accomplished, they were anesthetized with urethane and neural responses were recorded in SI during the application of flutter stimuli to the contralateral hindpaw. The fluorescence responses to the stimuli at unrewarded frequencies were selectively depressed in trained rats (Figure 7.6C,D), which had behaviorally neglected, unrewarded stimuli. The depression of cortical responses was not observed in the rats

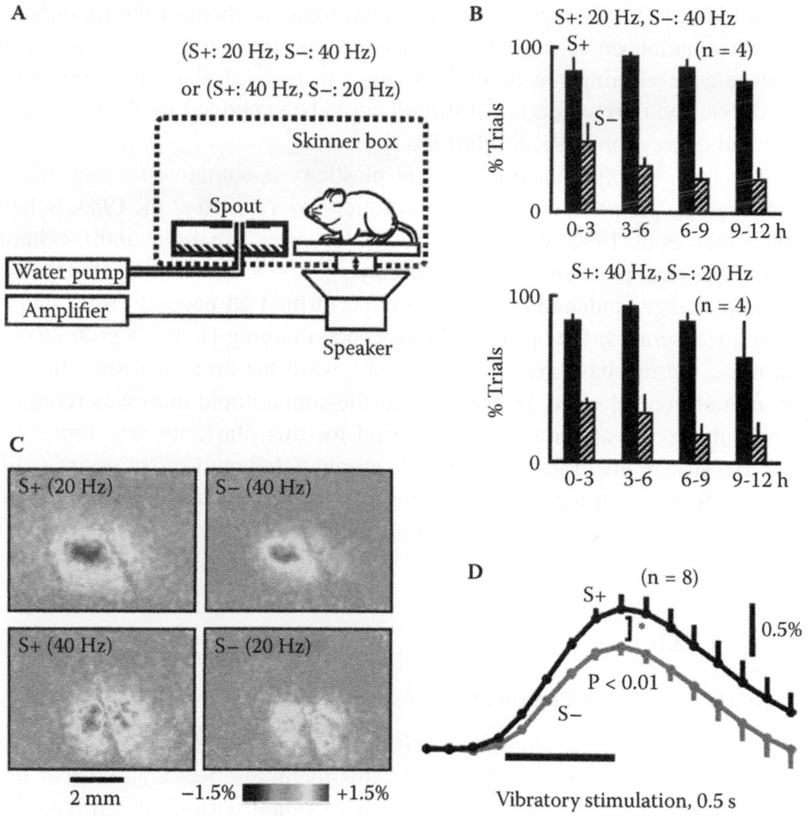

FIGURE 7.6 Learning-induced depression of cortical somatosensory activities. (A) Experimental setup for discrimination learning. (B) Percentage of trials with licking to rewarded stimuli (S+) or to unrewarded stimuli (S–) during every 3 h of the discrimination test. Results in 4 rats trained with S+ at 20 Hz and S– at 40 Hz are shown. Results in other 4 rats trained with S+ at 40 Hz and S– at 20 Hz are also shown. (C) Fluorescence responses to S+ and S– recorded in SI of a rat, which was trained with S+ at 20 Hz and S– at 40 Hz. Vibratory hindpaw stimulation for 0.5 s was used, and each image was taken immediately after cessation of the stimulation. Fluorescence responses in another rat, which was trained with S+ at 40 Hz and S– at 20 Hz, are also shown. (D) Time course of $\Delta F/F_0$ in 10 by 10 pixels in the eight trained rats. Means and S.E.M.s are shown. (Modified from Shibuki, K., Ono, K., Hishida, R., and Kudoh, M. 2006. *Neuroimage* 30: 735–744. With permission.)

trained with rewarded stimuli only. The correlation between the results of behavioral experiments and the stimulus-specific depression suggests that the latter might have some role in discrimination learning. Learning using reward or punishment is attributed to the functions of the limbic system including the amygdala (Baxter and Murray 2002, Rodrigues et al. 2004). Sophisticated discrimination learning in primates requires working memory in higher brain areas (Romo and Salinas 2003). However, cortical circuits in the adult primary sensory cortices show LTP and LTD (Kudoh and Shibuki 1997, Egger et al. 1999, Feldman 2000, Kudoh et al. 2002,

Birtoli and Ulrich 2004), suggesting functional roles for them in the modulation of sensory information processing. In this respect, stimulus-specific depression in SI could serve as a learning mechanism, because behavioral discriminating between rewarded (S+) and unrewarded (S−) stimuli could be explained by the stimulus-specific cortical depression at the level of SI.

Another form of experience-dependent plasticity is somatotopic map plasticity in S1 induced by peripheral afferent denervation (Merzenich et al. 1983, Schady et al. 1994, Elbert et al. 1995, Weiss et al. 2004). However, the molecular mechanisms underlying this map plasticity in S1 are largely unknown. We investigated the somatotopic map changes induced by tail cutting performed on neonatal mice and adult mice using transcranial flavoprotein fluorescence imaging (Kitaura et al. 2005). In tail-cut mice, the tail base area in S1 shifted toward the area corresponding to the tail tip area of control mice, indicating that the somatotopic map was reorganized after tail cutting. No apparent critical period for this plasticity was found before 8 weeks. Because many kinds of genetically manipulated mouse strains are available for experimental use, transcranial imaging of somatotopic map plasticity in mice might be useful for understanding the molecular mechanisms of somatotopic map plasticity in SI.

7.5 DISCUSSION

7.5.1 Merits of Transcranial Flavoprotein Fluorescence Imaging

In this chapter, flavoprotein fluorescence imaging combined with transcranial cortical imaging in mice has been introduced. This technique has a number of merits. Flavoprotein fluorescence provides a quantitative signal within all cell types of the brain, bypassing the difficulties introduced by the often inhomogeneous distribution of exogenous dyes or indicators. When exogenous dyes are used for optical imaging, bleaching of the dyes is inevitable during the recordings, making stable and reproducible recording difficult. Bleaching is also observed during flavoprotein fluorescence imaging when the recorded sites were exposed to strong blue light (Kubota et al. 2008). However, this bleaching is reversible, and flavoproteins are resistant to bleaching when exposure to excitation light is intermittently scheduled. These properties make flavoprotein fluorescence signals suitable for quantitative analysis of neural activities. Furthermore, transcranial imaging reduces surgical damages on the brain. Generally speaking, optical imaging of neural activities requires trained researchers. However, transcranial imaging of flavoprotein fluorescence in anesthetized mice is so easy that it is suitable for beginners or students of neuroscience. Transcranial imaging of flavoprotein fluorescence signals enables us to record experience-dependent cortical plasticity in the auditory, visual and somatosensory cortices, as demonstrated in this chapter. It is also applicable for visualizing cortical activities during audiogenic seizures in mice (Takao et al. 2006).

Flavoprotein fluorescence signals are applicable to functional imaging in animals other than mice. Short-term potentiation and depression induced by direct cortical stimulation with tetanic stimulation have been demonstrated in the somatosensory cortex of anesthetized rats (Murakami et al. 2004). Long-term cortical depression induced by discrimination learning in the rat somatosensory cortex has also been demonstrated (Shibuki et al. 2006). Flavoprotein fluorescence imaging of slice preparations is also useful due to its small variability in signal intensities (Hishida et al. 2007a, Kamatani et al. 2007). This technique is also applied to anesthetized cats (Husson et al. 2007). Taken together, flavoprotein fluorescence imaging is a widely applicable imaging technique to various problems in neuroscience.

7.5.2 Demerits of Transcranial Flavoprotein Fluorescence Imaging

One of the demerits of flavoprotein fluorescence imaging is that the temporal resolution may not be sufficient compared to other techniques using voltage-sensitive dyes or calcium indicators, because various metabolic reactions are required for the fluorescence changes. Synaptic transmission and neuronal excitation triggers an increase in intracellular Ca^{2+} level. The increased intracellular Ca^{2+} is partially taken up by the mitochondria (Budd and Nicholls 1996, Gunter et al. 1998, Rizzuto et al. 2000), and stimulates energy metabolism (McCormac et al. 1990, Hansford 1994, Territo et al. 2000, Kavanagh et al. 2000). These results can be observed as enhanced O_2 consumption after neural activities (Fein and Tsacopoulos 1988, Shibuki 1989, 1990, Vanzetta and Grinvald 1999). Because of the slow time course of fluorescence responses, later parts of the responses are masked by activity-dependent hemodynamic responses (Kitaura et al. 2007). However, the initial responses recorded within 0.5–0.8 s after the stimulus onset are hardly affected by vascular changes. Usually, the latency of cortical fluorescence responses after the onset of natural stimuli is about 0.2 s. When direct tetanic stimulation was given to the cortex, a fluorescence increase of about 70% of the maximal amplitude is achieved within 0.2 s after the stimulus onset. Therefore, a significant part of the latency is occupied by the transmission of neural information from the sensory organs to the cortex.

Another demerit of flavoprotein fluorescence imaging is that it may not be suitable for investigating large brains (Shibuki and Hishida 2004). In our preliminary experiments, clear fluorescence responses were not obtained in the sensory cortices of cats, dogs, or human subjects. This might be partly explained by species differences in metabolic rates per body weight. This parameter is inversely proportional to a quarter power of the body weight (Gillooly et al. 2001). Motion artifacts produced by breathing or heartbeats are more problematic in large animals. Further, the excitation or emission light used for fluorescence measurement can more easily penetrate through a small brain. The presence of baseline green fluorescence derived from connective tissue may also be crucial. In the case of cortical slices obtained from rats, about 50% of the green fluorescence is insensitive to the presence or absence of oxygen, so that this much is attributed to substances other than flavoproteins (Shibuki et al. 2003). Large brains, which require more mechanical support for the brain tissue, may have a high baseline fluorescence signal. However, the recent success of flavoprotein fluorescence imaging in cats (Husson et al. 2007) suggests that this demerit can be overcome by technical refinements.

7.6 FUTURE DIRECTIONS

7.6.1 Transcranial Control of Neural Activities

In transcranial imaging, a direct approach to the cortical surface is prevented by the presence of an intact skull. Physiological analyses of cortical functions sometimes require not only recordings, but also a suppression of cortical activities. Usually, a partial destruction of cortical areas using direct currents or a pharmacological blockade of cortical activities is used for this purpose. However, making a hole in the skull for this purpose reduces the merits of transcranial imaging. Furthermore, fine spatial control of the suppressed cortical area is difficult. Therefore, transcranial optical control of cortical activities, where possible, could be useful for investigating cortical functions when transcranial imaging is used. There are a number of optical methods for suppressing neural functions (Jay and Keshishian 1990, Takei et al. 1998, Kataoka et al. 2000, Cui et al. 2003, Dietrich et al. 1986). However, these methods require the application of photosensitive agents. Exposure to a strong infrared laser produces suppression of brain activities (Kataoka et al. 2000, Mochizuki-Oda et al. 2002). However, only transient suppression is obtained with a significant increase in tissue temperature by a few degrees. Mitochondria are abundantly present in presynaptic terminals (Billups and Forsythe 2002, Kosterin et al. 2005), and the metabolic load produced by neuronal excitation and synaptic transmission is efficiently covered by enhanced energy metabolism in mitochondria. Because flavoproteins are photosensitive, their inactivation caused by photobleaching may suppress cortical energy metabolism and lead to the production of photo-inactivation of cortical synaptic transmission. We tested this possibility in the mouse auditory cortex (Kubota et al. 2008).

Cortical slices were prepared from the mouse auditory cortex. Field potentials were recorded through a metal electrode placed in the supragranular layers of cortical slices. These field potentials were composed of early and late negative waves (Figure 7.7A). The former reflect the antidromic firing of pyramidal neurons and axonal activities, whereas the latter are produced by the trans-synaptic responses of pyramidal neurons (Kudoh and Shibuki 1997). To test the effects of photo-irradiation, the recording side was exposed to a blue ($\lambda = 475$ nm) laser. The amplitude of the late trans-synaptic component gradually decreased, and 50 min after the onset of exposure, this amplitude was only one third of that recorded immediately before the exposure (Figure 7.7A). The amplitudes of the early antidromic components of the field potentials were stable. After the cessation of light exposure, the late trans-synaptic components of field potentials slowly recovered in a few hours.

We tested the photo-inactivation of cortical synaptic transmission in anesthetized mice. The mouse auditory cortex includes the anterior auditory field (AAF) and the primary auditory cortex (AI) (Willott et al. 1993, Stiebler et al. 1997). In previous studies using electrophysiological recordings, neurons in AAF were activated by tonal stimuli with a latency shorter than that in neurons in AI (Linden et al. 2003). There are extensive corticocortical connections in the auditory cortex, including connections between AAF and AI (Budinger et al. 2000, de la Mothe et al. 2006), and positively correlated firing activities were found between neurons in AAF

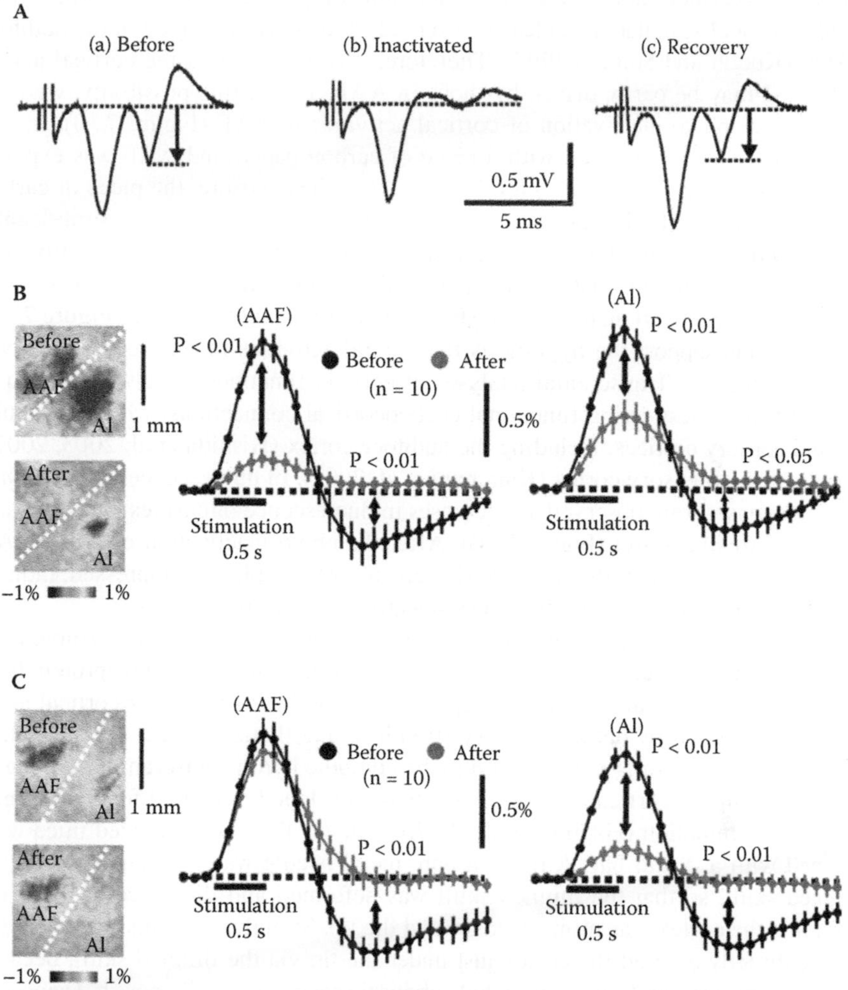

FIGURE 7.7 Transcranial photo-inactivation of cortical activities. (A) Supragranular field potentials elicited by layer VI stimulation recorded before (a) and immediately after (b) exposure to blue laser light for 50 min. Field potentials recorded 2 h after cessation of light exposure are also shown (c). The second negativity representing trans-synaptic potentials (arrows) in (a) and (c) was markedly suppressed in (b). (B) Fluorescence responses before and after selective photo-inactivation of AAF, recorded 0.5 s after stimulus onset. Temporal profiles of $\Delta F/F_0$ in AAF and AI before and after selective photo-inactivation of AAF are also shown. Statistical differences were evaluated by Wilcoxon signed rank test 0.5 and 1.5 s after stimulus onset for evaluating the effects of photo-inactivation on the neural responses (0.5 s) and vascular responses (1.5 s). (C) Fluorescence responses before and after selective photo-inactivation of AI, recorded 0.5 s after stimulus onset. Temporal profiles of $\Delta F/F_0$ in AAF and AI before and after selective photo-inactivation of AI are also shown. (Modified from Kubota, Y., Kamatani, D., Tsukano, H., Ohshima, S., Takahashi, K., Hishida, R., Kudoh, M., Takahashi, S., and Shibuki, K. 2008. *Neurosci. Res.* 60: 422–430. With permission.)

and AI (Eggermont 2000). The horizontal spreading of field potentials elicited by local electrical stimulation is clearly observed in the slices obtained in the auditory cortex (Kudoh and Shibuki 1997). Therefore, it is possible that the cortical activities in AI may be partly driven by those in AAF. To test this possibility, we used transcranial photo-inactivation of cortical activities in AAF (Figure 7.7B). In this experiment, AI was covered with a piece of carbon paper, and AAF was exposed to blue laser light ($\lambda = 475$ nm) for 50 min. After the exposure, the piece of carbon paper was removed. Fluorescence responses in AAF were clearly and significantly suppressed ($P < 0.01$). Fluorescence responses in AI were also significantly suppressed ($P < 0.01$). In contrast, when neural activities in AI were suppressed by direct photo-inactivation, those in AAF were not significantly affected (Figure 7.7C). These results support the hypothesis that neural activities in AI are partly driven by those in AAF. Transcranial analyses of cortical functions are also expected to be useful for investigating functional corticocortical connections, which are found around sensory cortices, including the auditory cortex (Hishida et al. 2003, 2007a) and the somatosensory cortex (Kamatani et al. 2007). In fluorescence imaging, vascular responses were observed as decreases in fluorescence intensities a few seconds after the stimulus onset (Figure 7.7B). After the photo-inactivation of AI or AAF, vascular responses in both AI and AAF were almost completely suppressed, indicating that neurovascular coupling is very sensitive to photo-inactivation.

Neural activities elicited by sensory stimuli in the primary auditory, somatosensory and visual cortices can be easily visualized using transcranial flavoprotein fluorescence imaging in anesthetized mice. However, activities in higher cortical areas are hardly elicited by natural sensory stimuli in anesthetized mice. If transcranial electrical stimulation is possible, it may be combined with transcranial imaging to investigate higher cortical functions. We have developed a method for transcranial electrical stimulation (Hishida et al. 2007b). The skulls of anesthetized mice were thinned with a dental bar. A tip of a bare metal needle was gently pushed on the thinned skull, so that the thinned skull was deformed and the space between the skull and the cortex was minimized around the tip. Stimulus currents applied to the needle directly entered the cortex just under the tip via the thinned skull, because current escape through the minimized subdural space was small enough. Using this method, we visualized the functional corticocortical connections between the primary and higher cortices, and between motor and somatosensory cortices in anesthetized mice. Thus, transcranial imaging combined with transcranial control of cortical activities is expected to be useful for investigating various aspects of higher cortical functions in mice.

7.6.2 Transcranial Imaging Using Two-Photon Microscopy

In this chapter, we focused on visualizing cortical activities in wide cortical areas. However, the depth profile of fluorescence signals in cortical layers is unknown, when single photon microscopy with a cooled CCD camera is used. Two-photon microscopy would be useful to investigate this problem. There is no theoretical limitation to prevent single cell imaging of neural activities using fluorescence signals (Kazama et al. 2008). In fact, activity-dependent changes in flavoprotein or NADH signals have been demonstrated at the single cell level using two-photon microscopy (Kasischke et al. 2004). Transcranial imaging using two-photon microscopy has already been used to visualize fine morphological changes at the single neuron level (Grutzendler et al. 2002, Ohki et al. 2005, Zuo et al. 2005). Therefore, a combination of transcranial flavoprotein fluorescence imaging and two-photon microscopy might be applicable to the investigation of the fine cellular mechanisms underlying experience-dependent plasticity in the mouse sensory cortices.

7.6.3 Transcranial Imaging in Awake Mice

Transcranial imaging is applicable to awake mice for investigating higher cortical functions suppressed by anesthesia (Shibuki et al. 2004). In these experiments, the superficial layers of the skull were removed and the remaining skull was covered with transparent dental resin to prevent infection and maintain the transparency of the skull. A few days after this surgery, the head of an awake mouse was fixed by screwing a metal piece attached to the skull. The responses elicited by monotonous tonal stimuli were rapidly attenuated after repeated exposure to the stimuli. However, the responses elicited by mixed sound stimuli that were slightly different in each trial were well maintained. These findings suggest the importance of the novelty or meaning of stimuli for eliciting cortical activities in sensory cortices including the surrounding higher association areas. However, there are a number of problems in these experiments. The cortical responses in awake mice are variable from trial to trial, so that reproducible results are not easily obtained. Furthermore, nonspecific spreading of cortical activities is frequently observed. Cortical responses later than 0.5 s after stimulus onset may be masked by vascular responses, which are fast and marked in awake mice. Fortunately, vascular responses are very sensitive to photo-irradiation (Figure 7.7B,C), so that photo-inactivation of vascular responses might be useful for investigating slow cortical activities in awake mice. Regardless of these difficulties, analysis of cortical activities in awake animals is one of the most interesting and challenging problems in optical imaging of neural activities.

REFERENCES

Antonini, A., Fagiolini, M., and Stryker, M.P. 1999. Anatomical correlates of functional plasticity in mouse visual cortex. *J. Neurosci.* 19: 4388–4406.

Aubin, J.E. 1979. Autofluorescence of viable cultured mammalian cells. *J. Histochem. Cytochem.* 27: 36–43.

Bakin, J.S., Kwon, M.C., Masino, S.A., Weinberger, N.M., and Frostig, R.D. 1996. Suprathreshold auditory cortex activation visualized by intrinsic signal optical imaging. *Cereb. Cortex* 6: 120–130.

Bao, S., Chang, E.F., Woods, J., and Merzenich, M.M. 2004. Temporal plasticity in the primary auditory cortex induced by operant perceptual learning. *Nat. Neurosci.* 9: 974–981.

Baxter, M.G., and Murray, E.A. 2002. The amygdala and reward. *Nat. Rev. Neurosci.* 3: 563–573.

Bennett, B.D., Jetton, T.L., Ying, G., Magnuson, M.A., and Piston, D.W. 1996. Quantitative subcellular imaging of glucose metabolism within intact pancreatic islets. *J. Biol. Chem.* 271: 3647–3651.

Benson, R.C., Meyer, R.A., Zaruba, M.E., and McKhann, G.M. 1979. Cellular autofluorescence—is it due to flavins? *J. Histochem. Cytochem.* 27: 44–48.

Billups, B., and Forsythe, I.D. 2002. Presynaptic mitochondrial calcium sequestration influences transmission at mammalian central synapses. *J. Neurosci.* 22: 5840–5847.

Birtoli, B., and Ulrich, D. 2004. Firing mode-dependent synaptic plasticity in rat neocortical pyramidal neurons. *J. Neurosci.* 24: 4935–4940.

Budd, S.L., and Nicholls, D.G. 1996. A reevaluation of the role of mitochondria in neuronal Ca^{2+} homeostasis. *J. Neurochem.* 66: 403–411.

Budinger, E., Heil, P., and Scheich, H. 2000. Functional organization of auditory cortex in the Mongolian gerbil (*Meriones unguiculatus*). III. Anatomical subdivisions and corticocortical connections. *Eur. J. Neurosci.* 12: 2425–2451.

Cang, J., Kaneko, M., Yamada, J., Woods, G., Stryker, M.P., and Feldheim, D.A. 2005. Ephrin—as guide to the formation of functional maps in the visual cortex. *Neuron* 48: 577–589.

Chance, B., Cohen, P., Jöbsis, F.F., and Schoener, B. 1962. Intracellular oxidation-reduction states *in vivo*. *Science* 137: 499–508.

Chance, B., Schoener, B., Oshino, R., Itshak, F., and Nakase, Y. 1979. Oxidation-reduction ratio studies of mitochondria in freeze-trapped samples. NADH and flavoprotein fluorescence signals. *J. Biol. Chem.* 254: 4764–4771.

Chang, E.F., and Merzenich, M.M. 2003. Environmental noise retards auditory cortical development. *Science* 300: 498–502.

Coutinho, V., Mutoh, H., and Knöpfel, T. 2004. Functional topology of the mossy fibre-granule cell-Purkinje cell system revealed by imaging of intrinsic fluorescence in mouse cerebellum. *Eur. J. Neurosci.* 20: 740–748.

Cui, Y., Kataoka, Y., Li, Q.H., Yokoyama, C., Yamagata, A., Mochizuki-Oda, N., Watanabe, J., Yamada, H., and Watanabe, Y. 2003. Targeted tissue oxidation in the cerebral cortex induces local prolonged depolarization and cortical spreading depression in the rat brain. *Biochem. Biophys. Res. Commun.* 300: 631–636.

de la Mothe, L.A., Blumell, S., Kajikawa, Y., and Hackett, T.A. 2006. Cortical connections of the auditory cortex in marmoset monkeys: Core and medial belt regions. *J. Comp. Neurol.* 496: 27–71.

Diamond, I.T., and Neff, W.D. 1957. Ablation of temporal cortex and discrimination of auditory patterns. *J. Neurophysiol.* 20: 300–315.

Dietrich, W.D., Ginsberg, M.D., Busto, R., and Watson, B.D. 1986. Photochemically induced cortical infarction in the rat. 1. Time course of hemodynamic consequences. *J. Cereb. Blood Flow. Metab.* 6: 184–194.

Diez-Garcia, J., Akemann, W., and Knöpfel, T. 2007. In vivo calcium imaging from genetically specified target cells in mouse cerebellum. *Neuroimage* 34: 859–869.

Douglas, R.J., Koch, C., Mahowald, M., Martin, K.A.C., and Suarez, H.H. 1995. Recurrent excitation in neocortical circuits. *Science* 269: 981–985.

Dräger, U.C. 1978. Observations on monocular deprivation in mice. *J. Neurophysiol.* 41: 28–42.

Duchen, M.R. 1992. Ca^{2+}-dependent changes in the mitochondrial energetics in single dissociated mouse sensory neurons. *Biochem. J.* 283: 41–50.

Egger, V., Feldmeyer, D., and Sakmann, B. 1999. Coincidence detection and changes of synaptic efficacy in spiny stellate neurons in rat barrel cortex. *Nat. Neurosci.* 2: 1098–1105.

Eggermont, J.J. 2000. Sound-induced synchronization of neural activity between and within three auditory cortical areas. *J. Neurophysiol.* 83: 2708–2722.

Ehret, G., and Riecke, S. 2002. Mice and humans perceive multiharmonic communication sounds in the same way. *Proc. Natl. Acad. Sci. U.S.A.* 99: 479–482.

Elbert, T., Pantev, C., Wienbruch, C., Rockstroh, B., and Taub, E. 1995. Increased cortical representation of the fingers of the left hand in string players. *Science* 270: 305–307.

Fein, A., and Tsacopoulos, M. 1988. Activation of mitochondrial oxidative metabolism by calcium ions in Limulus ventral photoreceptor. *Nature* 331: 437–440.

Feldman, D.E. 2000. Timing-based LTP and LTD at vertical inputs to layer II/III pyramidal cells in rat barrel cortex. *Neuron* 27: 45–56.

Frenkel, M.Y., and Bear, M.F. 2004. How monocular deprivation shifts ocular dominance in visual cortex of young mice. *Neuron* 44: 917–923.

Fritz, J., Shamma, S., Elhilali, M., and Klein, D. 2003. Rapid task-related plasticity of spectrotemporal receptive fields in primary auditory cortex. *Nat. Neurosci.* 6: 1216–1223.

Frostig, R.D., Lieke, E.E., Ts'o, D.Y., and Grinvald, A. 1990. Cortical functional architecture and local coupling between neuronal activity and the microcirculation revealed by in vivo high-resolution optical imaging of intrinsic signals. *Proc. Natl. Acad. Sci. U.S.A.* 87: 6082–6086.

Gao, E., and Suga, N. 2000. Experience-dependent plasticity in the auditory cortex and the inferior colliculus of bats: Role of the corticofugal system. *Proc. Natl. Acad. Sci. U.S.A.* 97: 8081–8086.

Geissler, D.B., and Ehret, G. 2002. Time-critical integration of formants for perception of communication calls in mice. *Proc. Natl. Acad. Sci. U.S.A.* 99: 9021–9025.

Giffin, F., and Mitchell, D.E. 1978. The rate of recovery of vision after early monocular deprivation in kittens. *J. Physiol. (Lond.)* 274: 511–537.

Gillooly, J.F., Brown, J.H., West, G.B., Savage, V.M., and Charnov, E.L. 2001. Effects of size and temperature on metabolic rate. *Science* 293: 2248–2251.

Gordon, J.A., and Stryker, M.P. 1996. Experience-dependent plasticity of binocular responses in the primary visual cortex of the mouse. *J. Neurosci.* 16: 3274–3286.

Grutzendler, J., Kasthuri, N., and Gan, W.B. 2002. Long-term dendritic spine stability in the adult cortex. *Nature* 420: 812–816.

Gunter, T.E., Buntinas, L., Sparagna, G.C., and Gunter, K.K. 1998. The Ca^{2+} transport mechanisms of mitochondria and Ca^{2+} uptake from physiological-type Ca^{2+} transients. *Biochim. Biophys. Acta* 1366: 5–15.

Hansford, R.G. 1994. Physiological role of mitochondrial Ca^{2+} transport. *J. Bioenerg. Biomembr.* 26: 495–508.

Harel, N., Mori, N., Sawada, S., Mount, R.J., and Harrison, R.V. 2000. Three distinct auditory areas of cortex (AI, AII, and AAF) defined by optical imaging of intrinsic signals. *Neuroimage* 11: 302–312.

Harrison, R.V., Stanton, S.G., Ibrahim, D., Nagasawa, A., and Mount, R.J. 1993. Neonatal cochlear hearing loss results in developmental abnormalities of the central auditory pathways. *Acta Otolaryngol.* 113: 296–302.

Haselgrove, J.C., Bashford, C.L., Barlow, C.H., Quistorff, B., Chance, B., and Mayevsky, A. 1990. Time resolved 3-dimensional recording of redox ratio during spreading depression in gerbil brain. *Brain Res.* 506: 109–114.

Hashimoto, M., Takeda, Y., Sato, T., Kawahara, H., Nagano, O., and Hirakawa, M. 2000. Dynamic changes of NADH fluorescence images and NADH content during spreading depression in the cerebral cortex of gerbils. *Brain Res.* 872: 294–300.

Hassinen, I., and Chance, B. 1968. Oxidation-reduction properties of the mitochondrial flavoprotein chain. *Biochem. Biophys. Res. Commun.* 31: 895–900.

Hensch, T.K. 2004. Critical period regulation. *Annu. Rev. Neurosci.* 27: 549–579.

Hess, A., and Scheich, H. 1996. Optical and FDG mapping of frequency-specific activity in auditory cortex. *Neuroreport* 7: 2643–2647.

Hishida, R., Hoshino, K., Kudoh, M., Norita, M., and Shibuki, K. 2003. Anisotropic functional connections between the auditory cortex and area 18a in rat cerebral slices. *Neurosci. Res.* 46: 171–182.

Hishida, R., Kamatani, D., Kitaura, H., Kudoh, M., and Shibuki, K. 2007a. Functional local connections with differential activity-dependence and critical periods surrounding the primary auditory cortex in rat cerebral slices. *Neuroimage* 34: 679–693.

Hishida, R., Kudoh, M., and Shibuki, K. 2007b. Transcranial electrical stimulation for activating cortico-cortical connections in anesthetized mice. *Neurosci. Res.* 58 Suppl. 1: S215.

Hofer, S.B., Mrsic-Flogel, T.D., Bonhoeffer, T., and Hübener, M. 2006. Prior experience enhances plasticity in adult visual cortex. *Nat. Neurosci.* 9: 127–132.

Huang, S., Heikal, A.A., and Webb, W.W. 2002. Two-photon fluorescence spectroscopy and microscopy of NAD(P)H and flavoprotein. *Biophys. J.* 82: 2811–2825.

Hubel, D.H., and Wiesel, T.N. 1962. Receptive fields, binocular interaction and functional architecture in the cat's visual cortex. *J. Physiol. (Lond.)* 160: 106–154.

Hubel, D.H., and Wiesel, T.N. 1970. The period of susceptibility to the physiological effects of unilateral eye closure in kittens. *J. Physiol. (Lond.)* 206: 419–436.

Hubel, D.H., Wiesel, T.N., and LeVay, S. 1977. Plasticity of ocular dominance columns in monkey striate cortex. *Philos. Trans. R. Soc. Lond. B. Biol. Sci.* 278: 377–409.

Husson, T.R., Mallik, A.K., Zhang, J.X., and Issa, N.P. 2007. Functional imaging of primary visual cortex using flavoprotein autofluorescence. *J. Neurosci.* 27: 8665–8675.

Jay, D.G., and Keshishian, H. 1990. Laser inactivation of fasciclin I disrupts axon adhesion of grasshopper pioneer neurons. *Nature* 348: 548–550.

Jöbsis, F.F., O'Connor, M., Vitale, A., and Vreman, H. 1971. Intracellular redox changes in functioning cerebral cortex, I. Metabolic effects of epileptiform activity. *J. Neurophysiol.* 34: 735–749.

Kamatani, D., Hishida, R., Kudoh, M., and Shibuki, K. 2007. Experience-dependent formation of activity propagation patterns at the somatosensory S1 and S2 boundary in rat cortical slices. *Neuroimage* 35: 47–57.

Kasischke, K.A., Vishwasrao, H.D., Fisher, P.J., Zipfel, W.R., and Webb, W.W. 2004. Neural activity triggers neuronal oxidative metabolism followed by astrocytic glycolysis. *Science* 305: 99–103.

Kataoka, Y., Cui, Y.L., Maegawa, Y., Ito, T., and Watanabe, Y. 2000. Direct and suppressive action of photon on brain neurotransmission. *Jpn. J. Physiol.* 50 Suppl.: S166.

Kavanagh, N.I., Ainscow, E.K., and Brand, M.D. 2000. Calcium regulation of oxidative phosphorylation in rat skeletal muscle mitochondria. *Biochim. Biophys. Acta* 1457: 57–70.

Kazama, H., Ichikawa, A., Kohsaka, H., Morimoto-Tanifuji, T., and Nose, A. 2008. Innervation and activity dependent dynamics of postsynaptic oxidative metabolism. *Neuroscience* 152: 40–49.

Kilgard, M.P., and Merzenich, M.M. 1998. Plasticity of temporal information processing in the primary auditory cortex. *Nat. Neurosci.* 8: 727–731.

Kitaura, H., Ikeda, K., Takahashi, K., Tohmi, M., and Shibuki, K. 2005. Tail amputation reorganized somatotopic maps in mouse somatosensory cortex. *Neurosci. Res.* 52 Suppl.: S117.

Kitaura, H., Uozumi, N., Tohmi, M., Yamazaki, M., Sakimura, K., Kudoh, M., Shimizu, T., and Shibuki, K. 2007. Roles of nitric oxide as a vasodilator in neurovascular coupling of mouse somatosensory cortex. *Neurosci. Res.* 59: 160–171.

Klaidman, L.K., Leung, A.C., and Adams, J.D.Jr. 1995. High-performance liquid chromatography analysis of oxidized and reduced pyridine dinucleotides in specific brain regions. *Anal. Biochem.* 228: 312–317.

Kosterin, P., Kim, G.H., Muschol, M., Obaid, A.L., and Salzberg, B.M.. 2005. Changes in FAD and NADH fluorescence in neurosecretory terminals are triggered by calcium entry and by ADP production. *J. Membr. Biol.* 208: 113–124.

Kubota, Y., Kamatani, D., Tsukano, H., Ohshima, S., Takahashi, K., Hishida, R., Kudoh, M., Takahashi, S., and Shibuki, K. 2008. Transcranial photo-inactivation of neural activities in the mouse auditory cortex. *Neurosci. Res.* 60: 422–430.

Kudoh, M., Sakai, M., and Shibuki, K. 2002. Differential dependence of LTD on glutamate receptors in the auditory cortical synapses of cortical and thalamic inputs. *J. Neurophysiol.* 88: 3167–3174.

Kudoh, M., and Shibuki, K. 1997. Importance of polysynaptic inputs and horizontal connectivity in the generation of tetanus-induced long-term potentiation in the rat auditory cortex. *J. Neurosci.* 17: 9458–9465.

Kunz, W.S., and Gellerich, F.N. 1993. Quantification of the content of fluorescent flavoproteins in mitochondria from liver, kidney cortex, skeletal muscle, and brain. *Biochem. Med. Metab. Biol.* 50: 103–110.

LeVay, S., Wiesel, T.N., and Hubel, D.H. 1980. The development of ocular dominance columns in normal and visually deprived monkeys. *J. Comp. Neurol.* 191: 1–51.

Lewis, D.V., and Schuette, W.H. 1976. NADH fluorescence, $[K^+]_o$ and oxygen consumption in cat cerebral cortex during direct cortical stimulation. *Brain Res.* 110: 523–535.

Linden, J.F., Liu, R.C., Sahani, M., Schreiner, C.E., and Merzenich, M.M. 2003. Spectrotemporal structure of receptive fields in areas AI and AAF of mouse auditory cortex. *J. Neurophysiol.* 90: 2660–2675.

Lipton, P. 1973. Effects of membrane depolarization on nicotinamide nucleotide fluorescence in brain slices. *Biochem. J.* 136: 999–1009.

Lothman, E., Lamanna, J., Cordingley, G., Rosenthal, M., and Somjen, G. 1975. Responses of electrical potential, potassium levels, and oxidative metabolic activity of the cerebral neocortex of cats. *Brain Res.* 88: 15–36.

Mayevsky, A., and Chance, B. 1974. Repetitive patterns of metabolic changes during cortical spreading depression of the awake rat. *Brain Res.* 65: 529–533.

McCormack, J.G., Halestrap, A.P., and Denton, R.M. 1990. Role of calcium ions in regulation of mammalian intramitochondrial metabolism. *Physiol. Rev.* 70: 391–425.

McGann, J.P., Pirez, N., Gainey, M.A., Muratore, C., Elias, A.S., and Wachowiak, M. 2005. Odorant representations are modulated by intra- but not interglomerular presynaptic inhibition of olfactory sensory neurons. *Neuron* 48: 1039–1053.

Merrill, D.K., and Guynn, R.W. 1981. The calculation of the cytoplasmic free $[NADP^+]$/[NADPH] ratio in brain: effect of electroconvulsive seizure. *Brain Res.* 221: 307–318.

Merzenich, M.M., Kaas, J.H., Wall, J., Nelson, R.J., Sur, M., and Felleman, D. 1983. Topographic reorganization of somatosensory cortical areas 3b and 1 in adult monkeys following restricted deafferentation. *Neuroscience* 8: 33–55.

Mironov, S.L., and Richter, D.W. 2001. Oscillations and hypoxic changes of mitochondrial variables in neurons of the brainstem respiratory centre of mice. *J. Physiol. (Lond.)* 533: 227–236.

Mochizuki-Oda, N., Kataoka, Y., Cui, Y., Yamada, H., Heya, M., and Awazu, K. 2002. Effects of near-infra-red laser irradiation on adenosine triphosphate and adenosine diphosphate contents of rat brain tissue. *Neurosci. Lett.* 323: 207–210.

Murakami, H., Kamatani, D., Hishida, R., Takao, T., Kudoh, M., Kawaguchi, T., Tanaka, R., and Shibuki, K. 2004. Short-term plasticity visualized with flavoprotein autofluorescence in the somatosensory cortex of anesthetized rats. *Eur. J. Neurosci.* 19: 1352–1360.

Murphy, K.M., and Mitchell, D.E. 1987. Reduced visual acuity in both eyes of monocularly deprived kittens following a short or long period of reverse occlusion. *J. Neurosci.* 7: 1526–1536.

Nakahara, H., Zhang, L.I., and Merzenich, M.M. 2004. Specialization of primary auditory cortex processing by sound exposure in the "critical period." *Proc. Natl. Acad. Sci. USA* 101: 7170–7174.

Nelken, I., Bizley, J.K., Nodal, F.R., Ahmed, B., Schnupp, J.W.H., and King, A.J. 2004. Large-scale organization of ferret auditory cortex revealed using continuous acquisition of intrinsic optical signals. *J. Neurophysiol.* 92: 2574–2588.

Ohki, K., Chung, S., Ch'ng, Y.H., Kara, P., and Reid, R.C. 2005. Functional imaging with cellular resolution reveals precise micro-architecture in visual cortex. *Nature* 433: 597–603.

Ohl, F.W., Wetzel, W., Wagner, T., Rech, A., and Scheich, H. 1999. Bilateral ablation of auditory cortex in Mongolian gerbil affects discrimination of frequency modulated tones but not of pure tones. *Learn. Mem.* 6: 347–362.

Ono, K., Kudoh, M., and Shibuki, K. 2006. Roles of the auditory cortex in discrimination learning by rats. *Eur. J. Neurosci.* 23: 1623–1632.

Pantev, C., Oostenveld, R., Engelien, A., Ross, B., Roberts, L.E., and Hoke, M. 1998. Increased auditory cortical representation in musicians. *Nature* 392: 811–814.

Prusky, G.T., and Douglas, R.M. 2003. Developmental plasticity of mouse visual acuity. *Eur. J. Neurosci.* 17: 167–173.

Ragan, C.I., and Garland, P.B. 1969. The intra-mitochondrial localization of flavoproteins previously assigned to the respiratory chain. *Eur. J. Biochem.*10: 399–410.

Rajan, R., Irvine, D.R., Wise, L.Z., and Heil, P. 1993. Effect of unilateral partial cochlear lesions in adult cats on the representation of lesioned and unlesioned cochleas in primary auditory cortex. *J. Comp. Neurol.* 338. 17–49.

Recanzone, G.H., Merzenich, M.M., Jenkins, W.M., Grajski, K.A., and Dinse, H.R. 1992a. Topographic reorganization of the hand representation in cortical area 3b owl monkeys trained in a frequency-discrimination task. *J. Neurophysiol.* 67: 1031–1056.

Recanzone, G.H., Merzenich, M.M., and Schreiner, C.E. 1992b. Changes in the distributed temporal response properties of SI cortical neurons reflect improvements in performance on a temporally based tactile discrimination task. *J. Neurophysiol.* 67: 1071–1091.

Recanzone, G.H., Schreiner, C.E., and Merzenich, M.M. 1993. Plasticity in the frequency representation of primary auditory cortex following discrimination training in adult owl monkeys. *J. Neurosci.* 13: 87–103.

Reinert, K.C., Dunbar, R.L., Gao, W., Chen, G., and Ebner, T.J. 2004. Flavoprotein autofluorescence imaging of neuronal activation in the cerebellar cortex *in vivo*. *J. Neurophysiol.* 92: 199–211.

Rizzuto, R., Bernardi, P., and Pozzan, T. 2000. Mitochondria as all-round players of the calcium game. *J. Physiol. (Lond.)* 529: 37–47.

Rodrigues, S.M., Schafe, G.E., and LeDoux, J.E. 2004. Molecular mechanisms underlying emotional learning and memory in the lateral amygdala. *Neuron* 44: 75–91.

Romo, R., and Salinas, E. 2003. Flutter discrimination: neural codes, perception, memory and decision making. *Nat. Rev. Neurosci.* 4: 203–218.

Rosenthal, M., and Jöbsis, F.F. 1971. Intracellular redox changes in functioning cerebral cortex. II. Effects of direct cortical stimulation. *J. Neurophysiol.* 34: 750–762.

Sawtell, N.B., Frenkel, M.Y., Philpot, B.D., Nakazawa, K., Tonegawa, S., and Bear, M.F. 2003. NMDA receptor-dependent ocular dominance plasticity in adult visual cortex. *Neuron* 38: 977–985.

Schady, W., Braune, S., Watson, S., Torebjork, H.E., and Schmidt, R. 1994. Responsiveness of the somatosensory system after nerve injury and amputation in the human hand. *Ann. Neurol.* 36: 68–75.

Schuchmann, S., Kovacs, R., Kann, O., Heinemann, U., and Buchheim K. 2001. Monitoring NAD(P)H autofluorescence to assess mitochondrial metabolic functions in rat hippocampal-entorhinal cortex slices. *Brain Res. Protocols* 7: 267–276.

Schuett, S., Bonhoeffer, T., and Hubener, M. 2002. Mapping retinotopic structure in mouse visual cortex with optical imaging. *J. Neurosci.* 22: 6549–6559.

Shibuki, K. 1989. Calcium-dependent and ouabain-resistant oxygen consumption in the rat neurohypophysis. *Brain Res.* 487: 96–104.

Shibuki, K. 1990. Activation of neurohypophysial vasopressin release by Ca^{2+} influx and intracellular Ca^{2+} accumulation in the rat. *J. Physiol. (Lond.)* 422: 321–331.

Shibuki, K., and Hishida, R. 2004. Flavoprotein autofluorescence: an underestimated signal for functional brain imaging. *Physiology News* 54: 18–19.

Shibuki, K., Hishida, R., Kitaura, H., Takahashi, K., and Tohmi, M. 2007. Coupling of brain function and metabolism: Endogenous flavoprotein fluorescence imaging of neural activities by local changes in energy metabolism. In *Handbook of Neurochemistry and Molecular Neurobiology. 3rd edition, Vol. 5, Neural Energy Utilization,* Ed. G. Gibson, and G. Dienel, 322–342. New York: Springer.

Shibuki, K., Hishida, R., Kudoh, M., and Teduka, K. 2004. Trans-cranial imaging of awaking mouse cortical activities using flavoprotein autofluorescence. *Neurosci. Res.* 50 Suppl. 1: S159.

Shibuki, K., Hishida, R., Murakami, H., Kudoh, M., Kawaguchi, T., Watanabe, M., Watanabe, S., Kouuchi, T., and Tanaka, R. 2003. Dynamic imaging of somatosensory cortical activities in the rat visualized by flavoprotein autofluorescence. *J. Physiol. (Lond.)* 549: 919–927.

Shibuki, K., Ono, K., Hishida, R., and Kudoh, M. 2006. Endogenous fluorescence imaging of somatosensory cortical activities after discrimination learning in rats. *Neuroimage* 30: 735–744.

Shuttleworth, C.W., Brennan, A.M., and Connor, J.A. 2003. NAD(P)H fluorescence imaging of postsynaptic neuronal activation in murine hippocampal slices. *J. Neurosci.* 23: 3196–3208.

Stiebler, I., Neulist, R., Fichtel, I., and Ehret, G. 1997. The auditory cortex of the house mouse: left-right differences, tonotopic organization and quantitative analysis of frequency representation. *J. Comp. Physiol. A* 181: 559–571.

Tagawa, Y., Kanold, P.O., Majdan, M., and Shatz, C.J. 2005. Multiple periods of functional ocular dominance plasticity in mouse visual cortex. *Nat. Neurosci.* 8: 380–388.

Takahashi, K., Hishida, R., Kubota, Y., Kudoh, M., Takahashi, S., and Shibuki, K. 2006. Transcranial fluorescence imaging of auditory cortical plasticity regulated by acoustic environments in mice. *Eur. J. Neurosci.* 23: 1265–1276.

Takao, T., Murakami, H., Fukuda, M., Kawaguchi, T., Kakita, A., Takahashi, H., Kudoh, M., Tanaka, R., and Shibuki, K. 2006. Transcranial imaging of audiogenic epileptic foci in the cortex of DBA/2J mice. *Neuroreport* 17: 267–271.

Takei, K., Shin, R.M., Inoue, T., Kato, K., and Mikoshiba, K. 1998. Regulation of nerve growth mediated by inositol 1,4,5-trisphosphate receptors in growth cones. *Science* 282: 1705–1708.

Territo, P.R., Mootha, V.K., French, S.A., and Balaban, R.S. 2000. Ca^{2+} activation of heart mitochondrial oxidative phosphorylation: Role of the F0/F1-ATPase. *Am. J. Physiol.* 278: C423–435.

Tohmi, M., Kitaura, H., Komagata, S., Kudoh, M., and Shibuki, K. 2006. Enduring critical period plasticity visualized by transcranial flavoprotein imaging in mouse primary visual cortex. *J. Neurosci.* 26: 11775–11785.

Tomlinson, F.H., Anderson, R.E., and Meyer, F.B. 1993. Brain pHi, cerebral blood flow, and NADH fluorescence during severe incomplete global ischemia in rabbits. *Stroke* 24: 435–443.

Vanzetta, I., and Grinvald, A. 1999. Increased cortical oxidative metabolism due to sensory stimulation: Implications for functional brain imaging. *Science* 286: 1555–1558.

Versnel, H., Mossop, J.E., Mrsic-Flogel, T.D., Ahmed, B., and Moore, D.R. 2002. Optical imaging of intrinsic signals in ferret auditory cortex: Responses to narrowband sound stimuli. *J. Neurophysiol.* 88: 1545–1558.

Wang, X., Merzenich, M.M., Sameshima, K., and Jenkins, W.M. 1995. Remodelling of hand representation in adult cortex determined by timing of tactile stimulation. *Nature* 378: 71–75.

Weber, B., Burger, C., Wyss, M.T., von Schulthess, G.K., Scheffold, F., and Buck, A. 2004. Optical imaging of the spatiotemporal dynamics of cerebral blood flow and oxidative metabolism in the rat barrel cortex. *Eur. J. Neurosci.* 20: 2664–2670.

Weiss, T., Miltner, W.H., Liepert, J., Meissner, W., and Taub, E. 2004. Rapid functional plasticity in the primary somatomotor cortex and perceptual changes after nerve block. *Eur. J. Neurosci.* 20: 3413–3423.

Wiesel, T.N., and Hubel, D.H. 1965. Extent of recovery from the effects of visual deprivation in kittens. *J. Neurophysiol.* 28: 1060–1072.

Willott, J.F., Aitkin, L.M., and McFadden, S.L. 1993. Plasticity of auditory cortex associated with sensorineural hearing loss in adult C57BL/6J mice. *J. Comp. Neurol.* 329: 402–411.

Zhang, L.I., Bao, S., and Merzenich, M.M. 2001. Persistent and specific influences of early acoustic environments on primary auditory cortex. *Nat. Neurosci.* 4: 1123–1130.

Zuo, Y., Yang, G., Kwon, E., and Gan, W.B. 2005. Long-term sensory deprivation prevents dendritic spine loss in primary somatosensory cortex. *Nature* 436: 261–265.

8 Functional Imaging with Mitochondrial Flavoprotein Autofluorescence

Theory, Practice, and Applications

T. Robert Husson and Naoum P. Issa

CONTENTS

8.1 INTRODUCTION

While great strides have been made in the development of novel optical imaging techniques in the past two decades, these methods still either measure coarsely or require relatively invasive procedures. The ideal optical measure would faithfully represent local neural activity, both spiking and nonspiking, at subcellular resolution, with persistent, reliable responses without the use of exogenous agents. Practically, there are several opportunities to measure endogenous signals that could serve as indirect measures of neural activity: most commonly, metabolic processes which take advantage of the tight coupling between neural activity and ATP production. Unfortunately, the most robust metabolic signals (such as blood deoxygenation) are only weakly linked to neural activity, and can be spatially diffuse, putting strict limitations on the interpretation of these data.

Recently, the endogenous fluorescence of mitochondrial flavoproteins (FA, flavoprotein autofluorescence) has been exploited for functional imaging in a variety of preparations, both in vivo and in vitro. These flavoproteins are oxidized during aerobic metabolism, which is closely coupled to neuronal signaling, and their fluorescence has been shown to follow neural activity in response to pharmacological, electrical, and sensory stimulation. They have been used to confirm functional maps in several cortical areas, and are supported by a growing body of in vitro work. As these signals (1) are highly spatially localized, (2) can be measured by a variety of optical techniques both in vitro and in vivo, and (3) require no exogenous dyes, they offer substantial improvement in our ability to map functional activity and trace neural circuits. Here, we describe the underlying biochemistry of the flavoprotein signal, the experimental details of using it for functional imaging in vivo, and outline example studies and future directions for this promising new approach to visualizing neural activity patterns.

8.2 SOURCES OF FLAVOPROTEIN AUTOFLUORESCENCE SIGNALS

8.2.1 The Relationship between Neural Activity and Neuronal Metabolism

Neurons, from a metabolic perspective, are exceptionally expensive cells to maintain—so much so that metabolic demand has placed a significant constraint on brain evolution. The massive expansion in gray matter volume during the emergence of the hominids is thought to have occurred after the acquisition of a higher quality diet and the concomitant reduction in gut size (another metabolically expensive tissue) necessary to increase brain volume (the Expensive Tissue Hypothesis[1]). This energetic cost may even constrain neural processing. Some studies suggest that the way in which neurons encode information is designed to minimize energy demand (reviewed in Reference 2), and that an optimal coding scheme represents a balance between the metabolic cost of an action potential and the fixed cost of maintaining a neuron at resting potential: when this ratio is low the most energy-efficient system requires few neurons that each encode large amounts of information with a high average firing rate, but when it is high (spikes are metabolically expensive) the balance is shifted towards sparse coding schemes. It is clear that the metabolic needs of neurons are enormous and may be an important factor in the brain's development and function.

Even during a resting state (a period of relatively low-action potential generation), neurons must actively maintain precise ion gradients, preserve complicated morphologies, recycle neurotransmitter, and maintain postsynaptic receptor distribution and function, all of which require large amounts of ATP.[3] In order to meet this demand, neurons are equipped with a high density of mitochondria (17.3% of cellular volume, as compared to 11.0% in astrocytes and 11.3% in oligodendrocytes[4]), highest in regions involved with synaptic action such as dendrites (62% of neuronal mitochondria are found in dendrites) and axons/presynaptic terminals (36%).[3,5] To keep these mitochondria running, the brain must consume a disproportionately large quantity of metabolic substrates, specifically oxygen and glucose from arterial blood. Despite representing only 2% of body mass, the brain consumes 20% of available oxygen and 25% of the body's glucose.[4,6]

This enormous demand is exacerbated by increased neural signaling. Attwell and Laughlin's estimates for the rat gray matter suggest that as much as 81% of ATP production in neurons is devoted to electrical activity, most of which is accounted for by postsynaptic effects of glutamate (53%) and a substantial amount is allotted to axonal and presynaptic processes (42%).[3] They estimate that a 1 Hz increase in the firing rate of an excitatory cortical neuron (with a background rate of 4 Hz) requires a 16% increase in ATP production (though this relationship may not be linear at higher rates). Most of this ATP is consumed by the sodium-potassium ATPase, but many other electrochemical mechanisms consume significant amounts of ATP as well (including glutamate recycling, presynaptic calcium extrusion, etc).

Despite the large variability in metabolic demand, ATP levels in neurons are surprisingly constant in vivo[7]; thus, there must be a rapid, direct upregulation of ATP production in response to neural activity. Within active cerebral cortex, increased neural activity requires large amounts of glucose and oxygen. As a result, oxygen is drawn from the local blood supply, desaturating hemoglobin[8–10] and inducing a subsequent spatially diffuse and long-lasting reperfusion of the active area by local arterial vasodilation.[11–13] If the demand for oxygen following increased activity is not met, perhaps due to damage of the local vasculature as occurs during a stroke, ATP depletion causes failure of Na/K ATPase and loss of transmembrane ion gradients; in cortex, all electrical activity ceases within seconds of ischemic injury.[14,15] Clearly, neuronal signaling is immediately dependent on an adequate supply of oxygen and glucose.[14,16–18]

The rapid use of oxygen immediately after neural activity suggests that a substantial fraction of activity-induced metabolism is aerobic and involves activation of the mitochondrial tricarboxylic acid cycle (TCA). The TCA cycle activity is therefore well correlated with the energy demands of active neurons, and could serve as a highly sensitive metabolic reporter of electrical activity. Because several complexes within the TCA cycle and electron-transport chain contain flavoproteins whose fluorescence changes with oxidation state, the metabolic changes induced by activity are readily accessible in the modulation of neuronal autofluorescence. In the next section we review the sources and properties of the FA signal, highlighting its advantages for use in functional imaging.

8.2.2 Origin of the Flavoprotein Autofluorescence Signal

8.2.2.1 Review of Brain Metabolism and Glucose Utilization

Unlike other tissues in the body, the brain is nearly completely reliant on one form of energy production. Stoichiometric relationships between oxygen consumption (160 μmol/100 g/min), and carbon dioxide (identical to oxygen) suggest that the respiratory quotient of the brain is very nearly 1, indicative of pure carbohydrate metabolism.[19] Indeed, its primary energy substrates are glucose (31 μmol/100 g/min) and ketone bodies, and there is very little beta-oxidation of fatty acids.[7,20,21] Ketone bodies are thought to be metabolized only when glucose is unavailable, such as during starvation,[22–24] so efforts to understand brain metabolism have focused on glycolytic and glycogenolytic processes.[7] Glycogen, the only stored energy source in the brain, only exists in small amounts and is mostly stored in astrocytes; it is thought to only become available to neurons during increased activity or hypoglycemia.[25] Therefore, the only major energy substrate for neurons is arterial glucose, making the brain almost entirely reliant on the local vasculature to accommodate its large metabolic demands.

Glucose metabolism in the brain involves several stages where fluorescent molecules are oxidized or reduced, leading to putative imaging signals (Figure 8.1). Glucose is first converted to pyruvate in the cytoplasm during glycolysis, a process that reduces two fluorescent nicotinamide adenine dinucleotide (NAD+) molecules to NADH (uses of this fluorescence change are discussed elsewhere, see Reference 26 and Reference 27, and Section 8.2.3 below). This pyruvate is carried into the mitochondria and converted to acetyl coenzyme-A via the pyruvate dehydrogenase complex (PDC), a series of enzymes that includes a flavoprotein adenine dinucleotide (FAD) moiety in the dihydrolipoyl dehydrogenase active site, which when oxidized in the presence of blue visible light emits green fluorescence (characterized below, see Section 8.3.1). The activity of this enzyme involves the reduction of NAD+ and the double-oxidation of $FADH_2$, thus lowering NADH autofluorescence and increasing flavoprotein autofluorescence.[28–30] The resulting acetyl coenzyme-A can then enter the TCA cycle. The TCA cycle converts three more NAD+ molecules to NADH, which then enter the electron transport chain (ETC), donating their electrons to generate the chemiosmotic gradient that drives ATP synthase's conversion of ADP to ATP; it is this activity of the TCA cycle that directly corresponds to the aerobic production of ATP in cells.

The first site of the electron transport chain, complex I (NADH ubiquinone oxidoreductase), is a membrane-bound protein that converts two NADH to NAD+ and

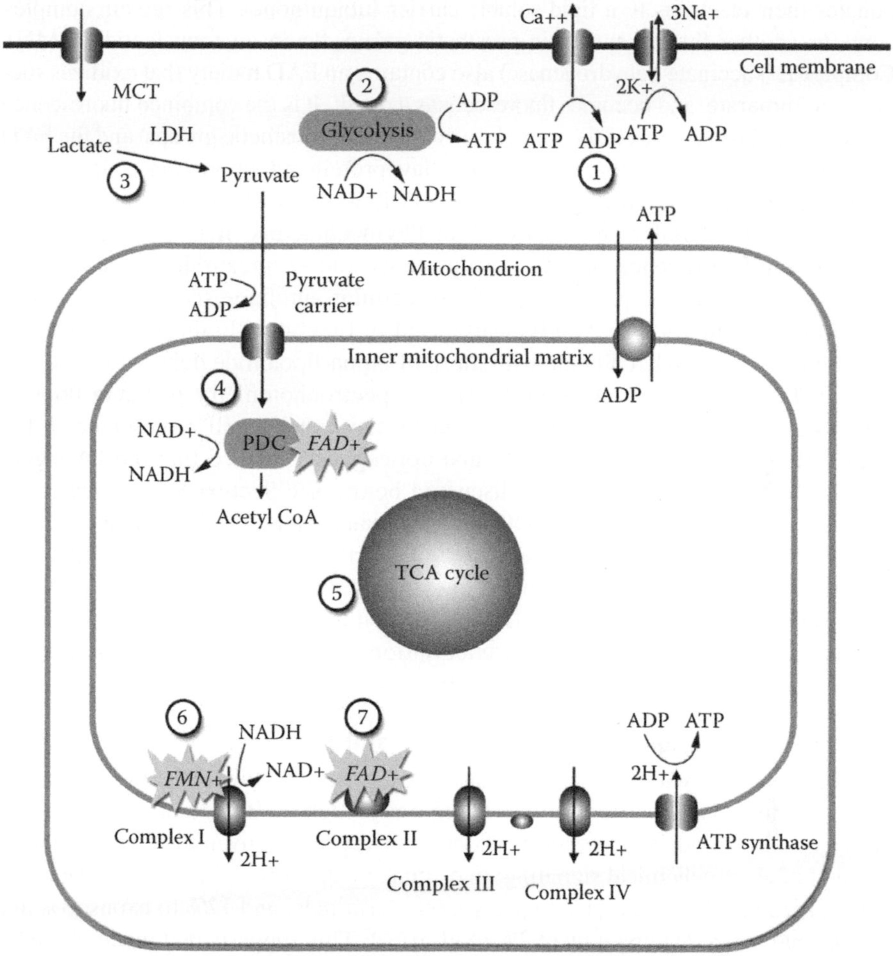

FIGURE 8.1 Mitochondrial sources of fluorescence. Fluorescent imaging signals derive from several stages of aerobic cellular metabolism. In response to increased neural activity, ATP hydrolysis (1) increases and the elevated ADP concentration upregulates the production of pyruvate through glycolysis (2); the concomitant formation of NADH reduces fluorescence from NAD^+. Pyruvate, either from glycolysis in the neuron or converted from lactate supplied by astrocytes (through monocarboxylate transporter 2 (MCT)) by pyruvate dehydrogenase (PDH) (3) is transported into the inner mitochondrial membrane where the pyruvate dehydrogenase complex (PDC) converts it to acetyl coenzyme-A (4); in the process NAD^+ fluorescence is reduced and FAD^+ fluorescence increases. This acetyl CoA enters the TCA cycle (5), eventually feeding into the electron transport chain. Within the electron transport chain Complex I activity increases both NAD^+ and FMN^+ fluorescence (6), and complex II increases FAD^+ fluorescence (7). The combined fluorescence of FMN^+ and FAD^+ in this pathway is the principle component of the flavoprotein autofluorescence signal.

donates their electrons to a lipid-soluble carrier (ubiquinone). This protein complex contains another fluorescent flavin prosthetic group, flavin mononucleotide (FMN). Complex II (succinate dehydrogenase) also contains an FAD moiety that oxidizes succinate to fumarate, and becomes fluorescent as a result. It is the combined fluorescence of these two flavoproteins (ETC complexes with flavin prosthetic groups) and the FAD associated with the PDC that underlies the flavoprotein autofluorescence signal used in brain imaging studies (though the FMN contribution is likely negligible[31]).

While many cellular enzymes contain flavin cofactors, it is well established that the autofluorescence of ETC flavoproteins and α-lipoamide dehydrogenase (from the PDC) are the primary sources of cellular autofluorescence.[28–30,32,33] The excitation and emission spectra typically used in functional brain imaging studies (Figure 8.3A,B) closely coincide with those of alpha-lipoamide dehydrogenase and ETC FAD fluorescence; although, based on spectrophotometry of flavin fluorescence signals, ETC flavoprotein fluorescence may not be well represented in the brain[31] (Figure 8.3C). When Shibuki and colleagues[34] showed that the FA signal is indeed a result of flavoproteins (discussed below, see Section 8.4.3), they could not distinguish between ETC or PDC flavoproteins because they used a nonspecific flavoprotein inhibitor (diphenyleneiodonium). Further study is needed to fully differentiate among these signals in brain in vivo; but, importantly, whether or not the flavoprotein signal is derived from ETC or PDC, it is restricted to the mitochondrial matrix and measures oxidative phosphorylation only, not glycolysis or any other activity-dependent cellular functions.

8.2.2.2 Localization of the Signal in Cells and Tissue

ATP diffuses slowly through the cytoplasm, so a general feature of cellular organization is that mitochondria are localized to regions of high ATP demand.[35,36] The distribution of mitochondria in neocortex is, then, unsurprising, given the high metabolic demand of electrochemical signaling: in primate visual cortex, 62% of mitochondria are localized to dendrites, 21% to presynaptic terminals, and 12% to axons (3% are in neuronal soma, leaving a mere 2% to glia).[5,6,37] This suggests that metabolic processes in neural tissue are mostly devoted to the maintenance of ionic gradients and the generation and propagation of electrical potentials. Attwell and Laughlin's energy budget for excitatory neurons estimated that 47% of ATP was consumed in processes related to the generation of action potentials, 34% to postsynaptic potentials, 13% to the resting membrane potential, and 3% each to glutamate recycling and presynaptic calcium flux.[3] These calculations strongly suggest that mitochondrial ATP production is spatially correlated with electrochemical activity and is primarily localized to synaptic sites.

Unfortunately, there is still no consensus on the pre- or post-synaptic source of the flavoprotein signal. Shibuki et al.[34] blocked postsynaptic potentials in cortical slices with CNQX and only partially abolished the FA signal, suggesting that presynaptic and postsynaptic activity contributed approximately equally to the overall signal. Reinert et al.[38] used CNQX in the cerebellum in vivo and found a 92% decrease in the amplitude of the light phase response, suggesting an almost entirely postsynaptic origin. This mismatch may reflect differences in glutamate receptor distribution in the cerebrum and cerebellum, or could be a result of metabolic differences in vivo versus in vitro. Both groups found that using a calcium-free bath reduced FA responses almost completely, which is consistent with a greater dependence on synaptic events than on action potentials, per se. It is therefore unclear whether the FA signal reflects presynaptic or postsynaptic activity or a combination of both.

Despite the much greater mitochondrial content of neurons, glial cells might also contribute to brain energy metabolism. There is evidence suggesting that the metabolic demands of sustained activity eventually outstrip a neuron's ability to generate ATP, and that astrocytes can be recruited to help sustain neural activity. While it is known that ~95% of ATP in the brain is produced by mitochondrial aerobic metabolism and only ~1–5% by glycolysis,[37] positron emission tomography (PET) studies by Fox et al.[39] noted a mismatch among signals interpreted as metabolic rate of oxygen consumption, glucose consumption, and blood flow, a phenomenon referred to as *metabolic uncoupling*. This suggested that, at rest, neurons are at or near their oxidative limit, so stimulus-evoked activity must very quickly recruit other sources of substrates for the TCA cycle. It has been proposed that astrocytes, which can detect increased activity through glutamate uptake at excitatory synapses, help sustain neural activity by generating excess lactate through glycolysis that is then "shuttled" to neurons, converted into pyruvate via lactate dehydrogenase, and fed into the TCA cycle.[19,40,41] While neural activity does lead to increased lactate production, whether this lactate can be taken up or used by neurons remains contentious.[42–45] This debate has important consequences for neurometabolic coupling, because it means that the relationship between the use of metabolic substrates and electrical activity may depend critically on glia. Astrocytes are now known to participate in functional properties of synaptic transmission, and there is evidence that their metabolic response to sensory stimulation has been greatly underestimated (reviewed in Reference 42). For example, it is known that glycogen stored in astrocytes is made available to neurons during periods of hypoglycemia or increased activity, and could be transported to neurons either as lactate or glucose.[46–48] Similarly, it is now well established that astrocytes respond to local neuronal activity by increasing aerobic glycolysis—a process that has been both measured chemically and imaged.[8,40] In general, it seems that astrocytes metabolically support neuronal activity by providing substrates for activity-dependent increases in oxidative metabolism.

The precise contribution from glia in metabolic optical signals, however, is not yet clear. As mentioned above, only 2% of mitochondria in the brain are glial, suggesting their contribution to mitochondrial signals is small, and these mitochondria are mostly restricted to cell bodies due to the narrow width of astrocytic filopodia.[42,49] Several studies indicate that increased neural activity induces increased glycolysis in astrocytes, not oxidative phosphorylation, though astrocytes certainly have oxidative capacity.[42] Because FA is a measure of oxidative phosphorylation only, it should not modulate in astrocytes during changes in neural activity; conversely, NADH autofluorescence will decrease because glycolysis, astrocytic or neuronal, results in the reduction of some NAD^+ to NADH. FA will also increase during any subsequent oxidative phosphorylation of "shuttled" lactate, and be directly related to any aerobic generation of ATP involved with the generation of action potentials (interestingly, Shibuki et al.[34] found that adding 50 mM L-lactate to the agar well used in vivo nearly doubled the FA response amplitude). Indeed, Kashiske et al. found that astrocytes emit high levels of background NADH autofluorescence and responded to neural activity with increased glycolysis, not oxidative phosphorylation.[8] In addition, it has been shown that in fibroblasts the NADH signal is not restricted to mitochondria, but show significant signal from other structures, especially lysosomes[26,50] (Figure 8.5). Thus, flavoprotein autofluorescence is a more direct indicator of neuronal ATP production, whereas NADH autofluorescence may represent a combination of both neuronal and glial metabolism.

Converging evidence suggests that there is a very close relationship between metabolic activity and neural signaling, and that neurons must allocate energy resources very carefully in order to participate in functional activity. Also, because the energy

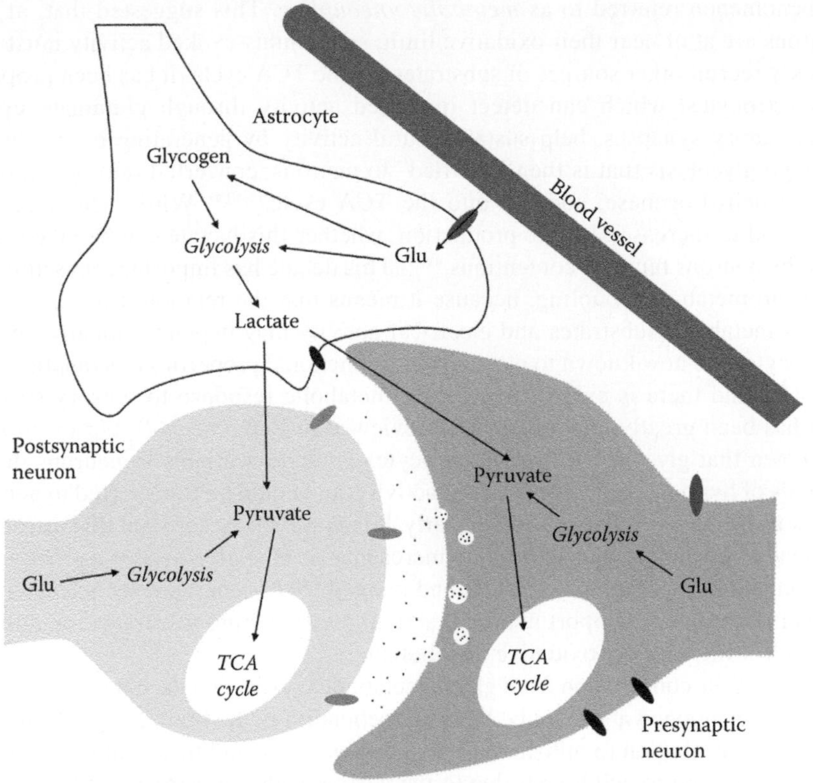

FIGURE 8.2 Cellular sources of flavoprotein autofluorescence. Flavoprotein autofluorescence signals derive from oxidative phosphorylation and therefore localize primarily to mitochondria in active neuronal tissue. Glucose is removed from arterial blood by the GLUT-1 transporter in astrocytes and the GLUT-3 transporter in neurons. Neuronal glucose undergoes glycolysis and pyruvate enters the mitochondria for oxidative phosphorylation of ADP. Astrocytic glucose can be converted to glycogen for storage or enter glycolysis. Crucially, astrocytic metabolism at synapses is glycolytic, not oxidative, as there are no mitochondria in astrocytic filopodia. Astrocytic glycolysis may produce lactate which could then be shuttled to neurons where it would be converted to pyruvate and fed into the TCA cycle.

demands of firing action potentials are immediate, increases in neuronal firing must be followed by proportional increases in ATP production in quick succession, or else the underlying homeostatic mechanisms required to keep neurons alive and functional will quickly fail. The question for physiologists then becomes, what are these underlying metabolic processes, and how can they be measured? As we will see, there are several opportunities for measuring local metabolic activity in the brain, which because of this tight neurometabolic coupling can be used as indirect measures of neural activity.

8.2.3 Imaging Opportunities from Neuronal Metabolism

Given the tight relationship between neural firing rate and energy metabolism, it has become increasingly popular to use the latter as a measure of the former. Ubiquitous well-established imaging techniques such as functional magnetic resonance imaging (fMRI) and positron-emission tomography (PET) rely on metabolic signals as measures of functional activity, but are limited in spatiotemporal resolution and have difficulty resolving precise neural activity patterns. There are several opportunities to measure the consumption of metabolic substrates in vivo using optical techniques; these offer much greater spatial and temporal resolution and can be used on both large (whole cortical area) and small (cellular circuit) levels. These techniques are all nondestructive, in that they preserve *in vivo* connections, do not require the use of exogenous agents, and can be used simultaneously with each other and in combination with standard electrophysiological techniques. Each has particular advantages and disadvantages, but either alone or in concert offers ways of examining neural activity on both large and small spatial scales with relatively little disruption of the underlying system.

Increased metabolic action requires a continuous and adequate supply of oxygen, and thus oxygenated hemoglobin. The rate at which tissue consumes oxygen can be measured optically due to the difference in absorption spectra between oxygenated and deoxygenated hemoglobin (the "intrinsic signal"[51,52]). However, the relationship between blood flow and neural activity is highly nonlinear. Directly after neural stimulation, oxygen tension in the stimulated area drops considerably, followed by a large, diffuse increase corresponding to wide-scale vasodilation.[11,12,53] This gives rise to the familiar "initial dip" seen in blood oxygen-level dependent (BOLD) imaging techniques such as intrinsic signal imaging (ISI) and fMRI, and the subsequent long, nonlocalized increase in signal that necessitates a great deal of spatial filtering and signal averaging.[54] Also, because local tissue perfusion takes place in the extracellular space, the theoretical limit to the spatial resolution of ISI is much larger than single cells. The size of the hemodynamic signal can be appreciated by considering spatial correlations in unfiltered intrinsic signal images; using this metric, we found that the hemodynamic signal encompasses an area approximately 30% larger than that delineated by flavoprotein autofluorescence.[54] There is also substantial variation in small-scale vasculature between cortical areas,[55] which could represent a problem for comparing functional activation between areas. However, ISI signals are very robust and require less illumination with high-energy light,[54] suggesting that this technique may be best suited for large-scale mapping of cortical regions in which illumination is limited or precise spatiotemporal information is not critical.

Another opportunity to use metabolic signals as a proxy for neural activity is the fluorescence of an electron-carrying molecule found in all cells, NADH. NAD^+ is reduced in both glycolysis and oxidative phosphorylation (see Section 8.2.2.1 above), causing small decreases in fluorescence that can be measured as a percent change from baseline. Importantly, this signal has been used to obtain high resolution (single-cell) images of functional activation in vivo by the use of two-photon microscopy, which was combined with both local field potential and oxygen tension measurements to produce a time-resolved view of spreading depression in hippocampal slices.[8] This study utilized at once many of the advantages offered by optical imaging—namely, simultaneous physiological measurements, single-cell resolution, and resolved temporal information. Unfortunately, we are not aware of any in vivo experiments using NADH autofluorescence with two-photon microscopy, so its usefulness for functional mapping has yet to be established. Also, as mentioned above (see Section 8.2.2.1), NADH fluorescence contains a significant glycolytic component, which may be useful for differentiating cell types on small scales, but which may overrepresent the metabolism of astrocytes over neurons over larger areas or in thicker tissue. This sentiment is echoed by Kann and Kovacs,[56] who argue that NADH signals must be interpreted with care due to astrocytic and extramitochondrial contamination.

A more direct measure of activity-dependent neuronal metabolism would only measure oxidative phosphorylation, which increases in neurons but not astrocytes during neural activity. FA represents the best opportunity in this regard, and its signal properties have been well established both in vitro and in vivo. While its signal strength is relatively small compared with ISI and NADH,[26,27,57] and appears to be weaker in larger mammals than in rodents,[54,58] it is measurable using standard optical imaging techniques in both mouse and cat. FA functional maps have been confirmed with ISI, and in fact show some superiority in spatial resolution even at large scales, and have been used to confirm maps of spatial frequency first measured with ISI but thought to be contaminated by hemodynamic artifacts (discussed below, see Section 8.4.2). The precise spatial resolution of FA signals has not been measured in neurons, but in prepared cardiomyocytes FA was able to resolve single mitochondria.[26] There is an immediate need for the characterization of the flavoprotein signal in neurons with two-photon microscopy, ideally in direct comparison with NADH autofluorescence. Evidence to date suggests that FA has the particular advantages of being restricted to subcellular compartments, representing neuronal (not glial) metabolism, and having the ability to measure both large-scale functional maps and small-scale neural circuits with the same optical signal.

It was initially thought that flavoprotein and NADH signals were exactly the same except opposite in sign, indicating that they both measured redox state with a strongly mitochondrial source.[59] While the flavoprotein time to peak and dark-phase temporally match the dip and prolonged light-phase of NADH autofluorescence, more precise measurements of these signals have revealed important differences[34,38,60]: First, the FA signal is more dependent on tissue oxygen availability than NADH signals, providing further evidence that flavoprotein signals represent oxidative and not glycolytic activity. Second, NADH and FA are differentially affected by metabolic uncouplers (which separate the ETC from the phosphorylation of ATP), suggesting that the signals can be pharmacologically isolated. Third, while both signals show increased time to peak and eliminated delayed phases in response to blocked Ca^{++} entry, simply decreasing extracellular calcium preferentially inhibits the flavoprotein signal. Finally, as mentioned above (sec Section 8.2.2.2), the NADH signal shows a large contribution from both astrocytes and nonmitochondrial cellular elements. Thus. while the two signals measure similar metabolic processes, they should not be considered substitutes, but rather strategic complements.

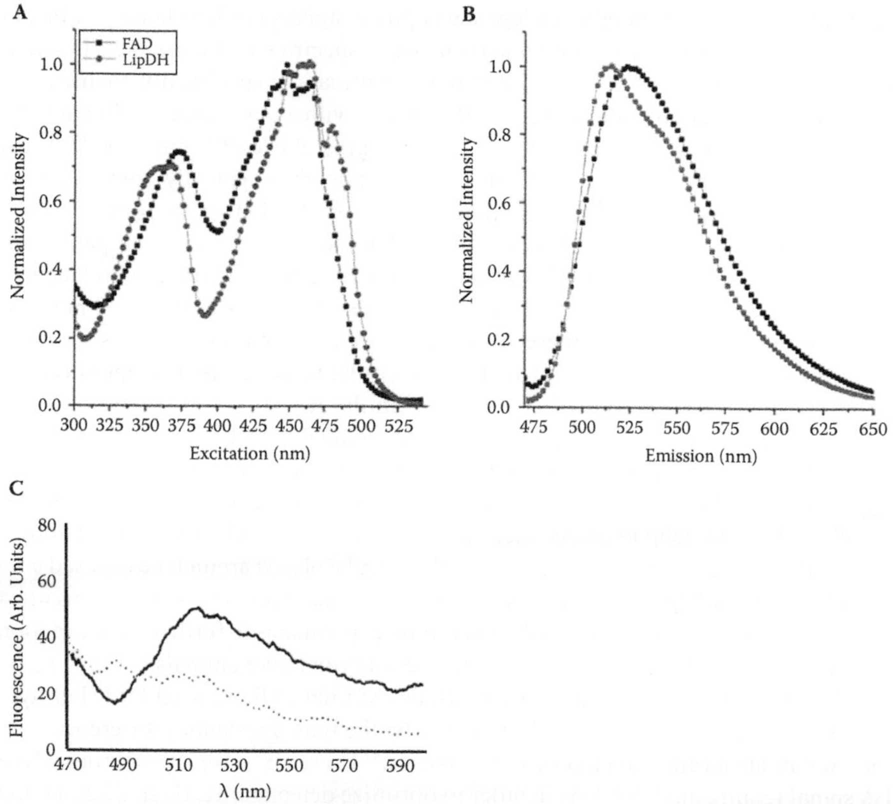

FIGURE 8.3 Excitation and emission spectra of flavoprotein autofluorescence. (A) The excitation spectra of electron transport chain flavins (FAD) and lipoamide dehydrogenase (from the PDC) are very similar, showing two peaks including a maximum at approximately 430–470 nm. (B) The emission spectra of these signals is also very similar, with a strong peak at approximately 520 nm and decaying emission through the visible spectrum. (From Huang, S. et al. [2002]. *Biophys. J.* 82 (5), 2811–2825. With permission.) (C) The contribution of ETC (dotted trace) and PDC flavins (black trace) to autofluorescence emission in brain. In the brain, PDC fluorescence, rather than ETC flavins, may be the dominant component of the flavoprotein autofluorescence signal. (From Kunz, W.S. and Gellerich, F.N. [1993]. *Biochem Med Metab Biol* 50 (1), 103–110. With permission.)

In general, endogenous metabolic signals are an underutilized tool in visualizing the dynamics and spatial organization of neural activity. Many opportunities exist both in vitro and in vivo to use these complementary techniques, especially together, to better understand neurometabolic coupling, the relationship between neurons and glia, local and distributed networks, functional aspects of neural circuits, and pathological conditions (discussed in Section 8.5.1 below).

8.3 METHODS AND SIGNAL PROPERTIES

8.3.1 EQUIPMENT AND METHODOLOGY

FA is distinguished from other endogenous optical signals (such as hemoglobin and NADH) by its particular excitation and emission spectra (see Figure 8.3). Loosely, flavoproteins emit green fluorescence in blue illumination; specifically, their excitation wavelengths are mainly between 420–490 nm with a maximum at 450 nm[26,56,61] and emit photons over a wide range of wavelengths above 500 nm.[31,32,62,63] Thus, for optical imaging purposes, excitation filter bandwidths typically emphasize the 450–490 nm range while emissions filters are typically long-pass above 500 nm (Takahashi et al.[64] use a 470–490 excitation filter and 500–550 nm band-pass filter, presumably to reduce the possibility of hemodynamic artifacts at longer wavelengths). It should be noted that some autofluorescence is emitted at lower than 500 nm, and is usually not captured with these filters. Both macroscopic tandem-lens arrangements (described in Reference 54) or standard epifluorescence microscopes (as used in Reference 34) have been used in vivo with similar results.

As with other optical imaging techniques, the animal must be secured in a stereotaxic apparatus, and every effort must be made to reduce mechanical disturbances. Typically, we stabilize the exposed brain area with 3% agarose in sterile saline and seal it with a glass coverslip to reduce mechanical disturbances and ensure a level optical plane. Alternatively, a metal or acrylic chamber can be placed around the exposed area and filled with carbogenated Ringer's solution (as in Reference 38). Some groups use a paralytic agent and artificial ventilation in mouse experiments to further reduce motion artifacts, but we have found this only to be necessary in larger animals.

The apparatus for FA imaging is remarkably similar to those used for ISI or dye-based optical imaging (Figure 8.4); practically, the only substantial difference is in the excitation and emission filter characteristics.[54] However, several properties of the FA signal require modifications in order to optimize detectability. First, illumination intensity must be relatively high, usually higher than that achieved by standard 12W halogen light sources. In our lab, we simply added an additional halogen light source and achieved suitable illumination, but other groups have reported using 75 W xenon and 100 W mercury–xenon lamps.[38,65] Second, due to the use of a long-pass emission filter and the low levels of fluorescence, light from external sources (such as a stimulus monitor used for vision studies) is a much larger contaminant than in ISI; the chamber surrounding the preparation must be completely opaque and every effort must be made to minimize ambient light. Finally, because of FA's relatively low signal strength, it is often useful to have long integration times (hundreds of ms), and therefore low frame rates (5–10 fps), as well as long interstimulus intervals to ensure

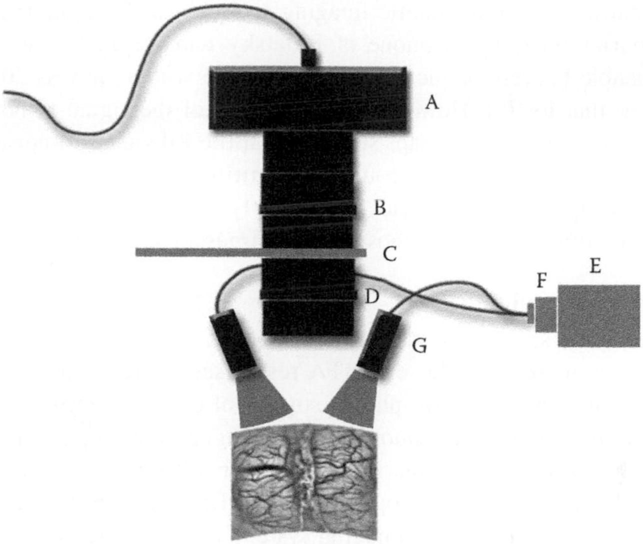

FIGURE 8.4 Imaging rig for FA. Standard tandem-lens combination for macroscopic imaging. (A) CCD camera attached to stereotaxic apparatus. (B) Top lens. (D) Bottom lens (typically, a 50 mm lens). The ratio of focal lengths of the top and bottom lenses sets the magnification factor. A 50 mm top lens is typically used for low-resolution imaging (1× magnification), and 135 mm lens for high-resolution imaging (2.7× magnification). The two lenses are mounted face-to-face. (C) Emission filter wheel (for multiple imaging modalities). (E) 12 W halogen light source. (F) Excitation filter wheel. (G) Fiber optic cables for illumination.

the late-phase reduction in fluorescence returns to baseline. The degree to which this is necessary will, like the previous two issues, depend greatly on the details of the preparation and are much more of a nuisance in vivo than in vitro, where the signal to noise ratio is much higher.

8.3.2 Properties of the FA Signal

8.3.2.1 Signal Strength

Signal strength also varies both by brain region and species. Electrical stimulation of the parallel fiber system in the cerebellum elicited strong responses (2% ΔF/F) and prominent inhibitory bands (see Section 8.3.2.5 below); responses in visual cortex are larger (5%) and easily detectable in response to sensory stimulation (0.25%). Initial reports suggested that the FA signal was extremely weak in large mammals and that functional imaging could only realistically be performed in rodents.[58] In fact, our lab has been able to generate and confirm strong FA responses in the cat, including maps of orientation and spatial frequency preference.[54,66] It is clear, however, that the signal strength is much weaker in the cat than in the mouse (0.5% ΔF/F as opposed to 5% ΔF/F in the mouse in response to electrical stimulation).

Because FA signals are relatively weak in vivo, most stimulation protocols must be repeated several times in order to improve signal-to-noise ratio with signal averaging, similar to hemodynamic imaging techniques. By applying the phase-encoding Fourier imaging technique of Kalatsky and Stryker[67,68] we are able to obtain reasonable FA retinotopic maps in the mouse with as few as 20 repetitions, comparable to that for ISI. However, our estimates of the signal-to-noise ratio for mapping cat spatial frequency maps suggests that the ISI signal-to-noise is approximately 1.3 times that of FA, which could be a limiting factor in these data.[66] While it is clear that the signal strength varies significantly by preparation, its responses are comparable and often superior to those of other imaging techniques.

8.3.2.2 Spatial Resolution

The question of the spatiotemporal resolution of FA is a complex one. On the spatial side, there is reason to believe that FA responses are restricted to single cells, or even single mitochondria: two-photon studies of cardiomyocytes reveal distinct mitochondrial flavoprotein autofluorescence[26] (Figure 8.5), matching our understanding of the localization of the signal to inner mitochondrial membranes (see Section 8.2.2.1). However, macroscopic studies are limited by optical scatter through neural tissue and the optics of the imaging system, and measurements of the optical point-spread function in vivo are exceptionally difficult (what constitutes a "point" of activity, and how can one be elicited without exciting surrounding tissue, is not at all clear). Our lab has found that iso-orientation domain maps generated with FA have an approximately 30% smaller autocorrelation peak than ISI, suggesting that FA has improved spatial resolution on the scale of orientation domains or smaller (~0.3 mm) (Figure 8.6D,E). Weber et al.[69] demonstrate that hemodynamic maps of barrel cortex

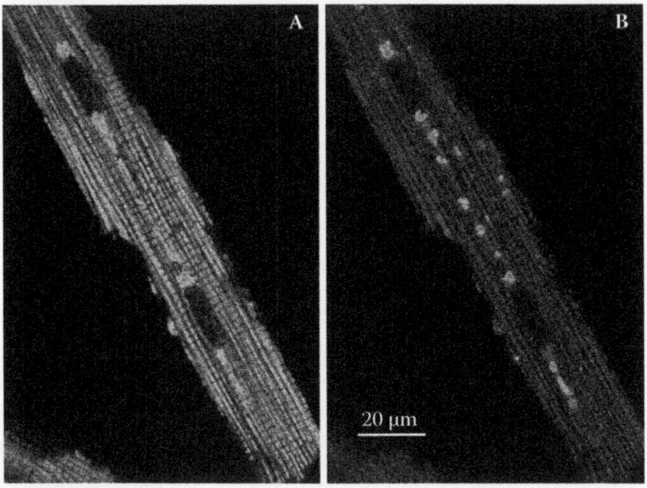

FIGURE 8.5 Multiphoton imaging of flavoprotein and NAD(P)H autofluorescence in cardiomyocytes. (A) NAD(P)H autofluorescence multiphoton imaging of canine cardiomyocyte (410–490 nm emission). (B) Flavoprotein autofluorescence multiphoton imaging of same cardiomyocyte (510–650 nm emission). The fluorescence in (A) is distributed throughout the tissue, whereas the responses in (B) are localized to individual mitochondria. (From Huang, S. et al. [2002]. *Biophys J* 82[5], 2811–2825. With permission.)

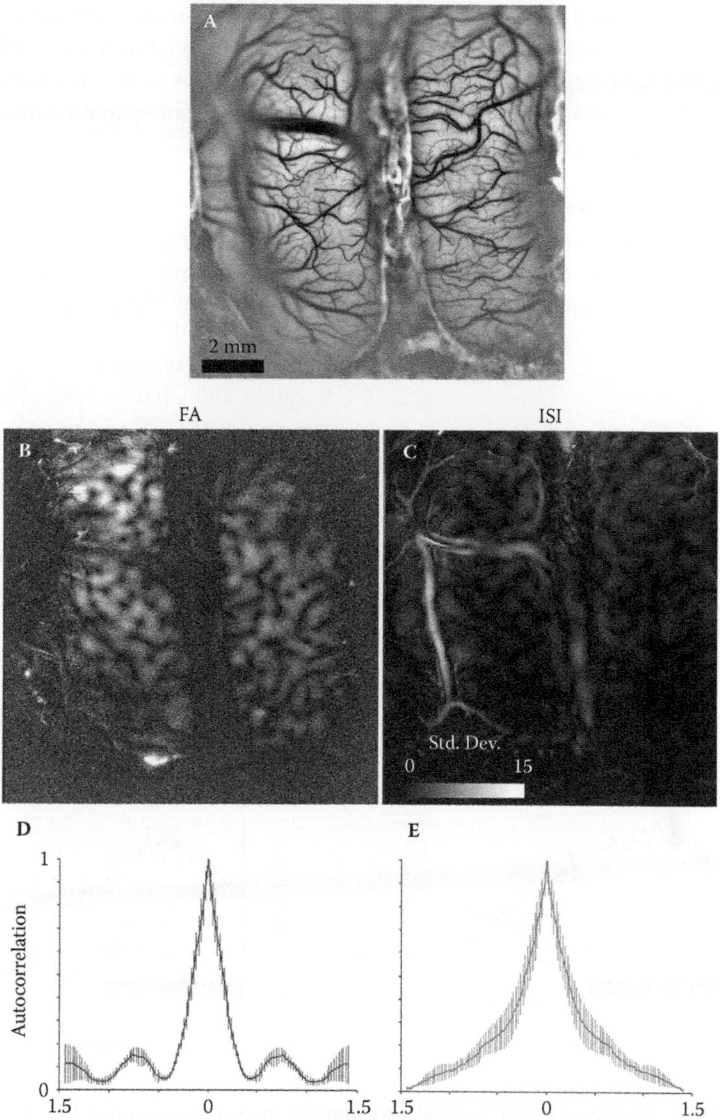

FIGURE 8.6 Comparison of FA and ISI signals over active cortex. (A) Image of cat primary visual cortex under green (530–550 nm) illumination to highlight vascular pattern. (B) Standard deviation of the FA signal over an orientation mapping session—the brighter the pixel, the more sharply it is tuned for stimulus orientation. (C) Standard deviation of the ISI signal over the same imaging session. The modulation of the ISI signal occurs primarily over blood vessels, whereas the FA modulation is strongest over active neural tissue (as revealed by the distinct orientation domain pattern). (D) Autocorrelation of FA iso-orientation map from the same region of cortex shows distinct peaks that represent orientation domains. (E) The autocorrelation function of an ISI iso-orientation map shows long range correlations on the order of orientation domain size. (Modified from Husson, T.R. et al. [2007]. *J. Neurosci* 27[32], 8665–8675. With permission.)

show an approximately 2.5 times larger response area than autofluorescence maps, suggesting that FA has comparatively fine spatial resolution. Given our understanding of the biochemistry of FA, and accepting in vitro studies as an indicator of signal source, we believe FA to offer a substantial improvement in spatial resolution over previous techniques for in vivo functional imaging studies.

8.3.2.3 Time Course and Temporal Resolution

Several groups have measured the optical impulse response of FA in a variety of in vivo preparations, summarized in Figure 8.7. In general, there is a large positive peak in fluorescence followed by a longer negative dip; the extent and duration

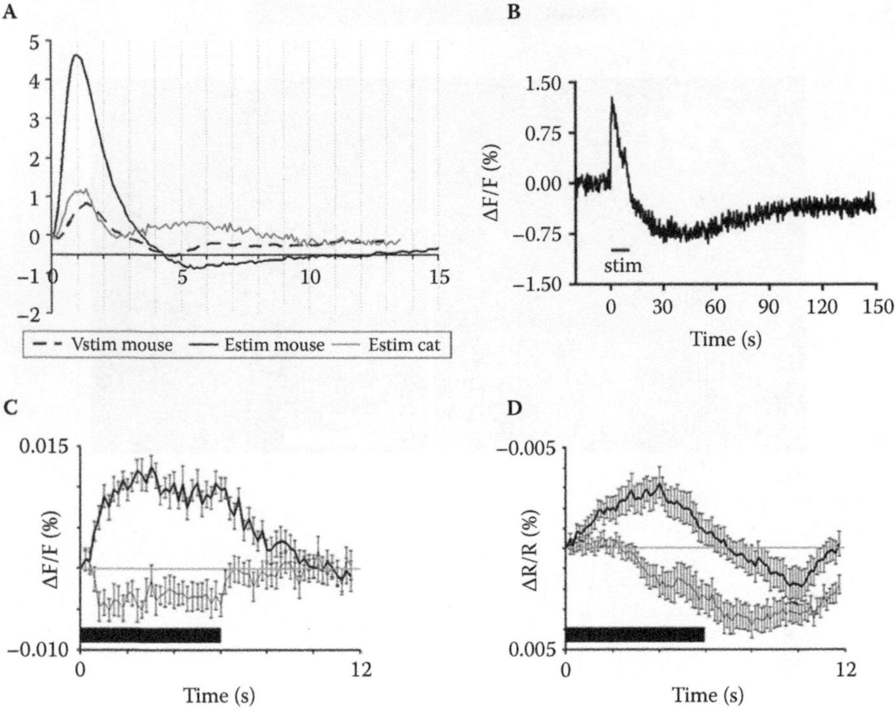

FIGURE 8.7 Time course of FA signal. (A) Impulse response function of FA in response to electrical stimulation (200 uA, 200 Hz, 200 ms) in mouse (black trace) and cat (grey trace) cortex. Also shown is the response in mouse to a brief (100 ms) flash of light (dotted trace). These time courses overlap with and confirm the time course originally described by Shibuki et al. 2003.[34] (B) Electrical stimulation (200 uA, 10 Hz, 10 s) in the mouse cerebellum. (C) FA of orientation domains in cat primary visual cortex in response to a drifting grating. The black trace shows the response in a domain that prefers the presented orientation, and the grey trace is the response in a domain that prefers the orthogonal orientation. (D) Responses in the same domains to the same stimuli, but measured with intrinsic signal imaging. The FA response is better correlated with the stimulus onset and offset, and is restricted to domains that prefer the presented orientation. ([A,C,D] from Husson, T.R. et al. [2007]. *J. Neurosci* 27[32], 8665–8675. With permission.) ([B] from Reinert, K.C. et al. [2004]. *J Neurophysiol* 92[1], 199–211. With permission.)

of these "light" and "dark" phases depends greatly on the particular preparation and stimulus protocol. In primary visual cortex of anesthetized mice, one second of electrical stimulation evokes a strong light phase response that peaks 1 second from the onset of the stimulus, and a dark phase with a minimum at approximately 5 seconds. Ten seconds of electrical stimulation of the cerebellar cortex evokes a similarly quick light phase but longer dark phase. Visual stimulation with a 200 ms flash of light produces similar results, but with a reduced signal amplitude and thus SNR (Figure 8.7A). In the cat, visual stimulation of an orientation domain with its preferred drifting grating showed unimodal FA responses with fast rise and decay times, in sharp contrast with ISI responses typically used to generate orientation preference maps. Electrical stimulation in cat cortex showed a very similar response time as seen in mice (Figure 8.7A).

Extrapolating temporal resolution from the impulse-response function requires an assumption of linearity (summation of optical signals), an assumption that may or may not hold in different parts of the CNS. Therefore, despite a well-characterized impulse-response function, it is not clear exactly how quickly oxidative phosphorylation can follow neural activity. In our hands we have seen FA responses modulate with a periodic visual stimulus up to 0.3 Hz, but due to low light levels at small integration times, faster changes may be visible with stronger illumination systems. It has been shown that background FA levels decrease with decreasing interstimulus intervals of electrical stimulation, in line with a more reduced redox potential over a longer time course.[70] It is therefore difficult to calculate a specific temporal resolution; but, given known comparisons between FA and ISI time courses, we can safely assume it can follow changes at least as quickly as hemodynamic responses.

8.3.2.4 Vascular Artifacts

One of the most significant problems with hemodynamic imaging techniques is their susceptibility to vascular artifacts.[10,51,55,71] Such arguments have been made against maps of spatial frequency preference in primary visual cortex, suggesting that these responses do not represent the organization of spatial frequency preference but are artifacts of the underlying vasculature.[72] Usually, functional maps must be subjected to extensive templating (exclusion of pixels over blood vessels), but because hemodynamic signals modulate over blood vessels they often report functional selectivity to stimuli despite being nonneural tissue, making templating very difficult and the possibility of inaccurate mapping very real. This represents a major constraint on the use and interpretation of functional maps generated with BOLD intrinsic signals.

Flavoprotein autofluorescence, on the other hand, measures nonhemodynamic signals. The appearance of blood vessels does not modulate in FA images, and therefore blood vessels do not appear tuned for an arbitrary stimulus. Indeed, we have shown that the variation in the autofluorescence signal is greatest over neural tissue,[54] whereas the modulation in ISI signals is strongest over blood vessels (Figure 8.6B,C). This strongly suggests that FA functional maps do not represent the local vascular pattern but rather the response of cortical neurons. We have recently used this reduction in vascular artifacts to confirm the structure of spatial frequency maps in cat primary visual cortex.[66]

In mice, the lateral spread of hemodynamic signals can lead to significant challenges in functional mapping studies. It has been noted by several studies[54,65,69] that the retinotopic maps of visual cortex in mice are smaller with FA than with ISI, suggesting that the large nonspecific blood flow response is misrepresenting the detailed structure of this area. In studies of ocular dominance plasticity, it is crucial to make accurate measurements of the size of both monocular and binocular zones of visual cortex[73]; with ISI, these structures may appear much larger and more overlapped than they are. FA's spatially restricted response profile (Figure 8.7c,d) may be much better suited to make these fine spatial measurements.

8.3.2.5 Linearity

Estimates of the metabolic cost of increased firing rate suggest a linear relationship, with an approximate production of 3–5 × 10^8 ATP molecules per action potential.[3] Thus, as a direct measure of ATP production, we would expect FA to also show a linear increase in fluorescence in response to increased neural activity. To assess this, we have used electrical stimulation to increase the amount of neural activity over an imaged region of cortex and found that this relationship is approximately linear for reasonable stimulus ranges (under 100 uA), and plateaus at very high amplitudes (data not shown). Reinert et al.[38] also find that FA is approximately linear with stimulus amplitude in the cerebellum, and Shibuki et al.[34] find that FA response amplitudes (but not size of responsive area) is linear with stimulation frequency. These findings suggest that FA can linearly follow increases in neural activity over a wide range of activity levels.

Moreover, autofluorescence can also decrease when inhibitory circuits are activated. Gao et al. [74] studied autofluorescence in the cerebellum, in which Purkinje cells have very high baseline firing rates that can be modulated both upward by parallel fiber input and downward by inhibitory inputs from GABAergic molecular layer interneurons. With electrical stimulation of these parallel fibers at 50–200 μA two regions change their fluorescence: there is a strong increase in fluorescence along the parallel fiber input, and a pronounced decrease in fluorescence in parasagittal zones that would be inhibited by molecular layer interneurons. The reduction in fluorescence in parasagittal zones was abolished by a $GABA_A$ receptor blockade, suggesting that it depends on inhibitory neurotransmission. Immunostaining of the fixed cerebellar tissue collected after autofluorescence imaging showed that the parasagittal regions in which fluorescence decreased corresponded to parasagittal zebrin-II compartments,[75] suggesting that these neighboring functional zones respond oppositely to parallel fiber inputs. This was the first demonstration that FA could be used to image not only excitatory but also inhibitory activity.

8.3.2.6 Analysis Techniques

Sensory-driven response amplitudes are small compared to baseline fluorescence, so some sort of normalization is needed to visualize fluorescence responses. Many of the normalization techniques in standard use with other imaging methods, like intrinsic signal imaging, can be easily adapted to FA.[67,76] For example, we use the same analysis tools for processing FA and ISI images from cat and mouse visual cortex, with the obvious caveat that one or the other signal must be inverted if the two signals are to be compared directly.

However, because FA does not modulate over blood vessels, templating procedures are much more straightforward than with ISI, but care must be taken in the construction of stimulus preference maps. With FA, the response over blood vessels in mean-subtracted images is generally negative for any stimulus that activates neural tissue.[66] The maximal response for pixels over blood vessels, therefore, appears as the condition that stimulates neural tissue the least, and is therefore least affected by high-pass spatial filtering. Thus, if blood vessels are not excluded from analysis, the pixels over them could be mistakenly assigned to low-responsive conditions. Fortunately, in highly responsive conditions, blood vessels appear as dark regions, clearly distinguished from neural tissue. Using these conditions to construct templates (e.g., by a threshold) prevents the mischaracterization of pixels over blood vessels as can occur in hemodynamic preference maps.

Because the autofluorescence signal has been known to reversibly bleach,[77] for repetitive stimulation protocols it may be advantageous to extrapolate a "bleaching curve" and subtract it from the optical trace.[60] This can be accomplished by measuring baseline responses during each interstimulus interval and using the changes in these blanks to fit an exponential decay. In addition, for direct comparison between flavoprotein and either absorption (ISI) or NADH signals, the signals must be inverted; in the case of a periodic representation (such as that for retinotopic maps in the mouse), this can be accomplished by phase-shifting the response map by 180° (as performed in Reference 78). Some studies using NADH autofluorescence in vivo[79] have attempted to eliminate hemodynamic artifacts by subtracting fluorescence at an isobestic wavelength of oxy/deoxyhemoglobin, but we have not found this to be necessary even in cats.

8.3.2.7 Possible Pitfalls

As a relatively new technique for functional imaging, FA still requires some care in properly obtaining and interpreting data. Several obstacles that have been encountered remain unexplained, and further study is needed to both understand the details of the FA signal and to appreciate its applicability to various preparations.

Until recently there appeared to be a strong discrepancy between in vivo and in vitro measures of neuronal redox state with optical techniques. Specifically, in vivo signals seemed to follow sustained activity much more accurately than in vitro measures, which consisted of a much more profound dark phase recovery period. Turner et al.[27] reviewed this issue at length and concluded that differences in tissue oxygenation between intact brains (which have local vasculature) and slices (which rely entirely on perfusion by the bath media) explain these differences and bring the two signals in line with one another. Interestingly, the Ebner group has found that inhibiting astrocytic TCA activity with fluoroacetate caused a stronger attenuation of the dark phase than the light phase,[80] suggesting that the dark phase is mostly nonneuronal. Given the exciting work done in slice preparations and the outstanding spatial resolution of FA in prepared tissue (discussed below, see Section 8.4.3), this reconciliation between in vitro and in vivo signals means they can be used as complementary methods in the same model system.

A more practical concern has also arisen due to the nature of flavin fluorescence. Lower excitation wavelengths (blue), corresponding to higher energy light, can cause photodamage in neural tissue at much lower intensities than other wavelengths such as those used for ISI. High energy light generates singlet oxygen, a strong oxidizing agent, which can lead to damage of lipid membranes and cell death.[81,82] While no reports exist of photodamage to neural tissue in macroscopic applications of FA, and responses persist even after several hours of imaging, in these and microscopic studies (including two-photon microscopy) the intense illumination requirements of FA need to be balanced against the risk of photodamage and photobleaching.[83–86] Given that FA time courses in response to electrical stimulation require a long dark phase recovery period (see Figure 8.7A,B), there seems to be some indication that bleaching of the flavoprotein signal can occur; also, a recent study has used this illumination light to focally inactivate neural tissue.[77] It thus seems advisable to control the excitation light source such that it only illuminates the tissue while imaging and at the lowest possible intensity.

Another potential problem with FA is that by using a long-pass emission filter hemodynamic signals could contaminate images. It is known, for example, that hemoglobin can absorb excitation light and quench emitted fluorescence.[87] Our group has done a systematic comparison between hemodynamic and autofluorescence time courses in vivo (Figure 8.8) and found that, as expected, in response to electrical stimulation the

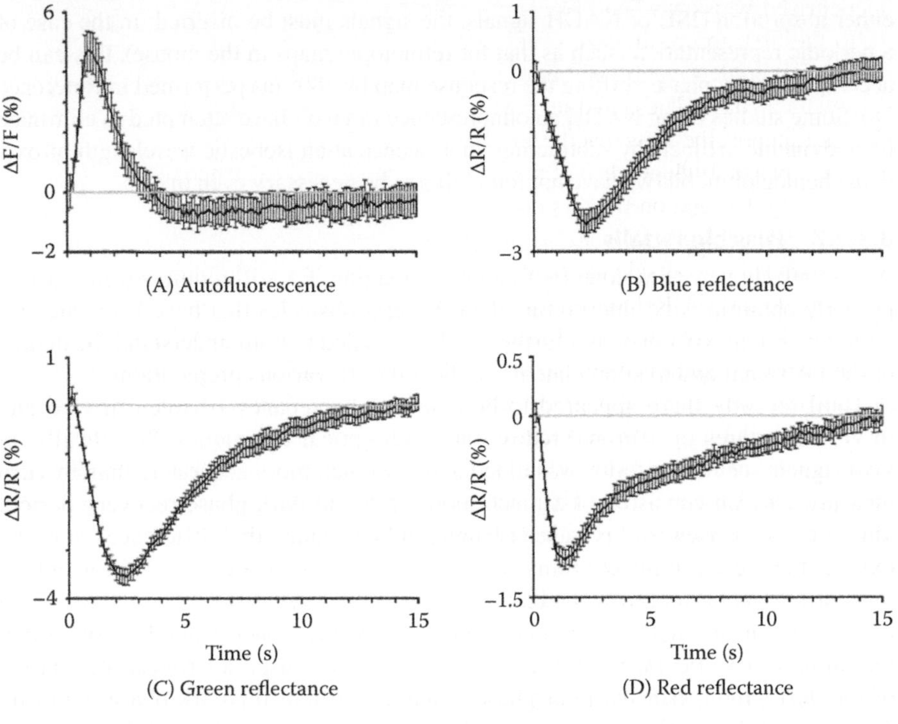

FIGURE 8.8 Autofluorescence and intrinsic signal time courses. FA and absorption signal time courses averaged over 22 cycles of electrical stimulation in mouse. Each plot is the response of one pixel measured using (A) the FA signal (420–490 nm excitation filter; 515+ nm emission filter), (B) the blue absorption signal (420–490 nm excitation filter, no emission filter), (C) the green absorption signal (530–550 nm excitation filter; no emission filter), and (D) the red absorption signal (600–620 nm excitation filter; no emission filter). Error bars are SEM. The FA response is opposite in sign and slightly faster than the reflectance signals, suggesting that it is distinct from hemodynamic absorbance signals. Results were similar in cat. (From Husson, T.R. et al. [2007]. *J. Neurosci* 27[32], 8665–8675. With permission.)

hemodynamic (intrinsic) signals are opposite in sign, corresponding to light absorption rather than fluorescence, and are slightly slower in time-to-peak than FA. The responses are similar when stimulating visually with a bright flash, though much lower in intensity. Also, Reinert et al. blocked hemodynamic signals by inhibiting nitric oxide synthase with L-NAME and showed that FA responses were preserved.[38] This suggests that the FA signals are not due to hemodynamic artifacts, and that well-separated filters can adequately restrict the signal to flavoprotein autofluorescence.

While FA has been reliably elicited in anesthetized mice and cats, it is not clear what effect anesthesia has on these responses. It is our experience that urethane or ketamine/urethane mixtures are very reliable in mice,[54,64,65] but both ketamine/xylazine and pentobarbital have also been successfully used.[38,65] Our lab has only rarely imaged responses under isoflurane anesthesia, and have had little success with alpha-cholorolase induction protocols, though it is not clear what effect these substances would have on mitochondrial activity. In cats, we used sodium thiopental regularly, and were able to see activation with isoflurane in ferrets. A systematic study of the effects of anesthesia protocols on FA responses would not only better inform researcher's choice for in vivo studies, but especially in combination with other metabolic imaging techniques could enlighten the investigation of these anesethetics' mechanisms of action, as many common choices are known to be vasoactive or metabolic depressants (for example, barbiturates inhibit the oxidation of NADH, see Reference 88 and references therein).

8.4 EXAMPLE STUDIES

8.4.1 Early Studies

The pioneering studies of fluorescent intracellular molecules and their relationship to energy utilization began almost 50 years ago by Britton Chance, who first used flavin, NADH, and cytochrome fluorescence to trace the process of oxidative metabolism.[29,62,63,89–95] These studies found that both ADP and calcium ions increased the redox activity of these fluorophores, suggesting a strong link between their activity and mitochondrial ATP production. These signals were later used to study both functional activation (mostly in response to electrical stimulation) in cell cultures and slices from the hippocampus and cerebral cortex[59,96,97] and pathological responses to spreading depression and epilepsy.[87,98–100] The development and availability of high quality CCD cameras allowed these relatively small signals to be used in a variety of preparations, beginning with the work of Mironov and Richter[96] in the brainstem, Shuttleworth et al.[57] in the hippocampus, Shibuki et al.[34] in the somatosensory cortex, and Reinert et al.[38] in the cerebellum. These studies not only showed reliable and robust responses in several preparations, but also extended FA's functional imaging abilities to in vivo preparations.

8.4.2 The Development of FA Imaging In Vivo

Shuttleworth et al.[57] imaged NAD(P)H (both NADH and NADPH) and flavin fluorescence in response to electrical and chemical stimulation in slice preparations of mouse hippocampal area CA1. They characterized their respective time courses, both consisting of an initial, stimulus-linked component and an aftershoot (as in Figure 8.7A,B). These responses were abolished by the ionotropic glutamate receptor blockade, suggesting that the signal was predominantly postsynaptic, and the authors concluded that while these signals were not explicitly coupled to calcium ion flux they were reliable measures of neural activity.

Dr. Katsuei Shibuki and colleagues extended these findings to the rat somatosensory cortex (see the chapter by Shibuki et al. in this book). In slices, they found large increases in FA (27 +/– 2%) in response to electrical stimulation using a highly sensitive cooled CCD camera and wide-view epifluorescence microscope. Taking this robust setup forward, they were the first to show stimulus-evoked activity patterns with FA in anesthetized mice. They used both electrical and vibratory skin stimulation to characterize the FA responses under urethane anesthesia and distinguished them from hemodynamic signals under the same conditions. This was a crucial step in proving the difference between ISI and FA responses and showing the viability of the signal in vivo.

The Ebner Lab at the University of Minnesota followed these results by showing strong in vivo responses in the mouse cerebellum.[38] They used a well-understood neuronal circuit, the parallel fiber/Purkinje cell system, which upon electrical stimulation showed a well-defined band of fluorescence along the parallel fiber tract (Figure 8.9A,B; these responses were confirmed shortly thereafter by Reference 101). Their signal had the same time course shown previously by Shuttleworth and Shibuki, with an initial bright phase followed by a longer dark phase, and signal strength was linearly related to stimulus amplitude and frequency. They also used pharmacological treatments similar to those used by Shibuki et al. in slice (see Section 8.4.3 below) to show that the in vivo fluorescence was a result of flavoprotein activity, and found that the optimal wavelengths of excitation and emission light matched those previously reported for flavoproteins. In subsequent studies, this group has shown that electrical stimulation of parallel fibers evokes parasagittal inhibitory bands that are abolished by $GABA_A$ antagonists (Figure 8.9), reflected in FA imaging as dark bands that run alongside the bright parallel fibers.[74] Their use of FA in this robust and stereotyped system demonstrates the ability of FA to identify both excitatory and inhibitory activity patterns and to trace neural circuits over large areas in vivo.

One of the primary advantages of optical imaging techniques is that they allow the visualization of changes in spatial activity patterns over time. The Shibuki lab quickly followed up their characterization of the FA signal by showing that it can be used to study short-term plasticity in the somatosensory cortex in vivo.[102] They showed that optical responses to somatosensory stimulation were reliably enhanced if they were followed by deep-layer (1.5–2 mm from cortical surface) electrical stimulation in the responsive region; likewise, more superficial (<0.5 mm) electrical stimulation attenuated responses. Also, they could greatly potentiate ipsilateral responses with ipsilateral stimulation, but could also potentiate contralateral responses, suggesting a commissural projection affecting this plasticity. Subsequently, they mapped tonotopic organization of primary auditory cortex and the anterior auditory field, and demonstrated differences in FA response durations between mice raised in sound-deprived versus normal acoustic environments and used the in vitro slice preparation to find specific response differences between the vertical layers of cortex.[64] Because

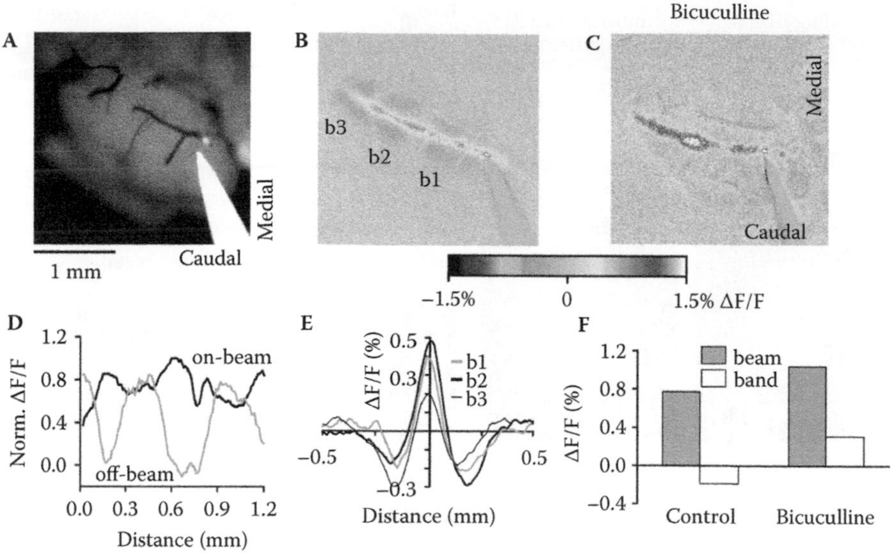

FIGURE 8.9 **(See color insert following page 224)** Imaging inhibition with FA. (A) Image of mouse cerebellar cortex; the stimulating electrode is apparent. (B) Pseudocolor image of FA response to electrical stimulation in cerebellum. Fluorescence increased along the linear band (red), but decreased in the parasagital zones (blue). (C) Response amplitude along longitudinal transects through the on-beam excitatory region (black) and the parasagittal off-beam region (grey). (D) Response amplitude across the on- and off-beams. (E) Amplitude of the on-beam and parasagittal bands in control and bicuculline-administered conditions. (F) Pseudocolor image of FA response after bicuculline administration. Note that the inhibitory bands are abolished with bicuculline. (Modified from Gao, W. et al. [2006]. *J Neurosci* 26[32], 8377–8387. With permission.)

they could measure the same flavoprotein fluorescence signal both in vivo and in vitro, they could use the complementary advantages of each to create a wholistic view of these experience-dependent changes. Finally, they have shown that FA can reliably measure ocular dominance plasticity in the mouse visual cortex,[65] one of the most studied assays of experience-dependent changes in neural circuitry. In sum, the Shibuki group has demonstrated that FA can be used to functionally map several areas of the mouse brain, measure experience-dependent plasticity, and link large-scale activity to specific circuits in vitro.

It was thought, however, that the only viable in vivo preparation for FA was the mouse. Indeed, the difference in signal strength between sliced (~20% changes in ΔF/F) and anesthetized mouse cortex (~2%) is enormous, and preliminary reports indicated that the difference between small and large mammals is also large.[58] Our group, however, found that FA could be reliably used to image not only the retinotopic organization of mouse visual cortex (Figure 8.10A–D) but also the classic orientation maps in cat (Figure 8.10E–H).[54] These maps were very similar to those generated using intrinsic signal imaging, but were more spatially localized and required less spatial filtering of raw data (due to the correlation structure shown in Figure 8.6D–E). Moreover, we have used this greater signal specificity to confirm the structure of spatial frequency

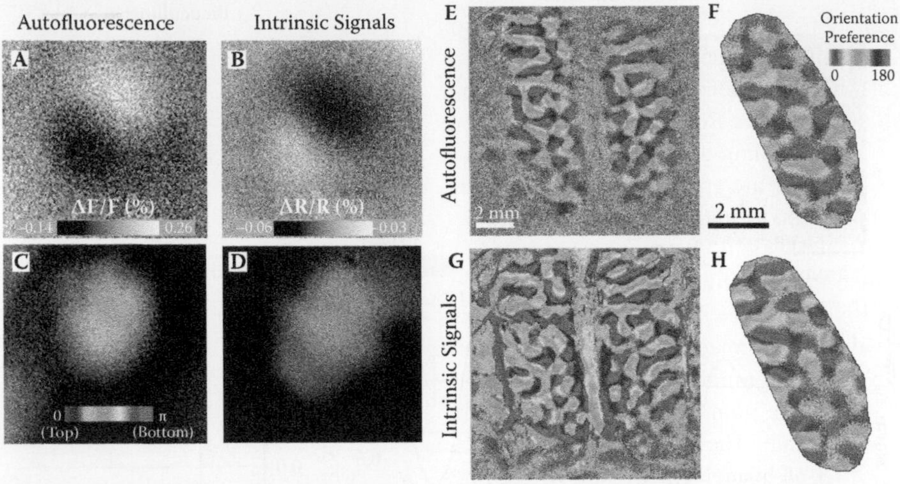

FIGURE 8.10 (See color insert following page 224) Functional imaging of primary visual cortex with FA. (A,B) Responses over visual cortex in mouse in response to a drifting visual bar. Regions that responded to the stimulus showed increased brightness with FA imaging, and reduced brightness (increased absorbance) with intrinsic signal imaging. (C,D) Retinotopic maps generated using the phase-encoding method of Reference 67 for FA (C) and ISI (D) (ISI images have been inverted before calculating map of retinotopy for direct comparison with FA). (E,F) Orientation maps of cat primary visual cortex generated using FA in two separate experiments. (G,H) Orientation maps of the same two experiments in (E,F) generated with ISI. ISI images report responses over blood vessels as tuned for orientation (see, for example, the large vessel in the left margin of the experiment in (G), whereas FA does not. After templating, the organization of orientation preference over neural tissue has been shown to be the same in both FA and ISI maps. (From Husson, T.R. et al. [2007]. *J. Neurosci* 27[32], 8665–8675. With permission.)

maps in the cat,[66] which have been called into question due to the possibility of vascular artifacts in ISI maps. Because FA is a nonhemodynamic measure of neural activity, the similarity between FA and ISI spatial frequency maps suggests that the patchy structure of spatial frequency domains observed by functional imaging is a genuine feature of the organization of primary visual cortex.

Over the past 5 years, FA has been shown to be a reliable and relatively noninvasive measure of neural activity patterns in several brain regions that is linear with stimulus intensity and sensitive to evoked activity. Experiments are currently under way to bring FA's spatial specificity and tighter link to neural activity into more complicated systems, such as primate sensory and motor cortex. There is a great deal of hope that this technique will be an improvement on previous techniques, such as hemodynamic and voltage sensitive dye imaging, and that new opportunities will emerge to study the details of neuron–glia and neurometabolic coupling in intact animals.

8.4.3 Functional Imaging In Vitro

One of the fundamental advantages offered by FA is its ability to measure functional activity both in vivo and in vitro with the same underlying signal. This allows an unprecedented degree of interchange between preparations—findings at the level of cortical areas can be "zoomed in" on by further slice preparations, and findings in slices or cultures can be extended to intact brains. Because of this advantage and the extensive work done on FA in slices, we here review several studies that have been performed in vitro, and discuss their relationship with findings in live animals. In general, these techniques greatly improve our understanding of the FA signal and have outstanding potential for revealing the detailed physiology that underlies larger-scale brain activity.

Huang et al. have characterized the autofluorescence of both NAD(P)H and flavoproteins in cardiomyocytes using two-photon laser scanning microscopy (2P).[26] Their 2P excitation and emission spectra match previously reported characteristics (see Section 8.3.1) and showed appropriate polarity (blocking oxidative phosphorylation via cyanide increases flavoprotein fluorescence and decreases NAD(P)H fluorescence). Importantly, they were able to identify individual mitochondria with limited signal averaging (five trials) and limited photodamage (due to the longer wavelengths used in 2P). At this resolution, they were able to differentiate between cytosolic and mitochondrial NAD(P)H signals, with the latter appearing as distinct "hot spots" amongst weaker background fluorescence; measuring flavoprotein autofluorescence, they saw only these "hot spots" (which were aligned with those seen in NAD(P)H imaging), providing direct evidence that FA is restricted to single mitochondria (Figure 8.5). This differentiation allows for better calculation of FAD/NAD(P)H ratios, which could be done over individual mitochondria (and may be valuable for measuring disease states). While isolated cardiomyocytes are a far cry from in vivo brain preparations, the mitochondrial energy demand of the heart is very large and is comparable to that in brain, so results in neurons may prove similar [6]. More recently, they have employed 2P techniques in rat hippocampal slices to study the metabolic response to hypoxia, but only using NADH autofluorescence.[103] This work provides at least a proof of concept that FA can spatially resolve individual mitochondria.

The Shibuki group has contributed a wide variety of studies using flavoprotein autofluorescence in cortical slices. In order to definitively isolate the endogenous signal, they performed pharmacological treatments and compared local field potential recordings to optical traces.[34] Abolishing all activity with tetrodotoxin blocked both the optical and field potential recording. Reduced Ca^{++} in the bath eliminated postsynaptic activity in the field potential as well as optical responses. Reducing both glucose and oxygen content also restored neural activity but not FA, and finally the flavoprotein blocker diphenyleneiodonium was shown to completely abolish the optical responses while completely preserving electrical activity in the slice (though diphenyleneiodonium is a widely used flavin inhibitor, there is controversy over its precise effects, see Reference 104). These treatments strongly implicate flavoproteins in the signals obtained measuring green fluorescence with blue excitation light (these findings were subsequently confirmed by Reference 38, except the finding of mixed pre- and postsynaptic origin [see Section 8.2.2.2]). They have since studied intracortical activity patterns in several preparations, and have demonstrated functional, experience-dependent connections between primary and secondary somatosensory cortex[105] and between primary auditory and surrounding areas.[106] Their repeated success in applying FA to several brain regions and many functional assays highlights the robustness of the method.

8.5 CONCLUSIONS AND FUTURE DIRECTIONS

The use of mitochondrial flavoprotein autofluorescence in the last few years has demonstrated substantial advantages for functional imaging, yet its full utilization has yet to begin. There is reason to believe that further improvements in spatiotemporal resolution are possible, and work towards this end can benefit from recent advances in imaging technology. Even with established techniques and preparations, FA opens the door to new lines of investigation that could have widespread implications for fields ranging from neurochemistry and neuro-metabolic coupling to functional imaging of local and global circuits.

However, several investigations are immediately required to fully understand the source and resolution of the FA signal. Preliminary reports with two-photon microscopy in cardiac muscle suggest subcellular resolution (Figure 8.5), but these results have not been extended to neurons either in vitro or in vivo. In theory, FA could offer an improvement due to its subcellular localization and endogenous source, and thus might be particularly suited to long-term study of synaptic changes (such as in Reference 107). Though the underlying signal is small, FA may stand to benefit from recent advances in optical imaging technology that allow for whole-field illumination with minimal photobleaching.[108] As the mechanical modifications to traditional two-photon imaging systems for FA are small, and because FA responses have been measured with such techniques in other cell types,[50,109–111] these applications may not be far away.

FA could also be used to map functional circuits in vivo. For example, the demonstration that FA can be used to identify fine cortical structures in large animals suggests that FA could be used to visualize functional interactions between columnar structures such as orientation domains in visual cortex with either pharmacological, electrical, or sensory stimulation, or several in combination. The lack of long-range spatial correlations in FA maps[54] makes clear the improvement over ISI for these studies. In addition, FA could also help unravel the highly contentious issue of neurometabolic coupling. As it serves as a faithful indicator of redox state and ISI measures local tissue deoxygenation, a combination of the two techniques could help visualize the changes that occur in neural tissue as a result of activity or insult.

Fortunately, recent advances in our understanding of neurometabolic coupling suggest that metabolic signals may be very useful in studying various types of functional plasticity (reviewed in Reference 112). It has been shown that neuron–glia interactions are modified with changes in synaptic plasticity, suggesting that FA, with its presumptive restriction to synapses (see Section 8.2.2.2) may be able to indirectly monitor these changes. Specifically, long-term changes in the functional properties of synapses such as LTD and LTP are supported by related changes in the metabolic processes at those synapses, as there is a great density of mitochondria at synapses, FA in combination with two-photon microscopy may be able to build on previous long-term imaging studies of synaptic plasticity in vivo. On larger scales, FA may prove a more subtle measure of developmental plasticity than ISI, which has been used, for example, to investigate the effects of monocular deprivation in developing mice and cats.[65,73,113–118]

Perhaps the most promising application of FA is in the realm of translational neuroscience. It is now appreciated that a large number of neurological conditions are associated with mitochondrial dysfunction, including amyotrophic lateral sclerosis, Alzheimer's, Parkinson's, and Huntington's diseases.[119–122] It may be possible to use FA in transgenic mouse models of these diseases to better assess the regional and functional deficits across neural tissue. For example, Parkinson's disease can be modeled by specific inhibition of complex I of the mitochondrial ETC.[123–125] As FA is a direct measure of ETC activity, these mice can be used both in vivo and in vitro to study local or global alterations in functional activity as well as changes in activity brought about by pharmacological treatments. Also, studies have already exploited the ratio between FAD and NADH autofluorescence to grade precancerous epithelial cells[126]; perhaps similar approaches could be applied to the study of gliomas or neuromas in cortex, which already have acceptable animal models.[127,128] In these and many other ways, FA could serve as a powerful assay of metabolic disease.

The imaging of mitochondrial flavoprotein autofluorescence is a powerful yet underappreciated tool for measuring many aspects of brain function. It adds another way of visualizing the interactions between metabolism and neural activity, and serves as a complement to preexisting techniques that measure other relevant signals. In a perfect world, we would be able to measure every aspect of neural activity simultaneously with precise resolution; while this is still very much a fantasy, we now have a variety of techniques that can be used in combination to understand the complex interactions and changes in the brain. FA brings us one step closer to a direct visualization of these processes.

ACKNOWLEDGMENTS

We would like to thank Dr. Murray Sherman and members of the Sherman lab for helpful discussions, and members of the Issa lab, especially Atul Mallik and Ari Rosenberg, for extensive discussion and review. Work in our lab was supported by grants from the Brain Research Foundation and Mallinckrodt Foundation (NPI).

REFERENCES

1. Aiello, L.C. and Wheeler, P. (1995). The expensive-tissue hypothesis: The brain and the digestive system in human and primate evolution. *Curr Anthropol* 36(2), 199.
2. Laughlin, S.B. (2001). Energy as a constraint on the coding and processing of sensory information. *Curr Opin Neurobiol* 11(4), 475–480.
3. Attwell, D. and Laughlin, S.B. (2001). An energy budget for signaling in the grey matter of the brain. *J Cereb Blood Flow Metab* 21(10), 1133–1145.
4. Foster, K.A. et al. (2006). Optical and pharmacological tools to investigate the role of mitochondria during oxidative stress and neurodegeneration. *Prog Neurobiol* 79(3), 136–171.
5. Wong-Riley, M.T. (1989). Cytochrome oxidase: An endogenous metabolic marker for neuronal activity. *Trends Neurosci* 12(3), 94–101.

6. Rolfe, D.F. and Brown, G.C. (1997). Cellular energy utilization and molecular origin of standard metabolic rate in mammals. *Physiol Rev* 77(3), 731–758.
7. McKenna, M.C. et al. (2006). Energy metabolism of the brain. In *Basic Neurochemistry: Molecular, Cellular, and Medical Aspects* (Siegel, G.J. et al., Eds.), Elsevier Academic Press, pp. 531–557.
8. Kasischke, K.A. et al. (2004). Neural activity triggers neuronal oxidative metabolism followed by astrocytic glycolysis. *Science* 305(5680), 99–103.
9. Thompson, J.K. et al. (2003). Single-neuron activity and tissue oxygenation in the cerebral cortex. *Science* 299(5609), 1070–1072.
10. Malonek, D. and Grinvald, A. (1996). Interactions between electrical activity and cortical microcirculation revealed by imaging spectroscopy: Implications for functional brain mapping. *Science* 272(5261), 551–554.
11. Logothetis, N.K. (2002). The neural basis of the blood-oxygen-level-dependent functional magnetic resonance imaging signal. *Philos Trans R Soc Lond B Biol Sci* 357(1424), 1003–1037.
12. Logothetis, N.K. and Wandell, B.A. (2004). Interpreting the BOLD signal. *Annu Rev Physiol* 66, 735–769.
13. Vanzetta, I. and Grinvald, A. (1999). Increased cortical oxidative metabolism due to sensory stimulation: Implications for functional brain imaging. *Science* 286(5444), 1555–1558.
14. Dugan, L.L. and Kim-Han, J.S. (2006). Hypoxic-Ischemic Brain Injury and Oxidative Stress. In *Basic Neurochemistry: Molecular, Cellular, and Medical Aspects* (Siegel, G.J. et al., Eds.), Elsevier Academic Press, pp. 559–572.
15. Kristian, T. and Siesjo, B.K. (1997). Changes in ionic fluxes during cerebral ischaemia. *Int Rev Neurobiol* 40, 27–45.
16. Schiff, S. and Somjen, G. (1987). The effect of graded hypoxia on the hippocampal slice: An in vitro model of the ischemic penumbra. *Stroke* 18(1), 30–37.
17. Schurr, A. and Rigor, B.M. (1989). Cerebral ischemia revisited: New insights as revealed using in vitro brain slice preparations. *Cellular and Molecular Life Sciences (CMLS)* 45(8), 684–695.
18. Schurr, A. et al. (1997). Brain lactate, not glucose, fuels the recovery of synaptic function from hypoxia upon reoxygenation: An in vitro study. *Brain Res* 744(1), 105–111.
19. Magistretti, P.J. and Pellerin, L. (1996). Cellular bases of brain energy metabolism and their relevance to functional brain imaging: Evidence for a prominent role of astrocytes. *Cereb Cortex* 6(1), 50–61.
20. Page, M.A. and Williamson, D.H. (1971). Activity of ketone-body utilization pathway in brain of suckling and adult rats. *Biochem J* 121(1), 16P.
21. Yang, S. et al. (1987). Fatty acid oxidation in rat brain is limited by the low activity of 3- ketoacyl-coenzyme A thiolase. *J Biol Chem* 262(27), 13027–13032.
22. Nehlig, A. and Pereira de Vasconcelos, A. (1993). Glucose and ketone body utilization by the brain of neonatal rats. *Progr Neurobiol* 40(2), 163–220.
23. Hawkins, R.A. et al. (1986). Regional ketone body utilization by rat brain in starvation and diabetes. *Am J Physiol Endocrinol Metab* 250(2), E169–178.
24. Ruderman, N.B. et al. (1974). Regulation of glucose and ketone-body metabolism in brain of anaesthetized rats. *Biochem J* 138(1), 1–10.
25. Oz, G. et al. (2007). Human brain glycogen content and metabolism: Implications on its role in brain energy metabolism. *Am J Physiol Endocrinol Metab* 292(3), E946–951.
26. Huang, S. et al. (2002). Two-Photon Fluorescence Spectroscopy and Microscopy of NAD(P)H and Flavoprotein. *Biophys J* 82(5), 2811–2825.
27. Turner, D.A. et al. (2007). Differences in O(2) availability resolve the apparent discrepancies in metabolic intrinsic optical signals in vivo and in vitro. *Trends Neurosci* 30(8), 390–398.

28. Kunz, W.S. and Kunz, W. (1985). Contribution of different enzymes to flavoprotein fluorescence of isolated rat liver mitochondria. *Biochim Biophys Acta* 841(3), 237–246.
29. Hassinen, I. and Chance, B. (1968). Oxidation-reduction properties of the mitochondrial flavoprotein chain. *Biochem Biophys Res Commun* 31(6), 895–900.
30. Voltti, H. and Hassinen, I.E. (1978). Oxidation-reduction midpoint potentials of mitochondrial flavoproteins and their intramitochondrial localization. *J Bioenerg Biomembr* 10(1), 45–58.
31. Kunz, W.S. and Gellerich, F.N. (1993). Quantification of the content of fluorescent flavoproteins in mitochondria from liver, kidney cortex, skeletal muscle, and brain. *Biochem Med Metab Biol* 50(1), 103–110.
32. Chorvat, D., Jr. et al. (2005). Spectral unmixing of flavin autofluorescence components in cardiac myocytes. *Biophys J* 89(6), L55–L57.
33. Aubin, J.E. (1979). Autofluorescence of viable cultured mammalian cells. *J Histochem Cytochem* 27(1), 36–43.
34. Shibuki, K. et al. (2003). Dynamic imaging of somatosensory cortical activity in the rat visualized by flavoprotein autofluorescence. *J Physiol* 549(pt. 3), 919–927.
35. Ames, A. (2000). CNS energy metabolism as related to function. *Brain Res Rev* 34(1–2), 42–68.
36. Jones, D.P. (1986). Intracellular diffusion gradients of O2 and ATP. *Am J Physiol Cell Physiol* 250(5), C663–675.
37. Erecinska, M. and Silver, I.A. (1989). ATP and brain function. *J Cereb Blood Flow Metab* 9(1), 2–19.
38. Reinert, K.C. et al. (2004). Flavoprotein autofluorescence imaging of neuronal activation in the cerebellar cortex in vivo. *J Neurophysiol* 92(1), 199–211.
39. Fox, P. et al. (1988). Nonoxidative glucose consumption during focal physiologic neural activity. *Science* 241(4864), 462–464.
40. Pellerin, L. and Magistretti, P.J. (1994). Glutamate uptake into astrocytes stimulates aerobic glycolysis: A mechanism coupling neuronal activity to glucose utilization. *Proc Natl Acad Sci USA* 91 (22), 10625–10629.
41. Pellerin, L. et al. (1998). Evidence Supporting the Existence of an Activity-Dependent Astrocyte-Neuron Lactate Shuttle. *Dev Neurosci* 20(4–5), 291–299.
42. Hertz, L. et al. (2007). Energy metabolism in astrocytes: High rate of oxidative metabolism and spatiotemporal dependence on glycolysis/glycogenolysis. *J Cereb Blood Flow Metab* 27(2), 219–249.
43. Mangia, S. et al. (2003). The aerobic brain: Lactate decrease at the onset of neural activity. *Neuroscience* 118(1), 7–10.
44. Mangia, S. et al. (2003). Issues concerning the construction of a metabolic model for neuronal activation. *J Neurosci Res* 71(4), 463–467.
45. Caesar, K. et al. (2008). Glutamate receptor-dependent increments in lactate, glucose and oxygen metabolism evoked in rat cerebellum in vivo. *J Physiol* 586(5), 1337–1349.
46. Brown, A.M. et al. (2004). Energy transfer from astrocytes to axons: The role of CNS glycogen. *Neurochem Int: Role of Non-synaptic Communication in Information Processing* 45(4), 529–536.
47. Brown, A.M. (2004). Brain glycogen re-awakened. *J Neurochem* 89 (3), 537–552.
48. Dringen, R. et al. (1993). Glycogen in astrocytes: Possible function as lactate supply for neighboring cells. *Brain Res* 623(2), 208–214.
49. Wolff, J.R. and Chao, T.I. (2004). Cytoarchitectonics of non-neuronal cells in the central nervous system. *Adv Mol Cell Biol* 31, 1–52.
50. Andersson et al. (1998). Autofluorescence of living cells. *J Microscopy* 191, 1–7.
51. Frostig, R.D. et al. (1990). Cortical functional architecture and local coupling between neuronal activity and the microcirculation revealed by in vivo high-resolution optical imaging of intrinsic signals. *Proc Natl Acad Sci USA* 87(16), 6082–6086.

52. Grinvald, A. et al. (1986). Functional architecture of cortex revealed by optical imaging of intrinsic signals. *Nature* 324(6095), 361–364.
53. Takano, T. et al. (2007). Cortical spreading depression causes and coincides with tissue hypoxia. *Nat Neurosci* 10(6), 754–762.
54. Husson, T.R. et al. (2007). Functional imaging of primary visual cortex using flavoprotein autofluorescence. *J. Neurosci* 27(32), 8665–8675.
55. Harrison, R.V. et al. (2002). Blood capillary distribution correlates with hemodynamic-based functional imaging in cerebral cortex. *Cereb Cortex* 12(3), 225–233.
56. Kann, O. and Kovacs, R. (2007). Mitochondria and neuronal activity. *Am J Physiol Cell Physiol* 292 (2), C641–657.
57. Shuttleworth, C.W. et al. (2003). NAD(P)H fluorescence imaging of postsynaptic neuronal activation in murine hippocampal Slices. *J Neurosci* 23(8), 3196–3208.
58. Shibuki, K. and Hishida, R. (2004). Flavoprotein autofluorescence. In *Physiology Online* (Vol. 2005), The Physiological Society.
59. Duchen, M.R. (1992). Ca(2+)-dependent changes in the mitochondrial energetics in single dissociated mouse sensory neurons. *Biochem J* 283(pt. 1), 41–50.
60. Kosterin, P. et al. (2005). Changes in FAD and NADH Fluorescence in Neurosecretory Terminals Are Triggered by Calcium Entry and by ADP Production. *J Membr Biol* 208(2), 113–124.
61. Benson, R.C. et al. (1979). Cellular autofluorescence—is it due to flavins? *J Histochem Cytochem* 27(1), 44–48.
62. Chance, B. et al. (1962). Intracellular oxidation-reduction states in vivo. *Science* 137, 499–508.
63. Chance, B. et al. (1962). Metabolically linked changes in fluorescence emission spectra of cortex of rat brain, kidney and adrenal gland. *Nature* 195, 1073–1075.
64. Takahashi, K. et al. (2006). Transcranial fluorescence imaging of auditory cortical plasticity regulated by acoustic environments in mice. *Eur J Neurosci* 23 (5), 1365–1376.
65. Tohmi, M. et al. (2006). Enduring critical period plasticity visualized by transcranial flavoprotein imaging in mouse primary visual cortex. *J Neurosci* 26(45), 11775–11785.
66. Mallik, A.K. et al. (2008). The organization of spatial frequency maps measured by cortical flavoprotein autofluorescence. *Vision Res.* 48: 14, 1545–1553.
67. Kalatsky, V.A. and Stryker, M.P. (2003). New paradigm for optical imaging: Temporally encoded maps of intrinsic signal. *Neuron* 38(4), 529–545.
68. Kalatsky, V.A. et al. (2005). Fine functional organization of auditory cortex revealed by Fourier optical imaging. *Proc Natl Acad Sci USA* 102: 37, 13325–13330.
69. Weber, B. et al. (2004). Optical imaging of the spatiotemporal dynamics of cerebral blood flow and oxidative metabolism in the rat barrel cortex. *Eur J Neurosci* 20(10), 2664–2670.
70. Brennan, A.M. et al. (2007). Modulation of the amplitude of NAD(P)H fluorescence transients after synaptic stimulation. *J Neurosci Res* 85(15), 3233–3243.
71. Erinjeri, J.P. and Woolsey, T.A. (2002). Spatial integration of vascular changes with neural activity in mouse cortex. *J Cereb Blood Flow Metab* 22(3), 353–360.
72. Sirovich, L. and Uglesich, R. (2004). The organization of orientation and spatial frequency in primary visual cortex. *Proc Natl Acad Sci USA* 101(48), 16941–16946.
73. Cang, J. et al. (2005). Optical imaging of the intrinsic signal as a measure of cortical plasticity in the mouse. *Vis Neurosci* 22(5), 685–691.
74. Gao, W. et al. (2006). Cerebellar cortical molecular layer inhibition is organized in parasagittal zones. *J Neurosci* 26(32), 8377–8387.
75. Sillitoe, R.V. et al. (2003). Zebrin II compartmentation of the cerebellum in a basal insectivore, the Madagascan hedgehog tenrec Echinops telfairi. *J Anat* 203(3), 283–296.
76. Bonhoeffer, T. and Grinvald, A. (1996). Optical imaging based on intrinsic signals. In *Brain Mapping: The Methods* (Toga, A.W. and Mazziotta, J.C., Eds.). Academic Press, pp. 55–97.

77. Kubota, Y. et al. (2008). Transcranial photo-inactivation of neural activities in the mouse auditory cortex. *Neurosci Res* 60(4), 422–430.
78. Husson, T.R. et al. (2006). Autofluorescence imaging in mouse and cat visual cortex. In *Soc. Neurosci.* 32: 640–649.
79. Harbig, K. et al. (1976). In vivo measurement of pyridine nucleotide fluorescence from cat brain cortex. *J Appl Physiol* 41(4), 480–488.
80. Reinert, K.C. et al. (2007). Flavoprotein autofluorescence imaging in the cerebellar cortex in vivo. *J Neurosci Res* 85(15), 3221–3232.
81. Kolosov, M.S. et al. (2003). Photodynamic injury of isolated neuron and satellite glial cells: Morphological study. *IEEE J. Selected Topics in Quantum Electronics* 9(2), 337–342.
82. Oleinick, N.L. and Evans, H.H. (1998). The Photobiology of Photodynamic Therapy: Cellular Targets and Mechanisms. *Radiation Res* 150(5), S146–S156.
83. Sacconi, L. et al. (2006). Overcoming photodamage in second-harmonic generation microscopy: Real-time optical recording of neuronal action potentials. *Proc Natl Acad Sci USA* 103(9), 3124–3129.
84. Hopt, A. and Neher, E. (2001). Highly nonlinear photodamage in two-photon fluorescence microscopy. *Biophys J* 80(4), 2029–2036.
85. Patterson, G.H. and Piston, D.W. (2000). Photobleaching in two-photon excitation microscopy. *Biophys J* 78(4), 2159–2162.
86. Schonle, A. and Hell, S.W. (1998). Heating by absorption in the focus of an objective lens. *Opt Lett* 23(5), 325–327.
87. Jobsis, F.F. et al. (1971). Intracellular redox changes in functioning cerebral cortex. I. Metabolic effects of epileptiform activity. *J Neurophysiol* 34(5), 735–749.
88. Choi, I.Y. et al. (2002). Effect of deep pentobarbital anesthesia on neurotransmitter metabolism in vivo: On the correlation of total glucose consumption with glutamatergic action. *J Cereb Blood Flow Metab* 22(11), 1343–1351.
89. Chance, B. and Williams, G.R. (1955). A method for the localization of sites for oxidative phosphorylation. *Nature* 176(4475), 250–254.
90. Chance, B. et al. (1955). Respiratory enzymes in oxidative phosphorylation. V. A mechanism for oxidative phosphorylation. *J Biol Chem* 217(1), 439–451.
91. Chance, B. and Williams, G.R. (1955). Respiratory enzymes in oxidative phosphorylation. IV. The respiratory chain. *J Biol Chem* 217(1), 429–438.
92. Chance, B. and Williams, G.R. (1955). Respiratory enzymes in oxidative phosphorylation. III. The steady state. *J Biol Chem* 217(1), 409–427.
93. Chance, B. and Williams, G.R. (1955). Respiratory enzymes in oxidative phosphorylation. II. Difference spectra. *J Biol Chem* 217(1), 395–407.
94. Chance, B. and Williams, G.R. (1955). Respiratory enzymes in oxidative phosphorylation. I. Kinetics of oxygen utilization. *J Biol Chem* 217(1), 383–393.
95. Chance, B. et al. (1962). Intracellular oxidation-reduction states in vivo. *Science* 137, 499–508.
96. Mironov, S.L. and Richter, D.W. (2001). Oscillations and hypoxic changes of mitochondrial variables in neurons of the brainstem respiratory centre of mice. *J Physiol* 533(pt. 1), 227–236.
97. Schuchmann, S. et al. (2001). Monitoring NAD(P)H autofluorescence to assess mitochondrial metabolic functions in rat hippocampal-entorhinal cortex slices. *Brain Res Brain Res Protoc* 7(3), 267–276.
98. Hashimoto, M. et al. (2000). Dynamic changes of NADH fluorescence images and NADH content during spreading depression in the cerebral cortex of gerbils. *Brain Res* 872(1–2), 294–300.
99. Mayevsky, A. and Chance, B. (1974). Repetitive patterns of metabolic changes during cortical spreading depression of the awake rat. *Brain Res* 65(3), 529–533.

100. O'Connor, M.J. et al. (1973). Effects of potassium on oxidative metabolism and seizures. *Electroencephalogr Clin Neurophysiol* 35(2), 205–208.
101. Coutinho, V. et al. (2004). Functional topology of the mossy fibre-granule cell—Purkinje cell system revealed by imaging of intrinsic fluorescence in mouse cerebellum. *Eur J Neurosci* 20(3), 740–748.
102. Murakami, H. et al. (2004). Short-term plasticity visualized with flavoprotein autofluorescence in the somatosensory cortex of anaesthetized rats. *Eur J Neurosci* 19(5), 1352–1360.
103. Vishwasrao, H.D. et al. (2005). Conformational Dependence of Intracellular NADH on Metabolic State Revealed by Associated Fluorescence Anisotropy. *J Biol Chem* 280(26), 25119–25126.
104. Riganti, C. et al. (2004). Diphenyleneiodonium inhibits the cell redox metabolism and induces oxidative stress. *J Biol Chem* 279(46), 47726–47731.
105. Kamatani, D. et al. (2007). Experience-dependent formation of activity propagation patterns at the somatosensory S1 and S2 boundary in rat cortical slices. *Neuroimage* 35(1), 47–57.
106. Hishida, R. et al. (2007). Functional local connections with differential activity-dependence and critical periods surrounding the primary auditory cortex in rat cerebral slices. *Neuroimage* 34(2), 679–693.
107. Trachtenberg, J.T. et al. (2002). Long-term in vivo imaging of experience-dependent synaptic plasticity in adult cortex. *Nature* 420 (6917), 788–794.
108. Holekamp, T.F. et al. (2008). Fast three-dimensional fluorescence imaging of activity in neural populations by objective-coupled planar illumination microscopy. *Neuron* 57(5), 661–672.
109. Mulder, D.J. et al. (2006). Skin autofluorescence, a novel marker for glycemic and oxidative stress-derived advanced glycation endproducts: An overview of current clinical studies, evidence, and limitations. *Diabetes Technol Ther* 8(5), 523–535.
110. Rocheleau, J.V. et al. (2004). Quantitative NAD(P)H/flavoprotein autofluorescence imaging reveals metabolic mechanisms of pancreatic islet pyruvate response. *J Biol Chem* 279(30), 31780–31787.
111. Evans, N.D. et al. (2003). Non-invasive glucose monitoring by NAD(P)H autofluorescence spectroscopy in fibroblasts and adipocytes: A model for skin glucose sensing. *Diabetes Technol Ther* 5(5), 807–816.
112. Magistretti, P.J. (2006). Neuron-glia metabolic coupling and plasticity. *J Exp Biol* 209(pt. 12), 2304–2311.
113. Smith, S.L. and Trachtenberg, J.T. (2007). Experience-dependent binocular competition in the visual cortex begins at eye opening. *Nat Neurosci* 10(3), 370–375.
114. Kaneko, M. et al. (2008). TrkB kinase is required for recovery, but not loss, of cortical responses following monocular deprivation. *Nat Neurosci* 11(4), 497–504.
115. Crair, M.C. et al. (1998). The role of visual experience in the development of columns in cat visual cortex. *Science* 279(5350), 566–570.
116. Crair, M.C. et al. (1997). Relationship between the ocular dominance and orientation maps in visual cortex of monocularly deprived cats. *Neuron* 19(2), 307–318.
117. Issa, N.P. et al. (1999). The critical period for ocular dominance plasticity in the Ferret's visual cortex. *J Neurosci* 19(16), 6965–6978.
118. Frank, M.G. et al. (2001). Sleep enhances plasticity in the developing visual cortex. *Neuron* 30(1), 275–287.
119. Beal, M.F. (1998). Mitochondrial dysfunction in neurodegenerative diseases. *Biochimica et Biophysica Acta (BBA)—Bioenergetics* 1366(1–2), 211–223.
120. Rego, A.C. and Oliveira, C.R. (2003). Mitochondrial dysfunction and reactive oxygen species in excitotoxicity and apoptosis: Implications for the pathogenesis of neurodegenerative diseases. *Neurochem Res* 28(10), 1563–1574.

121. Beal, M.F. (2004). Mitochondrial Dysfunction and Oxidative Damage in Alzheimer's and Parkinson's Diseases and Coenzyme Q10 as a Potential Treatment. *J Bioenerg Biomembr* 36(4), 381–386.
122. Lin, M.T. and Beal, M.F. (2006). Mitochondrial dysfunction and oxidative stress in neurodegenerative diseases. *Nature* 443(7113), 787–795.
123. Schapira, A.H. et al. (1990). Mitochondrial complex I deficiency in Parkinson's disease. *J Neurochem* 54(3), 823–827.
124. Betarbet, R. et al. (2000). Chronic systemic pesticide exposure reproduces features of Parkinson's disease. *Nat Neurosci* 3(12), 1301–1306.
125. Dauer, W. and Przedborski, S. (2003). Parkinson's disease: Mechanisms and models. *Neuron* 39(6), 889–909.
126. Skala, M.C. et al. (2007). In vivo multiphoton microscopy of NADH and FAD redox states, fluorescence lifetimes, and cellular morphology in precancerous epithelia. *Proc Natl Acad Sci USA* 104(49), 19494–19499.
127. Holland, E.C. (2001). Brain tumor animal models: Importance and progress. *Curr Opin Oncol* 13(3), 143–147.
128. Gutmann, D.H. et al. (2006). Mouse models of human cancers consortium workshop on nervous system tumors. *Cancer Res* 66(1), 10–13.

9 Visualizing Adult Cortical Plasticity Using Intrinsic Signal Optical Imaging

Ron D. Frostig and Cynthia H. Chen-Bee

CONTENTS

9.1 INTRODUCTION

Our understanding of the functional organization and plasticity of the adult sensory cortex has been transformed in the last 25 years. The view that cortical functional representations of the sensory surface in adult animals are fixed, especially for the primary sensory cortices, has been replaced by the view that they are dynamic and continuously modified by the animal's experience. Such "experience-dependent" plasticity in adult cortical functional representations has been demonstrated in a range of mammalian species across many primary cortices and has been implicated in a range of fundamental processes including rehabilitation following peripheral sensory loss or damage, recoveries from central nervous system damage, improvements in sensory-motor skills with practice, and learning and memory, thus underscoring its importance to an organism's survival. For recent reviews on adult cortical plasticity see References 1,2.

While much progress has been made in understanding the phenomenology and underlying mechanisms, research on the plasticity of adult cortical functional representations is still in its infancy. So far, several techniques have been available for the study of adult cortical plasticity. Ideally, a technique should have the ability to sample: (1) neuronal spiking activity, (2) noninvasively, (3) simultaneously from large cortical regions, (4) with high, 3D spatial resolution, (5) and high temporal resolution, (6) in the awake animal. This ideal technique would enable the direct, noninvasive (hence, long-term and nondamaging) assessment of a cortical functional representation (comprised of many neurons collectively occupying a volume of cortex) with sufficient spatial and temporal resolution to track real-time changes in the functional representation in the awake, behaving animal. In recent years, there has been a growing interest also in studying evoked subthreshold activity, and therefore the ability to record subthreshold activation should be added to the above list of abilities of the ideal technique for studying cortical plasticity. Unfortunately, such an ideal technique does not exist but each of the most widely used current techniques for studying cortical plasticity offers a particular subset of the just-described advantages.

The most popular technique is the use of a single microelectrode to record from the cortex. It offers the advantage of recording action potentials and subthreshold activity directly from cortical neurons with high spatial (point) and temporal (ms) resolution sufficient to follow real-time changes in neuronal activity at any location along a volume of cortex, with the disadvantage that recordings are invasive to the cortex. In order to assess the functional representation of a sensory organ (e.g., a finger, a whisker), neurons are recorded from different cortical locations and the functional representation of the organ is then defined as the cortical region containing neurons responsive to stimulation of that organ (i.e., neurons that have receptive fields localized at the sensory organ). A change in the spatial distribution of neurons responsive to a given sensory organ and/or in their amplitude of response is typically taken as evidence for plasticity in the functional representation of that sensory organ.[3] As a cortical functional representation could be comprised of thousands to millions of neurons distributed over a volume of cortex, the use of a single microelectrode to characterize the plasticity of a functional representation requires many recordings across a large cortical region, recordings that can only be obtained in a serial fashion and require many hours to complete; thus, the animal is typically anesthetized. Because the number of recordings will be limited due to cortical tissue damage incurred during the experiment and time constraints, the characterization of a functional representation is dependent on the extrapolation between recording locations. Also, because of its invasiveness, this technique is not ideal for the assessment of the same functional representation before and after a manipulation within the same animal. The recent development of simultaneous recordings from a chronically implanted array of microelectrodes (see review[4]) provides relief to some of the challenges described above, and has great promise for its potential for long-term recordings from the same animal, although recordings still damage the cortical tissue, the sample size is still small, and extrapolation is still needed between recording locations. Therefore, for better understanding of large-scale cortical functional organization and its plasticity, microelectrode arrays are still limited in fulfilling the need for visualizing mass-action of millions of neurons at different points in time before, during, and after an experimental manipulation.

In this chapter we discuss in detail how visualizing the activity from a large population of neurons can be achieved with our principal technique for studying plasticity of adult cortical functional representations known as intrinsic signal optical imaging (ISOI),[5,6,7] with examples of its applications for functional imaging of rat primary somatosensory cortex. ISOI lacks the ability to directly measure neuronal activity (spiking or subthreshold) and track data in millisecond temporal resolution, but is capable of visualizing functional representations via the simultaneous sampling of a large cortical region with high spatial resolution. We find the combined use of ISOI with rat primary somatosensory cortex particularly appealing because: (1) this animal model offers many advantages for the study of adult cortical plasticity, (2) it can be noninvasive to the cortical tissue (by imaging through the skull) and thus offers the opportunity to image chronically from the same animal, and (3) it is capable of visualizing functional representations in real time (albeit providing only an aerial, i.e. two-dimensional, view of functional representations) and exploits a signal source that appears to be more closely related to underlying spiking activity. The remainder of the chapter is devoted to describing the background, theory and methodological issues relevant to imaging cortical functional representations, the successful application of ISOI for studying the plasticity of whiskers functional representations in the adult rat, and potential future directions for ISOI.

9.2 BACKGROUND AND THEORY

9.2.1 Advantages of Applying ISOI to the Rat Primary Somatosensory Cortex

One main advantage is the relative ease in imaging cortical activity in the rat. Unlike imaging experiments in cats and monkeys, the surgical preparation for imaging in rats is minimally invasive to the animal[8] because there is no longer a need to (1) control for respiration and heart rate artifacts via paralysis and artificial ventilation in addition to the synchronization of data collection to heart and respiration; (2) remove the skull and dura in order to image cortical activity, therefore leaving the brain intact and in optimal condition throughout the experiment; and (3) build and affix an elaborate "recording chamber" to the skull with screws and cement as typically done for imaging in cats and monkeys. Instead, only removal of the scalp and muscle tissue along with thinning of the skull with a drill above the imaged area is needed for better light penetration. In addition, because the skull and the dura are left intact, the opportunity is available to repeatedly image the same rat over an extended time period (chronic imaging), which can be advantageous for the study of cortical plasticity within the same animal. It is worthwhile to note that, in contrast to rats, the imaging of mice somatosensory cortex can be achieved through the intact skull (no thinning needed) because it is sufficiently transparent.[9]

The other main advantage is related to the use of the posterior medial barrel subregion (PMBSF), the largest subdivision within the primary somatosensory cortex, which processes sensory information from the large facial whiskers (or vibrissae), as an animal model of cortical function and plasticity. The body of literature accumulated thus far on the PMBSF suggests that this system shares many features with sensory cortex in higher mammals—such as topographical organization of whisker inputs to the cortex, and functional "columns" dedicated to processing tactile information from the whiskers—and thus provides a rich source of information from which hypotheses can be formulated about cortical plasticity.[10,11] The popularity of the PMBSF for the study of cortical plasticity stems in part from the relative ease in manipulating the whiskers: (1) precise stimulation can be delivered to a specific whisker or a specific combination of whiskers; (2) sensory deprivation can be easily achieved by trimming and/or plucking of whiskers; and (3) sensory deprivation can be easily reversed because of whisker regrowth following the termination of whisker trimming and/or plucking. Thus, because of the above-mentioned advantages, the combination of ISOI with rat PMBSF can be a powerful means for investigating adult cortical plasticity.

9.2.2 Intrinsic Signals and Their Sources

ISOI is based on the finding that when the brain is illuminated neuronal activity causes changes in the intensity of the light that is reflected from the brain. Accordingly, patterns of evoked activity can be detected and quantified by measuring reflectance patterns from the living brain.[5] These stimulus-evoked optical changes are referred to as *intrinsic signals* to differentiate them from other optical signals obtained using extrinsic probes such as voltage sensitive dyes or calcium indicators (see other chapters in this book for imaging with extrinsic probes). It should be noted that in recent years the term *intrinsic signals* has also been used to describe new classes of intrinsic signals such as fast scattering responses and auto-fluorescence responses, which are beyond the scope of the present chapter and will not be mentioned further (see other chapters in this book describing these.) The intrinsic signals originate from activity-dependent changes in several intrinsic sources in the brain including: oxygen consumption affecting hemoglobin saturation level (i.e., oxygenated versus deoxygenated hemoglobin, or HbO2 versus Hbr, respectively); changes in blood volume affecting tissue light absorption; changes in blood flow affecting hemoglobin saturation level; and changes in tissue light scattering related to physiologic events accompanying neuronal activity such as ion and water movement in and out of neurons, changes in the volume of neuronal cell bodies, capillary expansion, and neurotransmitter release (reviewed in References 12 and 13; see also chapter by Rector et al. in this book). The relative contribution of these various sources to the intrinsic signals is dependent on the wavelength of light used to illuminate the cortex. Roughly, blood-related sources dominate the intrinsic signals in the blue-to-red parts of the spectrum while tissue light scattering dominates the intrinsic signals in the infrared part of the spectrum.[7] In terms of signal-to-noise and better contrast between active/nonactive regions, the best images of cortical activity are routinely obtained using red or orange light illumination,[7] and thus the remainder of the present chapter is restricted to imaging of intrinsic signals using red light illumination.

9.2.3 Stimulus-Evoked Intrinsic Signals

Figure 9.1 contains representative stimulus-evoked intrinsic signals obtained from a cortex illuminated with red light (630 nm). Intrinsic signals evoked by a short-lasting

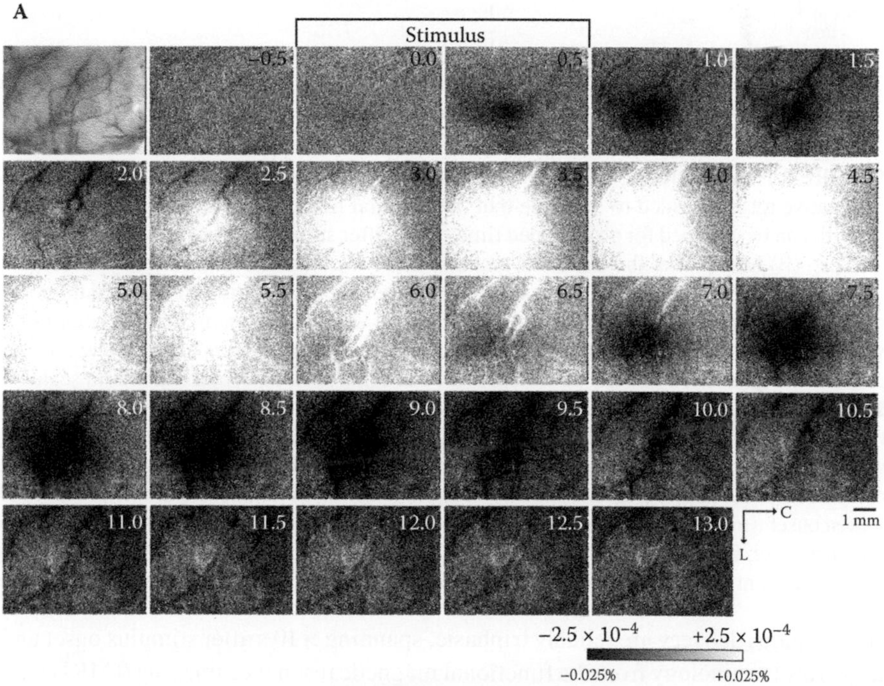

FIGURE 9.1 Visualization of evoked ISOI intrinsic signal. Data from a representative rat is provided to illustrate that high-spatial resolution images of intrinsic signal activity can be obtained for an extended time epoch after stimulus onset (up to 13.5 s; 500 ms frames). (A) A photograph is provided of the vasculature present on the surface of the imaged cortical region. Images of poststimulus data are created by first converting data into fractional change values relative to the 500-ms frame collected immediately prior to stimulus onset on a pixel-by-pixel basis before applying an 8-bit linear grayscale to the processed data such that middle gray = no change relative to prestimulus data, darker gray values = larger undershoots, lighter gray values = larger overshoots, and darkest (black) and lightest (white) values are thresholded at $\pm 2.5 \times 10^{-4}$ or ±0.025%. Prestimulus data can be visualized by creating an image in the similar manner, where the 500-ms frame immediately preceding stimulus onset (−0.5 s time point) is converted relative to the −1.0 s time point. Stimulus bar indicates when the 1-s stimulus delivery occurred; 1-mm scale bar and neuroaxis apply to all images. (Modified from Chen-Bee, C.H. et al., *J. Neurosci.*, 27, 4572, 2007. With permission.)

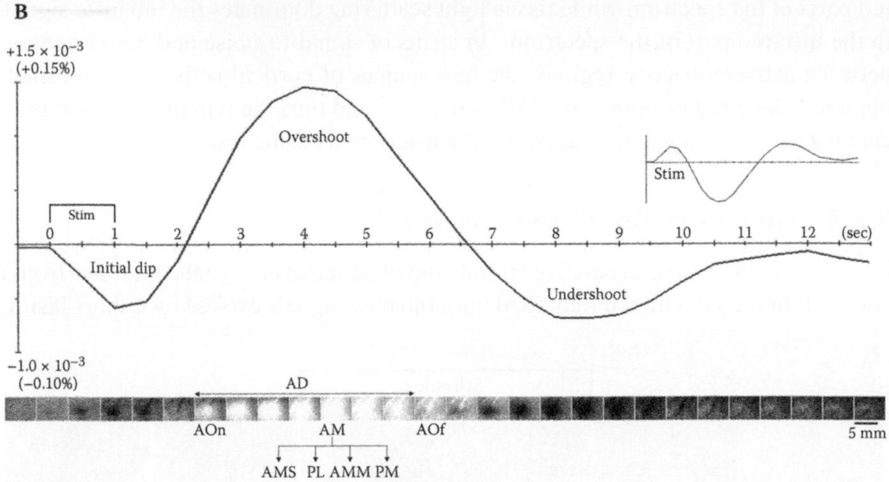

FIGURE 9.1 (continued) Visualization of evoked ISOI intrinsic signal. Data from a representative rat is provided to illustrate that high-spatial resolution images of intrinsic signal activity can be obtained for an extended time epoch after stimulus onset (up to 13.5 s; 500 ms frames). (B) Line plot of the fractional change values is obtained from a single binned pixel located centrally within the imaged cortex, with the inset containing the same data points plotted upside down as is traditionally done in ISOI studies. The same images seen in (A) were reduced in size and redisplayed below the line plot for easier comparison along the temporal domain (reduced images are also provided in Figures 9.2, 9.4, and 9.7). 5 mm scale bar applies to all images; neuroaxis remains the same. Note that stimulation of a single whisker for 1 s evokes a triphasic intrinsic signal consisting of an initial dip (dark patch) followed by an overshoot (bright patch) and then an undershoot (dark patch). Each intrinsic signal phase can be comprehensively characterized by assessing such spatiotemporal parameters as: AOn = area onset time; AM = area max time; AOf = area offset time; AD = area duration; AMS = area max size; PL = peak location; AMM = area max magnitude; and PM = peak max time. (Modified from Chen-Bee, C.H. et al., *J. Neurosci.*, 27, 4572, 2007. With permission.)

(1 s) stimulus delivery are reliably triphasic, spanning > 10 s after stimulus onset and, to borrow terminology from the functional magnetic resonance imaging (fMRI) field, consists of an initial dip below baseline, followed by the overshoot above baseline and then the undershoot below baseline (Figure 9.1B). High-spatial resolution images can be generated from the evoked triphasic signal to map the spatiotemporal profile of each signal phase (Figure 9.1A). To comprehensively characterize a given signal phase, various parameters are available for assessment pertaining to the presence/absence of an evoked signal area (e.g., onset, offset, total duration) and maximum evoked area (e.g., areal extent, peak magnitude, peak location) (see Figure 9.1B). Figure 9.2 summarizes the high-spatial resolution mapping of the triphasic signal across 60 rats and Figure 9.3 summarizes in detail how the three signal phases compare with each other. Relative to stimulus onset, the evoked initial dip area typically appears within 0.5 s (area-onset time), peaks 1–1.5 s (area max time), disappears by 2–2.5 s (area offset time), and thus lasts 1.5–2.5 s (area-duration; Figure 9.3C), followed by the overshoot area (2–3 s area onset, 3.5–4.5 s area max, 6–9 s area offset,

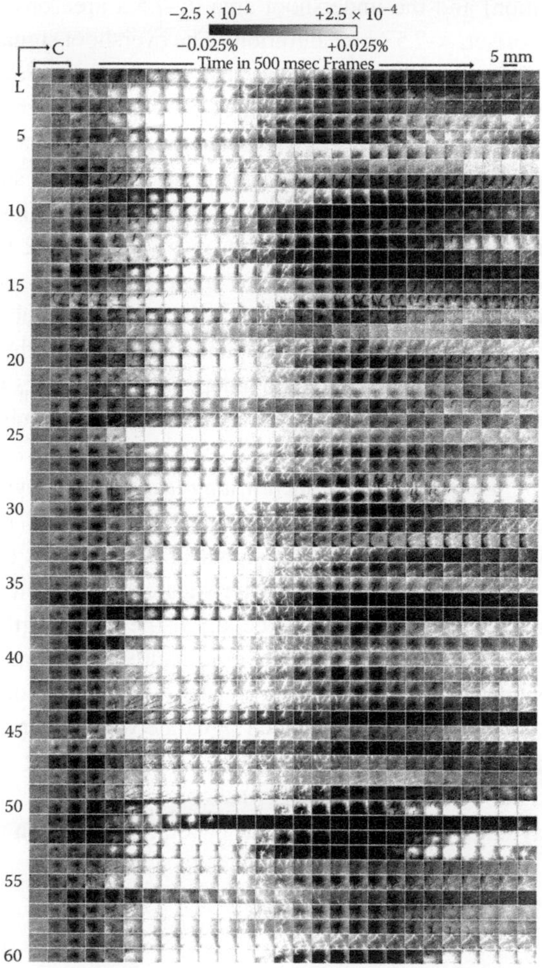

FIGURE 9.2 Visualization of the evoked ISOI intrinsic signal across 60 rats. Images of evoked intrinsic signal in 500-ms frames are provided through 13.5 s after onset of 1-s stimulus delivery to a single whisker (one row per rat). A total of 64 stimulation trials were collected per rat for the first 24 rats, whereas 128 trials were collected for the remaining 36 rats. Stimulus bar in the upper left corner indicates that the 1-s stimulus delivery occurs during the first two frames and applies to all rows; grayscale $\pm 2.5 \times 10^{-4}$, 5 mm scale bar, and neuroaxis apply to all images. Across all rats, the evoked intrinsic signal is reliably triphasic, consisting of an initial dip (first prolonged presence of a black activity area), followed by an overshoot (first prolonged white activity area), and then an undershoot (second prolonged black activity area). While some variability existed across rats, consistent differences between the three signal phases were apparent including time after stimulus onset when an activity area first appeared (area-onset time or AOn), total duration of activity area presence (area-duration or AD), and spatial extent size of the maximum activity area (area-max-size or AMS). Note that by the end of the 13.5 s poststimulus time epoch, an additional overshoot and/or undershoot was observed in the majority of rats. (Modified from Chen-Bee, C.H. et al., *J. Neurosci.*, 27, 4572, 2007. With permission.)

3.5–5 s area duration) and the undershoot area (6–7.5 s area onset, 7.5–9.5 s area max, 10–12 s area offset, 3–7.5 s area duration). The overshoot signal phase is largest in spatial extent (Figure 9.3A), is strongest in peak magnitude (Figure 9.3B), and lasts the longest (Figure 9.3C), followed by the undershoot and then the initial dip, with poor co-localization of peak activity between the three signal phases (Figure 9.3D). For every signal phase, only a few parameters appear correlated, such as larger spatial extents with stronger peak magnitudes or longer total duration of evoked area presence with later offset of evoked area presence (see Table 2 of Reference 14). Correlations between signal phases are even more sparse (e.g., no correlation in peak magnitude or spatial extent between any two phases; see Table 3 of Reference 14), demonstrating the large degree of independence between the different signal phases with respect to their attributes. Such independence impedes using the various attributes of the initial dip to predict the attributes of the subsequent overshoot and undershoot signal phases.

It remains to be determined how such a long-lasting signal with three distinct phases is dependent on the wavelength of illumination and/or stimulus delivery duration used. While a similar tri-phasic signal is present with the use of orange illumination,[14] it is possible that the characteristics of the detected signal can differ between illumination wavelengths, especially if the wavelengths differ substantially in what underlying sources contribute dominantly to the intrinsic signals. For example, in following the evoked intrinsic signals over an extended time period, it would be interesting to compare the detected signal between green, red, and infrared illumination. Duration of stimulus delivery is also likely to affect the characteristics of the detected signal; indeed, using spectroscopic optical imaging to specifically follow the hemodynamic signals Hbr, HbO2, or total hemoglobin over an extended

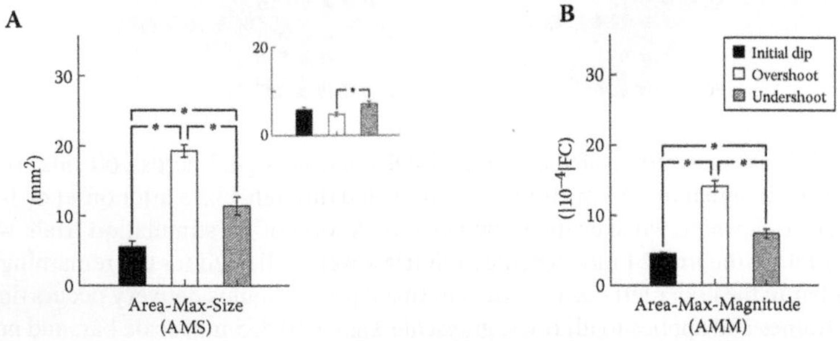

FIGURE 9.3 Comparison between the three ISOI signal phases. (A,B) Interphase comparison of area-max-size (AMS) and area-max-magnitude (AMM). The spatial extent size of the maximum activity area (AMS) and the magnitude of peak activity within the quantified area (AMM) were assessed for all three signal phases. Means and SEs of the quantified data are provided for 60 rats. Statistically significant differences between all possible pairs of signal phases exist for both AMS and AMM (indicated by asterisks). Inset in (A): The maximum activity area as quantified using a normalized threshold of 50% AMM. In striking contrast, note that the smallest maximum activity area is actually observed for the Overshoot signal phase. (Modified from Chen-Bee, C.H. et al., *J. Neurosci.*, 27, 4572, 2007. With permission.)

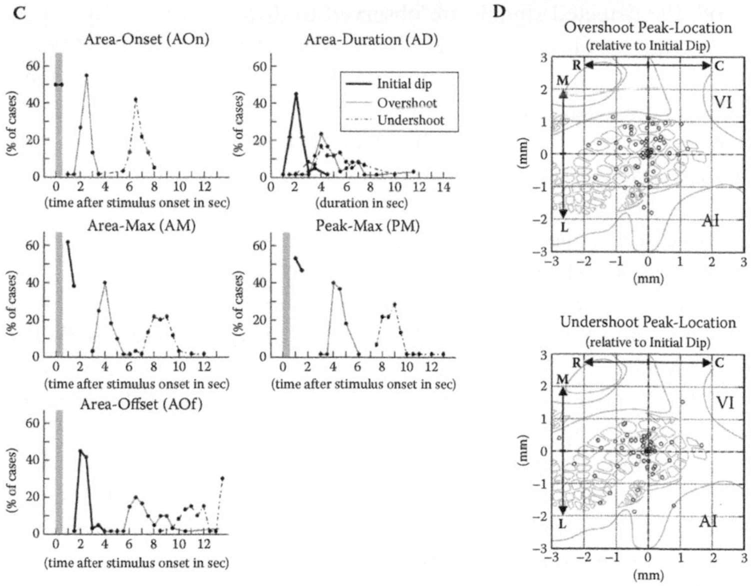

FIGURE 9.3 (continued) Comparison between the three ISOI signal phases. (C) Temporal profile of the evoked activity area per signal phase. Regarding the entire period that an evoked activity area is present, the following temporal characteristics were assessed per signal phase: AOn = area onset time; AM = area max time; AOf = area offset time; AD = area duration; and PM = peak max time. Occurrence of the 1-s stimulus delivery to a single whisker is indicated in all but the fourth panel with a vertical gray bar. While some variability existed across rats, note that the initial dip, overshoot, and undershoot phase of the evoked intrinsic signal exhibited stereotypical time points with respect to when the area-onset, area-max, and area-offset occurred, as well as when peak-max occurred. Also, the total duration that the evoked activity area was present (area-duration) was about twice as long for the overshoot and undershoot as compared to the initial dip, and that area max and peak max occurred at similar time points. (D) Interphase comparison of peak activity location. Interphase comparison of peak location is meaningful especially relative to the initial dip given that the initial dip peak response to stimulation used here (single whisker C2 stimulation) has previously been shown to localize above the appropriate anatomical location (whisker C2 anatomical representation in layer IV of primary somatosensory cortex). In order to summarize the spatial registry between the initial dip and overshoot peak locations across all 60 rats, the overshoot peak location was first converted to relative coordinates with respect to those of the initial dip such that a relative coordinate of (0,0) would indicate a perfect match in peak location for the two phases. Thus, a single rat can be plotted with a single point and a total of 60 rats can be summarized in a scatterplot. To better appreciate the anatomical significance of the variability in the plotted relative coordinates, the scatterplot is superimposed on an appropriately scaled schematic of tangential cortical layer IV cytochrome oxidase labeling that includes portions of primary somatosensory, visual (VI), and auditory (AI) cortex, with the relative origin centered above whisker C2 anatomical representation. Note that the overshoot peak location did not co-localize well with that of the initial dip; only 22% of the cases exhibited a margin of error that would still be considered to successfully lie above whisker C2 anatomical representation. The overall findings were the same when comparing peak locations for the initial dip versus the Undershoot signal phase. Lastly, only a slight improvement in colocalization of peak activity was observed between the overshoot and undershoot (data not shown). (Modified from Chen-Bee, C.H. et al., *J. Neurosci.*, 27, 4572, 2007. With permission.)

time period, the detected signals are observed to differ substantially depending on whether the stimulus duration is brief (2 s) or long (16 s[15]).

Similarly to ISOI with red illumination, functional magnetic resonance imaging (fMRI) also measures hemodynamic-related changes and interestingly it, too, detects an evoked signal that is tri-phasic and exhibit similar attributes to those of ISOI[16] across various data collection conditions, suggesting that a tri-phasic signal with attributes as described here may to some extent transcend factors such as the technique used (albeit one based on underlying hemodynamic signal sources), stimulus parameters, cortical region, quantification method, etc. For ISOI, it is thought that the initial dip is dominated by a rapid Hbr increase and HbO2 decrease, localizes best with activated neurons, and thus is routinely exploited in ISOI imaging studies; the overshoot is dominated by HbO2 increase, Hbr decrease, and volume changes in the vascular system due to influx of blood to the activated area, and is less localized to activated neurons[17]; as with the initial dip, the undershoot is dominated by Hbr increase, although the initial dip and the undershoot likely differ in the exact mechanisms by which the Hbr increase occurs. At the very least, involvement of reoccurring evoked neuronal activity, supra- or subthreshold, is ruled out in whatever mechanisms underlie the ISOI undershoot (Figure 9.4).

The ISOI initial dip is particularly relevant for the imaging of brain activity because of its tight spatial coupling to suprathreshold neuronal activity.[17] Note that the initial dip signal amplitude is quite small, typically between $1–10 \times 10^{-4}$ of the total amount of light that is reflected from the cortex (Figure 9.1A). Thus, on first inspection, the initial dip does not seem very promising for imaging brain activity because of its weak amplitude and slower time course (compared to ms domain of action potentials). Nevertheless, by simple manipulations of the intrinsic signal data as described later, clear images of evoked initial dip are obtained with very high spatial resolution (~50 μm) from large areas of the cortex.

9.2.4 Biological Noise

As mentioned earlier, there is no longer a need to control for heart rate and respiration artifacts when imaging rats. The absence of heart rate artifacts is presumably due to the faster heart rate of rats (relative to cats and monkeys) in combination with the lower temporal resolution (e.g., 500 ms frames) of typical ISOI experiments. At higher temporal resolutions, it is likely that heart rate artifacts will be a concern even when rats are imaged. The reason respiration artifacts are absent when imaging in rats versus cats or monkeys is less clear; it may be due to the opportunity specifically in rats to image through the thinned skull, thereby providing dampening of brain movement during breathing, and/or in general greater brain movement during breathing in cats and monkeys versus rats prior to any surgical preparation.

Two types of biological noise are of particular interest when imaging the ISOI initial dip, both of whose magnitudes are much larger than that of the initial dip: (1) global, spontaneous fluctuations and (2) local contributions overlying surface blood vessels. Spontaneous fluctuations in intrinsic signals can be ten times greater and occur on a slower time scale (oscillations of ~ 0.05–0.1 Hz or one complete cycle

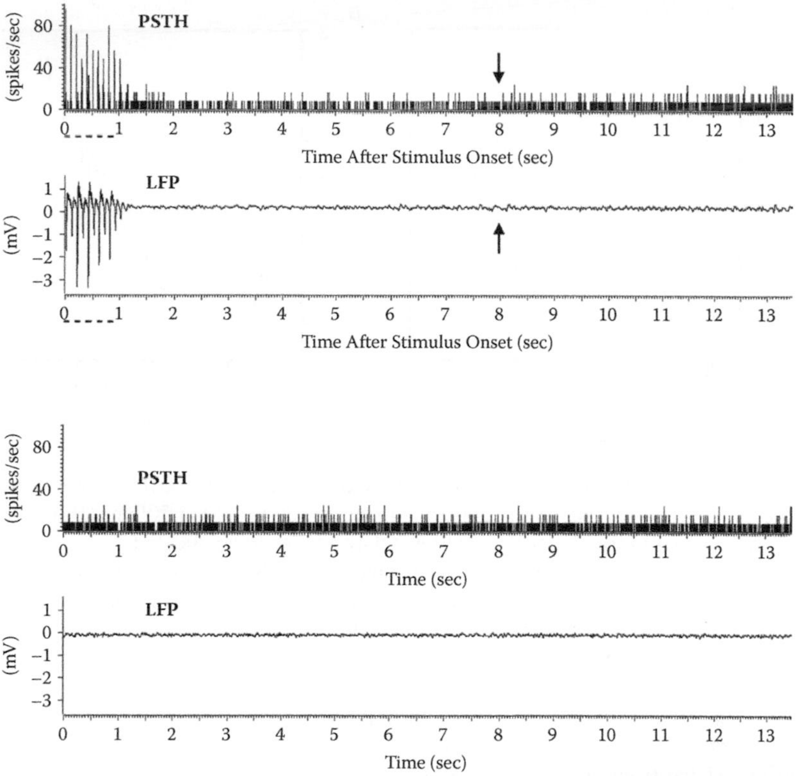

FIGURE 9.4 Suprathreshold and subthreshold neuronal activity recorded through 13.5 s after onset of 1-s stimulus. *Top half:* Data from the supragranular layer at the initial dip peak location of a representative rat are provided to illustrate that suprathreshold (PSTH) and subthreshold (LFP) neuronal activity, recorded simultaneously from the same electrode, was followed for many seconds after stimulus delivery. Stimulus bars indicate the 1-s delivery of 5 Hz whisker C2 stimulation, and arrows indicate the approximate time after stimulus onset when area max is achieved for the ISOI undershoot phase. Other than the obvious round of evoked suprathreshold and subthreshold neuronal activity occurring during stimulus delivery, note the lack of a second round of increased activity for both the PSTH and LFP for the interval between stimulus offset and up to 13.5 s after stimulus onset, including within a few seconds prior to the undershoot area max time point. *Bottom half:* Control data from the same recording location in the same rat are provided to illustrate that no occurrences of spontaneous activity were observed with similar magnitudes as those of evoked activity for either the suprathreshold (PSTH) or subthreshold (LFP) recordings. (Modified from Chen-Bee, C.H. et al., *J. Neurosci.*, 27, 4572, 2007. With permission.)

every 10–20 s) as compared to the stimulus-evoked initial dip (Figure 9.5). Because stimulus delivery evokes only a small change on top of the large spontaneous intrinsic signals fluctuations, the successful imaging of stimulus-evoked intrinsic signals requires that these spontaneous fluctuations are somehow averaged out. As they are not time-locked to stimulus delivery, spontaneous intrinsic signals fluctuations can be minimized by averaging a set of stimulation trials. The number of imaging trials

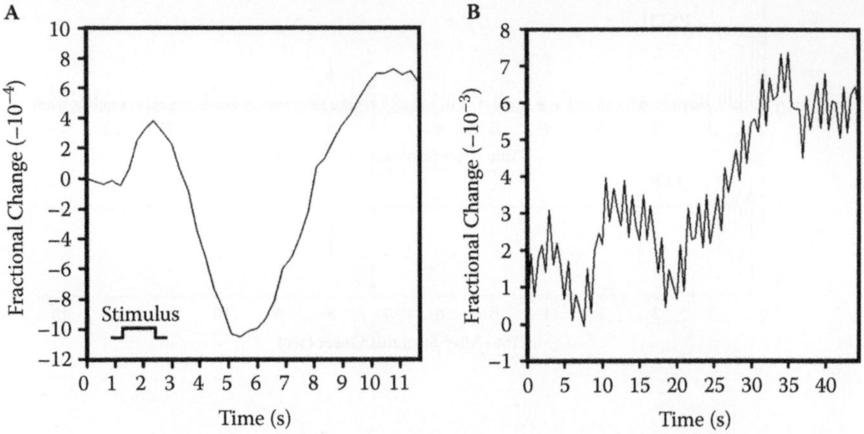

FIGURE 9.5 Temporal profile of evoked and spontaneous ISOI intrinsic signal obtained from the rat barrel cortex. Data were collected in either 300 ms (A) or 500 ms (B) frames and plotted on the y-axis in fractional change units relative to the first collected frame. By convention, decreasing light reflectance is plotted as upgoing. (A) Stimulus-evoked intrinsic signal from the left cortex of a rat. Intrinsic signal was sampled over a 0.16 mm^2 cortical area in 300 ms frames after averaging 32 trials. On the x-axis, the timepoint of 0 contains intrinsic signal collected during 0–299 ms, with timepoints 1.2–2.4 s containing intrinsic signal collected during the 1.2 s stimulation of whisker C2 at 5 Hz. As compared to the slow changes in the spontaneous intrinsic signal shown in (B), the amplitude of the faster changes in the stimulus-evoked cortical intrinsic signal was smaller by an order of magnitude. (B) Spontaneous cortical intrinsic signal from the right PMBSF of a rat. Intrinsic signal was sampled over a 0.16 mm^2 cortical area in 500 ms frames during a single 45.5 s trial. On the x-axis, the timepoint of 0 contains intrinsic signal collected during 0–499 ms while the timepoint of 5 s contains intrinsic signal collected during 5000–5499 ms, and so forth. The magnitude of changes in intrinsic signal remained similar when the frame duration is increased to 5 s and the trial interval is increased to 455 s (data not shown). (Modified from Chen-Bee, C.H. et al., *J. Neurosci. Methods*, 68, 27, 1996. With permission.)

needed (32–128 trials) for sufficient capturing of the initial dip, as well as the other two signal phases, is comparable to that of single unit recording experiments. Any residual presence of spontaneous fluctuations can then be addressed at the level of data analysis. To better understand these spontaneous fluctuations, we have collected and analyzed control trials in the same manner as stimulation trials and found that their presence can be substantial in magnitude and areal extent despite the averaging across many trials (Figure 9.6). Fortunately, these spontaneous fluctuations are nonspecific both in the temporal and spatial domains, making them distinguishable from stimulus-evoked intrinsic signal (see also imaging of spontaneous fluctuations with voltage sensitive dyes in Chapter 6 of this book). Spontaneous fluctuations have also become a topic of major interest in the fMRI field,[18] highlighting again the similarities between these two imaging methods.

The averaging of stimulation trials is not effective in minimizing local contributions that overlie surface blood vessels because they are also time-locked to stimulus delivery. However, as they typically follow a slower time course as compared to the

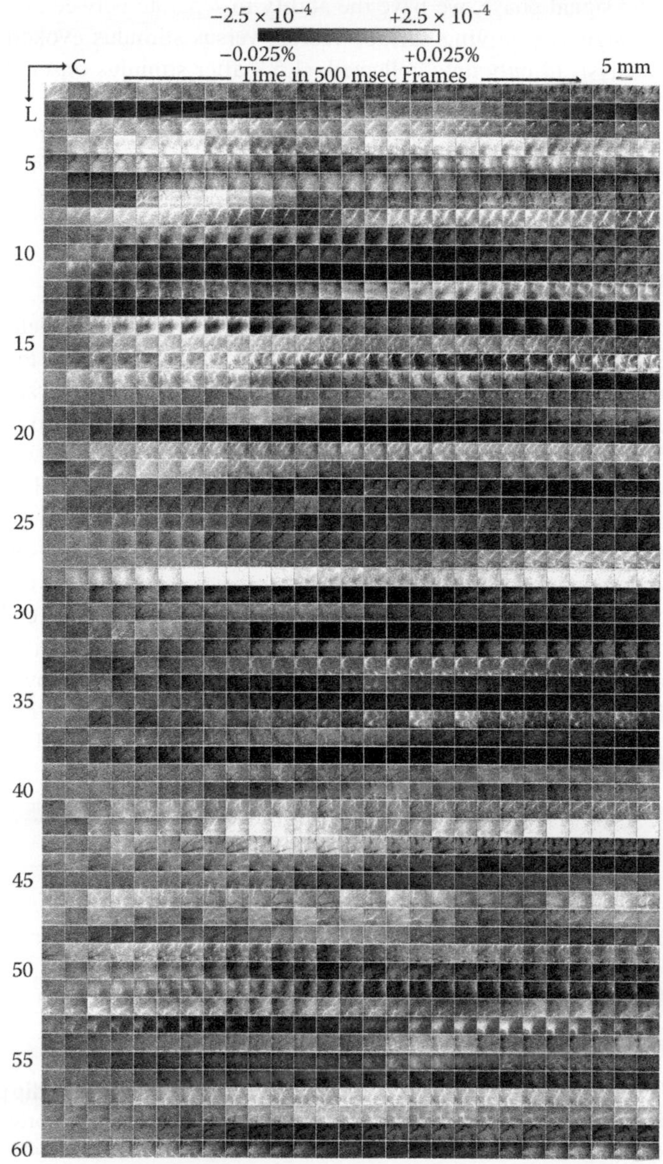

FIGURE 9.6 Visualizing 13.5 s of ISOI control (no stimulation) data across 60 rats. Images of ISOI intrinsic signal in 500-ms frames are provided for control trials (one row per rat) that were randomly interlaced with stimulation trials. A total of 64 control trials were collected per rat for the first 24 rats, whereas 128 trials were collected for the remaining 36 rats. Grayscale, 5 mm scale bar, and neuroaxis apply to all images. Overshoot and/or undershoot fluctuations in intrinsic signal area were observed in control trials that were collected randomly with stimulation trials, although these fluctuations did not exhibit stereotypical characteristics such as time of area onset, area max, or area offset. (Modified from Chen-Bee, C.H. et al., *J. Neurosci.*, 27, 4572, 2007. With permission.)

ISOI initial dip signal phase, we have the ability to separate between contributions from intrinsic signals overlying surface vessels versus stimulus-evoked initial dip by limiting analysis to only data collected <1.5 s after stimulus onset (Figure 9.7). An alternative method to separate these artifacts is to use a periodic stimulus and Fourier analysis in order to filter out artifact contributions with a different peak location in the frequency domain as compared to the evoked intrinsic signals (see more details in Chapter 10 of this book).

9.2.5 The Imaging Camera

Successful imaging of intrinsic signals requires a camera capable of detecting small changes in light reflectance, i.e., a camera system with a good signal-to-noise ratio. As no fluorescent probes are involved, high light levels can be used to achieve imaging conditions that are "shot-noise" limited, which is advantageous to any imaging technique. Shot-noise is an unavoidable noise that emerges from the statistical nature of light quanta. Because the shot-noise depends on the square root of light intensity, it is desirable to work as close as possible to the saturation level of the camera and therefore reduce its relative contribution. Thus, cameras that can register high light intensities without compromising their linearity are advantageous for ISOI.

We use cooled, digital, charge-coupled device (CCD) cameras that contain pixels with large well capacities (several hundred thousand photons per pixel); the well capacity indicates the number of electrons that can be accumulated by one pixel of the CCD chip before an overflow of charge. Such large well capacities enable the use of high-intensity illumination, effectively reducing shot-noise and consequently

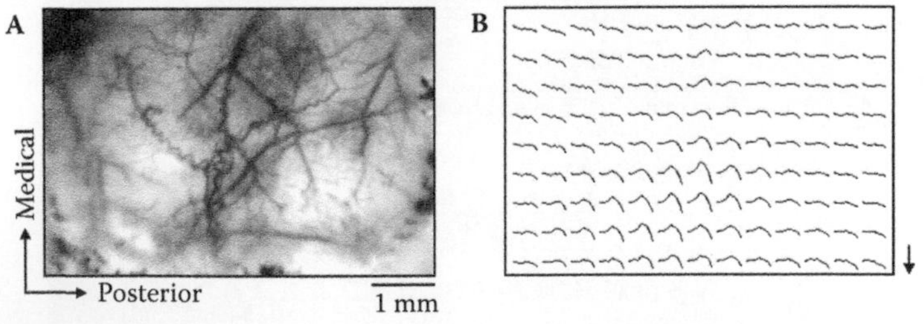

FIGURE 9.7 Separating intrinsic signal contributions from the ISOI initial dip phase versus those overlying surface blood vessels. (A) The imaged area within the left cortex of an adult rat as imaged through a thinned skull. Dark streaks correspond to large blood vessels found on the cortical surface or dura matter. Orientation and scale bar apply to all panels. (B) Array of intrinsic signals evoked by 5 Hz stimulation of whisker D1 for 1 s from a collection of 128 trials. Increasing intrinsic signal (i.e., decreasing light reflectance) is plotted as upgoing (downward pointing scale bar = 1×10^{-3}), with each plot corresponding to the 4.5 s time course as averaged for the underlying 0.46 mm × 0.46 mm area. The region exhibiting an increase in the initial dip phase of the intrinsic signals overlying the cortical tissue is located in the lower center of the total imaged area, with a concurrent but slower increase in intrinsic signals overlying nearby large surface blood vessels (upper center). (Modified from Chen-Bee, C.H. et al., *J. Neurosci. Methods*, 97, 157, 2000. With permission.)

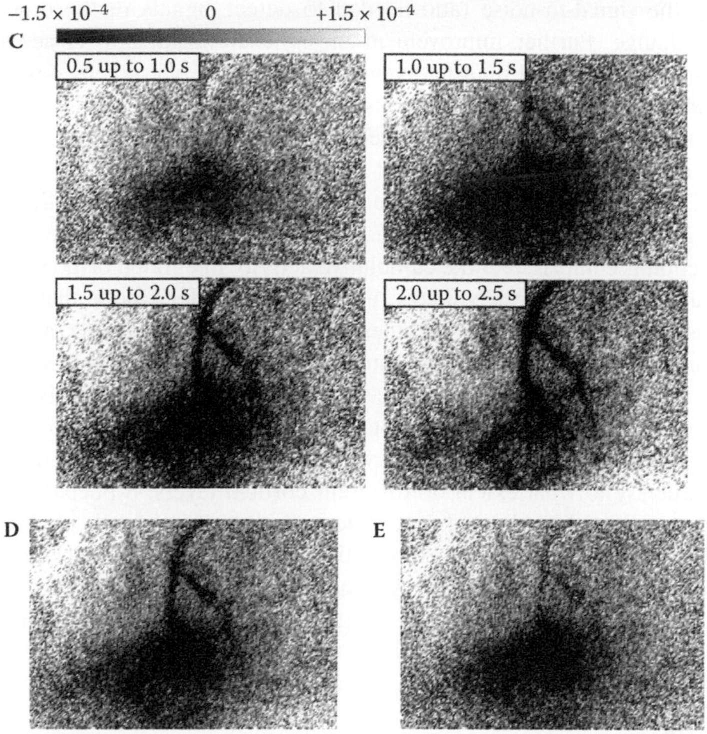

FIGURE 9.7 (continued) Separating intrinsic signal contributions from the ISOI initial dip phase versus those overlying surface blood vessels. (C) Images of evoked intrinsic signals in 500 ms frames from same data presented in (B). Each image was generated after dividing a given poststimulus frame (indicated at top left of each image; 0 s = stimulus onset) by a frame collected immediately prior to stimulus onset and applying an 8-bit, linear grayscale map to the processed data so that increased intrinsic signal greater than -1.5×10^{-4} is mapped to a grayscale value of black. Evoked ISOI initial dip intrinsic signal overlying the cortical tissue (black patch in lower center) is present starting 0.5 s poststimulus onset and remains elevated through 2.0 s poststimulus onset before diminishing in strength. In contrast, evoked intrinsic signals overlying large surface blood vessels (black streaks in upper center) follows a slower time course, with minimal activity present 1.0 s poststimulus onset that increases and remains elevated past 2.5 s poststimulus onset. (D,E) Visualization of evoked initial dip intrinsic signals collected either 0.5 up to 2.5 s (D) or 0.5 up to 1.5 s (E) poststimulus onset. The resultant image for 0.5 up to 2.5 s (D) or 0.5 up to 1.5 s (E) poststimulus onset is equivalent to averaging all four frames or the first two frames shown in (C), respectively (each frame has been divided by the frame collected immediately prior to stimulus onset). When data processing includes the time epoch of 0.5 up to 2.5 s poststimulus onset (D), evoked intrinsic signals overlying both the cortical tissue (black patch in lower center) and large surface blood vessels (black streaks in upper center) are visualized. However, data processing of the shorter poststimulus time epoch (E) generates an image of evoked initial dip intrinsic signals overlying the cortical tissue with limited presence of evoked intrinsic signal overlying the large surface blood vessels. Gray scale bar applies to all figures (C–E). (Modified from Chen-Bee, C.H. et al., *J. Neurosci. Methods*, 97, 157, 2000. With permission.)

improving the signal-to-noise ratio needed to detect signals in the order of 10^{-4} fractional change. Further improvement in signal-to-noise ratio is accomplished by performing 2 × 2 pixel binning as well as averaging of trials as described in a later section (in recent years, considerable improvements in CMOS-based cameras have made them an attractive alternative for CCD-based imaging; for example, see Chapter 6 of this book.)

The CCD camera is positioned above the cortex to be imaged and thus offers an aerial, or 2D, collection of light reflection from the illuminated cortex. Therefore, data collection is comprised of the cumulative activity integrated orthogonally to the cortical surface. The camera operates in a "frame-transfer" mode that enables the continuous capture of sequence of frames for imaging before, during, and after sensory stimulation. Penetration of light into the cortex depends on the wavelength of the illumination; longer wavelengths penetrate deeper into the cortical tissue and thus the use of red illumination is also advantageous in terms of penetration. However, at present it is still difficult to estimate to what extent the red light penetrates in vivo and consequently to what extent do different cortical layers, especially the deeper layers, contribute to the collected data. Due to the exponential loss of photons with cortical depth and due to a further exponential loss of photons in the return path from cortical depth to the surface, at the very minimum ISOI is inherently biased for functional measurements of activity in the upper layers of the cortex, although the upper layers of the cortex include large dendritic trees of neurons located in deep cortical layers and therefore activity in the upper layers can be influenced or modulated by neurons in deep cortical layers.

9.2.6 Imaging Plasticity in Sensory Cortex

In recent years most of ISOI research has been focused on imaging functional columns—groups of vertically organized neurons that exhibit preference in responding to a specific stimulus (e.g., orientation columns, ocular dominance columns, and other columnar organizations in the visual cortex), which are believed to be an important module of the functional organization of cortex. Thus, the study of cortical plasticity can be conducted at the level of functional columns. Imaging plasticity of the columnar organization requires differential imaging of a cortical area after the application of different stimuli that activate different columns of cortical neurons. Division or subtraction of differences in activation patterns by orthogonal stimuli (e.g., horizontal versus vertical grating; left versus right eye) is used to highlight the cortical location of columns with a preferred response to a specific stimulus. Plasticity is then typically assessed by quantifying changes in the maps of cortical columns following sensory manipulation (see, for example, Reference 19). An alternative to differential imaging is the use of "single condition" maps where the response to one stimulus is not divided by its opposite stimulus but by either a "cocktail" that included cortical activation by all other stimuli, or divided by the cortical response to a blank screen.

Imaging the functional representation of a peripheral organ provides an alternative means to study plasticity in the adult cortex. An organ functional representation is defined as the total cortical area that is activated stimulation of the sensory organ and is visualized by subtracting or dividing background (spontaneous) activity that precedes the stimulation from the poststimulus activity rather than employing a differential imaging or "single condition" strategy to create the columnar preference maps described above. This approach is preferable in cortical regions such as the somatosensory cortex in which the sensory representations of the organs overlap and do not exhibit clear columnar organization for stimulus submodality (e.g., orientation) to the extent seen in visual cortex. Plasticity of a sensory organ's functional representation is typically assessed by quantifying changes in the cortical area activated by the sensory organ following manipulation. The focus of the present chapter is on the imaging of functional representation for the study of cortical plasticity.

9.2.7 Relationship between Initial Dip and Neuronal Activity

ISOI with red illumination provides an indirect measure of neuronal activity, and thus it is important to determine the relationship between the initial dip and the underlying neuronal activity. In the rat PMBSF, we have found that the pattern of initial dip activity is very similar to that of the underlying spiking neurons. Specifically, the peak of the initial dip activity corresponds spatially to the location where neurons respond in the strongest fashion to the same sensory stimulus. Sampling at progressively larger distances away from the peak, both initial dip and spiking activity become progressively weaker with increasing distance from peak activity, and areas with no optical activity do not exhibit any spiking activity either.[20,21,22] In recent years, however, correlation between intrinsic signals and local field potentials (LFPs) have also been described,[19] especially correlation with certain frequencies of the LFP.[23] In addition, based on imaging in the olfactory bulb, it has also been suggested that the initial dip can be influenced by glial rather than neuronal underlying source.[24] Further research is needed for establishing the exact correspondence between intrinsic signals, action-potentials, LFPs, and glial sources to refine our ability to interpret functional imaging findings.

9.3 METHODOLOGY

9.3.1 Anesthesia

At the start of an experiment, rats are inducted with a bolus injection of sodium pentobarbital (Nembutal). As an option, rats may be first inducted with an inhalant (Isoflurane) for easier delivery of the Nembutal bolus injection. Rats are then maintained for the remainder of the experiment with supplemental injections of Nembutal as needed. Changes in the anesthesia level can affect the quality of imaging data; too light a level can lead to imaging data with increased contributions from surface vasculature while too deep a level can lead to a decrease in the ratio of stimulus-evoked signal to spontaneous background activity. This relationship is presumably the outcome of imaging with red light illumination where intrinsic sources dominating the intrinsic signals are metabolic mechanisms related to oxygen metabolism. Thus, maintaining a constant level of anesthesia is essential for a successful imaging experiment. To assess the level of anesthesia, we examine various physiological indicators including withdrawal and corneal reflexes, body temperature, respiration rate, and the level of oxygenated blood as indicated by the coloring of the eyes and extremities.

9.3.2 Surgical Preparation

The scalp is retracted above the skull overlying the cortical region of interest, and the exposed skull carefully thinned to a thickness of approximately 150 μm using a high-speed drill. To maintain its transparency, the thinned skull is then bathed in saline that is contained within a simple petroleum jelly chamber erected around the thinned skull and sealed with a glass coverslip. Alternatively, warm 2% agar gelatin may be applied to the thinned skull region and topped with a glass coverslip to ensure an even surface. When cool, the agar holds its form while maintaining the moistness and transparency of the thinned skull. If the rat is revived for chronic imaging, all exogenous materials are removed from the skull, the scalp sutured, and topical and systemic antibiotics administered to decrease the risk of infection immediately after the first imaging session. Rats are able to resume normal behaviors (e.g., grooming, feeding, drinking) within a day after surgery despite the removal of some muscle tissue during the initial surgery. We have found that the thinned skull remains in excellent condition throughout the interim between imaging sessions, provided that the retracted scalp is kept moist with saline throughout the duration of the initial imaging experiment. In nearly every case we have found that the thinned region of the skull and underlying vessels in the dura and pia matter appear largely unchanged from the first imaging session.

9.3.3 Imaging Set-Up

A schematic of our basic set up is shown in Figure 9.8. The anesthetized rat is held in a stereotactic frame (not shown). A stable light source is essential as magnitude changes in stimulus-evoked intrinsic signals is in the order of ~10^{-4}. One means of illumination is provided by a standard 100 W tungsten-halogen light source stabilized with a regulated power supply in order to minimize voltage ripples to less than 10^{-5}. The light then passes a protective heat filter and a red band-pass filter (630 ± 15 nm) before being directed by flexible light guides to illuminate the cortex through the thinned skull. Another light source option is the use of superbright LEDS. Although the LED output intensity is in the order of a few milliwatts and declines gradually over time with battery drainage, these limitations are not concerning for our typical imaging experiments and in recent years, we have routinely used a superbright red LED connected in series to a resistor and 6 V battery for a simple, excellent, and cost-effective source of bright and completely non-fluctuating illumination. A sensitive (12- or 16-bit) cooled CCD camera is used to capture the light reflected from the cortex. To achieve the optimal viewing area and working distance, the CCD camera is fitted with a 50 mm lens that is inverted and attached to an extender. The camera is focused several hundred micrometers below the cortical surface to minimize contributions from surface blood vessels. A computer controls data collection and stimulus delivery for a set of trials. The collected data are then

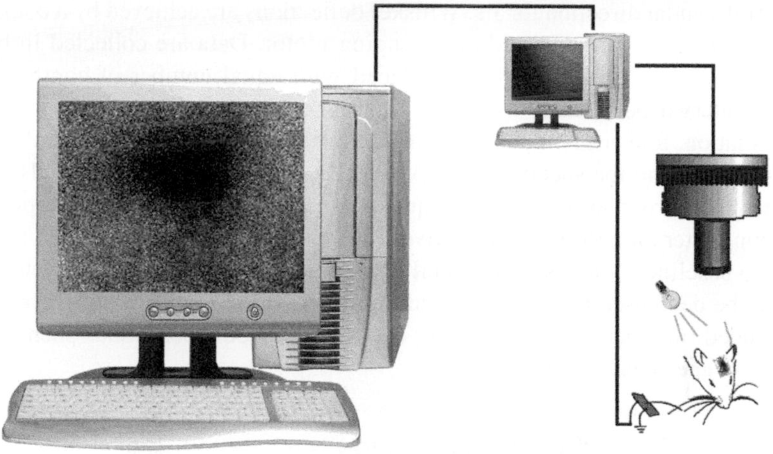

FIGURE 9.8 Set-up for in vivo intrinsic signal optical imaging of cortical functional representations. Images are taken through the animal's thinned skull. The cortex is illuminated with a 630 nm light source. The CCD camera captures a sequence of images before, during and after a computer controlled mechanical stimulation to a whisker. The sequence of digital images is sent to a computer that also controls the experiment. After averaging of 32–128 image sequences, analyzed data can be displayed as an image using an 8-bit linear grayscale on a computer monitor. The black patch represents activity evoked by 5 Hz stimulation to a single whisker. Typically, one computer acquires the data and controls the experiment, while the data are sent to a second computer for off-line analysis.

transferred to a second computer so that incoming data can be analyzed concurrently with the collection of the next set of trials.

9.3.4 Data Collection

For each trial, the CCD camera collects an uninterrupted sequence of frames. The total field of view and pixel resolution of a given frame can vary depending on the pixel array size of the camera's CCD chip and the lens magnification used. The size of the pixel array (or subset thereof) along with the speed of the camera dictates the shortest frame duration that can be used for capturing a sequence of frames in real time. Thus far, for imaging of the initial dip from rat primary somatosensory cortex, we routinely use frames that view ~6.5 mm × 5.0 mm cortical area with ~190 × 140 pixel array (1 pixel ~ 12 μm^2 cortical area). Furthermore, at the single trial level, frames with durations as short as 100 ms have been collected but are typically collapsed to a final temporal resolution of 500 ms frames, and the total number of frames captured per trial sequence is sufficient to follow at least several seconds' worth of data after stimulus onset. To compensate for CCD chips with lower well capacities, faster cameras have the option to collect and subsequently collapse more frames to a final temporal resolution of 500 ms frames as a means to increase the signal-to-noise ratio without increasing the total number of collected trials. In a stimulation trial, one second of spontaneous activity is collected prior to stimulus delivery. Stimulation consists of 5 Hz deflections of a single whisker in the rostral-caudal direction for 1 s. Whisker deflections are achieved by a computer-controlled copper wire attached to a stepping motor. Data are collected in blocks of 32 stimulation trials randomly interlaced with equal number of control trials, with an interval between consecutive stimulus deliveries of ~21 s. For each type of stimulation, four trial blocks are then summed together to form a complete data file. Depending on the specific needs of a study, collection of control trials can be eliminated in order to double the acquisition rate of stimulation trials, provided that a long interval between consecutive stimulus deliveries is maintained to allow return to baseline. Conversely, the total number of trials needed to complete a data file may be decreased by half, provided that the collected trials are dispersed over an extended time period to sufficiently overcome periodic confounds such as fluctuations in anesthesia level.

9.3.5 Data Analysis and Quantification

A summary for analyzing and quantifying the ISOI initial dip is provided in Figure 9.9. To determine the overall baseline level of activity from the imaged cortex prior to stimulus onset, a prestimulus ratio value is calculated for each pixel within the pixel array via dividing the latter of the two prestimulus frames by the earlier prestimulus frame (see Figure 9.9B). The median, rather than the mean, ratio value (referred to hereafter as prestimulus baseline) is then used to indicate the baseline level of activity because it is less sensitive to outlier values due to random and/or biological noise. Analysis of stimulus-evoked initial dip is restricted to 1 s of data collected 0.5 up to 1.5 s after stimulus onset to ensure that only the rising phase of the initial dip is captured with minimal contribution from surface blood vessels. A poststimulus ratio value is assigned to each pixel within the pixel array by first averaging this 1 s of poststimulus data and then dividing the averaged data by 0.5 s of data collected immediately prior to stimulus onset (see Figure 9.9C). As more active cortex reflects less light than less active cortex, a pixel whose poststimulus ratio value is smaller than the prestimulus baseline would correspond to a cortical region exhibiting increased initial dip activity. Prior to quantifying the evoked initial dip activity area, a two-step Gaussian filter (half-width = 5) is applied to the poststimulus ratio values to remove high spatial frequency noise likely attributable to shot noise. The filtered poststimulus ratio values are then thresholded at absolute increments away from the prestimulus baseline, the number of qualifying pixels counted, and the total pixel count translated to an unit of area (mm^2; see Figure 9.9D–F). The peak location and magnitude within the quantified area of evoked initial dip activity can be obtained, with peak location reliably coregistering with the correct anatomical location in the cortex.[8,21] For 2D visualization of the evoked initial dip activity area as a dark patch, an 8-bit linear gray scale can be applied to the poststimulus ratio values such that those pixels with poststimulus ratio values similar to the prestimulus baseline are assigned a middle shade of gray, those with smaller poststimulus ratio values (i.e., corresponding to increased initial dip activity) are assigned darker

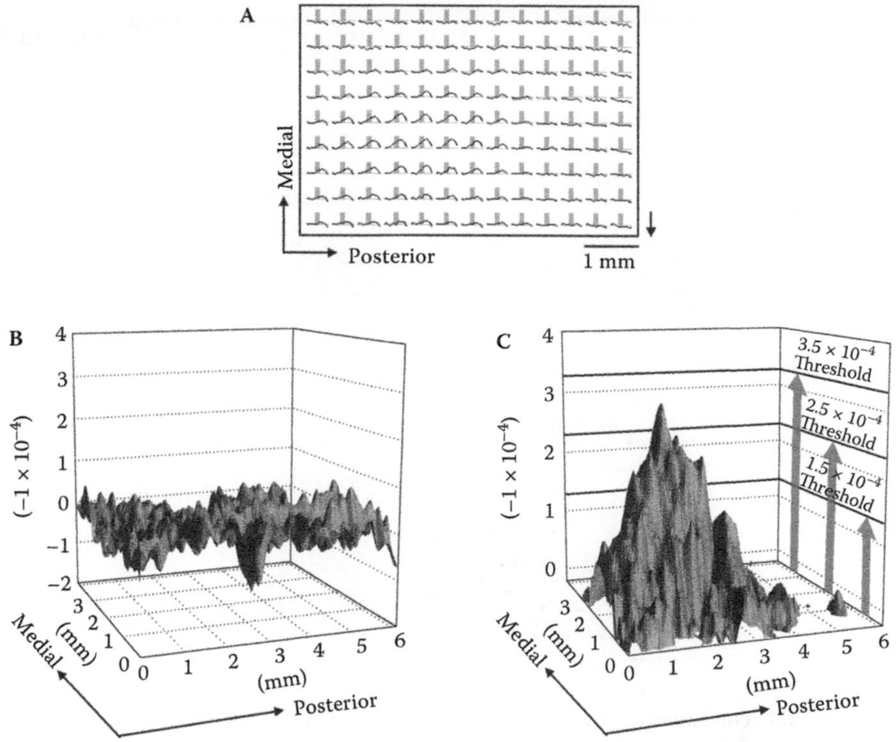

FIGURE 9.9 Areal extent quantification with the use of absolute thresholds combined with the prestimulus baseline. (A) Array of intrinsic signals evoked by stimulation of a single whisker E2 after collection of 128 trials (downward pointing scale bar = fractional change of 1 × 10^{-3}). Each plot corresponds to the 4.5 s time course as averaged for the underlying 0.46 mm × 0.46 mm area, with the time epoch of 0.5 up to 1.5 s poststimulus onset highlighted in gray. Note the region of evoked intrinsic signals is located in the left center of the total imaged area. (B) Determination of prestimulus baseline. Data collected from 1.0 up to 0.5 s prior to stimulus onset are converted to ratio values relative to data collected from 0.5 up to 0 s prior to stimulus onset, with the x- and y-axes indicating cortical location and the z-axis indicating strength of prestimulus intrinsic signals. The median ratio value (0.21 × 10−4 for this example) is used as a measure of the average prestimulus activity over the entire imaged area. Ratio values are filtered (Gaussian half width 5) prior to the 3D plotting. (C) Setting absolute thresholds when used in combination with the prestimulus baseline. Data collected 0.5 up to 1.5 s poststimulus onset are converted to ratio values relative to prestimulus data such that the processed data may be thought of as a "mountain of evoked activity," with the x- and y-axes indicating cortical location and the z-axis indicating strength of poststimulus intrinsic signals. Ratio values are filtered (Gaussian half-width 5) prior to the 3D plotting. Thresholds are set at absolute increments away from the prestimulus baseline (0.21 × 10^{-4} as indicated by z-axis minimum). Three arbitrary increments are illustrated here: 1.5, 2.5, and 3.5 × 10^{-4} away from pre-stimulus baseline. (Modified from Chen-Bee, C.H. et al., J. Neurosci. Methods, 97, 157, 2000. With permission.)

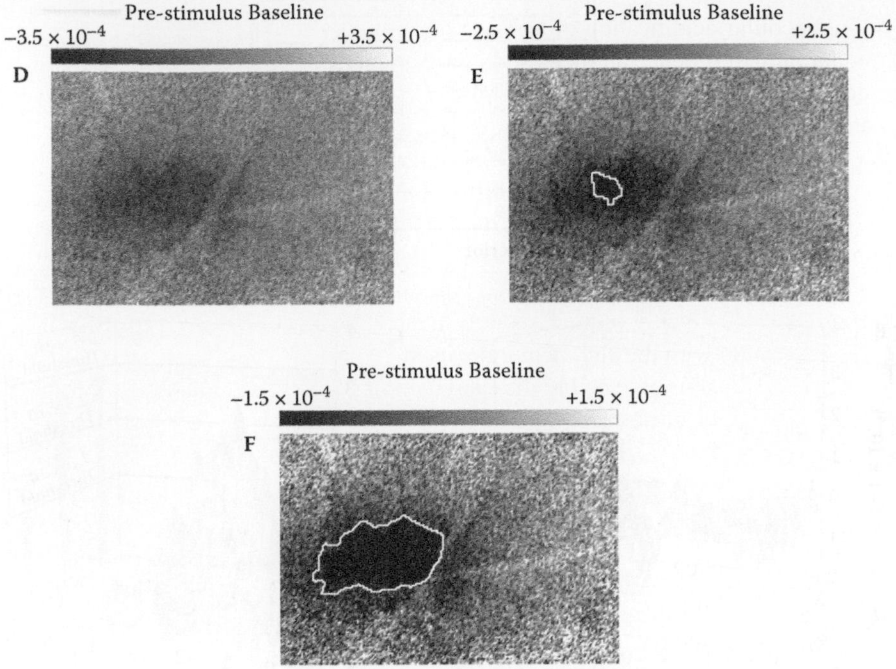

FIGURE 9.9 (continued) Areal extent quantification with the use of absolute thresholds combined with the prestimulus baseline. (D–F) Visualizing the quantified area of evoked intrinsic signals using absolute thresholds combined with the poststimulus baseline. An 8-bit, linear grayscale mapping function is applied to the ratio values so that the quantified area (enclosed by a white border) is visualized as a black patch within the total imaged area for each of the three absolute thresholds. Note that an area was not quantified with the highest (3.5×10^{-4}) threshold (D) as the peak ratio value for this example is 3×10^{-4}. Orientation and horizontal scale bar in (A) also apply to (D–F). (Modified from Chen-Bee, C.H. et al., J. Neurosci. Methods, 97, 157, 2000. With permission.)

shades of gray (darkest shade = black), and those with larger poststimulus ratio values are assigned lighter shades of gray (see Figure 9.9D–F). For more details on data analysis and quantification see Reference 25 and 26.

9.3.6 Additional Considerations for Data Analysis and Quantification

Other considerations have arisen during our recent characterization of the three ISOI signal phases,[14] where analysis and quantification is no longer restricted to just the initial dip. Although discussed within the context of characterizing three signal phases, these considerations are also relevant for other imaging studies where a comparison between different signal phases can be exchanged for research questions such as comparing between different treatment conditions, before versus after a manipulation, etc.

When trying to quantitatively characterize and compare the three phases of the ISOI signal, one consideration is what criteria to use in deciding which poststimulus time epoch to use per phase for quantification. As explained in a later section, a set of criteria is used to identify the typical time epoch used for the quantification of the ISOI initial dip, which took into consideration such aspects as time of signal onset and peak relative to stimulus onset as well as total signal duration. In the case where the three ISOI signal phases are quantified, it is more challenging to identify one set of criteria that can be readily applied to all the three phases because these phases substantially differ in their times to signal onset and peak as well as their total durations. A relatively unambiguous option is to analyze the data collected within the single 500 ms frame containing the maximum evoked area for each of the three signal phases, respectively, for quantification of such parameters as the areal size of the maximum evoked area, peak magnitude within the maximum evoked area, and peak location within the maximum evoked area.[14] If it is necessary to redefine the time epoch to be used for quantifying the ISOI initial dip, then it should be noted that there are implications for analysis related to peak location. It may be that the location of peak activity can vary depending on which time epoch is used for quantification. At the very least, for the initial dip, we have explicitly confirmed that the location of peak activity is comparable when using the standard time epoch of 0.5 up to 1.5 s after stimulus onset versus the 500 ms frame containing the maximum evoked area (see Reference 14 for more details).

In regards to quantifying the size of the evoked area, another consideration is the set of absolute thresholds available for use that can be applied successfully to all three phases. In this particular situation, a large difference in amplitude exists between the three signal phases (strongest phase is ~3× that of the weakest phase). Hence, it is likely that only a narrow range of thresholds can be applied simultaneously to the three signal phases, where they are neither too high for the successful quantification of the weakest signal phase nor too low for the strongest signal phase. Interestingly, the use of a normalized threshold such as 50% peak magnitude may not be a viable option: we found that thresholding at 50% peak magnitude led to results where the smallest area was obtained for the signal phase (overshoot phase) with the strongest peak amplitude (Figure 9.3A inset), results which are in sharp contrast to the known largeness of the overshoot phase for both ISOI and fMRI signal[16]). Presumably, the conflicting results obtained when thresholding with the 50% peak magnitude is due to the large difference in amplitude between the three ISOI signal phases (for a more detailed description of the effects on areal extent quantification when using a threshold normalized to peak magnitude please refer to).[26]

Another consideration when quantifying the size of the evoked area is the usefulness of the prestimulus ratio value as an indicator of baseline activity level. The prestimulus ratio value is determined from a single 500-ms frame relative to its immediately preceding 500-ms frame as a means to extrapolate a trend in baseline activity. It has proved useful when quantifying the ISOI initial dip, a signal phase whose peak magnitude occurs within 2 s after stimulus onset. However, it is unlikely that the prestimulus ratio value is capable of correctly extrapolating baseline activity over a course of several to many seconds, as would be necessary in the case for quantifying the ISOI overshoot and undershoot phases whose peak magnitudes occur approximately 4 and 8 seconds, respectively, after stimulus onset. Indeed, as illustrated in Figure 9.6, large magnitude fluctuations in intrinsic signal activity are observed within 10+ seconds despite the averaging across many trials. When there is no acceptable alternative method to predicting baseline activity levels, an absolute value of 0 can be used for baseline purposes. In recent years, we have exclusively used the absolute value of 0 for baseline purposes even for imaging experiments when only the ISOI initial dip is quantified.

Lastly, an important consideration for ISOI as well as other functional imaging methods is how to define a reliable criterion for choosing an appropriate activity threshold level for areal extent quantification of evoked activity areas. This is a major concern as choosing different activity thresholds can result in dramatic differences in the final quantification of the exact same activity area (see Figure 9.10C–F). However defined, at the very least the criterion should help identify a threshold that will reliably reflect the underlying area of activated neurons. So far, current thresholding methods (including ours as described previously) are arbitrary and therefore not explicitly validated with some known correspondence with a specific level of neuronal activity. Functional imaging in general would benefit from a calibration process of functional images based on simultaneous imaging and neuronal recordings where the appropriate threshold could be identified based on maximal correspondence between the imaging-based versus neuronal-based activity area.

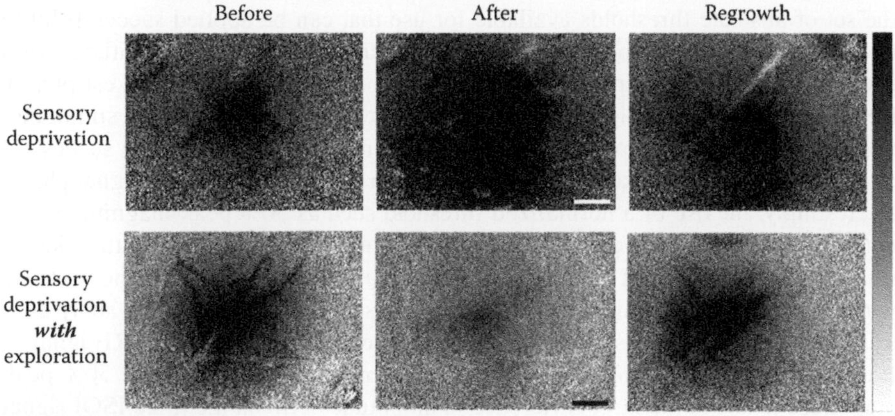

FIGURE 9.10 Plasticity of the spared whisker's functional representation is bidirectional depending on whisker use, and reversible upon restoration of normal sensory input. Unfiltered ratio images are provided from a sensory-deprived animal that remained in its home cage (top row) or was given an opportunity for behavioral assessment (bottom row), either before deprivation ("Before"), after 28 days of deprivation ("After"), and after 28 days of whisker regrowth ("Regrowth"). Ratio values are converted to grayscale values in which the prestimulus baseline is shown as gray and the black and white values on the grayscale bar are set to a decrease or increase of $\pm 2.5 \times 10^{-4}$ from baseline values respectively. Scale bar = 1 mm and applies to all images. Note that the two representative rats differed in the direction of plasticity of their whisker representations after sensory deprivation but were similar in that their whisker representations are nearly restored to predeprivation levels following regrowth of the whiskers. (Modified from Polley, D.B. et al., *Neuron*, 24, 623, 1999. With permission.)

9.4 SUCCESSFUL APPLICATION OF ISOI FOR STUDYING ADULT PLASTICITY OF FUNCTIONAL REPRESENTATIONS

In this section, we highlight a few examples of imaging adult cortical plasticity of whisker functional representation using ISOI. The first example is a case of adult plasticity following sensory deprivation (Figure 9.10). A given whisker functional representation is imaged before the deprivation to obtain a baseline. Next, all whiskers on the same snout, except for the imaged whisker, are removed for 28 days. After this deprivation period the functional representation of the same whisker is imaged again within the same animal, demonstrating an expansion of its representation. Next, for the same animals, the deprived whiskers were allowed to regrow for an additional period of 28 days and the functional representation of the same whisker was imaged for a third time. At this point, imaging revealed that once the previously deprived whiskers grew back, the functional representation of the nondeprived whisker returned to its predeprivation baseline size. Surprisingly, when the same experimental design was repeated on another group of adult rats except during the deprivation period the rats were allowed a few minutes of free navigation outside their cage, the results were opposite to the first group; namely, we found a major contraction of the remaining, nondeprived whisker when imaged after the deprivation period. Nevertheless, similar to the first group, the size of the functional representation of the same whisker returned to predeprivation baseline size once the deprived whiskers were allowed to regrow for 28 days. These findings were confirmed by postimaging single-unit recordings and support the view that the adult rat somatosensory cortex is highly plastic, and that such plasticity depends on the animal's behavior: if the animal is returned to its cage during deprivation then its nondeprived whisker representation expands, but if allowed during deprivation to use its nondeprived whisker in a behavioral context such as navigation outside its standard home-cage, then the representation contracts. For more details see Reference 20. This example highlights the advantage of chronic imaging of the same whisker functional representation before, during, and after a manipulation, where each animal serves as its own control, for studying brain function such as bi-directional plasticity and its reversal.

We studied further the relationship between whisker-related behavior and cortical plasticity in nondeprived, adult rats (Figure 9.11). Based on the results from the deprived rats, we were wondering whether an increase in whisker use over time will induce the same plasticity seen in nondeprived, behaving rats. To increase the use of whiskers in nondeprived rats, we designed a naturalistic habitat, an environment consisting of a large tank three-quarters filled with soil on top of which lies a system of plastic pipes emerging from a central hub—some of them leading to food and water dispensers—whose arrangement was regularly modified. The naturalistic habitat promotes underground tunnel digging, navigation, foraging, and social interactions with other rats, thereby providing a natural way to increase whisker use compared to caged-housed rats. In this experiment a "double control" approach was used where one group of rats served as their own control where their whiskers' functional representations were imaged before and after exposure for 28 days in the naturalistic habitat, whereas

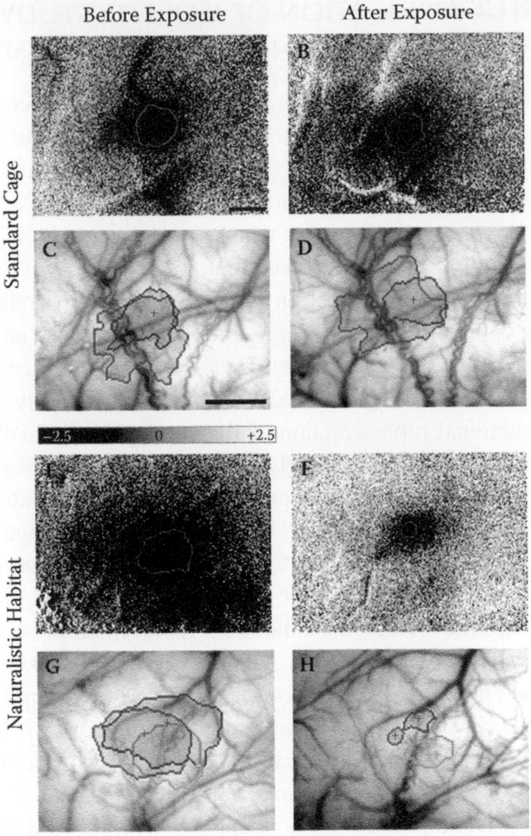

FIGURE 9.11 **(See color insert following page 224)** Natural whisker use outside the cage induces contraction of whiskers' representations as revealed with chronic optical imaging of nondeprived, adult rats. For all panels, color boundaries demarcate cortical areas with intrinsic signal activity $\pm 2.5 \times 10^{-4}$. (A,B,E,F) Ratio images of the C2 whisker representation collected before (A,E) and after (B,F) exposure to standard cage (A,B) or naturalistic habitat (E,F). Grayscale bar indicates intrinsic signal strength $\times 10^{-4}$. Black and white streaks correspond to large surface blood vessels. Scale bar in (A), which also applies to (B,E,F) = 1 mm. (C,D,G,H) Functional representations for whisker C1 (blue), C2 (red) and B2 (green; in [G,H]), obtained from two different standard cage (C and D) and naturalistic habitat (G,H) rats, superimposed onto images of the cortical surface as seen through the thinned skull before (C,G) and after (D,H) exposure. Functional activity peaks for each whisker are indicated by a cross. Scale bar in C, which also applies to (D,G,H) = 1 mm. (From Polley, D.B. et al., *Nature*, 429, 67, 2004. With permission.)

the whiskers' functional representations of the second group were imaged twice 28 days apart while living in the their standard cages located in the same room as the naturalistic habitat. After imaging several whiskers in each rat, the results clearly demonstrated a contraction of the functional representations of whiskers in animals that spent a month in the naturalistic habitat compared to no change in the representations of whiskers in animals living in standard cages. For more details see Reference 22 and for a review about such plasticity see Reference 27. This research adds further evidence to the idea of whisker use dependent contraction of cortical functional representations and illustrates how imaging can provide the means to move from the level of single whisker functional representations to multiwhisker level.

Another example of imaging adult plasticity that we have pursued targets plasticity induced by pharmacological interventions. While researching cortical plasticity induced by nerve growth factor (NGF), where rapid (minutes) expansion of a whisker functional representation was observed after topical NGF application,[28] we tested the hypothesis that the underlying mechanism for such plasticity is NGF-induced activation of the diffuse cholinergic modulatory system originating in the basal forebrain. One approach was to image cholinergic-induced plasticity and to compare it to that induced by NGF's. Because we wanted to study the local effects of cholinergic agents, rather than injecting them systemically or into the brain's ventricles, we directly applied them to the somatosensory cortex. To do so required the removal of the skull above the somatosensory cortex plus slight perforations of the dura to allow the cholinergic agents to reach the cortex directly. The results demonstrated that cholinergic agonists (e.g., carbachol and nicotine) induced expansion of a functional representation of a whisker whereas antagonists (e.g., scopolamine) contracted them (Figure 9.12).

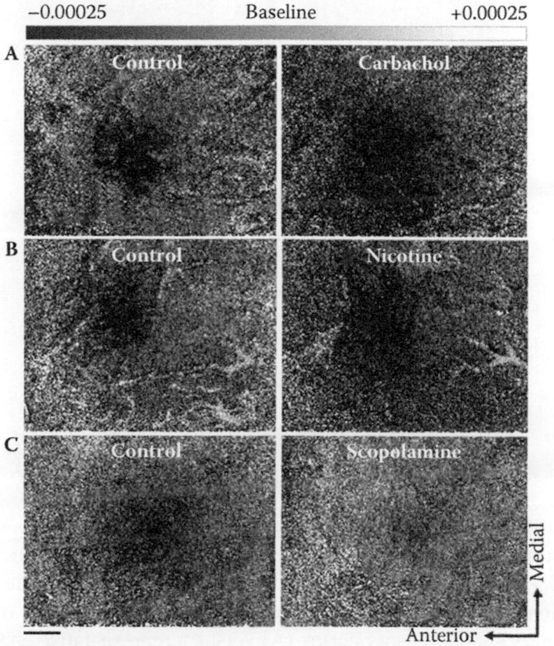

FIGURE 9.12 Effects of topical cholinergic agent application on the areal extent of whisker C2 functional representation. Data from representative animals are provided to illustrate the effects of carbachol (~69 minutes postapplication; (A), nicotine (~46 minutes postapplication; (B), or scopolamine (~46 minutes postapplication; (C) on the areal extent of whisker C2 functional representation compared with control images obtained within the same animals before topical cholinergic agent application. Orientation applies to all images. Scale bar = 1 mm in C (applies to A–C). (Modified from Penschuck, S. et al., *J. Comp. Neurol.*, 452, 38, 2002. With permission.)

For more details see Reference 29. This example illustrates the advantages offered by functional imaging for pharmacological studies interested in the dynamic response of an entire functional representation, in contrast to the challenges faced in trying to extract the same information via extrapolation of responses from a few single neurons especially with the time constraints imposed by the nature of such studies.

Lastly, another example of plasticity studies pursued in our laboratory focused on imaging plasticity induced by genetic manipulations of mice (Figure 9.13). Specifically,

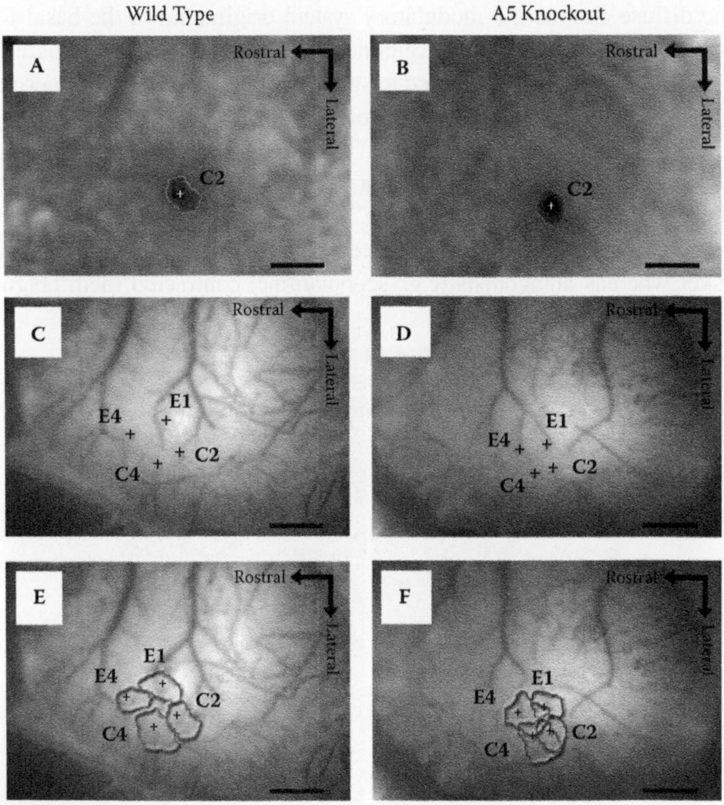

FIGURE 9.13 The functional representations of single whiskers in ephrin-A5 knock-out mice are qualitatively normal, but the organization of four medial whisker functional representations is malformed. (A,B) Grayscale functional images of filtered ratio values as obtained with 630 nm illumination depicting the C2 whisker functional representation of a representative control mouse primary somatosensory cortex (A) versus a representative ephrin-A5 knock-out mouse (B). The activity peaks of the C2 whisker functional representations are denoted by the crosses, the sizes measured as areas at half-height (50% of the peak) are denoted by the polygons. Dark pixels denote highest activity, and gray pixels denote baseline activity. All images presented here were obtained through the intact skull. All scale bars = 1 mm. (C,D) The same mice as in (A,B) with the activity peaks of the E4, E1, C4, and C2 whisker functional representations superimposed on the structural images of the cortex and vasculature taken with 540 nm illumination. (E,F) The same images as (C,D), but now with the areas at half-height added. (Modified from Prakash, N. et al., *J. Neurosci*, 20, 5841, 2000. With permission.)

we studied the effect of knocking out the gene for Ephrin-A5, a putative axonal guidance label, on the functional organization of the somatosensory cortex in adult mice as compared to wild-type mice. We found a malformation of the functional organization of the cortex in the knock-out mice manifested in the spacing between whisker representations but not in the whiskers' functional representations themselves. This study exemplifies another advantage of studying functional plasticity by imaging, namely that the consequences of a genetic manipulation on brain function can be assessed. For more details see References 30 and 31.

While a main goal of the present chapter is to emphasize the advantages offered by ISOI, we are appreciative of the advantages offered by the more traditional techniques used to study plasticity. In fact, there are several types of questions regarding cortical plasticity that can be answered only by using traditional techniques such as single-unit recording techniques. Whenever possible we attempt to apply as many techniques to the same animal when pursuing the characterization of plasticity and its underlying mechanisms. We feel that the use of an imaging technique such as ISOI is an optimal first step, as it provides a clear image of the "macro" or population vantage point of cortical activity and can follow cortical plasticity within the same animal. The functional images obtained with ISOI can then be used to guide the precise placement of microelectrodes for neuronal recordings, micropipettes for injection of neuronal tracers and/or iontophoresis of drugs and microdialysis probes at "hot spots" of activity, areas of reduced activity, and areas of no activity. Consequently, the investigator can benefit from using the optical activity imaging map to ask more targeted questions about the physiology, anatomy and/or pharmacology underlying cortical plasticity. Thus, the optical map of activity can be seen as a central component of a multidisciplinary approach to the study of cortical structure and function and their relationships in the normal and plastic cortex.

9.5 FUTURE DIRECTIONS

The most desirable direction for the study of plasticity is to be able to use ISOI in a noninvasive fashion in freely behaving animals. This will enable the imaging of changes in the cortex on a continuous basis without the need for anesthesia. Ideally, the camera chip is placed directly on animal's thinned skull and transmits the optical data to the host computer by telemetry, but flexible connections that do not restrain the animal's movement could also serve this purpose. As prices of the CCD chips drop and the trend of miniaturization of electronic circuits that can support such imaging continues, the ability for long-term real-time imaging of cortical function within the same animal may be realized in the near future, as it has already been realized for voltage-sensitive dye imaging (see Chapter 6 of this book). ISOI would also benefit from an improved ability to resolve signals from different depths in the cortex, so as to provide functional imaging of single cortical layers and consequently the ability to produce detailed 3D maps of cortical activation.

Another promising direction would be the development of a multipurpose imaging system that combines the advantages of ISOI with the advantages offered by other imaging techniques. Combining the high-resolution spatial maps obtained with ISOI, for example, with the high temporal resolution offered by dye-based optical imaging (e.g., voltage sensitive dyes), and the detailed images of neuronal structure and function provided with multiphoton imaging will allow researchers to address questions about cortex on multiple spatial and temporal scales. By visualizing cortical activity on many levels, from single dendritic spine to assemblies of thousands and millions neurons, one may gain the optimal insight into the mechanisms of cortical function and its plasticity.

ACKNOWLEDGMENTS

We thank our previous collaborators Susan Masino, Michael Kwon, Neal Prakash, Jonathan Bakin, Yehuda Dory, Barbara Brett-Green, Silke Penschuck, Daniel Polley, and Eugene Kvasnak for their original contributions. Supported by NIH grants NINDS NS-34519, NINDS NS-39760, NINDS NS-43165, NINDS NS-48350, NINDS NS-055832, and NSF IBN-9507636.

REFERENCES

1. Xerri, C. Imprinting of idyosyncratic experience in cortical sensory maps: Neural substrates of representational remodeling and correlative perceptual changes. *Behav Brain Res* **192**, 26–41 (2008).
2. Moucha, R. and Kilgard, M.P. Cortical plasticity and rehabilitation. *Progr Brain Res* **157**, 111–122 (2006).
3. Merzenich, M.M. et al. Somatosensory cortical map changes following digit amputation in adult monkeys. *J Comp Neurol* **224**, 591–605 (1984).
4. Nicolelis, M.A. and Ribeiro, S. Recordings: The next steps. *Curr Opin Neurobiol* **12**, 602–606 (2002).
5. Grinvald, A. et al. Functional architecture of cortex revealed by optical imaging of intrinsic signals. *Nature* **324**, 361–364 (1986).
6. Ts'o, D.Y. et al. Functional organization of primate visual cortex revealed by high resolution optical imaging. *Science* **249**, 417–420 (1990).
7. Frostig, R.D. et al. Cortical functional architecture and local coupling between neuronal activity and the microcirculation revealed by in vivo high-resolution optical imaging of intrinsic signals. *Proc Natl Acad Sci USA* **87**, 6082–6086 (1990).
8. Masino, S.A. et al.. Characterization of functional organization within rat barrel cortex using intrinsic signal optical imaging through a thinned skull. *Proc Natl Acad Sci USA* **90**, 9998–10002 (1993).
9. Prakash, N. et al. Malformation of the functional organization of somatosensory cortex in adult ephrin-A5 knock-out mice revealed by in vivo functional imaging. *J Neurosci* **20**, 5841–5847 (2000).
10. Armstrong-James, M. The nature and plasticity of sensory processing within adult rat barrel cortex. *Cereb Cortex* **11**, 333–373 (1995).
11. Simons, D.J. Neuronal integration in the somatosensory whisker/barrel cortex. *Cereb Cortex* **11**, 263–297 (1995).
12. Grinvald, A. et al. Optical imaging of neuronal activity. *Physiol Rev* **68**, 1285–1366 (1988).
13. Grinvald, A. et al. Intrinsic signal imaging in the neocortex: Implications for hemodynamic-based functional imaging. In *Imaging in Neuroscience and Development a Laboratory Manual* (Ed. R. Yuste and A. Konnerth) (Cold Spring Harbor Laboratory Press, New York, 2005).

14. Chen-Bee, C.H. et al. The triphasic intrinsic signal: Implications for functional imaging. *J Neurosci* **27**, 4572–4586 (2007).
15. Berwick, J. et al. Fine detail of neurovascular coupling revealed by spatiotemporal analysis of the hemodynamic response to single whisker stimulation in rat barrel cortex. *J Neurophysiol* **99**, 787–798 (2008).
16. Logothetis, N.K. and Pfeuffer, J. On the nature of the BOLD fMRI contrast mechanism. *Magn Reson Imaging* **22**, 1517–1531 (2004).
17. Malonek, D. and Grinvald, A. Interactions between electrical activity and cortical microcirculation revealed by imaging spectroscopy: Implications for functional brain mapping. *Science* **272**, 551–554 (1996).
18. Fox, M.D. and Raichle, M.E. Spontaneous fluctuations in brain activity observed with functional magnetic resonance imaging. *Nat Rev* **8**, 700–711 (2007).
19. Das, A. and Gilbert, C.D. Long-range horizontal connections and their role in cortical reorganization revealed by optical recording of cat primary visual cortex [see comments]. *Nature* **375**, 780–784 (1995).
20. Polley, D.B. et al. Two directions of plasticity in the sensory-deprived adult cortex [see comments]. *Neuron* **24**, 623–637 (1999).
21. Brett-Green, B.A. et al. Comparing the functional representations of central and border whiskers in rat primary somatosensory cortex. *J Neurosci* **21**, 9944–9954 (2001).
22. Polley, D.B. et al. Naturalistic experience transforms sensory maps in the adult cortex of caged animals. *Nature* **429**, 67–71 (2004).
23. Niessing, J. et al. Hemodynamic signals correlate tightly with synchronized gamma oscillations. *Science* **309**, 948–951 (2005).
24. Gurden, H. et al. Sensory-evoked intrinsic optical signals in the olfactory bulb are coupled to glutamate release and uptake. *Neuron* **52**, 335–345 (2006).
25. Chen-Bee, C.H. et al. Areal extent quantification of functional representations using intrinsic signal optical imaging. *J Neurosci Methods* **68**, 27–37 (1996).
26. Chen-Bee, C.H. et al. Visualizing and quantifying evoked cortical activity assessed with intrinsic signal imaging. *J Neurosci Methods* **97**, 157–173 (2000).
27. Frostig, R.D. Functional organization and plasticity in the adult rat barrel cortex: Moving out-of-the-box. *Curr Opin Neurobiol* **16**, 445–450 (2006).
28. Prakash, N. et al. Rapid and opposite effects of BDNF and NGF on the functional organization of the adult cortex in vivo. *Nature* **381**, 702–706 (1996).
29. Penschuck, S. et al. In vivo modulation of a cortical functional sensory representation shortly after topical cholinergic agent application. *J Comp Neurol* **452**, 38–50 (2002).
30. Prakash, N. et al. Basal forebrain cholinergic system is involved in rapid nerve growth factor (NGF)-induced plasticity in the barrel cortex of adult rats. *J Neurophysiol* **91**, 424–437 (2004).
31. Prakash, N. and Frostig, R.D. What has intrinsic signal optical imaging taught us about NGF-induced rapid plasticity in adult cortex and its relationship to the cholinergic system? *Mol Imaging Biol* **7**, 14–21 (2005).

10 Fourier Approach for Functional Imaging

Valery A. Kalatsky

CONTENTS

10.1 INTRODUCTION

Determining the function of the brain is one of the important goals of neuroscience and its associated clinical disciplines of neurology, psychiatry, neurosurgery, and psychology. Functional brain mapping or neuroimaging is a set of imaging modalities that permits localization and visualization of brain activity. Neuroimaging has the potential of allowing observation of the function of the human nervous system, and to aid in clinical diagnosis and monitor ongoing neural function.

There are at least a dozen of neuroimaging approaches that allow monitoring of neural functions. These approaches differ very widely in their spatial and temporal resolution, level of invasiveness, and costs. They can be divided into two groups: (1) ones that measure neural activity (emission of electromagnetic fields) directly and (2) ones that rely on indirect measurements of metabolic signals. The techniques that fall into the first group typically are not used for neuroimaging because they are either very invasive, for example, electrophysiology, or provide very low spatial resolution, for example, electroencephalography (EEG) or magnetoencephalography (MEG). The other group consists of a variety of techniques that either use radioactive labels such as positron emission tomography (PET) or use signals of endogenous nature and include functional magnetic resonance imaging (fMRI) and optical imaging of intrinsic signals (ISOI). The latter group is particularly attractive for brain mapping because it provides moderate to high spatial resolution and does not require introduction of labels. What is characteristic of this group is that it measures hemodynamic signals, that is, the signals related to changes in blood oxygenation, blood volume, and blood flow rate.[1,2] As a consequence the measured signals are contaminated by the cardiovascular artifacts such as heart beat, respiration, and vasomotion.[3] The conventional imaging protocols are implemented in the time domain where randomized stimulus presentation is followed by data collection and averaging. As the result the evoked activity cannot be straightforwardly disassociated from the noise. The standard approach to noise reduction requires massive averaging, which in turn leads to long experimental runs.

Recently, we developed a rather general methodology that effectively removes the physiological noise.[4] A simple observation that the components of the noise are cyclic suggested implementation of the imaging protocol in the frequency domain instead of the time domain. The noise components appear localized in rather narrow frequency bands in the frequency domain, thus periodic stimulation at a frequency far from these bands will evoke a periodic response which is nearly free of the cardiovascular contaminations. Finally, a straightforward Fourier analysis extracts the parameters of the evoked response.

In this chapter we present the basics of the Fourier approach, which can be applied to any imaging modality that suffers from cyclic contaminations, and demonstrate its use for rapid acquisition of cortical maps by optical imaging of intrinsic signals.

10.1.1 Neocortex and Cortical Maps

Neocortex, or simply cortex, is the most intricate and functionally diverse structure of the mammalian brain; humans have the most developed cortex, accounting for more than 50% of the brain's weight. The cortex is responsible for higher functions such as sensory perception, motor control, and consciousness. Even though functional brain mapping refers to 3D mapping of the whole brain, most of the attention is devoted to cortical imaging. The cortex is a nearly 2D sheet of neurons that measures about 2500 cm^2 in area. It is this two-dimensionality that makes brain maps corollaries to geographic maps.

Cortical maps can be divided into three categories with spatial resolution being a defining factor:

1. *A gross division of the cortical surface into modality-specific areas.* It has been known that localized damage to cerebral hemispheres due to a stroke or a wound may lead to a well-defined deficiency in the execution of specific tasks. Numerous studies, mostly done by neurologists, dating back more than 100 years have identified functional cortical areas responsible for specific functions. To name a few, these include John Hughlings Jackson, motor cortex; Pierre Paul Broca, the motor speech or Broca's area; and Carl Wernicke, language comprehension and production (Wernicke's area).[5] A few decades later Korbinian Brodmann used comparative cytoarchitectonic (histology) analysis to divide the human cortex into 52 distinct regions,[6,7] which are usually referred to as Brodmann areas. His classification is still widely used. For example the functional areas mentioned above correspond to the following areas in the Brodmann's taxonomy: area—primary motor cortex, area 44—Broca's area, area 22—Wernicke's area. Two visual areas frequently mentioned here are area 17—primary or striate visual cortex (V1) and area 18—secondary or extrastriate visual cortex (V2). The cortical parcellation defined by Brodmann has been refined since its creation. As anatomical and neuroimaging techniques become more and more precise the number of architectonic and functional subdivisions will steadily grow.[8]
2. *Topographic representations of inputs to the primary and lower secondary cortical areas and outputs from motor cortex.* The inputs of the major sensory modalities, i.e., sight, hearing, and touch, are relayed to the cortex via thalamic axonal projections. These inputs are organized in a manner that closely follows the spatial organization of the peripheral receptors; specifically, the retinal photoreceptors send inputs to the primary visual or area V1, the cochlear hair cells to the primary auditory cortex or area A1, and the skin mechanoreceptors to the primary somatosensory cortex or area S1.[5] Accordingly, the representations of the sensory inputs in the primary sensory areas are topologic maps; they may not preserve scales and directions across the cortical areas but they do preserve relationship between points. The names of these mappings are the names of the receptor layer or the function followed by the suffix "topy": retinotopy or visuotopy for vision, tonotopy or cochleotopy for hearing, and somatotopy for touch. In vision and touch, the mapping is somewhat natural from the 2D receptor space onto the 2D cortical space. Interestingly, in hearing, it is from 1D cochlear frequency space onto the 2D cortical space; the role played by the extra cortical dimension is not well understood. Typically, topographic mapping is observed in some higher cortical areas, for example, in those that receive direct inputs from the primary areas. The higher one goes along the hierarchy of cortical areas, however, the lower is the resemblance of the representations to those of the transduction levels, which is a signature of the more complex processing performed by the higher cortical areas.

3. *Complex features beyond the simple topographies described above.* Cerebral cortex has a layered structure similar to a stack of pancakes. The in-layer dimensions are referred to as horizontal or tangential, and the dimension perpendicular to that is referred to as vertical or radial. The layers contain different types of cells that are heavily interconnected in the radial direction. Just this observation alone allows for a conjecture: the cells along the radial dimension may have similar functional properties. In reality this conjecture was preceded by the discovery of Vernon Mountcastle[9] who showed that cat's somatosensory cortex is subdivided into small domains, about 500 μm, within which the cells' property tend to be similar. He named this domain *cortical columns* or simply "columns." The term has been widely used and abused thereafter. A few years later, in their classic paper of 1962, Hubel and Wiesel showed that cat's visual cortex is subdivided into domains selective for orientation, the orientation columns.[10] Also, they showed that ocular dominance, the propensity of cortical cells to prefer visual input from one eye to the other, has columnar organization.[10,11] Despite a significant effort,[12,13] even the masters failed to tell the global arrangement of orientation columns. Perhaps this was because it was very difficult to reconstruct the fast spatially varying orientation field from a small set of sparse microelectrode recordings. The global organization of the orientation columns was revealed about 20 years later by optical imaging.[14,15] Columnar organizations of orientation preference and ocular dominance, their global representations, dynamics, and development in visual cortex, have been a fascinating subject for many decades for both neuroscientists and physicists alike.[16–20]

In summary, tangentially the cortex is divided into many functional areas, and the cortical columns within many of the known areas are organized into maps. These maps have the property that similar stimuli or similar responses are represented by columns of cells that lie near to one another in the cortex. In the sensory cortex, such functional maps do not provide complete information about stimulus preferences of a single neuron. Rather, they average the common elements of the stimulus preferences of neurons in the column.

10.1.2 Conventional Method for Brain Mapping

Most methods of neural functional evaluation use randomly interleaved, episodic stimulus presentation (also known as block-design stimulation): either the responses following the many repetitions of each particular stimulus are averaged in relation to the time of onset of the stimulus, or else (in the approach referred to as reverse correlation) the stimuli preceding the neuronal responses are averaged in relation to the time of the response. Reverse correlation usually is not used for imaging data because the signal is slow and very noisy. These approaches are well suited to study systems with unstructured noise. They assume that the noise is random and will average out over many trials.

Conventional methods for data acquisition and analysis of the intrinsic optical responses are based on episodic presentation of stimuli.[21] That is, for construction of the orientation map in V1,[14,22] one collects images while presenting a moving square- or sine-wave grating (a series of parallel bars) of one orientation, followed by a blank screen or static image for relaxation of the response, after which the sequence is repeated for another randomly chosen orientation. The resulting transient responses to each stimulus are averaged, and the images are normalized by the response to a "blank" stimulus (or to the average "cocktail blank" response to the entire stimulus set) to produce the map of the cortical response to a given orientation. This approach irreversibly mixes the evoked responses with the signals of ongoing activity and has other major drawbacks. First, it assumes that there exists a stable base line image that can be used as a reference for the evoked responses. This assumption is flawed since the base line undergoes strong fluctuation of ubiquitous *1/f* character, where the power of the noise is inversely proportional to the frequency at which it is measured (Figure 10.1). Second, interlacing blocks of stimulation with relaxation blocks of a

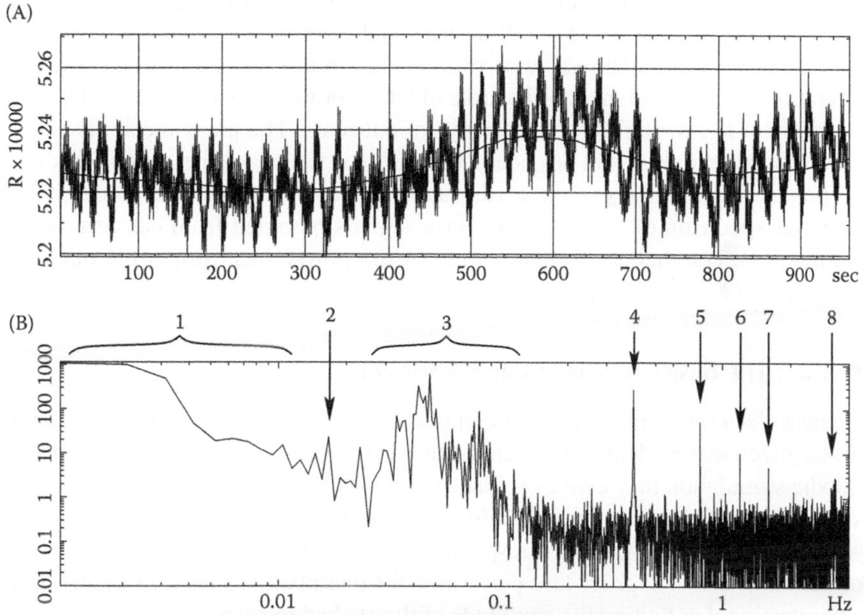

FIGURE 10.1 Time course and power spectrum of intrinsic optical signals in cat visual cortex. (A) Time course of light reflectance at a single pixel from a set of cortical images. The response is dominated by the vasomotor oscillations and slow variation of the base line. The latter is shown as the smooth line. The slow variation can be filtered out in the time domain before application of Fourier analysis. (B) The power spectrum of the reflectance signal plotted in (A). Components of the time series are indicated: 1—slow variation; 2—evoked response at the second harmonic of the frequency of stimulation (the fundamental component at 0.5 cycles/min cannot be resolved here), the amplitude and phase of which are used to construct cortical maps, 3—vasomotor signal; 4—ventilation artifact (fundamental frequency); 5,6,7—second, third, and fourth harmonics of the ventilation artifact; and 8—heart beat artifact. The subject was artificially ventilated; as a result the ventilation component is strongly periodic and sharp. The high frequency part of the spectrum (above 0.2 Hz) is dominated by shot noise, the Poisson-distributed statistics of photon counting.

static or blank screen of similar duration lengthens imaging time by a factor of two. The first shortcoming is relieved by the use of the periodic paradigm, which relies on the differential, rather than absolute, change of the signal. Imaging time is reduced by continuous stimulation; that is, the stimulus presentation traverses the stimulus parameter space in a continuous manner, which has additional benefits described below.

10.2 FOURIER APPROACH TO FUNCTIONAL BRAIN MAPPING

10.2.1 The Noise

As we already mentioned in the Introduction the hemodynamic intrinsic signals used for functional brain mapping, those related to changes in blood microcirculation are contaminated by the physiological activity of the cardiovascular system. Two of these processes are heart beat and respiration. Besides these obvious processes there is another one that is less understood; it is vasomotion. Vasomotion is the cyclic contraction and dilatation of small arteries and arterioles.[3] These oscillations are generated from within the vascular wall and are not a consequence of heartbeat, respiration, or neuronal input.[23] The role of vasomotion is not clear; it has been suggested that it is essential for the exchange of nutrients. The phenomenon is not specific to the brain only,[24] and has been observed in a variety of tissues and species.[25]

The three components of the cardiovascular activity can be readily identified in the power spectrum of the optical intrinsic signals recorded from cat visual cortex (Figure 10.1). Another prominent feature of the power spectrum is the ubiquitous 1/f-like noise.

10.2.2 The Basics of the Fourier Approach

A simple observation that the major components of the noise are cyclic in nature and thus localized in well-defined frequency bands leads to the following proposition: using periodic stimulation may evoke oscillatory changes in intrinsic signals that would be well-localized in the frequency domain and therefore could be disassociated from the cyclic contaminants. The evoked response then could be extracted by straightforward Fourier analysis. This procedure performed for one pixel is shown in Figure 10.1B, demonstrating that finding the amplitude of the evoked response is feasible.

What about the phase of the response? Fourier analysis provides two parameters for each harmonic: the amplitude and the phase. It appears that if one uses only a single-condition stimulus, the information about the phase of the response is not used at all. Actually, the phase of the response can be used to determine the delay between stimulus presentation and the lagging metabolic response; we will address this method for calculation of the delay below. The phase information, however, can be very valuable if one presents multiple single condition stimuli sequentially. In this case the phase of the response at different cortical sites will indicate which stimulus caused that response because the response will be phase-locked to the stimulus. Thus, one can temporally encode a sequence of stimuli, present it periodically, and then decode the responses, that is, assign the evoked activation back to the stimulus, according to their phases. This approach is particularly useful for mapping continuous representations such as the topographies or orientation preference described in Section 10.4. In this case the stimulus should traverse continuously the feature space; thus, one could map not just a single condition but the entire 1D feature space onto the cortex.

Implementation of this periodic continuous paradigm imposes the following requirements:

1. *Stimulus*: The feature space should be traversed periodically and continuously if possible. The period of stimulation should be chosen such that it does not fall into one of the noise bands.
2. *Acquisition*: The images of cortical surface, the frames, should be collected at a reasonably high rate and saved without averaging; having a well sampled temporal series is important for the following analysis. The frame acquisition must be synchronized with the stimulus.
3. *Analysis*: The required harmonics of the stimulation frequency, typically the first or second, are extracted by Fourier decomposition. The time series may be preprocessed, for example, by sliding-window averaging in the time domain to reduce the 1/f-like noise. The phase of the response will indicate which stimulus caused the response, and the amplitude will determine the significance or strength of that response. The resulting maps are 2D complex fields whose properties can be studied by the well-known complex analysis. Finally, the phases and amplitudes or both can be visualized as color coded cortical maps.

10.2.3 Stimulus Design

Stimulus design, perhaps, is the most important component of the periodic continuous imaging paradigm. It is impossible to give a recipe for all possible scenarios; here, we merely provide some guidelines. First of all, the stimulus must be presented periodically in time domain; that is, the temporal sequence of visual patterns should repeat itself. Typical periods are in the order of 10 seconds. Second, if the feature of interest is inherently continuous, then the stimulus should be designed so that the feature is continuously traversed. Continuity of the stimulus assures uniform coverage of the cortical neuronal populations sensitive to the feature under study, whereas the standard episodic stimuli provide a sparse sampling leading to weaker evoked responses. There is no straightforward suggestion that can be made for cortical mapping of representations of complex and/or highly abstract classes of objects or entities, for example, flowers, numerical digits, or letters of an alphabet. The columnar organization of the cortex, however, suggests continuous representation of preferred stimuli. Therefore, the representation of such classes in the cortex might not be organized as a spatial map with a representation of one object smoothly transforming into a representation of another. Instead, some distributed and/or transient representation scheme might be used.

The best feature spaces that meet these guidelines are orientation and direction of motion (referred to as direction); these are both innately cyclic and continuous. Unfortunately, these are about the only parameter spaces that have the desired properties. In a general case of noncyclic feature space one has to introduce at least one discontinuity in the stimulus "flow." Typically, distinguishing between the responses evoked by the two extremes bordering the discontinuity does not present a dilemma because they are segregated spatially on the cortical surface. To further resolve this ambiguity one can pad the discontinuity with a few seconds of the "blank" or null stimulus, which is a uniformly lit, typically grey, screen in vision.

Here we give details of the stimuli we have used to map retinotopic areas in the mouse visual cortex and to map orientation/direction preference in the cat visual cortex.

10.2.3.1 Stimulus for Retinotopy

The visual space is 2D. It can be parameterized by azimuth, i.e., angular deviation from the midline or vertical meridian, and by elevation, i.e., angular deviation from the line of the horizon (alternatively polar coordinates can be used for parameterization of the visual space). Thus, mapping of retinotopy requires two independent stimuli, one for the azimuth and the other one for the elevation, presented individually. Stimulus design is straightforward in this case: it is a vertical bar to map azimuth, and horizontal bar to map elevation, crossing the visual space as shown in Figure 10.2A,B. The stimuli have an apparent discontinuity since the positions of the bar at phase values π and $-\pi$ are not identical; this arrangement is shown in Figure 10.2A. Distinguishing between responses evoked by the two bars should not be difficult since the bars are well separated in visual space. This ambiguity vanishes for the elevation stimulus, Figure 10.2B, because the distance between consecutive bars is larger than the screen height; as a result, not more than one bar is displayed on the screen at any time. This is an example of a "blank" padded stimulus.

It is possible to multiplex these two stimuli into one by running each at a distinct frequency[26]; we will not address this extension of the Fourier approach here.

10.2.3.2 Stimulus for Orientation and Direction Preference

A typical stimulus for direction preference mapping is a full-field drifting square grating (other patterns can be used as well); the drift is unidirectional. Note that a full-field grating can be moved only perpendicular to itself; translations parallel to the lines of the grating have no effect. The stimulus used for mapping orientation preference is similar to that for direction preference except for the direction of the drift, which alternates a few times during stimulus presentation. A simple continuous rotation of the drifting grating creates stimulation suitable for Fourier mapping (Figure 10.2C). Note that rotation alone or rotation combined with a slow drift, would lead to nonuniform speed distribution of stimulus elements over the screen. To make this distribution more even, it is required to have a fast drift combined with a slow rotation. The drift velocity is usually around 10 degree/s, assuming a drifting grating of spatial frequency 0.2 cycle/degree and temporal frequency 2 cycle/s (typical values for cat; degrees here measure visual angle). The same velocity of stimulus elements due to rotation only is reached if the angular velocity is about 3 rotations per minute (here we used typical values for monitor size 40 cm and distance to the subject 40 cm). We usually use rotation speed 3–6 times lower than this (0.5–1 rotation per minute).

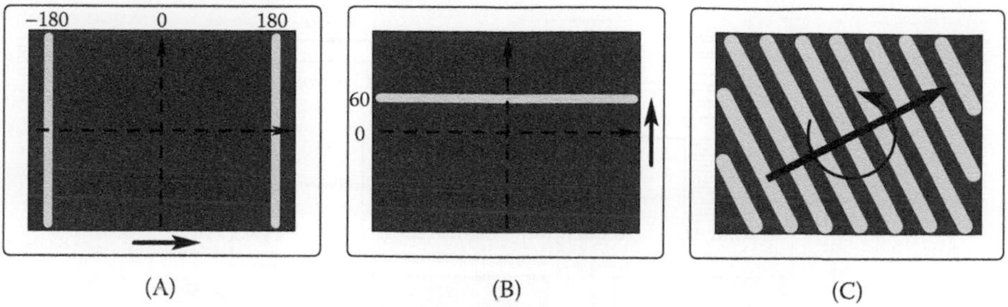

FIGURE 10.2 Stimulus design. (A and B) The retinotopy stimulus patterns. The stimuli are low duty-cycle square-wave gratings of low spatial frequency (0.01–0.02 cycle/degree) moving at a rate of 0.1–0.2 cycle/second. The brighter stripe has size of 1–2°. The vertical stripes are used to stimulate the lines of constant azimuth (a) and the horizontal to stimulate the lines of constant elevation (B). Therefore, a whole line of constant azimuth or elevation is stimulated at each moment. Different lines are stimulated at different phases of the stimulus cycle. The numbers at the top (A) and on the left (B) of the monitor are exemplar phases, measured in degrees, of the grating relative to the center of the screen. (C) Illustration of the stimuli employed to assess direction and orientation selectivity of the cat visual cortex. The stimulus is a drifting and rotating square-wave grating (spatial frequency 0.2–0.4 cycle/degree); thus, it has two degrees of freedom. The drift is the fast degree of freedom (temporal frequency 1–8 cycles per second) and the rotation is the slow degree of freedom (angle speed $f_r = 0.5$–2 rotations per minute). The stimulus appears to be drifting only for an observer looking at the stimulus for a short period of time (a few drifting cycles). A longer integration reveals the rotation. An important advantage of this stimulus versus a rotating star of radial stripes or radial sectors is its high uniformity of the stripe velocity as a function of the location on the screen, which makes it suitable for studying not only orientation selectivity but also direction selectivity. The response at $2f_r$ yields orientation selectivity; meanwhile the response at f_r (or sum of the responses at both f_r and $2f_r$ depending on the definition) gives directional selectivity.

10.2.4 Data Acquisition and Imaging System

The imaging system design has been previously described[4]; here, we address the key points only.

The complete imaging system (Figure 10.3) consists of the image acquisition computer and camera, a second computer that produces visual (or auditory or other sensory) stimulation, and optionally a third computer for online analysis and data storage (functions that could be performed on the image acquisition computer if analysis is done offline). We highly recommend using the three-computer configuration because the stimulus and acquisition computers run real-time (highest priority) applications (threads). Disrupting these applications by some other user-initiated processes may lead to lost events and corrupted acquired data. Since artifact removal is implemented by Fourier analysis, the imaging system can be assembled from off-the-shelf components and does not have any specialized or custom hardware, which is traditionally needed for synchronization with vascular and respiratory artifacts.

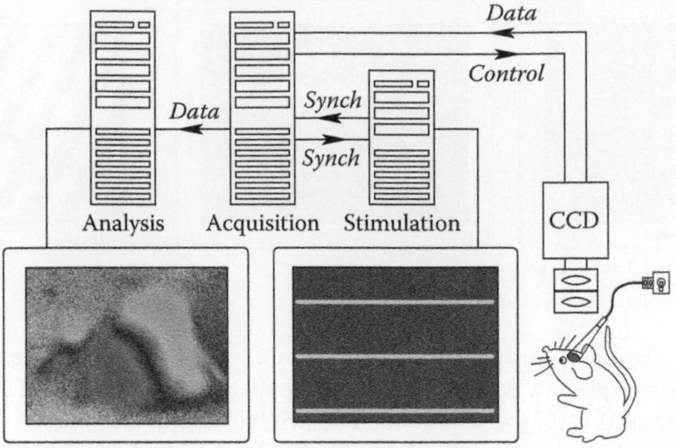

FIGURE 10.3 Imaging system. Images of the cortical surface illuminated by red light are acquired through a cranial window by a CCD camera. The images are transferred to the acquisition PC, which controls the camera's function, for preprocessing and storage. It is generally unnecessary to save the raw data stream (up to 60 MB/s for single tap cameras); therefore, the images are binned spatially and/or temporally, thus, reducing the data rate (typically to 3.75 MB/s). A frame header is attached to each preprocessed image and the final frame is stored locally or preferably remotely, as shown here, on a fast disk. Besides general experiment information, a frame header contains information about current state of stimulus. This synchronization information is received from the stimulation computer via a fast Ethernet link (we use fast UDP packets) or via a dedicated DIO link. The stimulus patterns are generated by the stimulation PC and displayed on a high refresh rate monitor. Analysis software cab be run on the analysis PC concurrently with acquisition, thus giving the experimenter immediate feedback about the experiment progress and the state of the subject.

10.2.5 Analysis and Visualization

After acquiring even a few stimulus cycles worth of data, the experimenter may start analysis. It consists of finding the real (cosine) and imaginary (sine) Fourier components at the frequency of stimulation for each individual pixel. In some cases, when a significant slow drift of the "baseline" is present such as that shown in Figure 10.1a, the time series can be preprocessed by sliding-window-averaging, temporal high-pass filtering, independently on each pixel to remove the slow activity and leave the evoked responses and the high frequency artifacts intact. These filters should have the window size measured in multiples of the period of stimulation, typically one or two full stimulus cycles. It is possible to come up with a more advanced filtering scheme, however, one should bear in mind that even a short 10-minute imaging experiment produces about 2 gigabytes of data. Our analysis scheme may not be the most accurate but it is nearly the most efficient one in dealing with large data sets.

The analysis described so far has been carried out on individual pixels. Further spatial high-pass filtering can be applied to the entire map if its quality is inadequate. For mouse imaging, we have not used spatial filtering at all. Occasionally, it was not possible completely to remove the effect of the global changes in reflectance, which left a constant DC bias in the phases even of unresponsive brain areas. In these cases, the DC bias evaluated from the unresponsive areas was subtracted from the entire map.

After preprocessing and Fourier extraction the components are converted to phase (φ) and amplitude (ρ) and the complex fields are ready for visualization. In contrast to the conventional approach to ISOI, where the amplitude of the response is used, the Fourier approach relies both on the phase and amplitude of the response—perhaps the phase being more important. The phase defines time of the response, which is used to find the state of the stimulus (equivalent to the stimulus condition if we were using episodic stimulation). The amplitude characterizes significance of the response. Typically, gray-scale coding is used to visualize changes in amplitude and hue (pure color) is used to visualize the phase. Hue is a cyclic variable in the HSL (hue, saturation, lightness) color space, which makes it a natural choice for coding the phase. Note that hue itself is suitable for continuous periodic mapping in species that have color vision.

10.3 EXPERIMENTAL PROCEDURES

10.3.1 Surgical Preparation

The surgical preparation and maintenance procedures are similar to those described previously.[4] All experimental procedures were approved by the University of Houston Committee on Animal Research.

Mice (C57BL/6 from Harlan) are anesthetized with an intraperitoneal injection of urethane (1.0 g/kg in 10% saline solution). A sedative, chlorprothixene (0.2 mg/mouse i.m.), is administered to supplement urethane. Atropine (5 mg/kg in mouse) and dexamethasone (0.2 mg/mouse) are injected subcutaneously. The animal's temperature is maintained at 37.5° by a rectal thermoprobe feeding back linearly to a heating pad on which the animal rested throughout the experiment. A tracheotomy is performed and electrocardiograph leads are attached to monitor the heart rate continuously throughout the experiment. Mice are placed in a stereotaxic instrument. A craniotomy is made over the area of interest; the dura mater is left intact. Imaging can be performed through intact skull for longitudinal studies. For acute experiments, however, it is desirable to have a craniotomy done for clearer view of the cortex. Low-melting point agarose (3% in saline) and a glass coverslip are placed over the exposed area.

Cats are initially anesthetized with ketamine (15–30 mg/kg i.m.), and implanted with a femoral or median catheter. Atropine (0.04 mg/kg) and dexamethasone (1.0–2.0 mg/kg) are injected subcutaneously to reduce tracheal secretions and minimize the stress response, respectively. To assess the anesthetic state of an animal, we continuously monitor its core temperature, electrocardiogram, expired CO_2, and peak airway pressure. A feedback-regulated heating pad maintains core body temperature at 37.5°. Pentobarbital (20–30 mg/kg, i.v.) is administered as needed to keep the animal at a surgical plane of anesthesia, determined by the animal's reflexes, heart rate and peak expired CO_2. A tracheotomy is performed and the animal is placed in a stereotaxic apparatus and ventilated with nitrous oxide:oxygen 2:1; 1% atropine sulfate and 10% phenylephrine hydrochloride are applied to the eyes. To protect from drying and focus the eyes on the stimulus monitor, contact lenses of the appropriate strength are fitted with the aid of a retinoscope. A craniotomy is made over the lateral gyrus of both hemispheres. Neuromuscular blockade is then induced by continuous infusion of pancuronium bromide (0.08–0.2 mg/kg/hr) mixed in 2.5% dextrose in lactated Ringer's solution (total volume of fluid infused, 5–10 ml/kg/hr), and ventilation is continued for the duration of the experiment. After neuromuscular blockade, anesthesia is delivered at a rate that maintains heart rate and peak expired CO_2 at their values prior to the blockade. The dura mater is reflected, and low-melting point agarose (3% in saline) and a glass coverslip are placed over the exposed cortex.

10.3.2 Imaging Procedures

Optical images of cortical intrinsic signal are obtained using a Dalsa 1M30 CCD camera (Dalsa, Waterloo, Canada) controlled by custom software, ContImage (Continuous Image). Using different tandem lens (Nikon, Inc., Melville, New York) configurations[27] images can be acquired at various magnifications: 1:1, 1:2.1, and 1:2.7. The surface vascular pattern or intrinsic signal images are visualized with illumination wavelengths set by a green (546 ± 10 nm) or red (610 ± 10 nm) interference filter, respectively. After acquisition of a surface image, the camera is focused 500–700 μm below the pial surface, an additional red filter is interposed between the brain and the CCD camera, and intrinsic signal images are acquired. A typical imaging run lasts for 10–12 minutes, with 2–3 minute breaks between runs (see Table 10.1). Frames are acquired at the rate of 30 fps and stored as 512 × 512 pixel images after binning the 1024 × 1024 camera pixels by 2 × 2 pixels spatially and by 4 frames temporally, reducing sampling rate for the stored images to 7.5 Hz and increasing well depth to 16 bit. The final pixel size is 8.9 μm for mouse and 24 μm for cat imaging. Images are analyzed using custom software kit ImAn (Image Analysis), MapAn (Map Analysis), and MapCorr (Map Correlator).

10.3.3 Visual Stimuli

Drifting bars and drifting-and-rotating full-field square-wave gratings (background 50% and contrast 40–60%) are generated by a generic video adaptor controlled by custom software, ContStim (Continuous Stimulus), and displayed on a high refresh rate monitor.

10.3.4 Data Processing and Analysis

Temporal processing of the individual pixel's time series consists of high-pass filtering (uniform kernel; length: 2 cycles of the stimulus) and extraction of the Fourier components of the first harmonic (fundamental) or second harmonic, depending on the stimulated feature space.

Spatial processing includes removal of DC shift (mouse maps) or high-pass filtering with large kernels, diameter 120 pixels or 2.6 mm (cat maps); the kernel size is 80 pixels for maps used for correlation analysis shown in Figure 10.6.

Similarity between maps is estimated by the standard normalized scalar product D. For instance, for two maps $\mathbf{a} = (a_x^i, a_y^i)$ and $\mathbf{b} = (b_x^i, b_y^i)$, where i enumerates the pixels, D is calculated as:

$$D = \frac{\mathbf{a} \cdot \mathbf{b}}{\|\mathbf{a}\| \|\mathbf{b}\|}, \ \mathbf{a} \cdot \mathbf{b} = \sum_i a_x^i b_x^i + a_y^i b_y^i, \ \|\mathbf{a}\|^2 = \mathbf{a} \cdot \mathbf{a}. \tag{10.A1}$$

Since many maps differed mainly by a uniform phase shift—for example, the "forward" and the complex conjugated "backward" maps (see Section 10.4.2)—we introduced an extension of the scalar product D, which we called correlation coefficient C, that allowed factorization of the common angular shift α. C and α are calculated as follows:

$$C = \frac{\sqrt{(\mathbf{a} \cdot \mathbf{b})^2 + (\mathbf{a} \times \mathbf{b})^2}}{\|\mathbf{a}\| \|\mathbf{b}\|}, \ \mathbf{a} \times \mathbf{b} = \sum_i a_x^i b_y^i - a_y^i b_x^i \tag{10.A2}$$

$$\alpha = \text{angle}(\mathbf{a}, \mathbf{b}) \equiv \arg(\mathbf{a} \cdot \mathbf{b}, \mathbf{a} \times \mathbf{b}). \tag{10.A3}$$

Note that $C = D\cos(\alpha)$.

To provide some reference values for maps' similarity we calculate autocorrelation coefficients. Autocorrelation coefficients are calculated as the average of the cross-correlation coefficients of a map with itself shifted by one pixel up, down, left, and right. The angles (10.A3) for all autocorrelation coefficients are very small (<0.5°), thus $C \approx D$.

10.4 RESULTS AND DISCUSSION

10.4.1 Relative Retinotopy

The responses to the drifting low-duty cycle gratings are shown in Figure 10.4. We emphasize the fact that the computation of each pixel in these images was done independently. Hence, no spatial filtering was applied to these images, and the quality of the phase and magnitude maps was not artificially enhanced. The maps clearly demonstrate high spatial resolution of this technique. The maps resolve not only the retinotopic organization of the two primary areas (the large central spot) but also the retinotopic structure of two small areas located laterally (on the left side of the central spot).

The phase maps in Figure 10.4A,D do not, however, show absolute retinal position on the monitor screen because they are shifted by the hemodynamic delay ϕ_d. These are thus maps only of relative retinotopy. The hemodynamic shift is evident from the fact that the blue region (coding for the central part of the screen) appears to be located in the lateral-caudal (lower-left) part of the excited area and not in the middle (Figure 10.4A). Another, perhaps a more clear, evidence of the delay is the presence of the yellow region at the rostral (top) end of the maps shown in Figure 10.4A. The

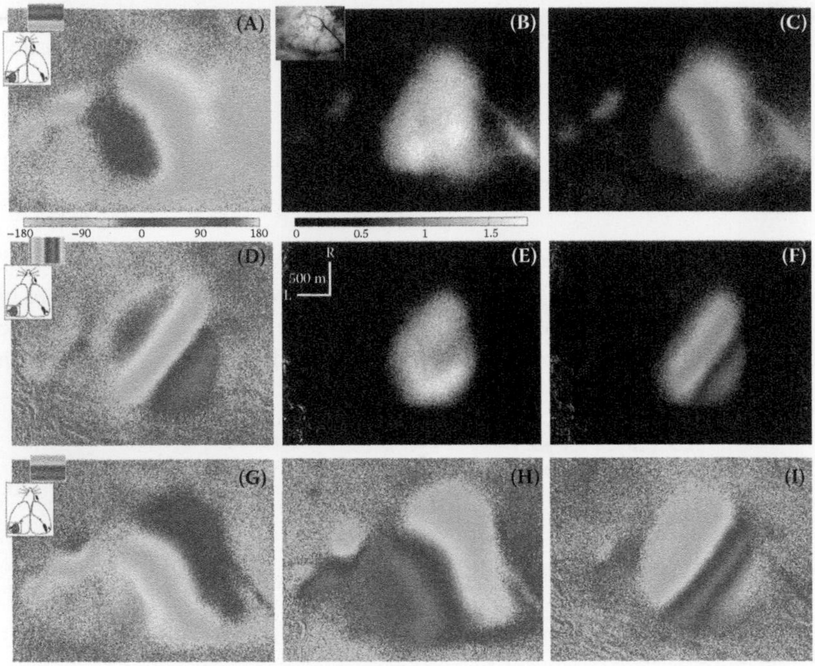

FIGURE 10.4 **(See color insert following page 224)** Maps of relative retinotopy, mouse visual cortex. (A) The phase map φ^+ of the actual response to the elevation stimulus shown on Figure 10.2B. (B) The magnitude ρ of the optical response the phase of which is illustrated in (A). Grey scale below shows response amplitude as fractional change in reflection $\times 10^{-4}$. (C) The polar map, combination of (A) and (B). (D–F) The phase, magnitude, and polar maps, respectively, of the response to the azimuth stimulus shown on Figure 10.2A. The maps show presence of two lateral retinotopic areas; it is clearer on the maps of elevation. (G) The phase map (φ^-) of the response to the reversed stimulus. Maps (A) and (G) appear to be shifted by the same hemodynamic delay 89°. (H,I) The phase maps $(\varphi^+ - \varphi^-)/2$ of absolute retinotopy, isoelevation and isoazimuth, respectively. Imaging time 10 minutes per dimension. The insets in (A,D,G) show the relative position of the monitor and its color code, as well as the location of the maps on the brain, top view. The inset in (B) shows the pattern of surface blood vessels. Color wedge applies to (A,D,G), units degrees of phase; the same wedge rescaled by a factor of 5 applies to (H,I), units degrees of visual angle. Scale in (E) applies to all images. R, rostral; L, lateral.

presence of this region contradicts the color coding (inset in Figure 10.4A), which does not have yellow regions at all because this color codes for the off-screen position of the stimulus bar.

Such maps of the relative retinotopy can still be very useful if one is interested in measuring differential quantities such as cortical magnification factors (the amount of cortical space used to represent 1° of the visual angle) or in studying relation of retinotopy to other cortical functional structures. For example, one can estimate the azimuth and elevation magnification factors. Also note that the color gradient is compressed by about a factor of 2 on the azimuth map (Figure 10.4d) relative to that on the elevation map (Figure 10.4a), which is in good agreement with previous observations by Reference 28.

10.4.2 Hemodynamic Delay

Maps of absolute retinotopy, the ones compensated for hemodynamic delay, can be computed using at least three strategies we havc described previously.[4,29] Here, we present the most elegant approach, a method of stimulus reversal.

The method of stimulus reversal is based on very simple observations stemming from time reversal symmetry. Of course, we cannot reverse time; instead, we reverse the stimulus. Let us imagine that we are running a hypothetical experiment using the stimulus shown in Figure 10.2b. We will assume that the stimulus whose position on the screen is defined by instantaneous phase $\varphi_s^+ \equiv \varphi_s$ will evoke a response whose phase measures φ^+. The two phases are related by a simple expression:

$$\varphi^+ = \varphi_s + \varphi_d, \tag{10.1}$$

where φ_d is the unknown hemodynamic delay. In the next hypothetical experiment we will present a stimulus identical to that used in the first experiment with one important exception: we will temporally reverse the stimulus; that is, if in the first experiment we had a horizontal bar sweeping the screen in the upward direction, in the second experiment the bar will be moving downward. A formula similar to Equation 10.1 applies here as well:

$$\varphi^- = \varphi_s^- + \varphi_d, \tag{10.2}$$

where φ^- is the phase of the response evoked by stimulus at φ_s^-. Here, we assumed that the hemodynamic delay φ_d is the function of neuro-vascular coupling that does not depend (or depends weakly) on the directionality of the stimulus. If φ_s^+ and φ_s describe identical configurations of the stimuli, they are related by the following equation:

$$\varphi_s^- = 2\pi - \varphi_s^+ = -\varphi_s^+, \tag{10.3}$$

where the second equality is justified by the cyclic nature of the phases. Substitution of Equation 10.3 into Equation 10.2 gives:

$$\varphi^- = -\varphi_s + \varphi_d. \tag{10.4}$$

Finally one can find φ_s and φ_d from Equation 10.1 and Equation 10.4:

$$2\varphi_s = \varphi^+ - \varphi^-, \tag{10.5a}$$

$$2\varphi_d = \varphi^+ + \varphi^-. \tag{10.5b}$$

Thus, by repeating the same stimulus presentation twice, in the "forward" and reversed "backward" versions, one can effectively remove hemodynamic delay simply by subtracting the phases of the two responses. In actual applications the procedures of subtraction and addition of the phases are substituted by the computationally efficient procedures of division and multiplication of the two complex fields.

The actual application of the stimulus reversal method is shown in Figure 10.4G. A good uniformity of the delay map (image not shown) justifies the use of the method of stimulus reversal. The average value of the delay calculated from the pixels where response magnitude was above 50% of the maximum response (N = 23700) was $\varphi_d = -89 \pm 7°$. Note that the phase is negative because the initial response is a decrease of the reflectance, the minimum of which is near 91°. We should point out that division by 2 (see Equations 10.5a,b) is not a well-defined operation for cyclic variables. In certain cases it may lead to maps that have a patchy appearance due to lines of discontinuities across which the phase changes by 180°. To circumvent this patchiness one can calculate $2\varphi_d$ from the delay map, note that the delay itself is double valued (typically $-180 < \varphi_d < 0$), and shift the entire map by this delay. The map of responses to the reversed stimulus should be complex conjugated before the shift in the opposite direction. For the discontinuous feature spaces, such as retinotopy, one can avoid the trouble of the two-valued division by 2. Instead of a map of φ_s one can use the difference map $2\varphi_s$ (Equation 10.5a) and a doubled color code. This approach should not be used for continuous cyclic spaces such as orientation because assignment of values to the difference map is always ambiguous there (see Figure 10.5). Finally, we would like to indicate that the delay φ_d needs to be measured only ones for a given period of stimulation; after that phase maps acquired from the same subject can be compensated using this known delay.

10.4.3 Absolute Retinotopy

Absolute isoelevation and isoazimuth maps of the mouse visual cortex corrected by the method of the stimulus reversal are shown in Figure 10.4H,I, respectively. Three distinct patches displaying retinotopic organization can be identified; they are clearer in the elevation map (Figure 10.4H). It is possible that each patch represents two visual areas. This hypothesis is supported by the fact that the larger patch consists of two visual areas V1 and V2. V1 or primary visual cortex occupies most of the area of the largest patch. V2 is a smaller mirror image of V1 located immediately laterally to V1. The V1-V2 boundary can be identified by the phase gradient reversal in Figure 10.4I; it corresponds to the representation of the central meridian.

Certain quantitative measures can be readily evaluated from the phase maps. For example the estimate of the average global azimuth magnification factor 0.016 mm/degree, which is in agreement with Reference 28. Local or differential magnification factors can be evaluated as well. For example, it is evident from the map of azimuth (Figure 10.4I) that the local azimuth magnification factor is constant over much of V1 and increases significantly near the V1–V2 border.[28] Also, it is evident that the elevation magnification factor is larger than the azimuth one by about a factor of 2, which is in good agreement with previous observations by Wagor et al.

10.4.4 Orientation and Direction Preference

Optical imaging using the traditional approach has been successful in revealing the structure of the domains of orientation preference in visual cortices of higher mammals.[14,22] Figure 10.5C,D shows orientation maps produced with the Fourier approach. With the new method, we can obtain high-quality orientation maps in as little as 1 minute. This is at least an order of magnitude acceleration of map acquisition, as compared to the conventional procedure (a typical block-design experiment lasts for about 30–60 minutes), without compromise of spatial resolution or

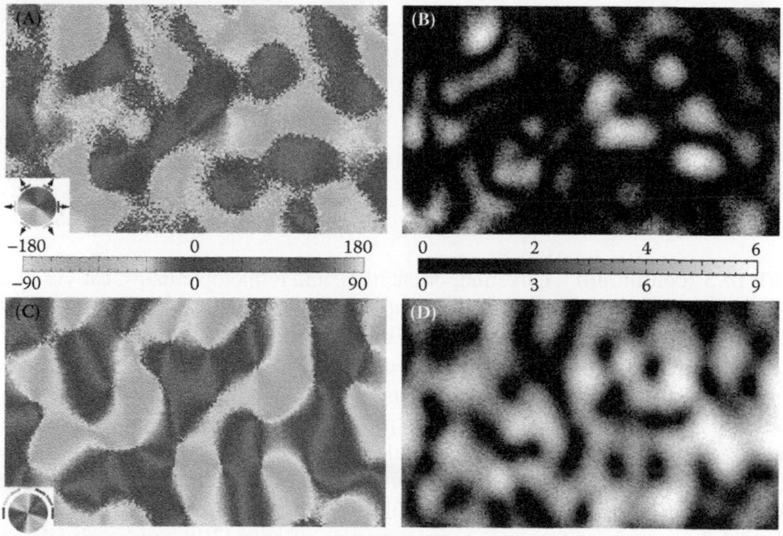

FIGURE 10.5 **(See color insert following page 224)** Direction, orientation, and retinotopic maps, cat visual cortex. (A,B) The phase φ and amplitude ρ maps of direction preference, respectively. These maps are constructed from the first harmonic of the response evoked by stimulus shown in Figure 10.2c. Stimulus parameters: rotation rate 1 cycle/min (30 seconds period for orientation), grating spatial frequency 0.2 cycle/degree and temporal frequency 2 cycle/s, background 50%, contrast 40%. Images were collected from the left hemisphere (areas 17/18), imaging time 10 minutes. The real and imaginary components of the map were spatially high-pass filtered, uniform circular kernel, diameter 120 pixels (the kernel diameter coincides with the height of the maps here). The phase map was compensated for the delay by shifting over 138°. While orientation maps exhibit point-like singularities the direction maps mostly have linear singularities, fractures. The fractures occupy a significant area leading to lower overall activity. The maximal activation was nearly half of that for orientation $\rho_{max} = 5.7$ and overall activation was only about 30% of that ($\rho_{mean} = 1.7 = 0.29\,\rho_{max}$). (C,D) The phase φ and amplitude ρ maps of orientation preference respectively. These maps are constructed from the second harmonic of the response evoked by stimulus shown in Figure 10.2C. The same data was used to construct the direction maps. The phase map was compensated for the delay by shifting over 120°. The point-like singularities of the complex fields, the so-called pinwheel centers, can be readily identified on both the phase and amplitude maps. Besides the pinwheel centers the activation is quite uniform $\rho_{max} = 9.4$, $\rho_{mean} = 4.8 = 0.51\,\rho_{max}$. Color wedge beneath (A) applies to (A,C), grey scale wedge (fractional change in reflection × 10^{-4}) applies to (B,D).

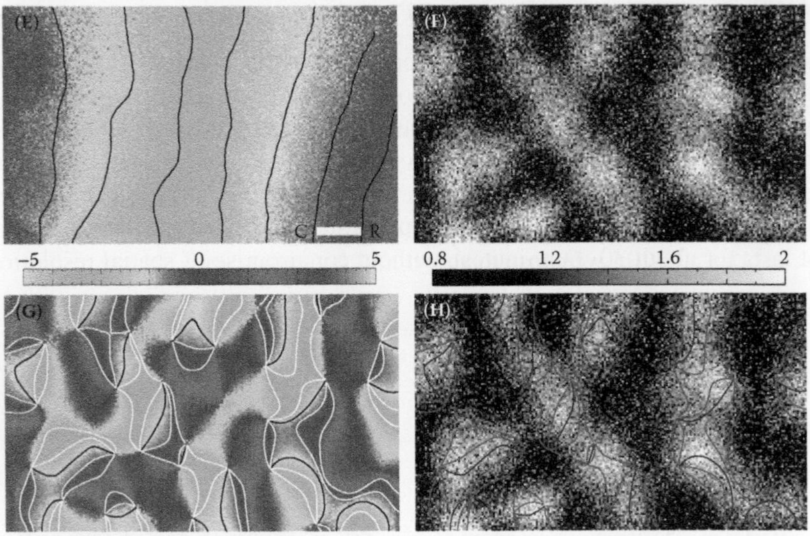

FIGURE 10.5 (continued) Direction, orientation, and retinotopic maps, cat visual cortex. (E) Retinotopic map: the phase difference map (2φ) of elevation; the same cortical location as in Figures (A–D). The stimuli were similar to those for mice (Figure 10.2B) except for the spatial frequency which was 0.05 cycles/degree. A full cycle of the color code corresponds to 10° of the visual angle since the phase gradient is doubled. The contours overlying the images have spacing of 1° of the visual angle. The representation of the horizontal meridian nearly coincides with left edge of the panel. The scale bar (0.5 mm) allows for an easy estimation of the elevation magnification factor 0.5 mm/deg (higher near the horizontal meridian), in agreement with Reference 36. The map is not filtered; the contours were drawn on a smoothed map, kernel 8 pixels. Note that only a small portion of area 17/18 can be visualized here, whereas in mice we could visualize nearly the entire visual cortex. (F) The amplitude of the response. The map clearly shows the activated orientation columns. The map was high-pass filtered (kernel diameter 80 pixels) to remove the uniform part of activation. (G) The phase map of orientation preference, same as in (a), with iso-orientaion contours corresponding to orientations of 90° (black), 90 ± 20 and 90 ± 40° (white). The 90 ± 20 contours outline the sites that should be strongly excited by a horizontal bar (20° is a typical half-width of the orientation tuning curves for single unit recordings[37]), while the 90 ± 40 contours are expected to encompass the excited region entirely. The half-width of the population orientation tuning measured by optical imaging can be estimated to be around 30°. The contours were drawn on a low-pass filtered map (kernel 4 pixels). (H) Overlay of the contours from (G) with the amplitude map from (B). The iso-orientaion domains perfectly match with the orientation columns. R, rostral; C, caudal. Scale bar 500 μm.

quality.[4] Thus, the new paradigm makes orientation maps much more quickly and with much less variability. The majority of orientation maps presented in the literature appear very smooth because they were low-pass filtered, whereas the maps presented here appear nearly as smooth without any filtering. We should point out that undue smoothing of maps, in particular orientation maps, may significantly alter their structure. For instance, low-pass filtering of orientation maps leads to pinwheel centers' annihilation (unpublished observations and private communication with Matthias Kaschube).

The representation of direction preference[30] can be calculated from the same data used to construct the orientation maps. The direction specific response should be extracted at the fundamental harmonic of the stimulus shown in Figure 10.2C, whereas the orientation specific response was extracted at the second harmonic. The direction preference maps are shown in Figure 10.5A,B. The direction maps have higher variability partially due to the weaker direction specific response, which measured about one-half of the orientation specific response for maximal activation and about one-third for the average activation (see caption to Figure 10.5). We should note that the evoked response depends on the period of stimulation: typically, for the periods we used, doubling the period of stimulation doubled the evoked response (this increase in evoked response may not be favorable since noise increases as well, see Figure 10.1B). Therefore, comparison of the evoked responses at the first and second harmonics of the stimulation frequency may be an easy comparison (the maps can be constructed from the same data) but not be a fair one (the period of stimulation are different). We will show in Section 10.4.6 that the direction and orientation specific responses differ by about a factor of 4 if evoked by stimuli of equal periodicities.

10.4.5 Validation of Orientation Maps

Although we presented a straightforward method for calculation of the hemodynamic delay and provided arguments for selecting the correct branch of the complex fields (maps) after taking their square root (division of the phase by 2 is equivalent to this procedure) in Section 10.4.2 and gave actual estimates of the phase delay in caption to Figure 10.5, the inquisitive reader may still question whether the orientation and direction maps shown in Figure 10.5 correspond to the true representations or the ones rotated by 180° in the phase space. One of the solutions to this ambiguity is to present a single condition stimulus that will activate only the cortical sites preferring that condition; for our particular problem, this stimulus is a drifting grating. Interestingly, this single-condition stimulus for orientation/direction preference can be used for retinotopic mapping as well, if the stimulus parameters are chosen appropriately. Alternatively, we can say that the stimulus used to map retinotopy (oriented bars, Figure 10.2A,B) should not only produce the expected fairly uniform retinotopic phase gradient but should also excite the orientation columns corresponding to the bar's orientation. Figure 10.5E–H shows our results confirming this hypothesis. The iso-orientaion domains from the orientation map match perfectly with the orientation columns revealed on the retinotopic map. Note that the retinotopy stimulus will evoke both the direction and orientation specific responses; however, the direction specific response is about four times weaker than the orientation specific response for equal stimulation period (see Section 10.4.6). The stimulation periods for maps shown in Figure 10.5a–d differ exactly by a factor of 2; that is why the difference in responses is not as strong (about 2 times).

10.4.6 Stability and Variation of Orientation and Direction Maps

The significant acceleration of map acquisition offered by the Fourier approach has a few apparent advantages of ethical and technical nature. First of all, each animal subject can be utilized much more efficiently since more experimental data can be collected per subject. Second, the variability between successive experimental runs, caused by changes in the physiological state or level of anesthesia, can be reduced simply because time between these runs can be shortened from about 1 hour to about 10 minutes or less. Finally, the experimenter can study a much broader class of stimuli than that allowed by the episodic approach.

We exploited these benefits of the Fourier approach to address the following questions: maps' stability and maps' variability as a function of intertrial time and as a function of stimulus perturbations. We collected a large number of orientation and direction maps evoked by the square-wave gratings with different parameters: spatial frequency, temporal frequency, rotation frequency, and stimulated eye. The correlation coefficients among these maps are presented in Figure 10.6, and the stimulus parameters and the strengths of the responses are given in Table 10.1. Our observa-

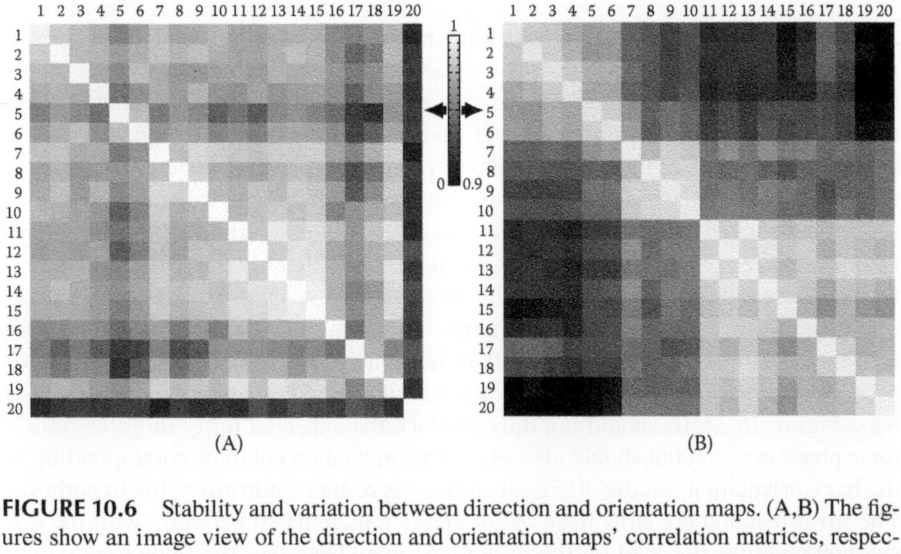

FIGURE 10.6 Stability and variation between direction and orientation maps. (A,B) The figures show an image view of the direction and orientation maps' correlation matrices, respectively, compiled for 20 independent experimental runs with various stimulus parameters. The region of interest (140 × 80 pixels) includes area 17 only and coincides with the lower-left corner of the region shown in Figure 10.5. A smaller region was chosen to exclude area 18, located in the upper-right corner in Figure 10.5,[38] from the correlation analysis because high spatial frequency stimuli evoked weak response there.[39] Stimulus parameters for each experiment are given in Table 10.1. The maps were spatially high-pass (kernel diameter 80 pixels) and low-pass (kernel diameter 4 pixels) filtered. Note the 10-fold difference in the grey-scale code.

TABLE 10.1
Stimulus Parameters and Mean (ρ_{mean}), Max (ρ_{max}), and Ratio ρ_{mean}/ρ_{max} of the Response Amplitude for Maps Used for Correlation Analysis in Figure 10.6

						Direction			Orientation		
	SF	TF	RF	E	T	Mean	Max	Ratio	Mean	Max	Ratio
1	0.4	2	1	L	0	0.65	2.10	0.31	1.59	3.84	0.41
2	0.4	2	1	L		0.76	2.29	0.33	1.89	4.09	0.46
3	0.4	2	1	L		0.72	2.58	0.28	1.70	3.48	0.49
4	0.4	2	1	L		0.61	2.22	0.28	1.52	3.56	0.43
5	0.4	2	−1	L		0.80	3.31	0.24	1.68	4.43	0.38
6	0.4	2	−1	L		0.90	3.05	0.29	2.07	4.60	0.45
7	0.2	2	−1	L	5	1.63	5.08	0.32	3.77	7.95	0.47
8	0.2	2	1	L		1.64	5.11	0.32	3.48	9.08	0.38
9	0.2	2	−1	L		1.49	5.68	0.26	3.15	7.26	0.43
10	0.2	2	1	L		1.62	5.83	0.28	3.52	8.27	0.42
11	0.2	2	1	R	6	1.46	4.52	0.32	5.10	10.68	0.48
12	0.2	2	−1	R		2.24	6.50	0.34	6.23	12.37	0.50
13	0.2	2	1	R		1.60	4.67	0.34	4.58	9.39	0.49
14	0.2	2	−1	R		2.07	6.40	0.32	5.22	11.05	0.47
15	0.2	2	−2	R	9	1.19	3.86	0.31	2.59	5.36	0.48
16	0.2	2	−0.5	R		1.81	5.43	0.33	7.48	15.33	0.49
17	0.2	1	1	R		1.44	5.21	0.28	5.85	11.82	0.50
18	0.2	2	1	R		1.38	5.26	0.26	5.10	11.05	0.46
19	0.2	4	1	R		1.95	7.00	0.28	4.27	9.75	0.44
20	0.2	8	1	R	11	1.22	4.28	0.29	4.60	10.66	0.43

Note: SF—spatial frequency, cycle/deg, TF—temporal frequency, cycle/s, RF—rotation frequency (negative values correspond to clockwise rotation), cycle/min, E—stimulated eye, L—left, R—right, T—time in hours. Experimental groups 7–10 (left eye) and 11–14 (right eye) had otherwise identical stimulus parameters. The varied parameters included spatial frequency, temporal frequency, rotation direction and frequency, and stimulated eye. The experiments are arranged in chronological order, the "T" column shows elapsed time (hours); less then 25% of all imaging sessions are shown here. The experiments without the time value were in the immediate succession to the previous ones; the runs lasted 7–10 minutes.

tions can be summarized as follows (we will mostly address the orientation maps' behavior; the numbers in parentheses refer to the maps in the correlation matrix):

1. All orientation maps displayed very high similarity: $C_{max} = 0.995$ (pair 11,13), $C_{min} = 0.899$ (4,20), the mean and standard deviation $C_{mean} = 0.950 \pm 0.026$ (the diagonal elements of the matrix were excluded from this and other statistical calculations).
2. Direction maps were much more variable. $C_{max} = 0.951$ (14,15), $C_{min} = 0.116$ (7,20), $C_{mean} = 0.664 \pm 0.192$. One of these maps (20) did not look like other

ones at all; it was contaminated by some stripy artifacts of technological origin (not related to physiology). Exclusion of this map from the analysis gave: $C_{min} = 0.209$ (5,18), $C_{mean} = 0.712 \pm 0.133$.

3. The maps with stronger response tend to have higher correlation (not a surprise since the noise level is about constant).
4. While the pattern of activation is stable, the amplitude of the response (both max and mean, see Table 10.1) may not be a stable measure of the evoked activation. For instance: (a) runs 4 and 16 had a nearly five-fold difference in the amplitudes, whereas the evoked patterns of activation were quite similar $C_{4,16} = 0.918$, (b) a block of identical stimuli (7–10), mean amplitude averaged across maps $\rho_{7\text{-}10} = 3.48 \pm 0.22$, $C_{7\text{-}10} = 0.986 \pm 0.004$, (c) a block of identical stimuli (11–14) $\rho_{11\text{-}14} = 5.28 \pm 0.60$, $C_{11\text{-}14} = 0.989 \pm 0.004$.
5. The ratio of the mean and maximum activity is stable: orientation $\rho_{mean}/\rho_{max} = 0.45 \pm 0.04$, direction 0.30 ± 0.03.
6. The correlation matrix, Figure 10.6b, has a clear block-diagonal structure. The following blocks can be readily identified: (a) (1–10) and (11–20) stimulation through left and right eye respectively, (b) (1–6) (7–20) stimulation by gratings with the same spatial frequency 0.4 and 0.2 cycle/degree, respectively, (c) (7–10) and (11–14) identical stimuli.
7. The response amplitude depends strongly on the period of stimulation. Runs 14–16, values given as pairs (period in minutes – ρ_{mean}), orientation: 0.25 – 2.59, 0.5 – 5.22, 1 – 7.48; direction: 0.5 – 1.19, 1 – 2.07, 2 – 1.81. The ratios of the responses (ρ^{OR}/ρ^{DI}) for equal periods are 4.4 (0.5 min) and 3.6 (1 min). The response peaks at around a period of 1 minute.
8. Stimulation through contralateral eye (7–10) evoked stronger activation than that through ipsilateral eye (11–14) $\rho_{contra}/\rho_{ipsi} = 1.52 \pm 0.20$. Correlation between the two groups $C_{(contra),(ipsi)} = 0.953 \pm 0.005$ was significantly lower than correlations within the groups (see above). The ratio of the maps can be used to create an ocular dominance map (not shown).
9. Stimulation with higher spatial frequency SF = 0.4 cycle/degree (1–6) evoked weaker responses than that with SF = 0.2 (7–10) $\rho_{SF=0.2}/\rho_{SF=0.4} = 2.00 \pm 0.25$; stimulation through ipsilateral eye in all maps. Correlation between the two groups $C_{(SF=0.4),(SF=0.2)} = 0.942 \pm 0.010$ was significantly lower than correlations within the groups $C_{SF=0.4} = 0.975 \pm 0.008$ and $C_{SF=0.2} = 0.986 \pm 0.004$.
10. Stimulation with various temporal frequencies TF = 1, 2, 4, and 8 cycle/s (17–20) evoked responses that were not significantly different from the reference responses at TF = 2 (11–14) $\rho_{(11-14)}/\rho_{(17-20)} = 1.07 \pm 0.18$; neither were different the correlation coefficients $C_{(11-14),(17-20)} = 0.981 \pm 0.005$ and $C_{(17-20)} = 0.980 \pm 0.005$. This result suggests that temporal frequency (or stimulus speed) may not be represented by means of a map in cat V1 or at least that it will be exceedingly difficult to measure such a representation.
11. Orientation maps obtained by the conventional episodic approach (data not shown) had considerably lower correlation coefficients. For example comparison among independent 28-minute episodic runs yielded correlations of

0.75^4, which is lower than any correlation value shown in Figure 10.6, even for distinctly different stimuli.

10.5 SUMMARY AND CONCLUSIONS

We have presented the basics of a general approach to functional brain mapping, and have given details on localization of instantaneous neuronal responses from slow hemodynamic signals using time-reversed stimuli. This approach should be widely applicable to brain imaging. Advantages of the new approach include removal of the noise, increase in spatial resolution, and a significant reduction in acquisition time. Along with theoretical considerations, we gave details of the practical implementation of the Fourier approach for optical imaging of intrinsic signals and demonstrated its use for construction of retinotopic maps in the mouse and retinotopic, orientation, and direction of motion preference maps in the cat visual cortex.

As a word of advice we would like to mention some important implications that the robustness of the orientation representation may lead to: Attempts to map some complex features, such as T-junctions, crossings, or more complex objects, may produce spurious results because even a small nonuniformity in the distribution of oriented edges may evoke a pattern of activity that is specific to that distribution of orientations rather than to the feature of interest. Alternatively, one can say that there are many nonidentical stimuli that have similar distributions of orientations; distinguishing the cortical representations of these stimuli will not be easy because the stimulus specific activation may be significantly weaker than the orientation specific component.

In addition to the results reported in this chapter we have obtained high-resolution maps in various cortices of different species. These include: mouse—auditory cortex (tonotopy), representation of whisker pad and digits in somatosensory cortex; rat—visual cortex (retinotopy), auditory cortex[29]; and ferret—visual cortex (retinotopy, ocular dominance, orientation, and direction selectivity). Applications of the proposed technique and optical imaging system are not limited to the cortical regions and species noted above. The same methodology should be applicable to intraoperative mapping of the human brain. Also, application of the methodology is not limited to optical imaging. It can be applied to functional MRI as well as to any other technique measuring metabolic or hemodynamic signals or even to electrophysiology (typically not required because of very high SNR for spike or evoked potential detection).

A similar method has been used in fMRI [31–34] for topographic mapping of human visual cortex. Surprisingly, it has not been used to map more complex features. Only very recently the Fourier approach was readopted for mapping orientation preference in the cat.[35] Moon et al. reported that they could not convincingly map orientation columns using conventional block-design stimulation and differential analysis method, however, they were successful in obtaining iso-orientation maps with high reproducibility using temporally encoded continuous cyclic orientation stimulation with Fourier data analysis. The capability of fMRI to register neuronal activity on the level of cortical columns should open new dimensions for the neuroscience community from systems to cognition and behavior.

ACKNOWLEDGMENTS

This work was supported by the Alfred P. Sloan Foundation and HFSP.

REFERENCES

1. Frostig, R.D., Lieke, E.E., Ts'o, D.Y., and Grinvald, A. Cortical functional architecture and local coupling between neuronal activity and the microcirculation revealed by in vivo high-resolution optical imaging of intrinsic signals. *Proc Natl Acad Sci USA* **87**, 6082–6 (1990).
2. Ogawa, S., Lee, T.M., Kay, A.R., and Tank, D.W. Brain magnetic resonance imaging with contrast dependent on blood oxygenation. *Proc Natl Acad Sci USA* **87**, 9868–72 (1990).
3. Colantuoni, A., Bertuglia, S., and Intaglietta, M. Quantitation of rhythmic diameter changes in arterial microcirculation. *Am J Physiol* **246**, H508–17 (1984).
4. Kalatsky, V.A. and Stryker, M.P. New paradigm for optical imaging: Temporally encoded maps of intrinsic signal. *Neuron* **38**, 529–45 (2003).
5. Kandel, E.R., Schwartz, J.H., and Jessell, T.M. *Principles of Neural Science* (McGraw-Hill, 2000).
6. Brodmann, K. *Vergleichende Lokalisationlehre der Grosshirnrinde in ihren Prinzipien dargestellt auf Grund des Zellenbaues* (Barth, Leipzig, 1909).
7. Garey, L.J. *Brodmann's Localization in the Cerebral Cortex* (Smith-Gordon, London, 1994).
8. Lewis, J.W. and Van Essen, D.C. Mapping of architectonic subdivisions in the macaque monkey, with emphasis on parieto-occipital cortex. *J Comp Neurol* **428**, 79–111 (2000).
9. Mountcastle, V.B. Modality and topographic properties of single neurons of cat's somatic sensory cortex. *J Neurophysiol* **20**, 408–34 (1957).
10. Hubel, D.H. and Wiesel, T.N. Receptive fields, binocular interaction and functional architecture in the cat's visual cortex. *J Physiol* **160**, 106–54 (1962).
11. Hubel, D.H. and Wiesel, T.N. Receptive fields and functional architecture of monkey striate cortex. *J Physiol* **195**, 215–43 (1968).
12. Hubel, D.H. and Wiesel, T.N. Shape and arrangement of columns in cat's striate cortex. *J Physiol* **165**, 559–68 (1963).
13. Hubel, D.H. and Wiesel, T.N. Sequence regularity and geometry of orientation columns in the monkey striate cortex. *J Comp Neurol* **158**, 267–93 (1974).
14. Blasdel, G.G. and Salama, G. Voltage-sensitive dyes reveal a modular organization in monkey striate cortex. *Nature* **321**, 579–85 (1986).
15. Grinvald, A., Lieke, E., Frostig, R.D., Gilbert, C.D., and Wiesel, T.N. Functional architecture of cortex revealed by optical imaging of intrinsic signals. *Nature* **324**, 361–4 (1986).
16. Wolf, F. and Geisel, T. Spontaneous pinwheel annihilation during visual development. *Nature* **395**, 73–8 (1998).
17. Swindale, N.V. A model for the formation of orientation columns. *Proc R Soc Lond B Biol Sci* **215**, 211–30 (1982).
18. Miller, K.D., Keller, J.B., and Stryker, M.P. Ocular dominance column development: Analysis and simulation. *Science* **245**, 605–15 (1989).
19. Koulakov, A.A. and Chklovskii, D.B. Orientation preference patterns in mammalian visual cortex: A wire length minimization approach. *Neuron* **29**, 519–27 (2001).
20. Lee, H.Y. and Kardar, M. Patterns and symmetries in the visual cortex and in natural images. *J Stat Phys* **125**, 1247–70 (2006).
21. Grinvald, A. et al. In-vivo Optical Imaging of Cortical Architecture and Dynamics. in *Modern Techniques in Neuroscience Research* (Eds. Windhorst, U. and Johansson, H.) 893–969 (Springer, New York, online version is available at http://www.weizmann.ac.il/brain/grinvald/html/pdf.html, 1999).

22. Bonhoeffer, T. and Grinvald, A. Iso-orientation domains in cat visual cortex are arranged in pinwheel-like patterns. *Nature* **353**, 429–31 (1991).
23. Nilsson, H. and Aalkjaer, C. Vasomotion: Mechanisms and physiological importance. *Mol Interv* **3**, 79–89, 51 (2003).
24. Mayhew, J.E. et al. Cerebral vasomotion: A 0.1-Hz oscillation in reflected light imaging of neural activity. *Neuroimage* **4**, 183–93 (1996).
25. Funk, W. and Intaglietta, M. Spontaneous arteriolar vasomotion. *Prog Appl Microcirc*, 66–82 (1983).
26. Kalatsky, V.A., O'Connor, E.M., and Tcheslavski, G.V. Concurrent multidimensional imaging of visual space representations in mouse visual cortex by Fourier optical imaging of intrinsic signals. *Soc. Neurosci. Abstr.* **503.9** (2006).
27. Ratzlaff, E.H. and Grinvald, A. A tandem-lens epifluorescence macroscope: Hundred-fold brightness advantage for wide-field imaging. *J Neurosci Methods* **36**, 127–37 (1991).
28. Wagor, E., Mangini, N.J., and Pearlman, A.L. Retinotopic organization of striate and extrastriate visual cortex in the mouse. *J Comp Neurol* **193**, 187–202 (1980).
29. Kalatsky, V.A., Polley, D.B., Merzenich, M.M., Schreiner, C.E., and Stryker, M.P. Fine functional organization of auditory cortex revealed by Fourier optical imaging. *Proc Natl Acad Sci USA* **102**, 13325–30 (2005).
30. Shmuel, A. and Grinvald, A. Functional organization for direction of motion and its relationship to orientation maps in cat area 18. *J Neurosci* **16**, 6945–64 (1996).
31. Engel, S.A. et al. fMRI of human visual cortex. *Nature* **369**, 525 (1994).
32. Engel, S.A., Glover, G.H., and Wandell, B.A. Retinotopic organization in human visual cortex and the spatial precision of functional MRI. *Cereb Cortex* **7**, 181–92 (1997).
33. Sereno, M.I. et al. Borders of multiple visual areas in humans revealed by functional magnetic resonance imaging. *Science* **268**, 889–93 (1995).
34. DeYoe, E.A. et al. Mapping striate and extrastriate visual areas in human cerebral cortex. *Proc Natl Acad Sci USA* **93**, 2382–6 (1996).
35. Moon, C.H., Fukuda, M., Park, S.H., and Kim, S.G. Neural interpretation of blood oxygenation level-dependent fMRI maps at submillimeter columnar resolution. *J Neurosci* **27**, 6892–902 (2007).
36. Tusa, R.J., Palmer, L.A., and Rosenquist, A.C. The retinotopic organization of area 17 (striate cortex) in the cat. *J Comp Neurol* **177**, 213–35 (1978).
37. Webster, M.A. and De Valois, R.L. Relationship between spatial-frequency and orientation tuning of striate-cortex cells. *J Opt Soc Am A* **2**, 1124–32 (1985).
38. Tusa, R.J., Rosenquist, A.C., and Palmer, L.A. Retinotopic organization of areas 18 and 19 in the cat. *J Comp Neurol* **185**, 657–78 (1979).
39. Movshon, J.A., Thompson, I.D., and Tolhurst, D.J. Spatial and temporal contrast sensitivity of neurons in areas 17 and 18 of the cat's visual cortex. *J Physiol* **283**, 101–20 (1978).

11 Optical Imaging of Neuronal Activity in the Cerebellar Cortex Using Neutral Red

Timothy J. Ebner, Gang Chen, and Wangcai Gao

CONTENTS

11.1 INTRODUCTION

Over the last decade our laboratory has developed and used optical imaging techniques to monitor neuronal activity in the cerebellar cortex *in vivo*. These techniques include the use of voltage sensitive dyes,[1–3] neutral red,[4,5] flavoprotein autofluorescence,[6,7] and

Ca^{++} imaging.[7] Optical imaging has allowed us to address a spectrum of questions about both normal and abnormal cerebellar cortical physiology.

This chapter focuses on the use of pH imaging, specifically neutral red, as a measure of neuronal activation in the cerebellar cortex. Our earliest attempts to use voltage sensitive dyes in the cerebellar cortex *in vivo* were disappointing due to the small size of the signal.[3] Therefore, we turned to other approaches to optically map neuronal activity in the cerebellum. The first of these approaches was neutral red imaging.[4] Neutral red imaging exploits the known close coupling between pH changes and neuronal activation.[8,9] In this chapter we discuss the theory underlying pH imaging with neutral red, describe the techniques used to image the cerebellar cortex *in vivo,* and provide examples of how neutral red imaging has generated insights into the functioning of cerebellar cortical circuits.

11.2 THEORY OF NEUTRAL RED IMAGING

A number of excellent reviews have extensively covered regulation of pH in the nervous system, pH and neuronal activity, and changes in pH during pathophysiological conditions.[9–11] Both a rich and complex topic, a general discussion of pH and the nervous system is beyond the scope of the present chapter. It suffices to state that neuronal activity is accompanied by extra- and intracellular pH shifts in neurons and glia across a wide range of species. In most preparations, including the cerebellar cortex and cerebral cortex, neuronal activity leads to neuronal acidification and glial alkalization. Extracellularly, the pH changes consist of a transient alkaline shift followed by a prolonged acidic shift. Also, large pH shifts, particularly acidification, occur in many pathological events in the central nervous system (CNS), including stroke, traumatic brain injury, epilepsy, and spreading depression.[11–14] Therefore, pH-based optical imaging is potentially an effective approach to map neuronal function and dysfunction *in vivo.*

A major issue is the choice of pH fluorescent dye for use in the CNS. Ideal properties of a pH indicator include maximal sensitivity within the physiological range in which pH shifts occur (~6.8–7.6), large signal-to-noise ratio, no toxicity, and deliverability. Another highly desirable property is sequestration to the compartment of interest (e.g., extracellular versus intracellular space). Characteristics, such as selectivity for specific cell classes (e.g., neurons versus glia) and for specific cell types (e.g., Purkinje cells versus granule cells), would be a major advance in pH imaging.

A number of fluorescent pH indicators have been developed for use in cells. The most widely used pH fluorescent dyes are BCECF (2′,7′-bis-(2-carboxyethyl)-5-(and-6)-carboxyfluorescein) and SNARF (seminaphtharhodafluor).[15–18] Both dyes have near-neutral *pKa* (7 and 7.5, respectively), in the physiological pH range of neurons (6.8–7.4). BCECF and SNARF have been primarily used in ratiometric imaging studies of extracellular or intracellular pH in vitro. We are unaware of published results using these pH indicators in the CNS *in vivo,* likely due to the difficulty of staining cells in the intact brain. Due to poor staining and toxicity, we were never successful in using these dyes to monitor neuronal activity in the rat cerebellar cortex *in vivo.* The only published pH optical imaging studies of activity patterns in the CNS have used the pH dye, neutral red.[4,19]

Neutral red (3-amino-*m*-dimethylamino-2-methylphenazine hydrochloride) has long been recognized as a vital dye that could be useful as an indicator of cellular pH.[20,21] Neutral red has a *pKa* in the physiological pH range (6.8 in saline, 7.0 in rat brain) and changes its absorbance inversely with changing pH.[22,23] Therefore, the intracellular acidification of neurons that results from depolarization/excitatory stimulation will produce an increase in fluorescence.[4] The optimal excitation is centered on 540 nm, and the maximal emission is 630 nm. Large variations in the concentration of ions that have major effects on the excitability of neurons and glia (Na^+, K^+, and Ca^{++}) have little effect on the emission spectra of neutral red.[4] Therefore, neutral red is relatively selective for shifts in pH.

Neutral red is highly lipophilic, uncharged at neutral pH, readily crosses the blood-brain barrier, and enters cells easily. Early uses of neutral red included as an assay of viability in cultured fish cells.[24] On entry into cells, neutral red is protonated, allowing it to be quickly trapped in cells and concentrated; as a consequence, little dye is associated with the cell membrane itself.[20,21] In slices of the cerebellar cortex, neutral red stains Purkinje cell bodies and their dendritic trees intensely.[20] However, the staining is not uniform, with interneurons exhibiting little staining while the molecular layer shows diffuse staining, presumably due to the densely packed parallel fibers.[20] Confocal microscopy of fresh sections of the cerebellar cortex stained with neutral red show that the dye is sequestered intracellularly in Purkinje cells (Chen and Ebner, unpublished data). Neutral red exhibits little toxicity *in vitro* and *in vivo*.[4,20] In agreement with La Manna's assessment,[21] neutral red is primarily located in the intracellular compartment and has many of the properties needed for measurement of intracellular pH, independent of extracellular pH.

Neutral red also has a measurable membrane potential dependent fluorescence that was attributed to changes in hydrophobicity.[25] It has been argued that this hydrophobic property could potentially be used to monitor faster changes in membrane potential, presumably by altering the association of neutral red with the plasma membrane.[22] However, this non-pH sensitivity is modest and unlikely to account for the large, slow fluorescent changes reported for neutral red *in vivo*.[4,25]

11.3 EXPERIMENTAL METHODS

11.3.1 Animal Preparation

All our neutral red imaging studies have used either adult rats or FVB mice. Anesthesia is induced by intramuscular injection of a cocktail solution of ketamine (60 mg/kg), xylazine (3 mg/kg), and acepromazine (1.2 mg/kg), and monitored by EKG and corneal and foot pinch reflexes. Supplemental injections of the ketamine and xylazine solution are given as needed. The animal is placed in a stereotaxic frame and body temperature is maintained at 37°C by a feedback-controlled homeothermic blanket system. The trachea is cannulated below the cricoid cartilage to allow for artificial respiration with 95% O_2 and 5% CO_2. A craniotomy exposes the cerebellar cortex, usually Crus I and II, and a watertight chamber of acrylic is constructed around the opening and filled with Ringer's solution containing (in mM): 123 NaCl, 3 KCl, 26 $NaHCO_3$, 2 $CaCl_2$, 2 $MgSO_4$, and 10 D-Glucose bubbled with 95% O_2 and 5% CO_2.

If needed, gallamine triethiodide (0.05 ml, 20 mg/mL, intramuscular injection) is used to paralyze the animal.

11.3.2 Dye Application

Because of its lipid solubility, neutral red readily penetrates the blood–brain barrier and cell membranes, easily staining the brain.[20,21] We have used two dye loading methods in the cerebellar cortex.[4,5] First, a 10 mM solution of neutral red is superfused over the exposed cortical surface for ~2 hours, supplemented with either intravenous (1 ml, 35 mM, for rats) or intraperitoneal (i.p.) injection of neutral red (0.3 ml, 35 mM, for mice). After staining, the neutral red solution in the chamber is thoroughly washed out and refilled with Ringer's solution. Second, we subsequently learned that two i.p. injections of neutral red work equally well (35 mM, 2 ml/injection; first injection, 30 m before imaging and during surgery; second injection, immediately prior to imaging). The i.p. staining method offers two distinct advantages. First, there is no precipitation of neutral red on the cerebellar surface that requires extensive washout. Second, the preparation time is shortened by 1–2 h, permitting a longer recording period.

11.3.3 Electrical Stimulations, Electrophysiological Recording, and Pharmacology

Stimulation of parallel fibers is achieved using a train of pulses (typically 50–200 μA, 100–200 μs pulse width, 5–35 Hz for 10 s) via an epoxylite-coated tungsten microelectrode (1–2 MΩ) placed at or just below the surface of the cerebellar cortex. To activate the climbing fiber projections to the cerebellar cortex a tungsten microelectrode is stereotaxically placed through the dorsal foramen magnum into the contralateral inferior olive and stimulated with a train of pulses (50–350 μA, 100 μs pulse width, 10 Hz for 5–10 s). Electrical stimulation of the ipsilateral face is delivered by two closely spaced, platinum needle electrodes placed subdermally using frequencies from 2 to 20 Hz for 5–20 s at 5–30 V with pulse widths of 100–500 μs. The voltage for face stimulation is adjusted to just below the threshold to evoke observable muscle twitches.

In some experiments, extracellular recordings of either field potentials or single cerebellar neurons are obtained with glass microelectrodes (2 M NaCl, 2–5 MΩ) using conventional electrophysiological techniques.[4,7] One of the advantages of optical imaging is that it allows precise placement of recording electrodes into the desired locations. The electrophysiological recordings are digitized (50 kHz), averaged on-line and stored for off-line analysis.

To examine the contributions of various receptors, neural signaling pathways, and pH modulation in the cerebellar cortex, various drugs were added to the Ringer's superfusing the chamber.

11.3.4 Optical Imaging

The animal in the stereotaxic frame is placed on an x-y stage mounted on a modified Zeiss or Nikon epifluorescence microscope. Either a 2× or 4× objective coupled with a 0.63× reducing lens is used to view the exposed cerebellar cortex. Images are acquired with either a PXL (512 × 512 pixels) or Quantix 57 (530 × 512 pixels) cooled charge coupled device (CCD) frame transfer camera with 12-bit digitization (Roper Scientific, Tucson, Arizona). The camera exposure time is 100–200 ms. A 100 W mercury–xenon lamp (Hamamatsu Photonics, Shizouka, Japan) with a direct current-controlled power supply (Opti Quip model 1600, Highland Mills, New York) was used as the excitation light source. The filter set for neutral red includes an excitation filter at 546 ± 10 nm, an emission filter at >620 nm, and a dichroic mirror of 580 nm.[4,21] Images are binned to yield resolutions of typically 10 × 10 μm^2 or 14 × 14 μm^2. Imaging flavoprotein autofluorescence used a band-pass excitation filter (455 ± 35 nm), an extended reflectance dichroic mirror (500 nm), and a >515 nm long-pass emission filter.[6] The Ca^{++} filter set was customized to exclude the majority of autofluorescence signal (excitation 490–510 nm), a long-pass dichroic mirror of 515 nm, and emission 520–530 nm.[7]

11.3.5 Data Processing and Analysis

The imaging protocol includes obtaining control images followed by a sequence of images with stimulation or other manipulations. The total number of frames acquired depends on the specific experimental question. Typically, the frame duration is 100–200 ms. Due to the high signal-to-noise ratio of neutral red imaging, the signal is easily detected and studied without averaging. The fluorescence change in each image, F_i, is quantified on a pixel basis by subtracting a background fluorescence frame (F_B) and then dividing by F_B, that is $\Delta F/F = (F_i - F_B)/F_B$. The average of 10–20 control frames was used as F_B. Next, the average ΔF/F in regions of interest is determined. For example, parallel fiber stimulation evokes a beam-like response and inferior olive stimulation evokes a band-like response that are defined as regions of interest. For spreading acidification and depression, the region of interest is the folium stimulated. To establish the significance of any changes in the amplitude of the optical response within a region of interest either paired student's t-test (within animals) or ANOVA (among animals) using a randomized complete blocked design followed by Duncan's post hoc is used.

More specialized analysis can be used as needed. For example, to quantify the optical responses evoked by face and inferior olive stimulation, a 2D Fourier analysis is used.[26] Based on the assumption that the parasagittal bands evoked are of lower spatial frequencies than the background noise and that the vasculature is primarily horizontally aligned, this analysis extracted the power in the parasagittal bands.

Two types of visualization methods are used. The first is simply based on the ΔF/F for a select image, in which either a grey or pseudocolor scale depicts the changes in fluorescence relative to background (examples shown in Figures 11.4A,B, 11.5C, and 11.6A). Given the large amplitudes of the optical responses, this first approach to visualization has a number of advantages. These include being relatively simple to implement, minimizing assumptions about the underlying statistics of the data, as well as showing the amplitudes of both the response and the background noise. The second involves an activation map constructed by statistical "thresholding." As described in detail previously,[4,26] statistically significant changes in fluorescence relative to background fluorescence are determined (examples shown in Figure 11.5A,B). Pixels with intensity levels greater than or equal to a threshold (e.g., 2–5 standard deviations of the mean background fluorescence) are defined as statistically significant, pseudocolored and superimposed on an image of the background. Statistical thresholding has the advantage of removing the background noise as well as visualization of the optical signal in relation to the underlying anatomy. However, this methodology has the intrinsic problem of selecting a threshold that distinguishes between the noise and the signal of interest. Ideally, both methods of visualization can be used to fully convey the nature of the optical responses.

11.4 NATURE AND ORIGIN OF THE NEUTRAL RED FLUORESCENCE SIGNAL

Our initial studies focused on understanding the pH dependence and origin of the neutral red signal. Fluorescence changes were determined when the rat cerebellar cortex stained with neutral red was superfused with Ringer's of varying pH (6.0–8.8). Bathing the cerebellar cortex in alkaline Ringer's results in a decrease in the mean fluorescence across the entire image. Conversely, bathing in acidic Ringer's results in an increase in the fluorescence. Mean fluorescence increased 5.6% with a decrease of 1 pH unit and decreased 7.1% with an increase of 1 pH unit. To exclude the possibility that changes in neuronal activity and/or excitability contributed to the optical signals, these experiments were repeated after blocking voltage-gated Na^+ and Ca^{++} channels, as well as AMPA receptors. The fluorescence changes were comparable, showing that the optical signals are primarily due to the manipulation of the pH.

A large number of experimental findings support the hypothesis that the pH dependent changes in fluorescence are primarily intracellular in origin. First, superfusion with Ringer's containing 20 mM sodium propionate produces an increase in fluorescence (Figure 11.1A). This increase in fluorescence is due to the uncharged form of the weak acid propionate acidifying the cytoplasm by carrying H^+ into cells.[27–29] Second, replacing normal Ringer's solution with a nominally CO_2-free Ringer's solution results in an abrupt decrease in fluorescence (Figure 11.1B). This decrease in fluorescence is consistent with a rapid efflux of CO_2 from the cytoplasm, equivalent to a net extrusion of protons.[30,31] The increase in intracellular pH results in a decrease in neutral red fluorescence. Third, bathing neurons in NH_4Cl produces an initial, brief intracellular alkalosis due to entry of NH_3 and is followed by a much longer intracellular acidification due to the passive influx of NH_4^+ and subsequent efflux of NH_3, leaving H^+ behind when the NH_4Cl is washed out.[28,29] Short duration perfusion with NH_4Cl results in the expected longer duration increase in fluorescence (Figure 11.1C). The recovery from this intracellular acid load can be

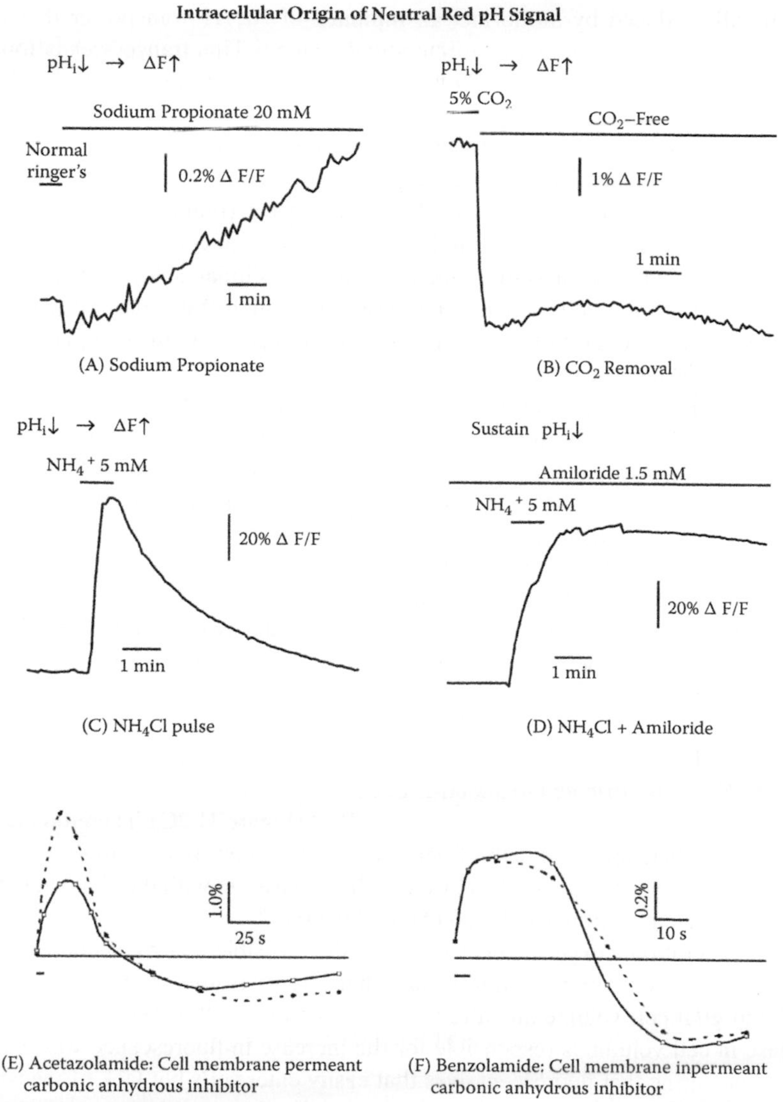

FIGURE 11.1 Neutral red fluorescence changes are due to intracellular pH shifts. A–F illustrate the changes in fluorescence in the cerebellar cortex as a function of addition of various drugs to the bath. (A) Sodium propionate (20 mM) produces intracellular acidification and an increase in fluorescence. (B) CO_2-free Ringer's evokes an intracellular alkalosis and a decrease in fluorescence. (C,D) A pulse of NH_4Cl (5 mM) produces an intracellular acidification that recovers over time and amiloride (1.5 mM) blocks the recovery with matching changes in fluorescence. (E,F) Parallel fiber stimulation denoted by the bar on the x-axis evokes a biphasic $\Delta F/F$ in the cerebellar cortex (solid line) and the evoked response increased after blocking carbonic anhydrous with the cell permeant inhibitor, acetazolamide (dotted line). The cell impermeant inhibitor, benzolamide, had no effect on the evoked response. Parallel fiber stimulation parameters were 200 μA, 150 μs, 20 Hz for 5 s (bar beneath x-axis). (From Chen, G. et al., *Neurosci.* 84: 645–668, 1998. With permission.)

substantially reduced by amiloride, an inhibitor of Na^+/H^+ transporter that facilitates the removal of protons to the interstitial space.[32] This transporter is found in Purkinje neurons and granule cells.[31,33,34] Accordingly, recovery from the fluorescence increase that follows a pulse of NH_4Cl was completely blocked by the addition of amiloride (Figure 11.1D). Finally, the cell permeant carbonic anhydrase inhibitor, acetazolamide, enhances the optical response to parallel fiber stimulation but the cell impermeant inhibitor, benzolamide, has no effect (Figure 11.1E,F). Blocking carbonic anhydrase intracellularly reduces the capacity of cells to buffer the excess protons generated,[9] and a similar increase in intracellular acidification has been observed in Purkinje cells.[35] These observations demonstrate that the neutral red signals in the cerebellar cortex reflect primarily intracellular shifts in pH, in agreement with LaManna's earlier observations.[21]

The next question we addressed is whether the changes in fluorescence arise from neurons or glia. Elevated extracellular potassium produces an intracellular acidosis in neurons and an extracellular alkalosis. The neuronal acidification due to depolarization involves a number of mechanisms, including production of metabolic acid, mitochondrial Ca^{++}/H^+ exchange, glutamate-mediated H^+ influx, and Ca^{++}-related acidification in Purkinje cell dendrites.[8,35–39] In astrocytes the depolarization activates an electrogenic Na^+-HCO_3^- cotransport that produces an alkaline shift.[40–42] The superfusion of the cerebellar cortex with Ringer's solution with 8 mM KCl evokes a rapid and dramatic increase in neutral red fluorescence (Figure 11.2A). The increase in fluorescence due to KCl is converted to a decrease in fluorescence when neuronal activity is blocked with a cocktail including glutamate receptor antagonists, blockers of transmitter release and voltage-gated Na^+ channels (Figure 11.2B). The decrease in fluorescence on depolarization with KCl was attributed to alkalosis of cerebellar glia and confirmed by blocking the glial depolarization evoked decrease with Ba^{++} (Figure 11.2C). Therefore, neurons and glia contribute optical signals of opposite sign to depolarizing stimuli (increase in fluorescence for neurons and a decrease for glia), consistent with the characteristics of activity dependent neuronal and glial pH modulation.[8,9]

Several additional experiments ruled out other possible factors as contributing to the signal observed with neutral red. Neuronal activity results in an increase in neuronal and glial cell volume and a reduction in the extracellular space.[43,44] That this increase in cell volume is responsible for the increase in fluorescence was excluded by demonstrating that fluorescent dyes that easily enter cells but lack pH sensitivity, show no measurable optical responses. Additionally, furosemide, an anion transport inhibitor that blocks extracellular space and cellular volume changes,[44,45] does not affect the optical responses.[46] Also, parallel fiber stimulation in the unstained cerebellar cortex does not produce detectable epifluorescence optical signals.

Finally, we demonstrated that the neutral red signal is neither the intrinsic hemodynamic signal [47,48] nor due to flavoprotein autofluorescence.[6,49] For differentiation from hemodynamic signals we showed that the reflectance signal at the wavelengths used to monitor neutral red epifluorescence is undetectable in the unstained cerebellar cortex.[4] In response to parallel fiber stimulation, the largest hemodynamic reflectance signal is on the order of 0.5% $\Delta R/R$, almost an order of magnitude smaller than the epifluorescence change obtained with neutral red. Finally, the neutral red signal consists of an initial increase in fluorescence, not a decrease, as expected for the intrinsic hemodynamic

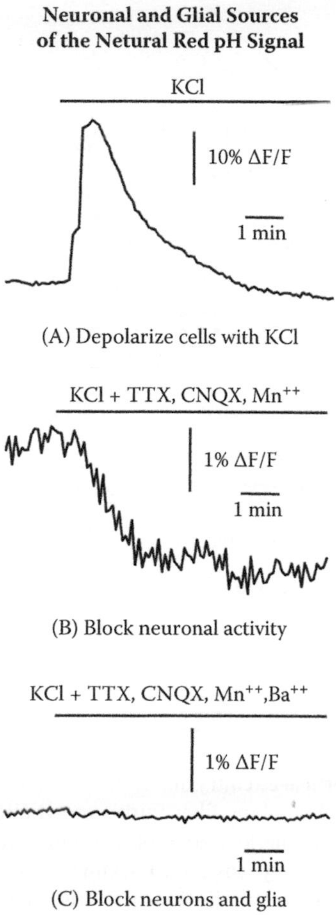

FIGURE 11.2 Neurons and glial cells respond differentially to depolarization. (A) Superfusion of the cerebellar cortex with 8 mM KCl evokes a large increase in fluorescence. (B) The addition of TTX (10 μM), CNQX (10 μM) and $MnCl_2$ (6 mM) to the Ringer's solution results in a decrease in fluorescence on superfusion with KCl. (C) The addition of $BaCl_2$ (4 mM) to the Ringer's solution blocked the decrease in fluorescence. (From Chen, G. et al., *Neurosci.* 84: 645–668, 1998. With permission.)

optical signal. To differentiate between neutral red and flavoprotein autofluorescence, we demonstrated there is no flavoprotein signal in the cerebellar cortex at the excitation and emission wavelengths used to monitor neutral red.[6] Therefore, activity-dependent signals generated by neutral red primarily reflect intracellular pH shifts.

11.5 PROPERTIES OF NEUTRAL RED IMAGING REVEALED BY PARALLEL FIBER STIMULATION

One of the intriguing aspects of the cerebellar cortex is its highly ordered and stereotypic circuitry and functional architectures (Figure 11.3). Optical imaging provides a

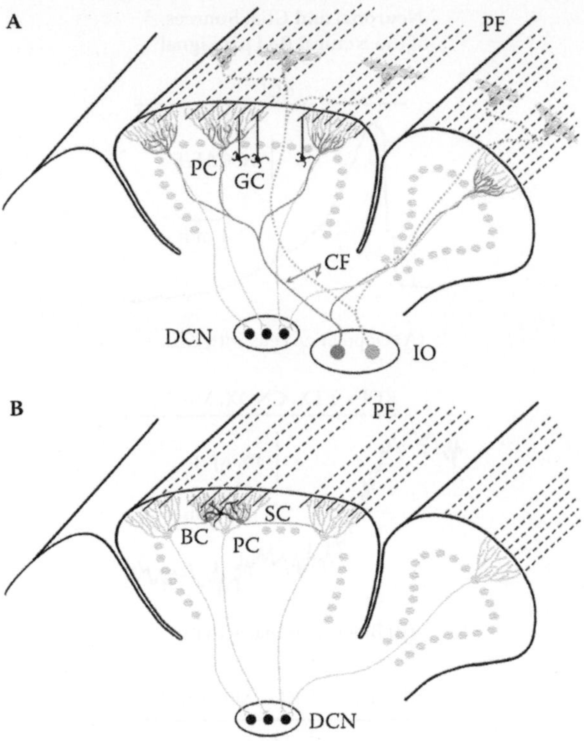

FIGURE 11.3 Cerebellar cortical circuitry and functional architectures. (A) Diagram of the parallel fiber-Purkinje cell and climbing fiber-Purkinje cell circuitry. Parallel fibers, which extend medial and lateral along the long axis of a folium, are the branched axons of the granule cells and make excitatory synapses on Purkinje cells. Also shown is the parasagittal organization of the climbing fiber projections from the inferior olive to Purkinje cells in parasagittal zones. (B) Molecular layer interneurons receive excitatory input from parallel fibers and in turn synapse on and inhibit Purkinje cells. PC = Purkinje cell, PF = parallel fiber, GC = granule cell, CF = climbing fiber, BC = basket cell, SC = stellate cell, DCN = deep cerebellar nuclei, IO = inferior olive. (From Ebner, T.J. et al., *Neuroscientist* 9: 37–45, 2003. (With permission.)

power tool to study these well-defined circuits and architectures. The parallel fibers define the first of these architectures. Parallel fibers are the molecular layer extension of granule cell axons and project medial and lateral along the long axis of a folium for several millimeters.[50,51] This massive system of fibers forms excitatory glutamatergic synapses on the dendrites of Purkinje cells and cerebellar interneurons (Figure 11.3A,B). Activating parallel fibers using a brief stimulus train evokes an optical response consisting of a longitudinal beam running parallel to the long axis of the folium.[4,5,46,52] The optical signal consists of an initial increase in fluorescence (acidic shift) that returns to baseline in approximately 60 s, followed by a beam of decreased fluorescence (alkaline shift) for up to 120 s. A major advantage of the neutral red optical signal is that its amplitude is 5–20 times larger (1–5% $\Delta F/F$) than the epifluorescence signals obtained from most voltage sensitive dyes (0.1–0.5% $\Delta F/F$)[3,53,54] or the reflectance change from the hemodynamic intrinsic signal (0.1–0.5% ΔR/R).[4,47,48] The optical signal obtained with neutral red in the cerebellar cortex is equal to or larger than the recently described flavoprotein autofluorescence signal or *in vivo* Ca^{++} imaging (Figure 11.4A).[6,7] The larger signal-to-noise ratio eliminates the need for averaging the responses to multiple stimulations. Clearly, a robust signal is a great advantage,

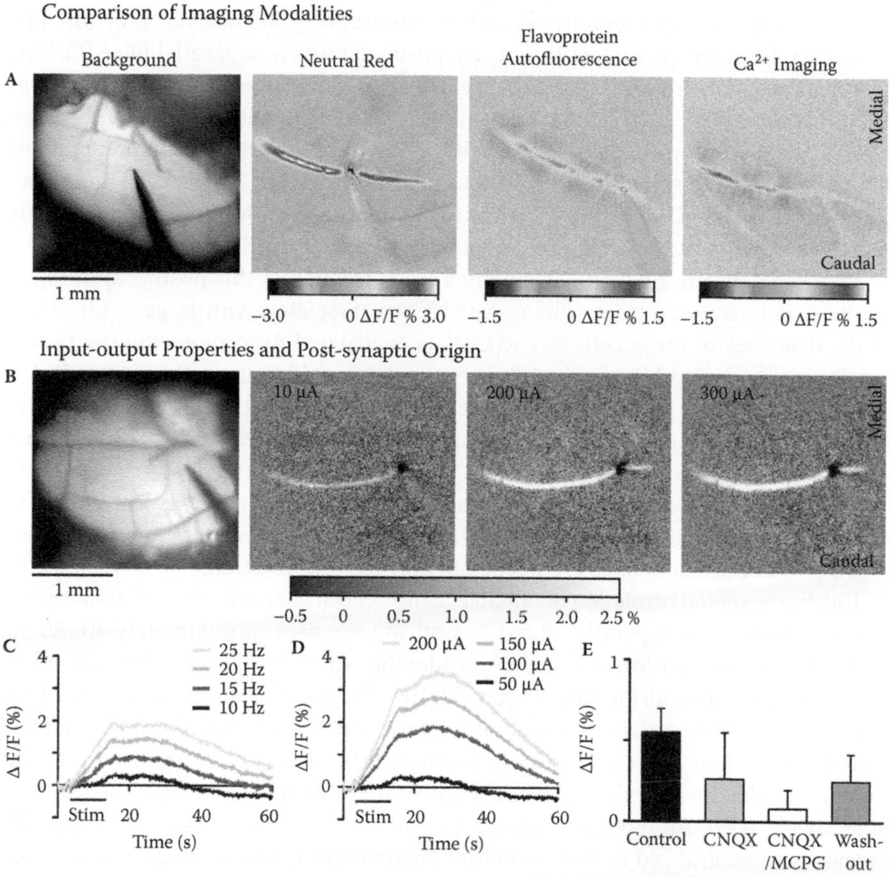

FIGURE 11.4 **(See color insert following page 224)** Properties of responses to parallel fiber stimulation. (A) Responses evoked by parallel fiber stimulation based on neutral red, flavoprotein autofluorescence and Ca^{++} imaging. (Figure 4a, panels 3,4 from Gao, W. et al., *J. Neurosci.* 26: 8377–8387, 2006. With permission.) (B) Optical responses evoked by surface stimulation of varying intensities. Note the beam width at 100 μA stimulation amplitude is still very well-defined. (Figure 4b from Ebner, T.J. et al., *Prog. Brain Research* 148: 125–138, 2005. With permission.) (C) Changes in optical response (ΔF/F) to parallel stimulation as a function of stimulation frequency (10 s train at 50 μA). The black bar beneath the x axis denotes the stimulation period. (D) Changes in optical response as a function of stimulation intensity (10 s train at 10 Hz). (E) Response to parallel fiber stimulation was reduced by AMPA and mGluR antagonists, CNQX (50 μM) and MCPG (1 mM), respectively, demonstrating that the signal is primarily post-synaptic in origin. (Figure 4c,d,e from Dunbar, R.L. et al., *Neuroscience* 126: 213–227, 2004. With permission.)

particularly in the intact CNS in which there are numerous sources of noise, including cardiovascular and respiratory motion artifacts.

The neutral red optical signal has several additional important characteristics. First, it is monotonically and linearly related to the stimulus parameters over a wide range of amplitudes and frequencies (Figure 11.4B–D). Second, the optical signal originates primarily from the postsynaptic elements (i.e., Purkinje cells and interneurons) and involves both iontotropic and metabotropic glutamate receptor (mGluR) activation. Both receptor types are present postsynaptically at parallel fiber-Purkinje cell synapses and provide for the vast majority of the Purkinje cells response to parallel fiber inputs.[55–57] Application of the non-NMDA glutamate receptor antagonist, CNQX, decreases the optical signals by 50–60%, and the mGluR receptor antagonist MCPG reduces the signal by an additional 15–25% (Figure 11.4E).[5,46,58] Therefore, approximately 80–85% of the increase in fluorescence is due to activation of the postsynaptic targets of the parallel fibers.

Purkinje cells are likely to contribute a large fraction of this postsynaptic signal as the depolarization opens voltage-gated Ca^{++} associated with large acidic shifts in the dendrites of these cells.[35] The Ca^{++} is exchanged for protons via the plasma membrane calcium ATPase, which is expressed at high density in Purkinje cell dendrites.[59] Neutral red preferentially stains Purkinje cells.[20] What remains unresolved is the exact contribution of cerebellar interneurons, including stellate, basket, and Golgi cells, all of which have dendrites in the molecular layer that are excited by parallel fibers.[50,51] Cerebellar interneurons were shown to be only weakly stained by neutral red,[20] suggesting these cells may make only a small contribution to the optical response. Further studies are needed to address this question.

Lastly, the spatial resolution obtainable using neutral red is excellent. Stimulation at low amplitudes evokes parallel fiber-like optical responses approximately 40–80 μm in width (anterior–posterior extent), considerably smaller than the ~300 μm width of a Purkinje cell dendritic tree (Figure 11.4B). The anterior–posterior extent of the optical response to surface stimulation is more constrained than the field potential recordings.[4,26] This is not unexpected, given that the neutral red response is primarily intracellular, and the field potentials are volume conducted through the extracellular space. Also, the optical response to parallel fiber stimulation is as spatially precise with neutral red as the responses obtained using other imaging techniques (Figure 11.4A and B). Therefore, the neutral red optical response is precise enough to monitor the activation of only a region of the dendritic tree of a Purkinje cell.

The above comments refer to the initial phase of increased fluorescence evoked by parallel fiber stimulation. What is the source of the subsequent decrease in fluorescence (Figure 11.1E,F)? One possibility is the decrease is due to an alkaline shift in neurons following the initial acidic shift. This would require that neuronal pH increase below resting levels following depolarization. The decreased fluorescence is most likely due to a depolarization-induced alkalosis occurring in cerebellar astrocytes.[40,60,61] Our studies did not tackle this important problem. One potential approach to examining this question would be to selectively block the Na^+-HCO_3^- cotransporters or mGluRs on astrocytes.

One limitation of neutral red imaging is the apparent inability to map postsynaptic inhibition. Stimulation of the parallel fibers activates molecular layer interneurons (stellate and basket cells) that in turn produce a powerful GABAergic inhibition of both Purkinje cells and other molecular layer interneurons as diagramed in Figure 11.3B.[62–64] The inhibition occurs within the beam of activated parallel fibers (on-beam inhibition) and lateral to the beam (off-beam inhibition). Both on- and off-beam inhibition is mediated via $GABA_A$ receptors.[64,65] On-beam inhibition can be detected indirectly as $GABA_A$ antagonists increase the amplitude and width of the activated beam.[5,46] However, there is no obvious signal lateral to the beam suggestive of off-beam inhibition (Figure 11.4). Recently, we demonstrated using flavoprotein and Ca^{++} imaging that off-beam inhibition can be detected and it is organized in parasagittal bands.[7,66] However, the failure to observe off-beam inhibition with neutral red is not unanticipated. The opening of $GABA_A$ receptors results in an intracellular acidification due to a HCO_3^- efflux, a finding that has been documented in a variety of preparations.[67,68] Therefore, the shifts in pH resulting from postsynaptic $GABA_A$ receptor mediated inhibition would not be distinguishable from postsynaptic glutamatergic excitation. Also, the pH effects are likely to be smaller and more difficult to detect.

The other major limitation of any form of pH imaging in the CNS is the relatively slow time course relative to electrophysiological time scales (Figure 11.1E). The pH changes, and therefore the fluorescence changes, occur over seconds and not milliseconds. As a result, neutral red imaging will be best for mapping spatial patterns of activity or processes with slower time courses, as detailed in the following sections. However, the latency of the fluorescence increase with neuronal activation is excellent, with changes detected in less than 100 ms. As a method for monitoring pH changes in the CNS, the response time is considerably faster than the 1–2 s response time of even the fastest pH microelectrodes.[9]

11.6 EXAMPLES OF USING NEUTRAL RED TO STUDY NORMAL AND ABNORMAL PROCESSES

11.6.1 Parasagittal Organization of the Cerebellar Cortex

In addition to the medial-lateral organization of the parallel fibers, another major functional architecture of the cerebellum is the parasagittal zonation of its afferent and efferent projections and molecular compartmentalization. Climbing fibers arise solely from the inferior olive, synapse monosynaptically on Purkinje cells and produce low frequency, all-or-none complex spikes.[51,69] The climbing fiber projection from the inferior olive makes excitatory, glutamatergic synapses on Purkinje cells in parasagittal zones in the cerebellar cortex (Figure 11.3A).[70] The output projections of the Purkinje cells to the deep cerebellar nuclei are also organized into parasagittal zones.[71] Further, numerous markers reveal a parasagittal molecular compartmentalization of Purkinje cells.[72,73] How this parasagittal architecture functions is a major question in cerebellar research.

We have used neutral red imaging to map the spatial patterns of activation in the rat and mouse cerebellar cortex evoked by peripheral stimulation.[19,26,74] Electrical stimulation of the vibrissae area of the ipsilateral face in the rat evokes optical responses in Crus I and II consisting of parasagittal bands (Figure 11.5A). The bands

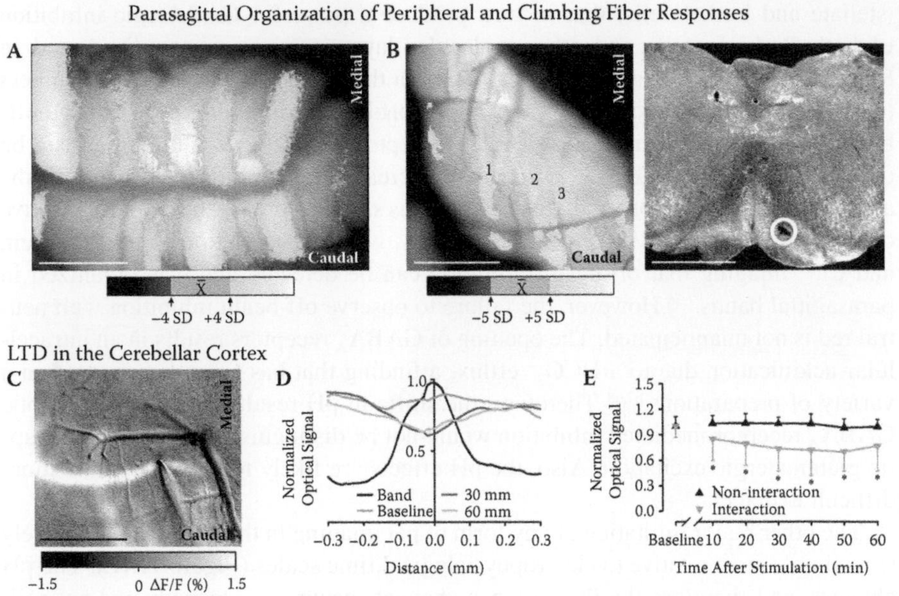

FIGURE 11.5 **(See color insert following page 224)** (A) Ipsilateral face stimulation evokes parasagittal bands in Crus I and II. Activation map shows statistically significant optical signals (> 4 SD above or below control) pseudocolored and superimposed onto a gray scale image of the folia. Color bar is calibrated in standard deviations relative to noise level of control image. Stimulation parameters were 500 μs pulses, 20 V, 5 Hz for 20 s. (From Chen, G. et al., *J. Neurophsiol.* 76: 4169–4174, 1996. With permission.) (B) Stimulation in the contralateral inferior olive also evokes parasagittal bands shown as a similar thresholded activation map (>5 SD above or below control). Stimulation parameters were 150 μA, 10 Hz, 10 s, 100 μs pulse width. On the right is an image of a section through the brainstem (right) showing an electrolytic lesion made using the stimulating electrode (white circle). (From Dunbar, R.L. et al., *Neuroscience* 126: 213–227, 2004. With permission.) (C) Intersection of the two responses evoked by conjunctive of parallel fibers and the contralateral inferior olive, used to induce LTD in the cerebellar cortex. (D) Average intensity profiles along the beam during baseline, 30 and 60 min after conjunctive stimulation. The intensity profiles were averaged after being centered on the band evoked by inferior olive stimulation. Intensity profile of the response to inferior olive stimulation shown in black. After conjunctive stimulation, the optical signal in the interaction region was reduced throughout the 60 min observation period, with the maximal decrease occurring at the peak of the inferior olive-evoked band. (E) Time course of the averaged, normalized optical signal (mean + SD) in the non-interaction and interaction regions compared with baseline (* indicates $P < 0.05$, Duncan's test, post hoc). Scale bar in A equals 500 μm and in B and C equals 1 mm. (Figures 11.5D,E from Gao, W. et al., *J. Neurosci.* 23: 1859–1866, 2003. With permission.)

are 100–500 μm in width, elongated in the anterior–posterior direction, commonly extend across at least two folia, and vary in number from 1–7. The optical responses evoked by peripheral stimulation are due to activation of postsynaptic elements, as they are essentially blocked by CNQX. At the mossy fiber-granule cell, parallel fiber-Purkinje cell and climbing fiber-Purkinje cells synapses AMPA receptors are major contributors to the post-synaptic responses.[57,75,76]

We hypothesized that these parasagittal responses are due to activation of the inferior olive and the climbing fiber projections to Purkinje cells. A number of observations are in agreement with this hypothesis. The optical bands evoked by peripheral stimulation are in register with the parasagittal compartmentalization of the olivocerebellar projection and Purkinje cells as delineated by immunostaining with anti zebrin II.[19,72,73] The parasagittal bands show a preferred frequency of face stimulation (6–8 Hz), consistent with the inherent rhythmicity of inferior olivary neurons and complex spike responses in Crus I and Crus II.[77–79] Contralateral inferior olivary stimulation evokes parasagittal bands with nearly identical spatial and frequency tuning characteristics to those evoked by peripheral stimulation (Figure 11.5B).[5,19,26,52] Finally, lidocaine injection into the inferior olive blocks the parasagittal bands evoked from the periphery.[26] These findings demonstrate that peripheral inputs activate climbing fibers to well-defined parasagittal zones in the rat cerebellar cortex. These observations support the hypothesis that the inferior olivary neurons are dynamically coupled, and that the frequency content of the stimulus is critical to engaging the inferior olive.

One unanswered question is why the peripheral stimulation evokes climbing fiber responses with little evidence of a contribution from mossy fiber activation of the granule cell-parallel fiber-Purkinje cell circuitry. Most likely the answer involves the powerful, all-or-none action of climbing fibers on Purkinje cells. The massive and long duration depolarization of Purkinje cells associated with the complex spike should result in large acidic shifts in Purkinje cell dendrites.[35,80] The mossy, fiber-granule, cell-Purkinje cell pathway has a smaller effect on Purkinje cell depolarization and simple spike firing,[62,81] and would likely generate smaller pH shifts. Potentially, pH indicators with greater sensitivity could be used to monitor the responses to mossy fiber inputs.

11.6.2 Optical Imaging of Plasticity at the Parallel Fiber-Purkinje Cell Synapse

We have also used neutral red imaging to examine the plasticity of the parallel fiber-Purkinje cell synapse *in vivo*. Conjunctive stimulation of climbing fiber and parallel fiber inputs onto cerebellar Purkinje cells results in long-term depression (LTD) at parallel fiber-Purkinje cell synapses.[82] Parallel fiber-Purkinje cell LTD has been hypothesized to play a major role in cerebellar motor learning.[83,84] The intracellular signaling mechanisms have been extensively studied *in vitro* and include the activation of group 1 metabotropic glutamate receptors (mGluR1) and protein kinase C (PKC) in Purkinje cells leading to internalization of AMPA receptors.[85,85–89] While parallel fiber-Purkinje cell LTD has been shown to exist in decerebrate animals,[82,90–92] there have been few studies in the intact animal and no characterization of the intracellular mechanisms or spatial aspects of LTD *in vivo*.

Using neutral red imaging, we optically mapped and characterized LTD of the parallel fiber-Purkinje cell synapse evoked by conjunctive stimulation of parallel fibers and the contralateral inferior olive.[52] The properties of parallel fiber-Purkinje cell LTD predict that the depression will be spatially specific and occur at the intersection of the evoked parallel fiber and climbing fiber responses,[93,94] as has been demonstrated *in vitro*.[95] At the site of the intersection of the parallel fiber-evoked beam and the climbing fiber-induced parasagittal band (Figure 11.5C), the subsequent responses to parallel fiber stimulation are depressed for at least 60 minutes (Figure 11.5E). The depression is spatially specific, confined to the region at which the parallel fiber and climbing fiber responses overlap (Figure 11.5D). Guided by the optical imaging, electrophysiological monitoring of the postsynaptic response to parallel fiber stimulation confirmed that the LTD occurred only at the intersection of the activated parallel fibers and climbing fibers. The generation of the LTD requires the conjunction of both inputs. Also, in agreement with the in vitro results, blocking mGluRs prevents the induction of LTD. Finally, we took advantage of a transgenic mouse that expresses an inhibitor of PKC selectively in Purkinje cells.[96] Parallel fiber-Purkinje cell LTD is abolished in this mouse as is vestibular-ocular reflex adaptation and learning-dependent timing of conditioned eyeblink responses.[96,97] As predicted, the optical counterpart of parallel fiber-Purkinje cell LTD is not present in this transgenic mouse. Conversely, LTD is preserved in the $mGluR_4$ knock-out mouse that is known to have intact parallel fiber-Purkinje cell LTD.[98] Therefore, neutral red imaging provided the first visualization of parallel fiber-Purkinje cell LTD *in vivo*, demonstrated its spatial specificity, and its dependence on $mGluR_1$ and PKC.

11.6.3 Spreading Acidification and Depression

A distinct advantage of optical imaging is the ability to study the spatial and temporal proprieties of cellular activity throughout a region of the brain. In addition to visualizing the activity of populations of cells or neuronal circuits, optical imaging methodologies have proven well-suited for characterizing the propagation of waves of activity in the CNS. For example, classical spreading depression of Leao[99] and its variants have been extensively studied with intrinsic signal optical imaging.[100–102] The discovery and investigation of calcium waves has relied on calcium imaging.[103,104] Spreading depolarization in the embryonic brain has been characterized using voltage sensitive dyes.[105] In this tradition, optical imaging with neutral red allowed us to discover and characterize a novel form of propagated activity in the cerebellar cortex *in vivo*.[46,106]

Surface stimulation of the cerebellar cortex normally evokes a highly constrained "beam" of activity, due to the activation of the parallel fibers and their postsynaptic targets. If the surface stimulation is sufficiently intense, the evoked beam of increase in fluorescence propagates anteriorly and posteriorly beyond the beam of activated parallel fibers and Purkinje cells (Figure 11.6A).[46,49,106] Essentially, this is an all-or-none event that is initiated when a threshold is reached. The increase in fluorescence is dramatic and attains levels as high as 30% above baseline. The propagation and increase in fluorescence continues for 1–2 minutes beyond the duration of the surface stimulation. The spread is initiated nearly simultaneously along the beam of activated parallel fibers and travels orthogonally to the beam. The average propagation speed is ~500 um/s with peak speeds up to 2000 μm/s. The wave of increased fluorescence spreads for considerable distances, traveling along the sulci to reach neighboring folia (Figure 11.6B).

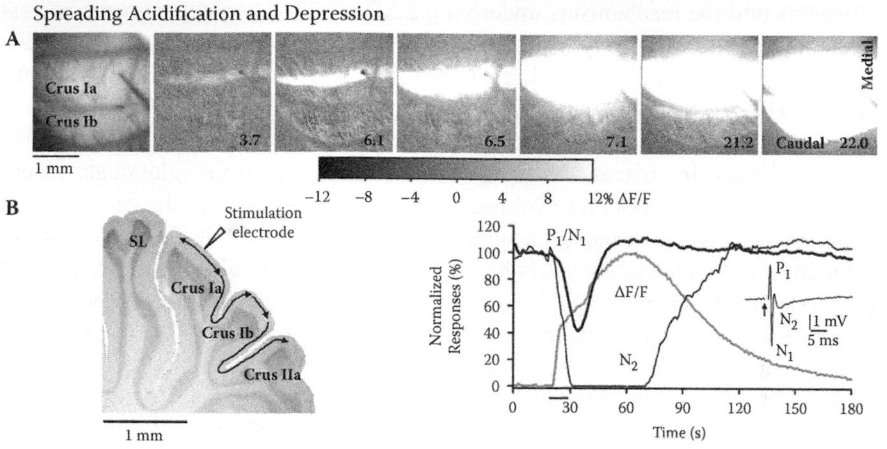

FIGURE 11.6 Spreading acidification and depression. (A) Series of optical images (stimulation minus background) illustrating the propagation of the optical response initiated by surface stimulation (150 μA, 150 μs pulses at 10 Hz for 10 s). The first image is a background image showing the position of the stimulating electrode. Subsequent images, marked with the time in seconds, are relative to the onset of stimulation. Various time intervals are shown to highlight different aspects of the propagation. (From Chen, G. et al., *J. Neurosci.* 21:9877–9887, 2001. With permission.) (B) Parasagittal section of the cerebellar cortex obtained after the imaging study showing the assumed pathway for propagation. (From Ebner, T.J. et al., *Prog. Brain Research* 148: 125–138, 2005. With permission.) (C) Time course of the normalized optical response (ΔF/F, grey) and the field potentials during SAD evoked by parallel fiber stimulation (bar beneath the x-axis, 10 Hz for 10 s). Inset shows an example of the field potentials. P_1/N_1 is the parallel fiber volley (thick black line) and N_2 is the postsynaptic response (thin black line). SAD is associated with a dramatic loss of cerebellar cortical excitability that recovers over several minutes. (From Chen, G. et al., *J. Neurophysiol.* 94: 1287–1298, 2005. With permission.)

The propagation of the increased fluorescence is accompanied by a transient but powerful depression of the cerebellar cortical circuitry (Figure 11.6C). Both the presynaptic and postsynaptic field potentials evoked by parallel fiber stimulation can be completely suppressed as the wave of increased fluorescence passes over a folium.[46,106] The presynaptic component recovers to normal levels within 30–60 s; however, recovery of the postsynaptic component takes considerably longer (60–300 s). Therefore, we named this propagation event *spreading acidification and depression* (SAD). The mechanisms underlying the suppression of cerebellar cortical circuitry during SAD are not known. The large pH shifts during SAD may directly decrease neuronal excitability, possibly by depressing voltage-gated sodium channels[107] and/or attenuating glutamate channel conductance.[108] A large number of factors differentiate SAD from classical spreading depression.[46,106] SAD can also be imaged with voltage sensitive dyes and flavoprotein autofluorescence, showing that large changes in depolarization and oxidative metabolism are occurring and potentially contribute to the decrease in excitability.[49]

Insights into the mechanisms underlying SAD are based on experiments assessing the contributions of presynaptic and postsynaptic elements of the cerebellar circuit. An increase in the excitability of the cerebellar circuits is needed to evoke SAD.[46,49] The greater the stimulation intensity, the greater the probability of evoking SAD. Both TTX and Ca^{++}-free Ringer's completely prevent evoking SAD by parallel fiber stimulation. In contrast, blocking AMPA and metabotropic glutamate receptors increases the threshold for evoking SAD but does not prevent the generation of SAD.[46] Increasing the stimulation intensity invariably results in SAD, even when combinations of glutamate receptor blockers are applied. We also demonstrated that the threshold for generating SAD is greatly lowered by blocking Kv1.1 potassium channels that play a major role in controlling neuronal excitability.[49] A single stimulus to the parallel fibers can evoke SAD in the presence of a Kv1.1 blocker, and spontaneous SAD is observed. Therefore, the activation of parallel fibers and the release of glutamate are essential for evoking SAD. Conversely, activation of Purkinje cells or molecular layer interneurons is not essential.

That the direction of SAD propagation is perpendicular to the parallel fibers raises the possibility that the molecular layer interneurons, with their axons extruding perpendicular to the parallel fibers, provide the underlying neuronal substrate (Figure 11.3B). However, cerebellar cortical interneurons are not critical elements of SAD as the threshold for SAD is actually decreased by $GABA_A$ blockers[46] and SAD occurs in mice lacking stellate cells.[109] Nor is the gaseous neurotransmitter, nitric oxide, critical, as SAD can be evoked in nitric oxide synthase deficient mice or in the presence of blockers of nitric oxide synthase.

Therefore, the mechanism underlying the generation and propagation of SAD remains unresolved. Our working hypothesis is that SAD is due to the unusual release of glutamate from a bundle of activated parallel fibers. In turn, the glutamate diffuses to and activates neighboring parallel fibers. Preliminary simulations show that this parallel fiber model accounts for the high speed, geometry and threshold properties of SAD.[110] We also hypothesized that SAD is primarily a pathophysiological process that can temporarily disrupt cerebellar function. One possibility is that a SAD-like process occurs in the Kv1.1 channelopathy, episodic ataxia type 1 (EA1).[111] EA1 is due to a mutation in the gene encoding for the Kv1.1 α-subunit[112,113] and EA1 patients have transient attacks of cerebellar dysfunction.[114,115] We have argued that the properties of SAD are consistent with the molecular and clinical features of EA1.[49,111]

11.7 DISCUSSION AND FUTURE DIRECTIONS

Improvements in pH indicators for imaging *in vivo* would be welcomed. As described in this chapter, the activity-dependent optical signals obtained using neutral red have an excellent signal-to-noise ratio and little toxicity. However, pH indicators that further increase the signal size will allow assessment of smaller perturbations in pH. Also, indicators with greater signal-to-noise ratios are needed for imaging in awake, behaving animals. Additionally, it would be useful to have a spectrum of pH dyes that allow imaging of different compartments, particularly dyes that sequester exclusively in the intracellular versus extracellular space. Indicators that are sensitive to different pH ranges and have different excitation/emission properties would allow monitoring multiple parameters in the same experiments (e.g., pH and Ca^{++}, pH and flavoprotein fluorescence, etc). There are a host of questions concerning the interactions between neurons and glia, cerebral metabolism, and the regulation of intracellular and extracellular pH that would benefit from improved pH indicators.

Ideally, the next generation of pH dyes would provide for cell specific imaging. In the cerebellum, having the capability to image specific populations of neurons, such as Purkinje cells or granule cells, or simply differentiate between neurons and glial cells, would constitute a major advance in optical imaging *in vivo.* Cell-specific imaging offers the chance of solving long-standing problems in cerebellar cortical physiology. For example, the controversy about whether parallel fibers are activated in "beam-like" patterns in response to natural inputs[116] could potentially be answered. A series of pH sensitive, GFP-based genetically encodable fluorescent probes have been developed,[117,118] and mice have been generated on a pH-sensitive variant of enhanced yellow-fluorescent protein.[119] An important caveat is the need to distinguish between the fluorescence changes due to intrinsic optical signals, such as flavoprotein autofluorescence, and the hemodynamic intrinsic signal.[6,47] In Section 11.1.4, Nature and Origin of the Neutral Red Fluorescence Signal, we discussed several approaches to differentiating these different optical signals. Also promising is the use of viral-vector techniques for gene transfer.[120,121] The viral-vector approach has the advantages of requiring less effort, time, and expense than generating transgenic animals, yet having the potential to deliver intrinsic pH sensors to specific cell types. Furthermore, the viral vector approach is not limited to a few species in which transgenic animals are available, but to a much wider range of species including nonhuman primates. Approaches based on the power of genetic manipulation will undoubtedly have a major impact on the optical imaging of activity dependent signals in the future, including pH imaging.

Concerning answering neurobiological questions, we would suggest using neutral red imaging (or other pH imaging modalities) in three areas. First, given its large signal size, limited toxicity, and ease of use, neutral red is an excellent tool for mapping neuronal patterns of activation in CNS circuits/structures. While our laboratory has been focused on the cerebellar cortex, regions of the nervous system in which spatial patterning is hypothesized to play a role in the information processing are also candidates for study. In addition, it may be possible to use activity-dependent pH signals to map neuronal activity in the awake mouse. Neutral-red- injected intraperitoneally readily crosses the blood–brain barrier and stains the cerebellum. With the transparency of the cranial bone in the mouse,[122] it may be possible to map neuronal activation in the awake, behaving mouse.

The second set of scientific questions are those of pH regulation in the CNS, how activity modulates neuronal and glial cells, and how shifting in pH can alter neuronal function.[9] The need to study pH-related changes in the intact nervous system is obvious and pH imaging *in vivo* will be a powerful tool.

Finally, the third area in which pH imaging can play an important role is studying abnormal and pathophysiological processes in the CNS. Acidosis plays a central role in neuronal damage during ischemic stroke through several mechanisms including nonspecific denaturation of proteins and nucleic acids,[123,124] triggering of cell swelling via Na^+/H^+ and Cl^-/HCO_3^- exchangers,[125] inhibition of mitochondrial energy metabolism,[12] stimulation of pathologic free radical formation,[126] and potentiation of glutamate receptor-medicated excitotoxicity.[127] Recent findings that acidosis activates a distinct family of membrane ion channels, the acid-sensing ion channels (ASICs) in both peripheral and central nervous systems has changed the view of acidosis-associated signaling in the brain.[128,129] Activation of these cation channels by protons plays an important role in a variety of physiological as well as acidosis-mediated neuronal injury. The acidosis due to brain ischemia causes an increase in glycolysis in both glia and neurons, leading to an increase in lactic acid production and a decrease of pH. This acidosis activates ASICs, resulting in Ca^{++} influx and subsequent neuronal damage. Similarly, epilepsy causes local acidosis due to increased energy demand and increased glycolysis.[13] These and other pathological processes are strong candidates for the use of pH imaging. While there were some early attempts using umbelliferone (7-hydroxycoumarin) as a fluorescence indicator of brain pH in cerebral ischemia,[130,131] clearly a renewed effort using modern optical imaging techniques is warranted.

REFERENCES

1. Elias, S. A. et al. Optical imaging of parallel fiber activation in the rat cerebellar cortex: Spatial effects of excitatory amino acids. *Neuroscience* 52, 771–786 (1993).
2. Kim, J. H. et al. Imaging of cerebellar surface activation in vivo using voltage sensitive dyes. *Neuroscience* 31, 613–623 (1989).
3. Ebner, T. J. and Chen, G. Use of voltage-sensitive dyes and optical recordings in the central nervous system. *Prog. Neurobiol.* 46, 463–506 (1995).
4. Chen, G. et al. Optical responses evoked by cerebellar surface stimulation *in vivo* using neutral red. *Neuroscience* 84, 645–668 (1998).
5. Dunbar, R. L. et al. Imaging parallel fiber and climbing fiber responses and their short-term interactions in the mouse cerebellar cortex in vivo. *Neuroscience* 126, 213–227 (2004).
6. Reinert, K. C. et al. Flavoprotein autofluorescence imaging of neuronal activation in the cerebellar cortex *in vivo*. *J. Neurophysiol.* 92, 199–211 (2004).
7. Gao, W. et al. Cerebellar cortical molecular layer inhibition is organized in parasagittal zones. *J. Neurosci.* 26, 8377–8387 (2006).
8. Chesler, M. and Kaila, K. Modulation of pH by neuronal activity. *Trends Neurosci.* 15, 396–402 (1992).
9. Chesler, M. Regulation and modulation of pH in the brain. *Physiol. Rev.* 83, 1183–1221 (2003).
10. Moody, W., Jr. Effects of intracellular H+ on the electrical properties of excitable cells. *Annu. Rev. Neurosci.* 7, 257–278 (1984).
11. Yao, H. and Haddad, G. G. Calcium and pH homeostasis in neurons during hypoxia and ischemia. *Cell Calcium* 36, 247–255 (2004).
12. Hillered, L. et al. Lactic acidosis and recovery of mitochondrial function following forebrain ischemia in the rat. *J. Cereb. Blood Flow Metab.* 5, 259–266 (1985).
13. Siesjo, B. K. et al. Extracellular and intracellular pH in the brain during seizures and in the recovery period following the arrest of seizure activity. *J. Cereb. Blood Flow Metab.* 5, 47–57 (1985).
14. Somjen, G. G. Acidification of interstitial fluid in hippocampal-formation caused by seizures and by spreading depression. *Brain Res.* 311, 186–188 (1984).

15. Sanchez-Armass et al. Regulation of pH in rat brain synaptosomes. I. Role of sodium, bicarbonate, and potassium. *J. Neurophysiol.* 71, 2236–2248 (1994).
16. Paradiso, A. M. et al. Na+-H+ Exchange in Gastric Glands As Measured with A Cytoplasmic-Trapped, Fluorescent Ph Indicator. *Proc. Natl. Acad. Sci. USA—Biol. Sci.* 81, 7436–7440 (1984).
17. Whitaker, J. E. et al. Spectral and photophysical studies of benzo[c]xanthene dyes: Dual emission pH sensors. *Anal. Biochem.* 194, 330–344 (1991).
18. Sanchez-Armass, S. et al. Spectral imaging microscopy demonstrates cytoplasmic pH oscillations in glial cells. *Am. J. Physiol.—Cell Physiol.* 290, C524–C538 (2006).
19. Chen, G. et al. Functional parasagittal compartments in the rat cerebellar cortex: An in vivo optical imaging study using neutral red. *J. Neurophysiol.* 76, 4169–4174 (1996).
20. Kado, R. T. Neutral red: A specific fluorescent dye in the cerebellum. *Jpn. J. Physiol.* 43(Suppl. 1), S161–S169 (1993).
21. LaManna, J. C. and McCracken, K. A. The use of neutral red as an intracellular pH indicator in rat brain cortex in vivo. *Anal. Biochem.* 142, 117–125 (1984).
22. Okada, D. Neutral red as a hydrophobic probe for monitoring neuronal activity. *J. Neurosci. Methods* 101, 85–92 (2000).
23. Griffith, J. K. et al. J. C. Distribution of intracellular pH in the rat brain cortex after global ischemia as measured by color film histophotometry of neutral red. *Brain Res.* 573, 1–7 (1992).
24. Babich, H. and Borenfreund, E. Applications of the neutral red cytotoxicity assay to in vitro toxicology. *ATLA—Alternatives to Laboratory Animals* 18, 129–144 (1990).
25. Cohen, L. B. et al. Changes in axon fluorescence during activity: Molecular probes of membrane potential. *J. Membr. Biol.* 19, 1–36 (1974).
26. Hanson, C. L. et al. Role of climbing fibers in determining the spatial patterns of activation in the cerebellar cortex to peripheral stimulation: An optical imaging study. *Neuroscience* 96, 317–331 (2000).
27. Chesler, M. The regulation and modulation of pH in the nervous system. *Prog. Neurobiol.* 34, 401–427 (1990).
28. Roos, A. and Boron, W. F. Intracellular pH. *Physiol. Rev.* 61, 296–434 (1981).
29. Thomas, R. C. Experimental displacement of intracellular pH and the mechanism of its subsequent recovery. *J. Physiol.* 354, 3P–22P (1984).
30. Boron, W. F. and De Weer, P. Intracellular pH transients in squid giant axons caused by CO_2, NH_3, and metabolic inhibitors. *J. Gen. Physiol.* 67, 91–112 (1976).
31. Gaillard, S. and Dupont, J. L. Ionic control of intracellular pH in rat cerebellar Purkinje cells maintained in culture. *J. Physiol.* 425, 71–83 (1990).
32. Rose, C. R. and Deitmer, J. W. Stimulus-evoked changes of extra- and intracellular pH in the leech central nervous system. II. Mechanisms and maintenance of pH homeostasis. *J. Neurophysiol.* 73, 132–140 (1995).
33. Ma, E. and Haddad, G. G. Expression and localization of Na+/H+ exchangers in rat central nervous system. *Neuroscience* 79, 591–603 (1997).
34. Orlowski, J. et al. Molecular cloning of putative members of the Na/H exchanger gene family. cDNA cloning, deduced amino acid sequence, and mRNA tissue expression of the rat Na/H exchanger NHE-1 and two structurally related proteins. *J. Biol. Chem.* 267, 9331–9339 (1992).
35. Willoughby, D. and Schwiening, C. J. Electrically evoked dendritic pH transients in rat cerebellar Purkinje cells. *J. Physiol.* 544, 487–499 (2002).
36. Siesjo, B. K. and Wieloch, T. Cerebral metabolism in ischaemia: Neurochemical basis for therapy. *Br. J Anaesth.* 57, 47–62 (1985).
37. Zhan, R. Z. et al. Intracellular acidification induced by membrane depolarization in rat hippocampal slices: Roles of intracellular Ca2+ and glycolysis. *Brain Res.* 780, 86–94 (1998).

38. Green, J. et al. Cytosolic pH regulation in osteoblasts. Regulation of anion exchange by intracellular pH and Ca2+ ions. *J. Gen. Physiol.* 95, 121–145 (1990).
39. Chen, J. C. and Chesler, M. Modulation of extracellular pH by glutamine and GABA in rat hippocampal slices. *J. Neurophysiol.* 67, 29–36 (1992).
40. Grichtchenko, I. I. and Chesler, M. Depolarization-induced alkalinization of astrocytes in gliotic hippocampal slices. *Neuroscience* 62, 1071–1078 (1994).
41. Brookes, N. and Turner, R. J. K(+)-induced alkalinization in mouse cerebral astrocytes mediated by reversal of electrogenic Na(+)-. *Am. J. Physiol.* 267, C1633–C1640 (1994).
42. Amos, B. J. and Chesler, M. Characterization of an intracellular alkaline shift in rat astrocytes triggered by metabotropic glutamate receptors. *J. Neurophysiol.* 79, 695–703 (1998).
43. Dietzel, I. et al. Transient changes in the size of the extracellular space in the sensorimotor cortex of cats in relation to stimulus-induced changes in potassium concentration. *Exp. Brain Res.* 40, 432–439 (1980).
44. Ransom, B. R. et al. Activity-dependent shrinkage of extracellular space in rat optic nerve: A developmental study. *J. Neurosci.* 5, 532–535 (1985).
45. Muller, M. and Somjen, G. G. Intrinsic optical signals in rat hippocampal slices during hypoxia-induced spreading depression-like depolarization. *J. Neurophysiol.* 82, 1818–1831 (1999).
46. Chen, G. et al. Role of calcium, glutamate neurotransmission, and nitric oxide in spreading acidification and depression in the cerebellar cortex. *J. Neurosci.* 21, 9877–9887 (2001).
47. Frostig, R. D. et al. Cortical functional architecture and local coupling between neuronal activity and the microcirculation revealed by in vivo high-resolution optical imaging of intrinsic signals. *Proc Natl Acad Sci USA* 87, 6082–6086 (1990).
48. Lieke, E. E. et al. Optical imaging of cortical activity: Real-time imaging using extrinsic dye-signals and high resolution imaging based on slow intrinsic-signals. *Annu. Rev. Physiol* 51, 543–559 (1989).
49. Chen, G. et al. Involvement of Kv1 potassium channels in spreading acidification and depression in the cerebellar cortex. *J. Neurophysiol.* 94, 1287–1298 (2005).
50. Eccles, J. C. et al. *The Cerebellum as a Neuronal Machine*. Springer-Verlag, Berlin (1967).
51. Ito, M. *The Cerebellum and Neural Control*. Raven Press, New York (1984).
52. Gao, W. et al. Optical imaging of long-term depression in the mouse cerebellar cortex *in vivo*. *J. Neurosci.* 23, 1859–1866 (2003).
53. Cohen, L. B. et al. Optical methods for monitoring neuron activity. *Annu. Rev. Neurosci.* 1, 171–182 (1978).
54. Grinvald, A. et al. Optical imaging of neuronal activity. *Physiol. Rev.* 68, 1285–1366 (1988).
55. Batchelor, A. M. et al. Synaptic activation of metabotropic glutamate receptors in the parallel fibre-Purkinje cell pathway in rat cerebellar slices. *Neuroscience* 63, 911–915 (1994).
56. Tempia, F. et al. Postsynaptic current mediated by metabotropic glutamate receptors in cerebellar Purkinje cells. *J. Neurophysiol.* 80, 520–528 (1998).
57. Konnerth, A. et al. Synaptic currents in cerebellar Purkinje cells. *Proc Natl Acad Sci U S A* 87, 2662–2665 (1990).
58. Gao, W. et al. Spatial properties of long-term depression in the mouse cerebellar cortex: An *in vivo* optical imaging study. *Soc. Neurosci. Abstr.* 27, 828.8 (2001).
59. Hillman, D. E. et al. Ultrastructural localization of the plasmalemmal calcium pump in cerebellar neurons. *Neuroscience* 72, 315–324 (1996).
60. Chesler, M. and Kraig, R. P. Intracellular pH transients of mammalian astrocytes. *J. Neurosci.* 9, 2011–2019 (1989).
61. Chesler, M. and Kraig, R. P. Intracellular pH of astrocytes increases rapidly with cortical stimulation. *Am. J. Physiol.* 253, R666–R670 (1987).
62. Eccles, J. C. et al. Parallel fibre stimulation and the responses induced thereby in the Purkinje cells of the cerebellum. *Exp. Brain Res.* 1, 17–39 (1966).

63. Eccles, J. C. et al. The inhibitory interneurones within the cerebellar cortex. *Exp. Brain Res.* 1, 1–16 (1966).
64. Hausser, M. and Clark, B. A. Tonic synaptic inhibition modulates neuronal output pattern and spatiotemporal synaptic integration. *Neuron* 19, 665–678 (1997).
65. Llano, I. and Gerschenfeld, H. M. Inhibitory synaptic currents in stellate cells of rat cerebellar slices. *J. Physiol.* 468, 177–200 (1993).
66. Moseley, M. L. et al. Bidirectional expression of CUG and CAG expansion transcripts and intranuclear polyglutamine inclusions in spinocerebellar ataxia type 8. *Nat. Genet.* 38, 758–769 (2006).
67. Voipio, J. et al. Pharmacological characterization of extracellular pH transients evoked by selective synaptic and exogenous activation of AMPA, NMDA, and GABAA receptors in the rat hippocampal slice. *J. Neurophysiol.* 74, 633–642 (1995).
68. Trapp, S. et al. Acidosis of hippocampal neurons mediated by a plasmalemmal Ca2+/H+ pump. *NeuroReport* 7, 2000–2004 (1996).
69. Eccles, J. C. et al. The excitatory synaptic action of climbing fibres on the purkinje cells of the cerebellum. *J. Physiol.* 182, 268–296 (1966).
70. Brodal, A. and Kawamura, K. Olivocerebellar projection: A review. *Adv. Anat. Embryol. Cell. Biol.* 64, 1–140 (1980).
71. Voogd, J. and Bigare, F. *The Inferior Olivary Nucleus*. Courville, J., deMontigny, C., and Lamarre, Y. (Eds.), pp. 207–234. Raven, New York (1980).
72. Hawkes, R. and Herrup, K. Aldolase C/zebrin II and the regionalization of the cerebellum. *J. Mol. Neurosci.* 6, 147–158 (1995).
73. Hawkes, R. et al. Structural and molecular compartmentation in the cerebellum. *Can. J. Neurol. Sci.* 20(Suppl. 3), S29–S35 (1993).
74. Ebner, T. J. et al. Optical imaging of cerebellar functional architectures: Parallel fiber beams, parasagittal bands and spreading acidification. *Prog. Brain Res.* 148, 125–138 (2004).
75. Llano, I. et al. Synaptic- and agonist-induced excitatory currents of Purkinje cells in rat cerebellar slices. *J. Physiol.* 434, 183–213 (1991).
76. Perkel, D. J. et al. Excitatory synaptic currents in Purkinje cells. *Proc. Biol. Sci.* 241, 116–121 (1990).
77. Llinas, R. and Yarom, Y. Oscillatory properties of guinea-pig inferior olivary neurons and their pharmacological modulation: An *in vitro* study. *J. Physiol.* 376, 163–182 (1986).
78. Welsh, J. P. et al. Dynamic organization of motor control within the olivocerebellar system. *Nature* 374, 453–457 (1995).
79. Sugihara, I. et al. Serotonin modulation of inferior olivary oscillations and synchronicity: A multiple-electrode study in the rat cerebellum. *Eur. J. Neurosci.* 7, 521–534 (1995).
80. Llinas, R. and Sugimori, M. Electrophysiological properties of in vitro Purkinje cell dendrites in mammalian cerebellar slices. *J. Physiol.* 305, 197–213 (1980).
81. Eccles, J. C. et al. The mossy fibre-granule cell relay of the cerebellum and its inhibitory control by Golgi cells. *Exp. Brain Res.* 1, 82–101 (1966).
82. Ito, M. et al. Climbing fibre induced depression of both mossy fibre responsiveness and glutamate sensitivity of cerebellar Purkinje cells. *J. Physiol.* 324, 113–134 (1982).
83. Eccles, J. C. An instruction-selection theory of learning in the cerebellar cortex. *Brain Res.* 127, 327–352 (1977).
84. Mauk, M. D. et al. Does cerebellar LTD mediate motor learning? Toward a resolution without a smoking gun. *Neuron* 20, 359–362 (1998).
85. Sakurai, M. Synaptic modification of parallel fibre-Purkinje cell transmission in in vitro guinea-pig cerebellar slices. *J. Physiol.* 394, 463–480 (1987).
86. Hartell, N. A. Induction of cerebellar long-term depression requires activation of glutamate metabotropic receptors. *NeuroReport* 5, 913–916 (1994).
87. Lev-Ram, V. et al. Synergies and coincidence requirements between NO, cGMP, and Ca2+ in the induction of cerebellar long-term depression. *Neuron* 18, 1025–1038 (1997).

88. Crepel, F. and Krupa, M. Activation of protein kinase C induces a long-term depression of glutamate sensitivity of cerebellar Purkinje cells. An *in vitro* study. *Brain Res.* 458, 397–401 (1988).
89. Linden, D. J. et al. A long-term depression of AMPA currents in cultured cerebellar Purkinje neurons. *Neuron* 7, 81–89 (1991).
90. Ito, M. and Kano, M. Long-lasting depression of parallel fiber-Purkinje cell transmission induced by conjunctive stimulation of parallel fibers and climbing fibers in the cerebellar cortex. *Neurosci. Lett.* 33, 253–258 (1982).
91. Ekerot, C. F. and Kano, M. Long-term depression of parallel fibre synapses following stimulation of climbing fibers. *Brain Res.* 342, 357–360 (1985).
92. Kano, M. and Kato, M. Quisqualate receptors are specifically involved in cerebellar synaptic plasticity. *Nature* 325, 276–279 (1987).
93. Marr, D. A theory of cerebellar cortex. *J Physiol.* 202, 437–470 (1969).
94. Albus, J. S. A theory of cerebellar function. *Math. Biosci.* 10, 25–61 (1971).
95. Augustine, G. J. et al. Local calcium signaling in neurons. *Neuron* 40, 331–346 (2003).
96. De Zeeuw, C. I. et al. Expression of a protein kinase C inhibitor in Purkinje cells blocks cerebellar LTD and adaptation of the vestibulo-ocular reflex. *Neuron* 20, 495–508 (1998).
97. Koekkoek, S. K. et al. Cerebellar LTD and learning-dependent timing of conditioned eyelid responses. *Science* 301, 1736–1739 (2003).
98. Pekhletski, R. et al. Impaired cerebellar synaptic plasticity and motor performance in mice lacking the mGluR4 subtype of metabotropic glutamate receptor. *J. Neurosci.* 16, 6364–6373 (1996).
99. Leao, A. A. P. Spreading depression of activity in the cerebral cortex. *J. Neurophysiol.* 7, 359–390 (1944).
100. Yoon, R. S. et al. Characterization of cortical spreading depression by imaging of intrinsic optical signals. *NeuroReport* 7, 2671–2674 (1996).
101. MacVicar, B. A. et al. Mapping of neural activity patterns using intrinsic optical signals: From isolated brain preparations to the intact human brain. *Adv. Exp. Med. Biol.* 333, 71–79 (1993).
102. Aitken, P. G. et al. Similar propagation of SD and hypoxic SD-like depolarization in rat hippocampus recorded optically and electrically. *J. Neurophysiol.* 80, 1514–1521 (1998).
103. Newman, E. A. and Zahs, K. R. Calcium waves in retinal glial cells. *Science* 275, 844–847 (1997).
104. Guthrie, P. B. et al. ATP released from astrocytes mediates glial calcium waves. *J. Neurosci.* 19, 520–528 (1999).
105. Momose-Sato, Y. et al. Optical mapping of the early development of the response pattern to vagal stimulation in embryonic chick brain stem. *J. Physiol.* 442, 649–668 (1991).
106. Chen, G. et al. Novel form of spreading acidification and depression in the cerebellar cortex demonstrated by neutral red optical imaging. *J. Neurophysiol.* 81, 1992–1998 (1999).
107. Tombaugh, G. C. and Somjen, G. G. Effects of extracellular pH on voltage-gated Na^+, K^+ and Ca^{2+} currents in isolated rat CA1 neurons. *J. Physiol.* 493(pt. 3), 719–732 (1996).
108. Traynelis, S. F. and Cull-Candy, S. G. Pharmacological properties and H+ sensitivity of excitatory amino acid receptor channels in rat cerebellar granule neurons. *J. Physiol.* 433, 727–763 (1991).
109. Huard, J. M. et al. Cerebellar histogenesis is disturbed in mice lacking cyclin D2. *Development* 126, 1927–1935 (1999).
110. Popa, L. S. et al. Activation of parallel fibers by extra-synaptic glutamate spillover mediates propagation of spreading acidification and depression. *Soc. Neurosci. Abstr.* 30, 535.6 (2004).
111. Ebner, T. J. and Chen, G. Spreading acidification and depression in the cerebellar cortex. *Neuroscientist* 9, 37–45 (2003).

112. Browne, D. L. et al. Identification of two new *KCNA1* mutations in episodic ataxia/myokymia families. *Hum. Mol. Genet.* 4, 1671–1672 (1995).
113. Harvey, A. L. Recent studies on dendrotoxins and potassium ion channels. *Gen. Pharmacol.* 28, 7–12 (1997).
114. Brandt, T. and Strupp, M. Episodic ataxia type 1 and 2 (familial periodic ataxia/vertigo). *Audiol. Neurootol.* 2, 373–383 (1997).
115. Eunson, L. H. et al. Clinical, genetic, and expression studies of mutations in the potassium channel gene *KCNA1* reveal new phenotypic variability. *Ann. Neurol.* 48, 647–656 (2000).
116. Bower, J. M. The organization of cerebellar cortical circuitry revisited: Implications for function. *Ann. N. Y. Acad. Sci.* 978, 135–155 (2002).
117. Kneen, M. et al. Green fluorescent protein as a noninvasive intracellular pH indicator. *Biophys. J.* 74, 1591–1599 (1998).
118. Elsliger, M. A. et al. Structural and spectral response of green fluorescent protein variants to changes in pH. *Biochemistry* 38, 5296–5301 (1999).
119. Metzger, F. et al. Transgenic mice expressing a pH and Cl-sensing yellow-fluorescent protein under the control of a potassium channel promoter. *Eur. J. Neurosci.* 15, 40–50 (2002).
120. Davidson, B. L. and Breakefield, X. O. Viral vectors for gene delivery to the nervous system. *Nat. Rev. Neurosci.* 4, 353–364 (2003).
121. Shevtsova, Z. et al. Promoters and serotypes: Targeting of adeno-associated virus vectors for gene transfer in the rat central nervous system in vitro and in vivo. *Exp. Physiol.* 90, 53–59 (2005).
122. Tohmi, M. et al. Enduring critical period plasticity visualized by transcranial flavoprotein imaging in mouse primary visual cortex. *J Neurosci.* 26, 11775–11785 (2006).
123. Kalimo, H. et al. Brain lactic acidosis and ischemic cell damage: 2. Histopathology. *J. Cereb. Blood Flow Metab.* 1, 313–327 (1981).
124. Nedergaard, M. et al. A. Acid-induced death in neurons and glia. *J. Neurosci.* 11, 2489–2497 (1991).
125. Kimelberg, H. K. et al. SITS-inhibitable Cl- transport and Na^+-dependent H^+ production in primary astroglial cultures. *Brain Res.* 173, 111–124 (1979).
126. Rehncrona, S. et al. Enhancement of iron-catalyzed free radical formation by acidosis in brain homogenates: Differences in effect by lactic acid and CO_2. *J. Cereb. Blood Flow Metab.* 9, 65–70 (1989).
127. Xiong, Z. G. et al. Acid-sensing ion channels (ASICs) as pharmacological targets for neurodegenerative diseases. *Curr. Opin. Pharmacol.* 8, 25–32 (2008).
128. Waldmann, R. et al. A proton-gated cation channel involved in acid-sensing. *Nature* 386, 173–177 (1997).
129. De La Rosa, D. A. et al. Distribution, subcellular localization and ontogeny of ASIC1 in the mammalian central nervous system. *J. Physiol.* 546, 77–87 (2003).
130. Anderson, R. E. et al. Focal cortical distribution of blood flow and brain pHi determined by in vivo fluorescent imaging. *Am. J. Physiol.* 263, H565–H575 (1992).
131. Anderson, R. E. and Meyer, F. B. Protection of focal cerebral ischemia by alkalinization of systemic pH. *Neurosurgery* 51, 1256–1265 (2002).

12 Quantitative In Vivo Imaging of Tissue Absorption, Scattering, and Hemoglobin Concentration in Rat Cortex Using Spatially Modulated Structured Light

David J. Cuccia, David Abookasis, Ron D. Frostig, and Bruce J. Tromberg

CONTENTS

12.1 INTRODUCTION

Significant changes in blood flow or in the integrity of cerebral vessels are believed to cause cerebrovascular disease (CVD) and to contribute to dementias including Alzheimer's disease.[1] Stroke, the most serious form of CVD, is one of the leading causes of death and adult disability worldwide. Acute treatments for stroke, however, are severely limited. Neuroprotective drugs under development show promise at halting the ischemic cascade, but as yet, no such compound has received federal approval in the United States. One of the biggest limitations to this development is the lack of understanding of the mechanisms by which cerebral vessels react to factors such as ischemia, inflammation, blood pressure changes, metabolic demands, and trauma.[2] In order to address these fundamental questions, functional brain imaging techniques such as fMRI and intrinsic signal optical imaging (ISOI) have emerged as tools to visualize and quantify cerebral hemodynamics.

In the neuroscience community, ISOI has long been used to study the organization and functional architecture of different cortical regions in animals and humans[3–5] (see other chapters in this book). Three sources of ISOI signals that affect the intensity of diffusely reflected light derive from characteristic physiologic changes in the cortex. For functional neuronal activation, these have been observed to occur over a range of timescales, including (1) light scattering changes, both fast (over 10 s of milliseconds) and slow (i.e., > ~0.5 s) (2) early (~0.5–2.5 s) absorption changes from alterations in chromophore redox status, i.e., the oxy/deoxy-hemoglobin ratio (known as the "initial dip" period), and (3), slower (~2–10 s) absorption changes due to blood volume increase (correlated with the fMRI BOLD signal). Light scattering changes have been attributed to interstitial volume changes resulting from cellular swelling, organelle swelling due to ion and water movement, capillary expansion, and neurotransmitter release.[6,7] The slower absorption factors have been demonstrated to correlate with the changes in metabolic demand and subsequent hemodynamic cascades following neuronal activation.[4,8,9]

Using animal models of acute and chronic brain injury, ISOI has been used to quantify the acute hemodynamic events in response to stroke, including focal ischemia and cortical spreading depression (CSD).[10–21] Researchers have also used ISOI to locate and quantify the spatial extent of the stroke injury, including ischemic core, penumbra, and healthy tissue zones.[18,22] CSD also plays a key role in migraine headache, and recent laser speckle imaging studies have revealed the neurovascular coupling mechanism to the transmission of headache pain.[23,24]

To fully understand the underlying mechanisms in vascular changes associated with cerebrovascular diseases such as stroke, an optical imaging technique that has the capability to rapidly separate absorption from scattering effects can enhance the information content of traditional ISOI, enabling (1) more accurate quantitation of hemodynamic function, (2) isolation of the electro-chemical changes characterized by light scattering, and (3) longitudinal chronic injury studies of function where structural reorganization due to neovascularization can cause significant alterations in scattering.[25,26]

Quantitative diffuse optical methods[27] such as spatially-resolved reflectance, diffuse optical spectroscopy (DOS), and tomography (DOT), and diffuse correlation spectroscopy (DCS) possess exquisite sensitivity to these functional and structural alterations associated with brain injury, and have been applied to the study of CSD.[11,15,28] DOS and DOT utilize the near-infrared spectral region (600–1000 nm) to separate and quantify the multispectral absorption (μ_a) and reduced scattering coefficients (μ_s'), providing quantitative determination of several important biological chromophores such as deoxy-hemoglobin (HbR), oxy-hemoglobin (HbO_2), water (H_2O), and lipids. Concentrations of these chromophores represent the direct metrics of tissue function such as blood volume fraction, tissue oxygenation, and edema. Additionally, the scattering coefficient contains important structural information about the size and density of scatterers and can be used to assess tissue composition (exctracellular matrix proteins, cell nuclei, mitochondria) as well as follow the process of tissue remodeling (wound healing, cancer progression). DOS utilizes a limited number of source-detector positions, e.g., 1–2, but often employs broadband content in temporal and spectral domains.[29] In contrast, DOT typically utilizes a limited number of optical wavelengths (e.g., 2–6) and a narrow temporal bandwidth, but forms higher resolution images of subsurface structures by sampling a large number of source-detector "views." To achieve maximal spatial resolution, the ideal DOT design would employ thousands of source-detector pairs and wavelengths. However, several engineering considerations including measurement time and instrument complexity currently limit the practicality of this approach.

In this chapter we present the basic principles of a new, noncontact quantitative optical imaging technology, modulated imaging (MI),[30–32] and provide examples of MI performance in 2 rat models of brain injury, cortical spreading depression (CSD) and stroke. MI enables both DOS and DOT concepts with high spatial (<1 mm) and temporal resolution (<1 s) in a simple, scan-free platform. MI is capable of both separating and spatially-resolving optical absorption and scattering parameters, allowing wide-field quantitative mapping of tissue optical properties. While compatible with time-modulation methods, MI alternatively uses spatially modulated illumination for imaging of tissue constituents. Periodic illumination patterns of various spatial frequencies are projected over a large area of a sample. The diffusely reflected image is modified from the illumination pattern due to the turbidity of the sample. Typically, sine-wave illumination patterns are used. The demodulation of these spatially modulated waves characterizes the modulation transfer function (MTF) of the material, and embodies the sample optical property information.

12.2 METHODS AND INSTRUMENTATION

12.2.1 Modulated Imaging Spectroscopy

The MI instrument platform was introduced originally by Cuccia et al.[30] Based on this design, we have developed a custom multispectral near-infrared (NIR) MI spectroscopy system capable of imaging between 650 and 1000 nm. A diagram of this system is shown in Figure 12.1.

Broadband NIR illumination is provided by an intensity-stabilized 250 W quartz-tungsten-halogen (QTH) lamp (Oriel QTH Source with Light Intensity Controller, Newport Corporation-Oriel Instruments, Stratford, Connecticut). Light is collimated and refocused with a pair of aspheric F/#0.7 optical lens systems (Oriel Aspherab). A custom-sized 3.5 in square hybrid hot mirror (Reynard Corporation, i.e., R00670-00) was placed between the lenses to limit the illumination to wavelengths below 1000 nm. Light engine optics taken from a digital projector (NEC HT1000) serve to homogenize and direct the light onto a 0.7 in digital micromirror device (DMD Discovery™ 1100 with ALP Accessory Package, ViALUX, Germany). Grayscale spatial sinusoid patterns are projected at 400 Hz using the ViALUX software development toolkit, which generates the necessary pulse-width modulation of binary sub-frames to produce a specified grayscale bit-depth (1–8 bits). Finally, a fixed focal length (f = 100 mm) projection lens illuminates the tissue at a slight angle from normal with a 15 × 25 mm illumination field. Detection was performed at normal incidence using a CRI Nuance™ camera system, which combines a 12-bit CCD camera and a liquid crystal tunable filter (LCTF; λ = 650–1100 nm, $\Delta\lambda$ = 10 nm). To avoid specular reflection, crossed linear polarizers are used in the illumination and detection arms. For this system, the former is a 1.5 in diameter NIR linear polarizer (Meadowlark Optics, VLM-200-IR-R) placed immediately after the projection lens, and the first stage of the Nuance LCTF serves as the latter. The DMD, CCD, and LCTF are controlled via USB by a laptop computer, and synchronized using

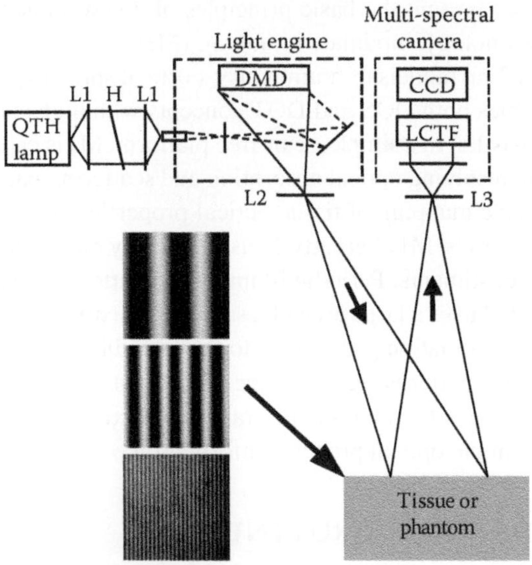

FIGURE 12.1 Modulated imaging platform. QTH—quartz tungsten halogen; L1—aspheric condenser; H—hybrid hot mirror; DMD—digital micromirror device; L2—projection lens; L3—camera lens; LCTF—liquid crystal tunable filter; CCD—charge-coupled device.

LabVIEW software (LabVIEW 8, National Instruments), enabling fast acquisition of a series of patterns with various spatial frequencies.

12.2.2 SFD Measurement, Calibration, and Modeling

A detailed description of SFD measurement, calibration, and diffusion modeling is provided by Cuccia.[32] In this work, we modeled diffuse reflectance using a transport-based White Monte Carlo (WMC) method.[33,34] Previously, we have found that compared with Monte Carlo, (1) diffusion predictions over- and underestimate low- and high-frequency diffuse reflectance, respectively, and (2) the quantitative accuracy of diffusion degrades with decreasing albedo.[32] Due to the moderate albedo of brain tissue (μ_s'/μ_a ~ 10–20), we chose to analyze all brain data with the WMC approach. This homogeneous tissue model is a significant simplification of the multilayered rat brain, and more work is necessary to accurately model this complex system. We discuss further the consequences of our simple model in Section 12.2.5.

12.2.3 Optical Property Inversion Methods

In this chapter, we use two inversion methods to calculate the absorption and reduced scattering from measurements of diffuse reflectance. When high measurement precision is desired, we use a "sweep" in spatial frequency space, producing an overdetermined set of diffuse reflectance measurements, which can be fitted to our WMC forward model predictions using least-squares minimization. This method is performed for all spatially averaged region analysis of optical properties and chromophores. When increased acquisition and/or processing speed is desired, we alternatively use a rapid two-frequency lookup table method based on cubic spline interpolation.[32] This data can be achieved with a minimal 3-phase, single frequency image set (by demodulating and averaging the images to obtain AC and DC amplitude maps, respectively). On typical personal computers this approach is capable of millions of inverse lookup calculations per second, and is therefore used to calculate all high-resolution images including time sequences. The signal-to-noise ratio (and thus the measurement precision) of either approach is limited by the data sampling, with the two-frequency method having a lower precision with the tradeoff of higher acquisition and processing speed.

12.2.4 Spectral Analysis–Chromophore Calculation

The quantitative absorption coefficient is assumed to be a linear (Beer's law) summation of individual chromophore absorption contributions:

$$\mu_a(\lambda) = 2.303\sum_{i=1}^{3} c_i \varepsilon_i(\lambda), \tag{12.1}$$

where c_i and $\varepsilon_i(\lambda)$ represent chromophore concentrations and molar extinction coefficients, respectively. Using reported extinction coefficients of HbO_2/HbR[35] and

H_2O,[36] we can invert Equation 12.1 and calculate tissue chromophore concentration separately at each pixel by linear least-squares fitting to the multispectral absorption images. Total hemoglobin (HbT) and oxygen saturation (S_tO_2) can then be calculated as HbT = HbR + HbO_2 and $S_tO_2 = HbO_2/(HbR + HbO_2) * 100$, respectively.

12.2.5 Optical Property Mapping: Resolution versus Quantitation

On a pixel-by-pixel basis, diffuse reflectance versus spatial frequency is fitted to the WMC forward model to extract the local absorption and reduced scattering optical property contrast. This process is repeated for each wavelength, resulting in multispectral absorption and scattering spectra at each pixel. The measured contrast from discrete absorbers and scatterers on millimeter and submillimeter spatial scales, however, will possess partial volume effects in all three spatial dimensions. This is due to the physical light transport length scales in tissue, limiting the true x-y resolution of optical property contrast to many detector pixels.[32] This phenomenon is not unique to MI, but present in all planar reflectance imaging measurements of turbid media. Absorption and scattering are calculated using a homogeneous reflectance model, extracting a locally averaged sampling of optical property contrast. Based on simulations of the tissue MTF for varying optical properties,[32] we expect the resulting image resolution to scale directly with the transport length, $l^* = (\mu_a + \mu_s')^{-1}$, and the spatial frequency of illumination. In this chapter, we place quantitative emphasis on *average* optical properties and chromophores measured over a field of view that is greater than l^*. Spatial maps and videos of these parameters are displayed and referred to as "contrast maps," with the caveat that high resolution features will exhibit degraded quantitative accuracy.

12.2.6 In Vivo Rat CSD Experiments

12.2.6.1 Animal Preparation

MI spectroscopy measurements were performed on an in vivo Wistar rat model with a thinned-skull preparation. All procedures were performed in accordance with approved IACUC protocol guidelines. The animals were anesthetized, placed in a stereotaxic frame, their skulls thinned and glass coverslip applied. This preparation is described in detail by Masino et al.[37] The resulting thinned skulls allowed direct imaging of the cortex over a 5 × 7 mm field-of-view (whisker barrel cortex, centered at the C2 location). In order to investigate the sensitivity of MI toward studying acute cortical injury, we induced cortical spreading depression (CSD) by applying 1 M KCl solution to the surface of the cortex through a perforated section of skull and dura, located approximately 3 mm above the camera's imaging field.

12.2.6.2 MI Measurement Protocol

For each of three animals, our MI measurement protocol was twofold. Prior to CSD induction, baseline spatial modulation data were acquired at 6 spatial frequencies (3-phase projections each) from 0 to 0.26 mm^{-1}, at 10 nm intervals over the entire range between 650 and 980 nm. Depending on the wavelength, image acquisition times ranged from 200 ms to 4 s, with total spectral imaging time of approximately 30 s per spatial pattern. The entire measurement (34 wavelengths, 3 phases, 6 frequencies) was repeated three times for statistical averaging yielding an entire measurement time of approximately 30 min.

Next, rapid dynamic measurements were performed, beginning 1 min prior to K^+Cl^- administration. Here, a significantly reduced data set was chosen in order to achieve high temporal resolution. Two spatial frequencies (0 and 0.26 mm^{-1}) were acquired with three phase projection images, as described in Section 12.2.2, at each of four wavelengths (680, 730, 780, and 830 nm). The resulting 12 images took in total 6 s, permitting a repetition rate of 10 measurements per minute. The animals were followed for a period of 10 min for rats 1 and 2, and a period of 30 min for rat 3.

All images in this study were smoothed by 2D convolution with a Gaussian filter function (FWHM = 3 pixels), and baseline repetitions were averaged prior to data processing. Additionally, time-series data were post-processed by smoothing slightly in time (Gaussian FWHM of 2 timepoints = 12 s).

12.2.6.3 Spatial Frequency Sensitivity Analysis

Because of the differential absorption sensitivity at low and high frequencies, optimal optical property separation is achieved when a large range of frequencies is used.[31] In Figure 12.2a, we depict this differential sensitivity using diffuse reflectance (MTF) predictions versus frequency, increasing μ_a by 100% from 0.02 (black line) to 0.04 mm^{-1} (gray line). This is done for two values of μ_s', 0.6 (solid lines) to 1.2 mm^{-1} (dashed lines), simulating a 100% change in scattering. Notice that the low frequencies have a significant reflectance change due to absorption, while high frequency reflectance remains nearly unchanged. Conversely, reflectance changes due to scattering are observed at all spatial frequencies. In Figure 12.2b, we further visualize this by plotting the reflectance sensitivity to 1% changes in absorption and

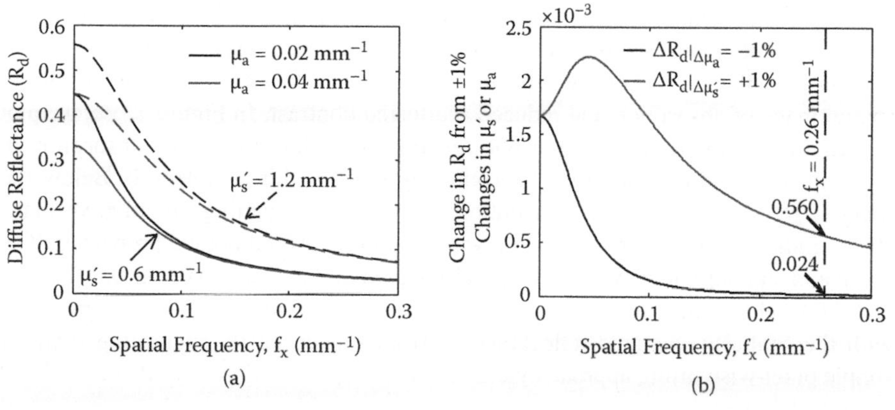

FIGURE 12.2 (a) Reflectance contrast in absorption and scattering covering a typical range of brain optical properties. (b) The frequency-dependent sensitivity to absorption (black line) and scattering (gray line), respectively. Reflectance at fx = 0.26 mm^{-1} is 24 times more sensitive to scattering contrast compared to that of absorption.

scattering. Whereas DC reflectance is equivalently sensitive to a fractional change in either absorption or scattering, at high spatial frequencies absorption contrast is lost while scattering contrast is retained. For instance, notice that at our maximum measurement frequency of 0.26 mm^{-1} the reflectance is roughly 24 times more sensitive to scattering compared to absorption ($\Delta R_d = 0.56\ \mu_s'$ versus $0.024 * 10^{-3}$ for μ_a). This plays an important role in Section 12.3.2 during our discussion of dynamic scattering measurement.

In realistic heterogeneous tissues, a tradeoff exists between maximizing the frequency range for optical property accuracy and obtaining similar sampling volumes. As tissue is a low-pass spatial filter, high frequencies are attenuated quickly with depth. Using diffusion-based forward modeling, we have estimated mean sampling depths at 650 nm using measured average background optical properties of brain tissue. This was done by predicting the depth sensitivity to contrast from a planar perturbation in absorption, given a background fluence profile from spatial frequencies 0 and 0.26 mm^{-1}. Based on these results, we observe qualitatively similar depth sampling, with mean depth sampling ranging between 2.5 mm and 1.2 mm (for $f_x = 0$ and 0.26 mm^{-1}, respectively). In all cases maximal sensitivity was found in the first 1–2 mm, where cortical hemodynamic changes occur.

12.3 RESULTS AND DISCUSSION

12.3.1 BASELINE MI SPECTROSCOPY

In Figure 12.3a we show a grayscale planar reflectance image of the cortical region of rat 1 at 650 nm. A dotted-line box denotes the region-of-interest (ROI) used for analysis, selected for its uniform illumination and the absence of cerebral bruising. The Monte Carlo-model fitting of spatial frequency data allows calculation of the absorption and reduced scattering coefficients. In Figure 12.3b we show the spatially averaged diffuse reflectance at 650 nm and the corresponding multi-frequency fit. Excellent agreement is observed between measurement data and the model-based fit, with derived μ_a and μ_s' coefficients of 0.033 and 0.70 mm^{-1}, respectively.

Analysis of multifrequency reflectance data separately at each pixel results in spatial maps of absorption and reduced scattering contrast. In Figure 12.3c, we plot the μa and μs′ maps recovered at 650 nm for rat 1. Note the strong absorption in the vein region, due to a large absorption by HbR at this wavelength. Below the images, we show histogram distributions of the corresponding quantitative maps above, indicating the degree of spatial variation in recovered optical properties. The mean and standard deviation for the pixel-wise μa and μs′ were 0.030 ± 0.007 mm^{-1} and 0.63 ± 0.13 mm^{-1}, respectively. These statistical results are in good agreement with the spatially averaged reflectance fit from Figure 12.3b, suggesting that our simple pixel-wise fitting approach yields optical properties similar to that calculated using a global analysis.

By mapping the absorption coefficient at multiple wavelengths, we can perform quantitative spectral imaging of tissue. In Figure 12.4, we summarize the baseline spectroscopy results for all three animals. In Figure 12.4a we show the μ_a (left) and μ_s' (right) coefficients versus wavelength (circles) recovered from spatially averaged

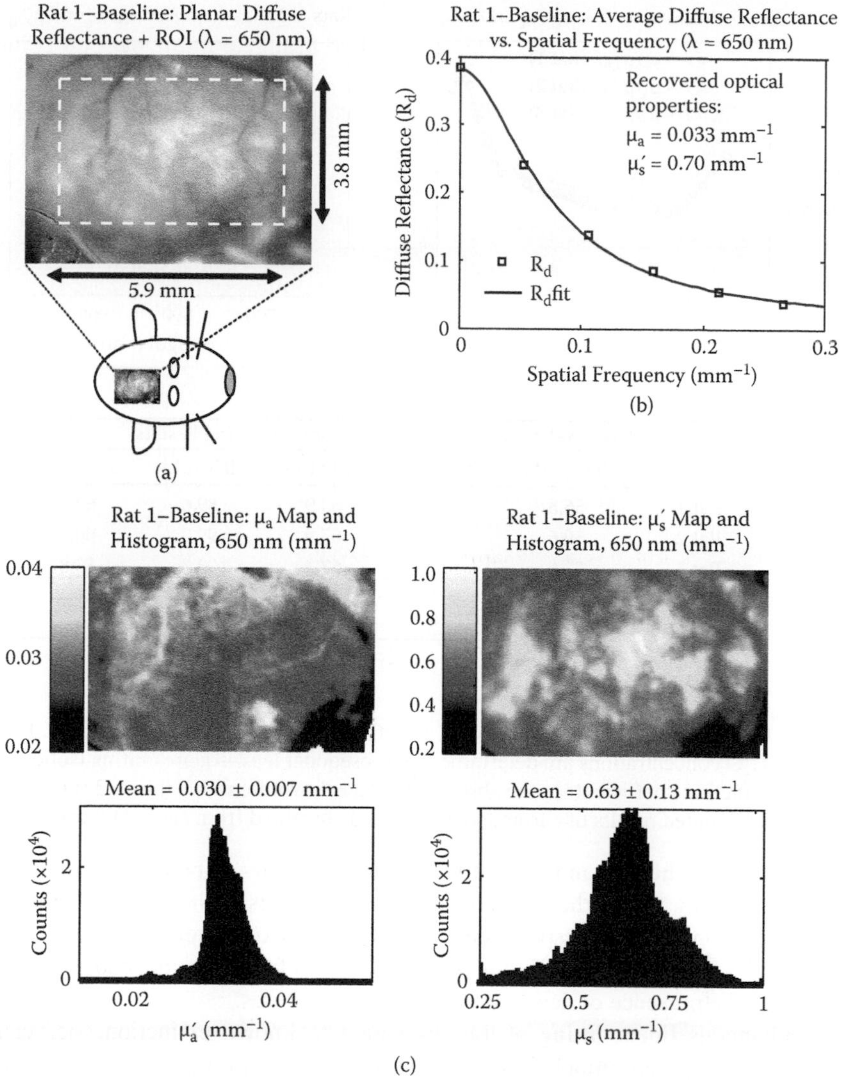

FIGURE 12.3 (a) Reflectance map for rat 1, showing the 3.8 × 5.9 mm region chosen for quantitative analysis. (b) Sample MTF reflectance data (squares) and fit (solid line) at 650 nm. (c) Recovered optical property maps (above) and corresponding image histogram over the analyzed ROI.

fitting. Data for rat 1 is shown in black (rat 2 in dark gray; rat 3 in light gray). Note the distinct spectral features in absorption, resulting from oxy- and deoxy-hemoglobin (HbO_2, HbR), and water (H_2O) absorption. The calculated scattering coefficient generally decays with increasing wavelength, and the results from a power law ($\mu_s' = A \cdot \lambda(\text{nm})^{-b}$, solid lines) fit are shown. A small residual coupling is observed between measured scattering and absorption spectral features. In particular, the

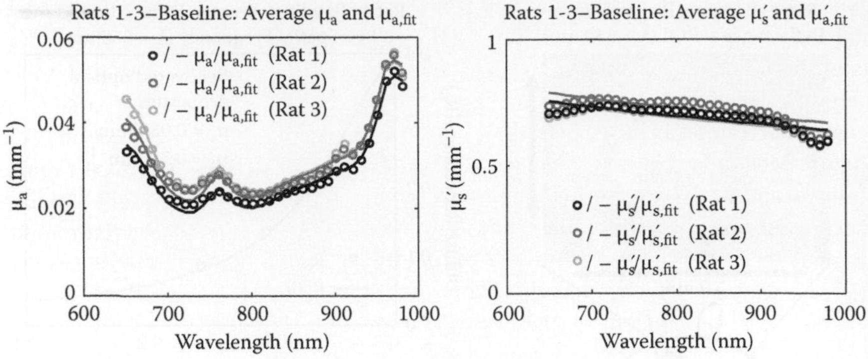

(a)

Rats 1-3–Baseline: Average Chromophore Fitting Results

Animal	HbO_2 (μM)	HbR (μM)	H_2O (%)	HbT (μM)	S_TO_2 (%)
Rat 1	56.3	33.2	63.9	89.6	62.9
Rat 2	59.6	39.7	67.9	99.3	60.0
Rat 3	57.6	45.7	67.4	103.3	55.8
Average	57.8	39.5	66.4	97.4	59.6
Std. Dev.	1.7	6.3	2.2	7.0	3.6

(b)

FIGURE 12.4 (a) Average μ_a (left) and μ_s' (right) spectra over entire ROI (circles). HbO_2, HbR, and H_2O concentrations are determined by subsequent least-squares fitting (solid lines) of molar extinction coefficients to the absorption. Data shown for rats 1 (blue), 2 (green), and 3 (red). (b) Tabulated results of chromophore values determined from spectral fitting.

scattering at the shortest and longest wavelengths appears to be underestimated by 5–10%, occurring where the corresponding absorption is highest (due to HbR and H_2O, absorption features, respectively). Based on our experiments in layered tissue phantoms,[38] we believe this effect is primarily due to frequency-dependent probing volumes in the presence of depth-heterogeneous structures.

Simultaneous linear fitting of the absorption to known extinction coefficients yields measures of chromophore concentration. Shown in Figure 12.4a, multispectral fitting (solid line) for rat 1 yields HbO_2, HbR, H_2O, HbT and S_tO_2 values of 56.3 μM, 33.2 μM, 63.9%, 89.6 μM, and 56.3%, respectively. Tabulated results of chromophore values for all three animals are shown in Figure 12.4b. Lipid absorption near 930 nm was not apparent in the μ_a spectrum, and when included in the spectral analysis was not found to significantly affect the results. The small absorption "bump" at 900–910 nm is an artifact of imperfect phantom calibration due to the presence of a sharp, strong silicone absorption peak that is present in the phantom.

We note that the solution for chromophore concentration is well-determined when the number of wavelengths is at least equal to the number of chromophores. Therefore, as few as two wavelengths can be used to separate HbO_2 and HbR (if a constant value of H_2O is assumed). Repeating the above analysis with 780 and 830 nm only (assuming $H_2O = 65\%$) yields results for HbO_2 and HbR within 10% of those from full spectral fitting. Repeating the above analyses using a simple diffusion-based model provided qualitatively similar results for absorption and scattering spectra, but in general was found to overestimate the absorption coefficient by 10–25%.

Absorption spectra at each pixel can be separately analyzed to yield spatial maps of local HbO_2, HbR, and H_2O distribution, shown in Figure 12.5. Notice the high concentration of HbR over the large superficial draining vessel (venous) regions, also reflected in the S_tO_2 image, highlighting the effect of tissue oxygen extraction. Conversely, notice that the high albedo regions with less structural detail are highly oxygenated, with S_tO_2 levels between 60 and 70%. Lastly, the H_2O map reveals a relatively homogeneous distribution of water.

12.3.2 Dynamic MI Spectroscopy of CSD

We performed measurements of CSD in each of the three rats, as described in Section 12.2.3. The results are presented as follows. We first present data for a single animal, choosing rat 3 for its long observation period of 30 minutes. Three ROIs are selected for analysis, and baseline MI spectroscopy results are reported for each of these regions. Next, the observed dynamic time courses of diffuse reflectance, optical properties, and chromophore concentrations are shown for each ROI. We then

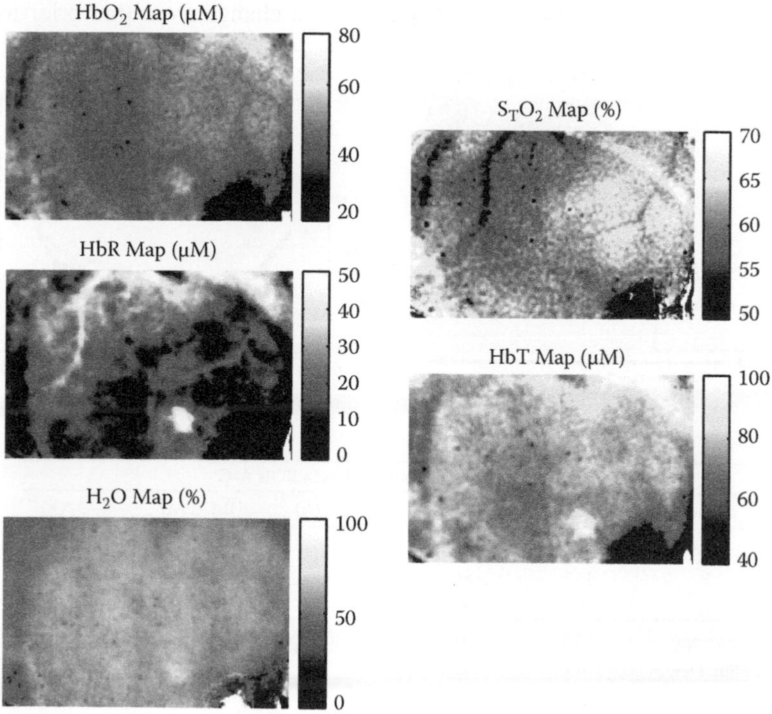

FIGURE 12.5 Chromophore fits to absorption spectra at each pixel yield maps of local HbO_2, HbR, and H_2O concentration (left). Total hemoglobin (HbT) and oxygen saturation (S_tO_2) maps can then be calculated from HbO_2 and HbR.

present the full spatio-temporal dynamic contrast data for rat 3 (2D + time) in the form of "snapshot" images.

Figure 12.6 summarizes the baseline spectroscopy measurements for rat 3. In Figure 12.6a, we show three regions of interest superimposed on the DC reflectance map, chosen to highlight three different characteristic temporal profiles observed within the field of view. In Figure 12.6b we show the baseline spectral fits for each of these regions, and in Figure 12.6c we tabulate the resulting calculated chromophore concentrations. In general, Region A (black) is a high albedo region lacking any large blood vessels, whereas Regions B (dark gray) and C (light gray) include high-absorption blood vessels and mild cerebral bruising from surgery. These differences are apparent in their recovered absorption spectra and fits, with on average 27% higher HbT, and 32% lower saturation in the vascular regions. Also, 7% higher H_2O is found in Regions B and C, which may indicate increased edema due to bruising.

In Figures 12.7–12.9 (for regions A–C, respectively), we present the temporal dynamics of CSD in each ROI of rat 3 as measured by MI. In part (a) of each figure, we plot the multispectral diffuse reflectance changes at $f_x = 0\ mm^{-1}$ (DC, top) and $f_x = 0.26\ mm^{-1}$ (AC, bottom). In part (b), we plot the recovered $\Delta\mu_a$ (top) and $\Delta\mu_s'$ (bottom) optical properties at each wavelength. While absolute values of diffuse reflectance and optical properties are measured separately at each time point, for visualization purposes all data are displayed as a change from that prior to KCl

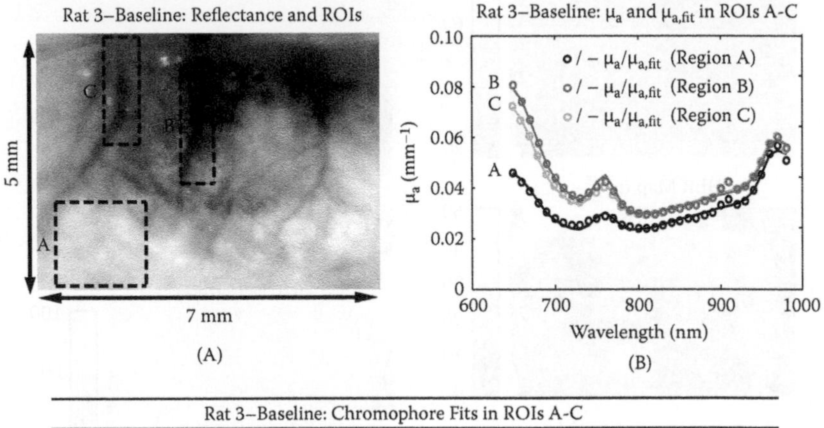

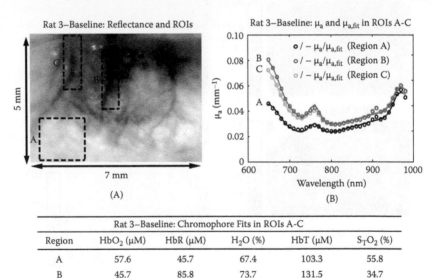

Rat 3–Baseline: Chromophore Fits in ROIs A-C

Region	HbO_2 (μM)	HbR (μM)	H_2O (%)	HbT (μM)	S_TO_2 (%)
A	57.6	45.7	67.4	103.3	55.8
B	45.7	85.8	73.7	131.5	34.7
C	54.2	75.8	70.5	129.9	41.7
Average	52.5	69.1	70.5	121.6	44.1
Std. Dev.	6.1	20.9	3.2	15.8	10.7

(C)

FIGURE 12.6 Regionwise spectral analysis of rat 1 baseline data including the respective (A) ROIs, (B) spectral absorption data (circles) and fit (lines), and (C) tabulated recovered chromophore data for each region

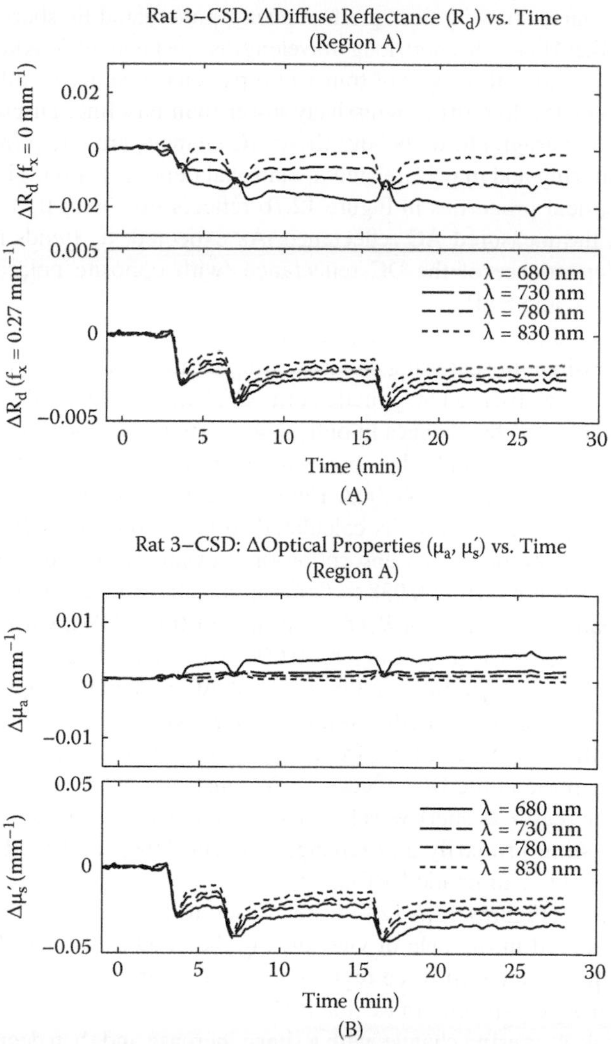

FIGURE 12.7 (A) Multispectral diffuse reflectance at DC ($fx = 0\ mm^{-1}$, top) and DC ($fx = .26\ mm^{-1}$, bottom) for Region A of rat 3 over approximately 30 min. (B) Corresponding recovered multispectral absorption (top) and reduced scattering (bottom) coefficients.

administration. Absolute optical property values at $t = 0$ (not shown) demonstrate excellent agreement (~5–10%) with full multifrequency baseline data.

Looking first at the reflectance time courses of Figure 12.7a (Region A), we see in general a series of three CSD events over the 30 minutes, with each transient event occurring for approximately 4.3 minutes. The first event occurs at minute 2.9 after KCl application, indicating an initial latency between the insult and the first resulting spreading depression wave. Reflectance contrast is present in both DC and AC frequency components, but with markedly different signatures. Generally, the DC time course shows a slow, gradual decay, punctuated by sharp, wavelength-dependent spikes/dips (for short/long wavelengths, respectively). Alternatively, the AC signature contains three sets of transient dips consistent across all wavelengths, with final values leveling off progressively lower than baseline. Discussed in detail in the following paragraph, we believe these AC changes are due primarily a result of optical scattering and may be related to neuronal depolarization. The corresponding derived optical properties in Figure 12.7b reflects this, with μ_s' trends tracking directly with the measured AC reflectance. As expected, μ_a trends reveal similar wavelength-dependence of the DC reflectance (with opposite polarity), reflecting changes in HbO_2 and HbR.

In Section 12.2.3.3 we noted that the diffuse reflectance at $f_x = 0.26$ mm^{-1} is 23 times more sensitive to scattering changes compared to absorption. In this context, we propose that the observed magnitude of the CSD-induced AC reflectance changes can only be explained by changes in optical scattering. To concretely illustrate this point, we pick as an example the observed 780 nm AC diffuse reflectance dip in Figure 12.7a at t = 3.7 min of −0.003. Here, the corresponding change in reduced scattering in Figure 12.7b, $\Delta\mu_s'$, is calculated to be −0.03 mm^{-1}. In order for this change to instead be due to an absorption-only event, μ_a would need to increase by 121% from baseline (from 0.038 to 0.084 mm^{-1}). This increase would also need to be accompanied by a drop in R_d ($f_x = 0$ mm^{-1}) of 0.12 (33%), whereas the *actual* observed DC reflectance only drops by 0.008 (<1%) and thus cannot explain the change. Secondly, we note that the three sets of AC reflectance dips occur consistently across all four wavelengths. While an approximate 120% increase in HbT could induce this decrease at high frequency, it would also require a large broad-wavelength decrease in the DC reflectance. We instead observe during these events that the DC increases at short wavelengths while the DC decreases at long wavelengths, suggesting primarily an exchange between HbO_2 and HbR volume fractions, as opposed to a dramatic HbT change.

Regions A–C (Figures 12.7–12.9) were chosen to highlight three different time signatures observed in the field of view during the CSD dynamics. The most contrasting feature between all three regions is the measured AC reflectance and the derived scattering coefficient. In Region B (Figure 12.8), each CSD event appears to cause a biphasic scattering change, with a sharp increase and then decrease, whereas a monophasic dip was observed in Region A (Figure 12.7). Region C (Figure 12.9) appears even more complex with a triphasic rise-dip-rise temporal profile. We observe that Regions A to C are located with increasing proximity to the CSD induction point (3 mm above the imaging field).

Because fractional changes in scattering and absorption have an equal (and opposite) effect on DC reflectance (see Section 12.2.3.3), any scattering (i.e., pathlength) changes measured here could be misinterpreted as absorption events with traditional ISOI analyses (i.e., DC reflectance only). In our observations, the measured scattering change of up to −0.05 mm^{-1} would be interpreted as an increase in absorption of up to +0.005 mm^{-1}, more than the maximum measured absorption change for wavelengths 730, 780, or 830 nm in any of the three regions. In order to account for differential pathlength changes, Kohl et al. proposed a multispectral model,[39] which they used to differentiate dynamic scattering and absorption changes using ISOI. This

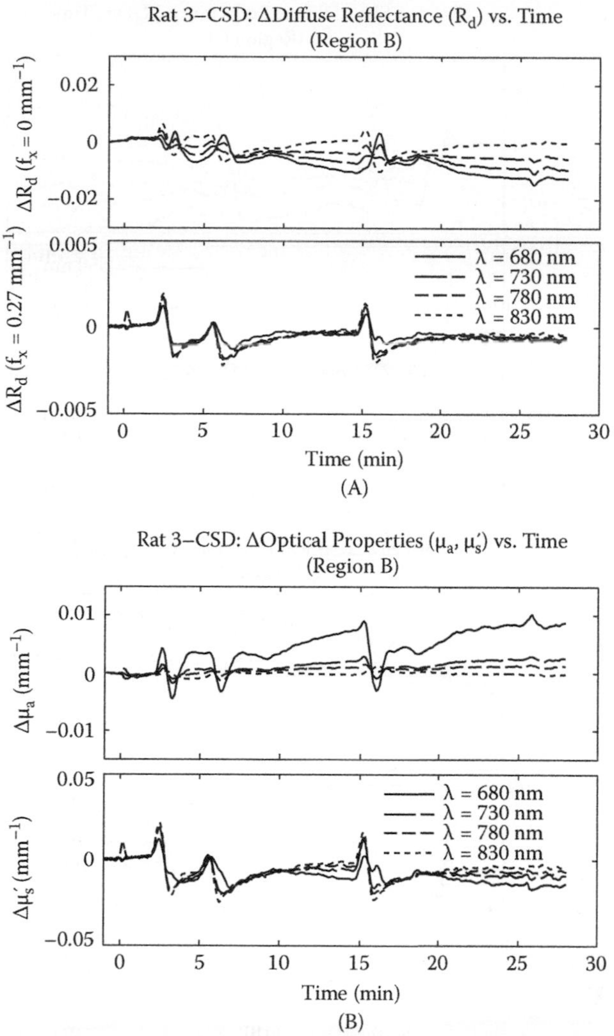

FIGURE 12.8 (A) Multispectral diffuse reflectance at DC (fx = 0 mm^{-1}, top) and DC (fx = 0.26 mm^{-1}, bottom) for Region B of rat 3 over approximately 30 min. (B) Corresponding recovered multispectral absorption (top) and reduced scattering (bottom) coefficients.

approach improves ISOI accuracy, and has been generally adopted as the method of choice for quantitative functional imaging. For dynamic measurements, we see MI as an improvement over this approach as it alternatively uses frequency domain measurements at a single wavelength to derive absolute scattering and absorption coefficients. This potentially provides a simplified single-wavelength measurement apparatus for detection of scattering, and also avoids potential mis-estimation of background optical properties.

Light scattering changes induced by spreading depression have been reported previously, and a comprehensive review is provided by Somjen. With in vivo spatially

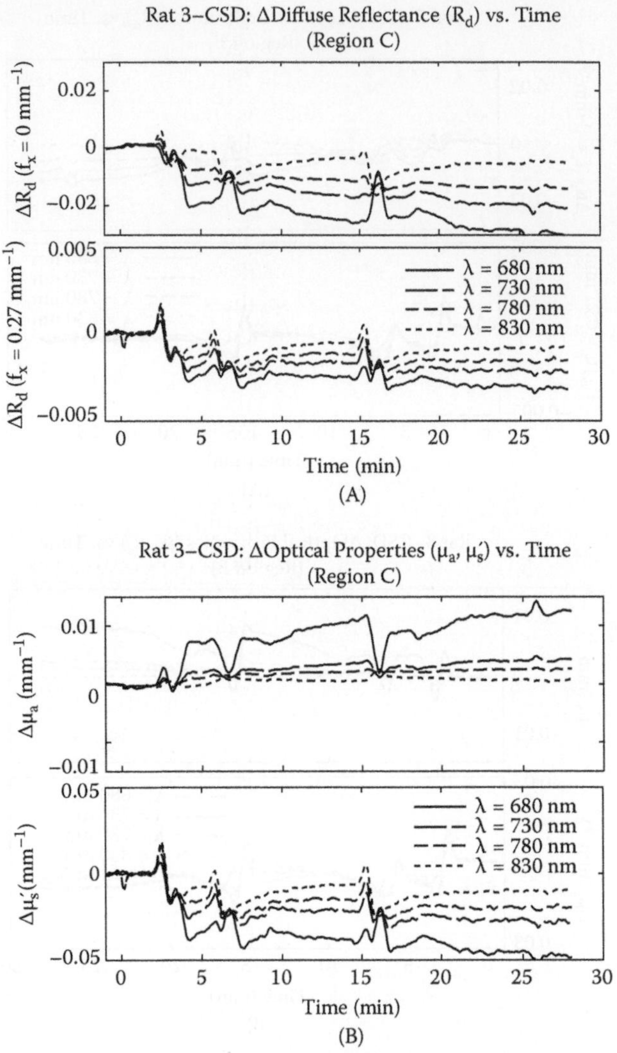

FIGURE 12.9 (A) Multispectral diffuse reflectance at DC (fx = 0 mm^{-1}, top) and DC (fx = 0.26 mm^{-1}, bottom) for Region C of rat 3 over approximately 30 min. (B) Corresponding recovered multispectral absorption (top) and reduced scattering (bottom) coefficients.

resolved reflectance measurements, Kohl et al.[11] separated absorption from scattering and observed a biphasic scattering response similar to that of Region A. With simultaneous laser scattering and electrophysiological measurements, both Jarvis et al. and Tao et al. found a strong correlation between electrical and optical scattering changes.[12,13,40] Tao et al. noted spatial heterogeneity in the dynamic spreading depression (SD) waveform related to the proximity to the SD induction site, similar to our results.

Using linear spectral analysis of absorption at all four wavelengths, we calculated the time-dependent chromophore concentration for Regions A, B, and C, presented in Figure 12.10A,B,C, respectively. In each region, the calculated baseline concentrations of H_2O were assumed to be constant. All three regions exhibit remarkably similar trends in HbR, HbO_2, HbT, and S_tO_2. This similarity is not clear in the DC traces of Figures 12.7–12.9, further highlighting the benefit of accurate separation of μa and μs′. Focusing on the first CSD event, there is a very consistent signature of: (1) a 2-minute latency post-KCl administration, (2) a 30-second period of decreasing S_tO_2 (3) a dramatic spike in both S_tO_2 (3–10%) and HbT (2–4 μM) with rise and decay times of approximately 1 minute each. For each region, the final S_tO_2 is approximately 5–10% lower than baseline, while the HbT restores to baseline values. This process repeats again twice more, except that the phase (2) desaturation appears to be absent. Additionally, in the "vessel" Region 3, we observe a gradual increase in HbT over the 30 minutes, indicating chronic blood pooling.

We show in Figure 12.11 the spatio-temporal evolution of both chromophore concentration and scattering changes from the first SD wave in rat 3. These are depicted in the form of a time derivative, i.e., (C(tn + 1) – C(tn))/(tn + 1 – tn), where C represents concentration/saturation/scattering values and tn represents time of acquisition for data point n. This visualization is appealing as it highlights the changes with high contrast.[18] From left to right, we show HbO_2, HbR, HbT, S_tO_2, and μ_s'. Notice the wave in scattering which propagates from top right to bottom left, at a rate of approximately 3 mm/min. An increase, or "spike" in scattering is observed initially

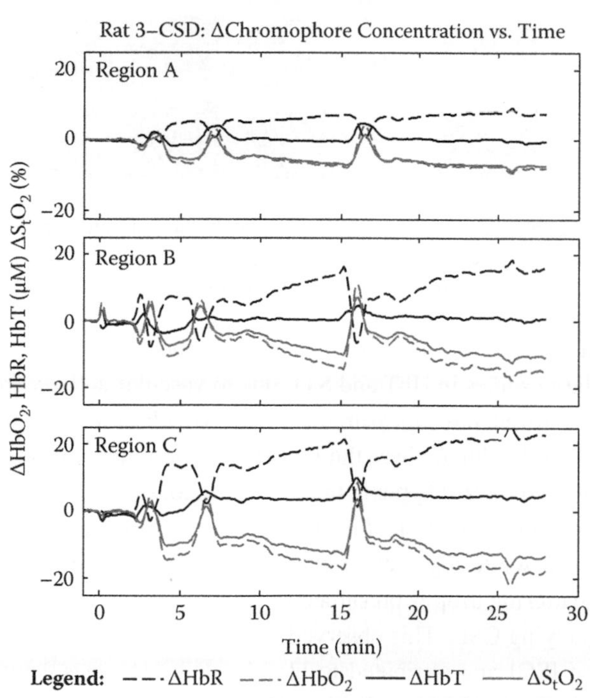

FIGURE 12.10 Recovered HbR, HbO_2, HbT, and STO2, for ROIs A, B, and C (top, middle, and bottom), recovered by analysis of the multispectral absorption coefficients from Figures 12.7–12.9b (top).

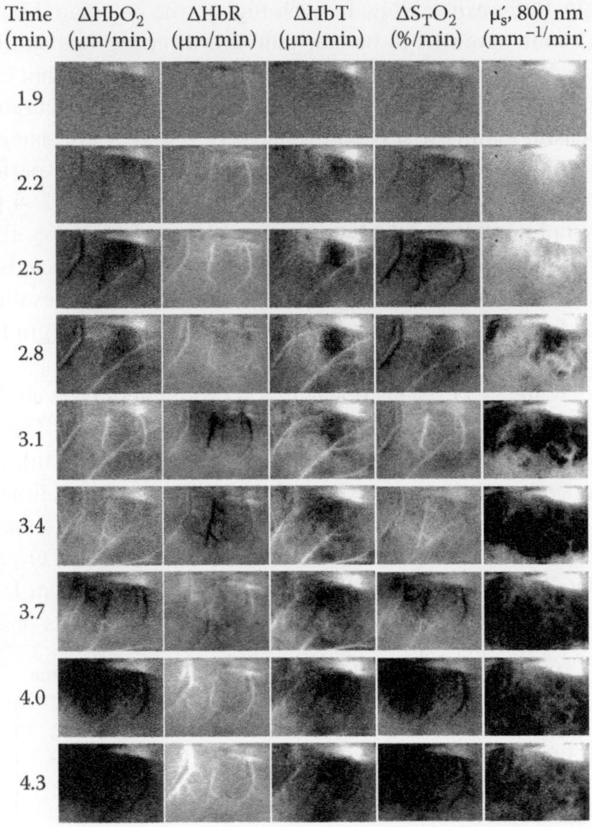

FIGURE 12.11 Spatio-temporal evolution of the hemodynamic and neural scattering response during a single spontaneous CSD event in rat 3. For visualization, a time derivative of the image sequence is displayed to highlight changes.

in the top right hand corner, in close proximity to the location of KCl administration. Note the large spikes in HbT and S_tO_2 due to vascular activity from depression wave propagation through the measurement field. We observe a transient increase in saturation and blood volume. Over the longer time periods, however, we observe a slow, sustained trend toward hypoxia in the vein regions.

The spatio-temporal evolution of the scattering coefficient in Figure 12.11 reveals a spatially defined scattering wave (reduction in μs′) that precedes hemodynamic changes. The scattering drop is presumed to be a consequence of neuronal depolarization accompanying CSD. This observed wave pattern has been shown previously with reflectance ISOI and attributed to blood volume changes.[18] Interestingly, the scattering depolarization wave is clearly followed in space and time by the increase in deoxyhemoglobin (HbR), decrease in saturation (S_tO_2), and drop in oxyhemoglobin (HbO_2); changes that are consistent with depolarization-induced neural tissue oxygen consumption.

12.3.3 Dynamic MI Spectroscopy of Stroke

In order to assess the sensitivity of MI to stroke, we conducted preliminary studies in a rat middle cerebral artery occlusion (MCAo) model, the most commonly involved artery in ischemic strokes. The left MCA was surgically cauterized using monopolar cautery or ligated to produce a permanent stroke. Figure 12.12 shows pre-versus post-MCAo results for a representative animal. Data were acquired at 5 wavelengths, 680, 800, 880, 920, and 980 nm in 3-minute intervals, and data points displayed every 6 minutes. Values for hemoglobin and saturation parameters were calculated from wavelength-dependent optical properties for each pixel, then averaged over the entire field of view (2 mm × 2 mm, 200 × 200 pixels) in the left hemisphere.

Stable baseline values for hemoglobin and S_tO_2 were measured for 27 minutes prior to MCAo. Dynamic changes were observed for each parameter within the first ~20 minutes of stroke and stabilized shortly thereafter with no significant variation between approximately 21 and 45 minutes post MCAo. The lack of new oxygenated blood supply following stroke resulted in a significant decrease in HbO_2 from ~50 μM to ~22 μM. In order to meet tissue metabolic demands, HbR was commensurately elevated from ~35 μM to ~49 μM due to oxygen extraction by ischemic neural tissues. This significant metabolic effect is also revealed by the dramatic drop in tissue saturation (S_tO_2) from baseline levels of ~60% to 30% after injury. Total hemoglobin (HbT), which roughly corresponds to blood flow, drops from 90 μM to 70 μM.

These results confirm previous observations of hemodynamic changes during cerebral ischemia in a rat model using diffuse optical tomography (DOT)[41] and

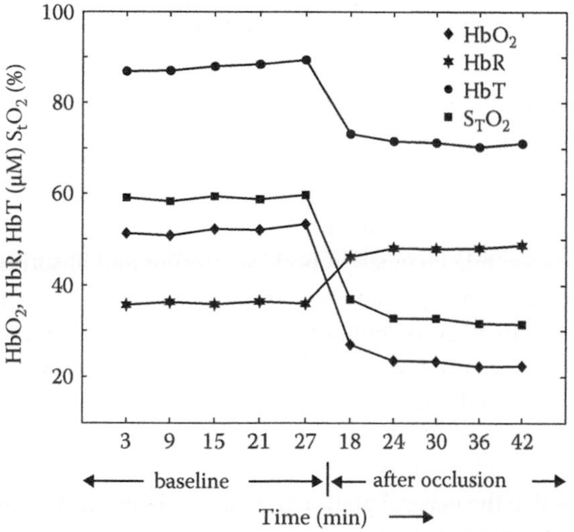

FIGURE 12.12 Mean values of cerebral components before and after MCA occlusion for a single representative animal. Hemoglobin and saturation parameters were calculated from wavelength-dependent optical properties for each pixel (680, 800, 880, 920, and 980 nm), then averaged over the entire field of view (2 mm × 2 mm, 200 × 200 pixels) in the left hemisphere.

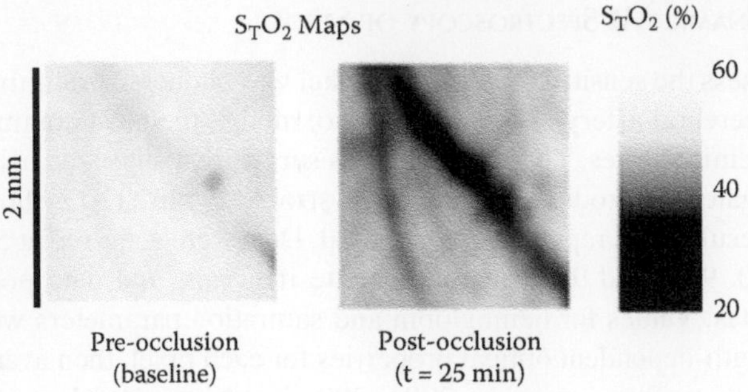

FIGURE 12.13 Maps of S_tO_2 before (baseline) and 25 minutes after MCA occlusion for representative animal. Field of view is 2 mm × 2 mm and 200 × 200 pixels in the left hemisphere.

multispectral reflectance imaging.[42] Although we report spatially-averaged cortical values in Figure 12.12, the MI technique also generates maps of cortical properties, making it a relatively simple approach for assessing hemodynamic and metabolic changes from different regions of the brain. For example, the S_tO_2 image for the same animal, shown in Figure 12.13, reveals uniform, low contrast tissue saturation levels of ~60% prior to MCAo. Twenty-five minutes following MCAo, clear spatial variations in S_tO_2 emerge, including the appearance of a high-contrast venous region where S_tO_2 values drop to ~20%. This strong tissue desaturation is a consequence of stroke-induced reductions in HbO_2 supply, accompanied by an increase in HbR as active neural tissue extracts oxygen. Similar spatio-temporal variations can be mapped and coregistered for all tissue chromophores.

12.4 CONCLUSION AND FUTURE DIRECTIONS

In conclusion we have presented a new method for quantitatively mapping tissue absorption and scattering properties, thereby allowing local sampling of in vivo concentrations of oxy- and deoxy-hemoglobin in CSD and stroke models of brain injury. Consistent dynamic changes in both scattering and absorption highlight the importance of optical property separation for quantitative assessment of tissue hemodynamics. Overall, these results demonstrate that MI can provide additional information content that enhances conventional ISOI methods. Future work involves further measurement of stroke and trauma, as well as epilepsy, migraine, and neuroplasticity. With continued technology development to optimize speed and spectral content, and advances in model-based reconstructions of tissue optical and physiological properties, we anticipate that the general platform of spatially modulated structured illumination can be easily integrated and coregistered with additional wide-field methods such as speckle correlation and fluorescence. In addition, MI can be used clinically in an intraoperative setting to provide feedback and guidance for surgeons. Because structured light introduces unique spatial and temporal selectivity, we expect that this approach will ultimately allow us to develop truly multidimensional methods for 3D tomography of neural tissue composition and metabolism.

ACKNOWLEDGMENTS

This work was supported by the National Institutes of Health under grants P41-RR01192 (Laser Microbeam and Medical Program: LAMMP), NS-43165 and NS-48350 (NIH-NINDS), by the U.S. Air Force Office of Scientific Research, the Medical Free-Electron Laser Program (F49620-00-2-0371 and FA9550-04-1-0101), and the Beckman Foundation. The authors would like to thank Jimmy Stehberg, Cynthia Chen-Bee, Teodora Agoncillo, Chris Lay, and Marlon Mathews for their assistance and suggestions related to this work.

REFERENCES

1. Pierce, M. et al., Advances in optical coherence tomography imaging for dermatology. *J. Investigative Dermatol.*, 123(3): 458–463, 2004.
2. Meeting Note: Cerebrovascular Biology and Disease, NHLBI Working Group. 2005.
3. Zepeda, A., C. Arias, and F. Sengpiel, Optical imaging of intrinsic signals: Recent developments in the methodology and its applications. *J. Neurosci. Meth*, 136: 1–21, 2004.
4. Frostig, R.D. et al., Cortical functional architecture and local coupling between neuronal activity and the microcirculation revealed by in vivo high-resolution optical imaging of intrinsic signals. *Proc Natl Acad Sci USA*, 87(16): 6082–6086, 1990.
5. Grinvald, A. et al., Functional architecture of cortex revealed by optical imaging of intrinsic signals. *Nature*, 324(6095): 361–364, 1986.
6. Cohen, L.B., Changes in neuron structure during action potential propagation and synaptic transmission. *Physiol Rev*, 53(2): 373–418, 1973.
7. Somjen, G.G., Mechanisms of spreading depression and hypoxic spreading depression-like depolarization. *Physiol Rev*, 81(3): 1065–1096, 2001.
8. Narayan, S.M., E.M. Santori, and A.W. Toga, *M*apping functional activity in rodent cortex using optical intrinsic signals. *Cereb Cortex*, 4(2): 195–204, 1994.
9. Narayan, S.M. et al., Imaging optical reflectance in rodent barrel and forelimb sensory cortex. *Neuroimage*, 1(3): 181–90, 1994.
10. O'Farrell, A.M. et al., Characterization of optical intrinsic signals and blood volume during cortical spreading depression. *Neuroreport*, 11(10): 2121–2125, 2000.
11. Kohl, M. et al., Separation of changes in light scattering and chromophore concentrations during cortical spreading depression in rats. *Opt Lett*, 23(7): 555–557, 1998.
12. Tao, L., Light scattering in brain slices measured with a photon counting fiber optic system. *J Neurosci Methods*, 101(1): 19–29, 2000.
13. Tao, L. et al., Light scattering in rat neocortical slices differs during spreading depression and ischemia. *Brain Res*, 952(2): 290–300, 2002.
14. Jones, M., J. Berwick, and J. Mayhew, Changes in blood flow, oxygenation, and volume following extended stimulation of rodent barrel cortex. *Neuroimage*, 15(3): 474–487, 2002.
15. Durduran, T., Non-Invasive Measurements of Tissue Hemodynamics with Hybrid Diffuse Optical Methods, in Dept. of Physics and Astronomy, University of Pennsylvania, 2004.
16. Ayata, C. et al., Pronounced hypoperfusion during spreading depression in mouse cortex. *J Cereb Blood Flow Metab*, 24(10): 1172–1182, 2004.
17. Jones, P.B. et al. Multimodal optical imaging of mouse Ischemic cortex. In *Optical Methods in Drug Discovery and Development, Proc SPIE Int Soc Opt Eng,* 2005. Boston, MA.
18. Chen, S. et al., In vivo optical reflectance imaging of spreading depression waves in rat brain with and without focal cerebral ischemia. *J Biomed Opt*, 11(3): 34002, 2006.

19. Chen, S. et al., Time-varying spreading depression waves in rat cortex revealed by optical intrinsic signal imaging. *Neurosci Lett*, 396(2): 132–136, 2006.
20. Ba, A.M. et al., Multiwavelength optical intrinsic signal imaging of cortical spreading depression. *J Neurophysiol*, 88(5): 2726–2735, 2002.
21. Yoon, R.S. et al., Characterization of cortical spreading depression by imaging of intrinsic optical signals. *Neuroreport*, 7(15–17): 2671–2674, 1996.
22. Chen, S. et al., Origin sites of spontaneous cortical spreading depression migrated during focal cerebral ischemia in rats. *Neurosci Lett*, 403(3): 266–270, 2006.
23. Dunn, A.K. et al., Dynamic imaging of cerebral blood flow using laser speckle. *J Cereb Blood Flow Metab*, 21(3): 195–201, 2001.
24. Bolay, H. et al., Intrinsic brain activity triggers trigeminal meningeal afferents in a migraine model. *Nat Med*, 8(2): 136–42, 2002.
25. Zepeda, A. et al., Reorganization of visual cortical maps after focal ischemic lesions. *J Cereb Blood Flow Metab*, 23(7): 811–820, 2003.
26. Manoonkitiwongsa, P.S. et al., Neuroprotection of ischemic brain by vascular endothelial growth factor is critically dependent on proper dosage and may be compromised by angiogenesis. *J Cereb Blood Flow Metab*, 24(6): 693–702, 2004.
27. Jacques, S.L. and B.W. Pogue, Tutorial on diffuse light transport. *J Biomed Opt*, 13(4): 041302, 2008.
28. Zhou, C. et al., Diffuse optical correlation tomography of cerebral blood flow during cortical spreading depression in rat brain. *Opt Express*, 14: 1125–1144, 2006.
29. Bevilacqua, F. et al., Broadband absorption spectroscopy in turbid media by combined frequency-domain and steady-state methods. *Appl Opt*, 39(34): 6498–6507, 2000.
30. Cuccia, D.J. et al., Modulated imaging: Quantitative analysis and tomography of turbid media in the spatial-frequency domain. *Opt Lett*, 30(11): 1354–1356, 2005.
31. Cuccia, D.J. et al. Quantitative recovery of tissue optical properties in the spatial frequency domain using modulated imaging. In *Optical Society of America Annual Meeting: Frontiers in Optics*. 2005. Tucson, Arizona.
32. Cuccia, D.J., Modulated Imaging: A Spatial Frequency Domain Imaging Method for Wide-field Spectroscopy and Tomography of Turbid Media in Biomedical Engineering. 2006, University of California: Irvine.
33. Kienle, A. and M.A. Patterson, Determination of the optical properties of turbid media from a single Monte Carlo simulation. *Phys Med Biol*, 41: 2221–2227, 1996.
34. Swartling, J. et al., Accelerated Monte Carlo models to simulate fluorescence spectra from layered tissues. *J Opt Soc Am A Opt Image Sci Vis*, 20(4): 714–727, 2003.
35. Prahl, S., Optical Absorption of Hemoglobin. Oregon Medical Laser Center Website (http://omlc.ogi.edu/spectra/hemoglobin/index.html), 1999.
36. Hale, G.M. and M.R. Querry, Optical constants of water in the 200 nm to 200 μm wavelength region. *Appl Opt*, 12: 555–563, 1973.
37. Masino, S.A. et al., Characterization of functional organization within rat barrel cortex using intrinsic signal optical imaging through a thinned skull. *Proc Natl Acad Sci USA*, 90(21): 9998–10002, 1993.
38. Weber, J.R. et al. Towards Functional optical imaging in layered tissues using modulated imaging. In *Optical Society of America Annual BIOMED Topical Meeting*. March 19–22, 2006, Ft. Lauderdale, FL.
39. Kohl, M. et al., Physical model for the spectroscopic analysis of cortical intrinsic optical signals. *Phys Med Biol*, 45(12): 3749–3764, 2000.
40. Jarvis, C.R., T.R. Anderson, and R.D. Andrew, Anoxic depolarization mediates acute damage independent of glutamate in neocortical brain slices. *Cereb Cortex*, 11(3): 249–259, 2001.

41. J. P. Culver et al., Diffuse optical tomography of cerebral blood flow, oxygenation, and metabolism in rat during focal ischemia. *J Cereb Blood Flow Metab*, 23: 911–924, 2003.
42. Jones, P.B. et al., Simultaneous multispectral reflectance imaging and laser speckle flowmetry of cerebral blood flow and oxygen metabolism in focal cerebral ischemia. *J Biomed Opt*, 13(4): 044007, 2008.

13 Intraoperative Optical Imaging

K.C. Brennan and Arthur W. Toga

CONTENTS

13.1 INTRODUCTION

Intraoperative optical intrinsic signal imaging (iOIS) provides an unparalleled opportunity to examine the basic physiology of the functioning human brain. At the same time, iOIS poses challenges for acquisition, analysis, and interpretation that are over and above those encountered in any other intact in vivo brain mapping. This may explain why, despite a profusion of optical imaging papers in the basic neuroscience literature, there have been relatively few publications on intraoperative optical imaging since its advent in 1992. Nevertheless, the clinical and research utility of this tool is considerable, and technological advances have made it even more so.

In terms of clinical utility, the most obvious benefits of intraoperative optical imaging are obtained when it is used to map "eloquent" cortex (i.e., language, sensory, motor, or visual areas) that are adjacent to or involved in pathology. This allows maximal surgical resection of pathological tissue while leaving essential areas intact. iOIS may also allow functional characterization of normal versus pathological tissue, providing an additional means to resect lesions while avoiding functioning brain.

The research utility of intraoperative OIS is equally large. Intraoperative OIS offers the only window into the investigation of human brain function with the combined spatial and temporal resolution of optical imaging. It provides the only direct way of confirming the insights made using optical imaging techniques in other mammals. And it serves as a crucial bridge to understanding the mechanisms of noninvasive techniques such as functional MRI (fMRI).

This chapter surveys the methodology of intraoperative OIS. We include comparisons between pre- and intraoperative maps along with methods to equate them in space. We also review how intraoperative OIS has been applied to address questions of basic physiology and cognitive function. Finally, we comment on the potential use of iOIS as a clinical brain mapping tool for neurosurgical guidance, and a research tool to understand basic mechanisms of cortical physiology and pathology.

13.1.1 History of the Technique

Haglund et al. were the first to observe optical intrinsic signals in humans, reporting activity-related changes in cortical light reflectance during seizure and cognitive tasks.[1] Since then, there have been reports describing the evolution of optical signals in the human cortex,[2] the registration of iOIS with fMRI,[3,4] the mapping of primary sensory and motor cortices,[1,5–9] the delineation of language cortices within[10,11] and across languages,[10] the imaging of epileptic foci and their treatment,[11–13] and the imaging of tumors, both with and without contrast dye.[11,14]

At the Laboratory of Neuro Imaging (LONI) at UCLA, we have successfully imaged 80 cases with iOIS: 41 seizure resection cases (16 language mapping, 4 motor mapping, 24 somatosensory mapping [in three cases more than one type of mapping was possible], 9 arteriovenous malformations, 2 hemangiomas, and 30 tumors).

13.2 SOURCES OF INTRINSIC SIGNALS

The origin of the optical intrinsic signal is still incompletely understood, even in reductionistic preparations such as brain slices. A comprehensive discussion of the subject is beyond the scope of this chapter, and the reader is referred to several excellent reviews,[15,16] as well as the other chapters of this book. However, it is instructive to briefly discuss candidate mechanisms because they highlight areas of potential opportunity for intraoperative imaging. In addition, Figure 13.1 provides a schematic showing known or potential sources of signal.

13.2.1 Blood Volume and Oxygenation

The largest changes in cortical reflectance in mammals come from the absorption of visible and near infrared light by hemoglobin. These changes provide an indirect index of underlying neural activity, as neuronal firing calls up increased perfusion to the active area. Fox and Raichle[17] first showed in human that neural activity results in both an increase in blood volume, and a net increase in hemoglobin oxygenation. Both of these changes can be visualized with OIS, with the use of filters. At visible light isosbestic points where oxy- and deoxyhemoglobin absorption is equal (~422, 452, 500, 530, 548, 570, 586, and 797 nm decreases in reflection correspond to increases in blood volume, and vice versa. This has been verified by a variety of techniques including intravascular dye imaging and optical spectroscopy.[18,19]

At points where absorption of oxy and deoxyhemoglobin differ, the signal is more sensitive to oxygen extraction. Typically, wavelengths between 600–660 nm are used, although near-infrared light below 800 nm can also be used. Imaging at wavelengths that are sensitive to oxygen extraction produces maps that are more spatially correlated with underlying neuronal activity than wavelengths that are influenced by blood volume changes.[20–23] This may be because fast changes in oxidative metabolism are more tightly coupled to electrical activity than the more delayed perfusion-related responses.[23] Imaging at 600–660 nm also offers the greatest signal-to-noise ratio, and, when focused beneath the cortical surface, least emphasizes blood vessel artifacts.[22] Due to the excellent spatial correlation of 610 nm signals with electrophysiological maps and its superior signal-to-noise ratio, our laboratory has focused on imaging optical signals in humans at 610 nm.[2,5,10,24]

13.2.2 Cytochromes

Hemoglobin is not the only absorber in the brain. Mitochondrial cytochromes, especially cytochrome *aa3* (or cytochrome oxidase) have also been visualized in animal preparations using spectroscopy.[25,26] Changes in reflectance were seen that were separable from large changes in the dominant absorber, hemoglobin. However, these experiments used techniques (ischemia, anoxia, hydrogen peroxide application) that would be expected to generate a large signal, and would be unacceptable

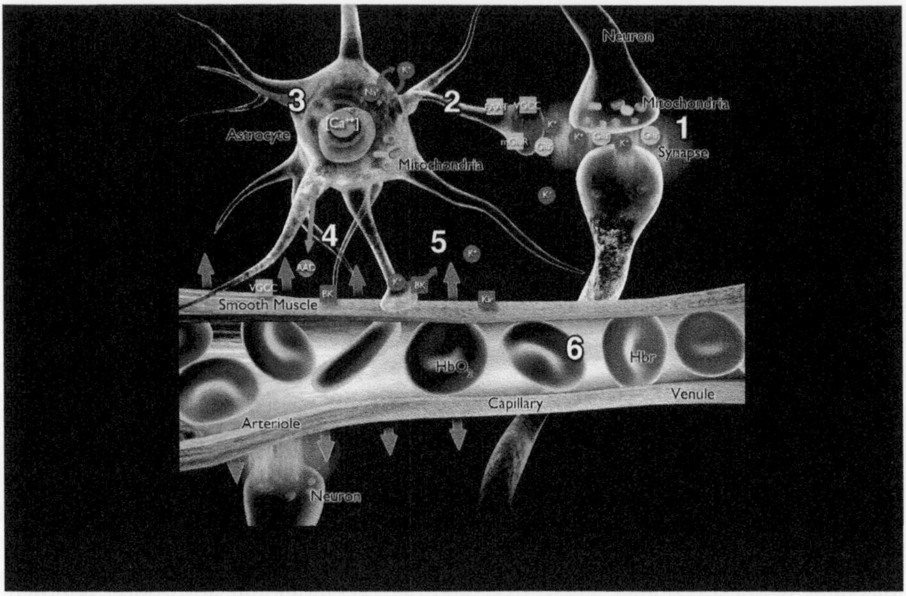

FIGURE 13.1 **(See color insert following page 224)** The neurovascular unit and its relation to intrinsic brain signals. Schematic shows the elements of the neurovascular unit, including neurons, astrocytes, and blood vessels (arteriole, capillary, and venule). (1) *Neurovascular coupling.* Synaptic activity results in release of glutamate into the synaptic cleft. Neuronal depolarization results in increase of extracellular K+. (2) Perisynaptic astrocytes respond with glutamate reuptake, spatial buffering of K+, and metabotropic glutamate receptor mediated calcium signaling. (3) Intracellular calcium elevation results in synthesis and release of arachidonic acid derivatives (AAD) (4), which diffuse to arteriolar smooth muscle cells causing vasodilatation (arrows; different AAD can add vasoconstriction). (5) K+ release at astrocyte endfeet is an additional mediator of arteriolar vasodilatation. (6) Initially, neuro-glial activity results in a net reduction of capillary and venous hemoglobin as a result of oxygen extraction (Hbr). Neurovascular coupling-mediated vasodilatation then leads to surplus oxygen delivery, resulting in a net oxygenation of capillary and venous hemoglobin, which is thought to be the basis of the fMRI signal. *Optically active elements of the neurovascular unit.* Oxygen extraction by metabolically active tissue (primarily neurons and glia) results in capillary and venous hemoglobin deoxygenation, which can be imaged with OIS at wavelengths sensitive to hemoglobin oximetry (generally 600–660 nm light is used). Vasodilatation results in net oxygenation of tissue, which can also be imaged with oximetric OIS. Changes in blood volume resulting from vasodilatation can be measured with OIS performed at wavelengths where oxy- and deoxyhemoglobin absorption is equal (isosbestic points; typically 530 or 570 nm). Neuro-glial activity causes intracellular changes in oxidative metabolism. Mitochondrial electron shuttles (NADH and flavoproteins) are intrinsically fluorescent, and their fluorescence changes with oxidation state. Mitochondrial flavoprotein fluorescence can be excited with blue light; emission of green light increases with flavoprotein oxidation. Finally, neuronal and glial swelling occurs with activity. Differences in light scatter caused by cellular swelling can be detected optically, although this is more difficult *in vivo* due to competing hemoglobin absorption changes. (Abbreviations: Glu-: glutamate; EAAT: excitatory amino acid transporter; mGluR: metabotropic glutamate receptor; VGCC: voltage-gated calcium channel; BK: calcium and voltage-sensitive potassium channel or BK channel; Kir: inwardly rectifying K+ channel.)[15,19,23,31,35,36,45,47,87] (Image courtesy of Amanda Hammond, LONI Scientific Visualization Group.)

in humans. Near-infrared spectroscopy studies investigating cytochrome oxidation in intact human skulls have been controversial.[27,28] It remains to be seen whether mitochondrial absorbers contribute to the human OIS signal. This question could be approached with intraoperative spectroscopy.

13.2.3 Light Scatter

The first demonstration of intrinsic signals in nervous tissue relied on changes in light scatter.[29] These changes are thought to be responsible for the optical signatures of neuronal activity in brain slices and isolated retina, as they are considered avascular preparations.[30–33,36] In isolated neuronal processes, changes in scatter can be shown to linearly correlate with membrane potential, and to occur with remarkably similar kinetics.[34–36] The origin of these signals is unclear but they are likely due to changes in the index of refraction of the plasma membrane.[34] In contrast, typical scatter changes detected in brain slices are much slower, with kinetics similar to the hemodynamic response in cortex (hundreds of milliseconds to seconds). The proposed mechanism for these changes is swelling of neurons and/or glia, as well as intracellular compartments such as mitochondria.[31,37]

Changes in scatter in the intact brain have been more difficult to quantify, due to the strong contribution of hemoglobin absorption. Spectroscopy experiments have inferred the presence of scatter-induced changes due to the presence of a residual signal after detection of oxy-, deoxy-, and total hemoglobin signal.[19] The salience of scatter-induced changes is greater above 600 nm, where hemoglobin absorption decreases. Invasive studies in cat at 660 nm and cat have shown signals with a temporal response similar to electrophysiology, which thus cannot be due to the slower hemodynamic response.[38] Although fast optical signals have been reported with near-infrared light through skull in intact humans, these changes have not been confirmed by other groups.[39,40] The potential utility of scatter-induced changes is that they can occur at much faster timescales than hemodynamic activity, and thus correlate more closely with neuronal activity. As yet investigation of fast scatter changes in the operating room has not been reported.

13.2.4 Intrinsic Fluorescence

Ultraviolet light excites NAD(P)H fluorescence, and this effect was first used for imaging purposes in the 1970s, in brain slices,[41] and in vivo.[42,43] The disadvantage of UV light is its phototoxicity. More recently, mitochondrial flavoprotein fluorescence, excited with blue light, has been used to generate functional signals, both in vitro[44,45] and in vivo.[46–48] Flavoprotein fluorescence signals have a similar time course to hemodynamic OIS signals, but they reportedly have greater signal to noise and are more spatially localized.[48,49] These methods involve simple modifications to conventional imaging apparatus (excitation and emission filters, dichroic), but have not to date been implemented for intraoperative imaging.

13.3 METHODS

13.3.1 Design Considerations Particular to the Operating Room

The operating room is a particularly challenging environment for experimentation, as the imperatives of patient safety impose constraints on experimental design. Firstly, there is typically no physical contact allowed with the subject due to the necessity of maintaining a sterile field; the operating neurosurgeons typically place all electrodes and stimulatory apparatus. Second, operating room lights and equipment generally need to remain on during recording, contributing to noise. Third, there is minimal to no control over anesthetic parameters and anesthetic agents, all of which can significantly affect the intrinsic signal. Fourth, the craniotomized human brain is highly susceptible to cardiac and respiratory induced motion artifact. Fifth, the field of view is necessarily limited to the craniotomy area chosen by the neurosurgeon. Much of this field of view likely includes pathologic tissue, which may or may not be of relevance to the experimental paradigm. Finally, time windows for image acquisition are typically very short, due to the need to minimize anesthetic exposure. Few institutional review boards will allow more than 40 minutes of experimentation, and typically the necessities of surgery allow much less.

13.3.2 Microscope and Light Source

For convenience, the operating microscope and its light source are usually used to collect images. The optics are comparable to most stereo microscopes, with variable magnification, long working distance, and wide field of view at the price of relatively low numerical aperture (NA). Higher NA systems can be used but their utility is limited by the fact that acquisition is taking place over a curved surface which needs a large depth of field (thus low NA) to be completely visualized. Typically a camera apparatus is mounted to the camera port of the operating microscope. Care needs to be taken to either reduce the weight of the camera-microscope apparatus, or to stabilize the arm of the operating microscope, to minimize vibration and drift. An alternative is to mount the camera directly to the skull via a bracket over the stereotactic head frame.[50]

The tungsten light sources typical to operating room microscopes are generally low-noise and offer a broad spectrum which can be filtered at the camera for the desired spectral region of interest. Filtering can also take place at the light source, but this regime cannot assure that the camera only receives light in the desired spectral region of interest.[14] Other light source options also exist. For example, light emitting diodes (LEDs, e.g., by Phillips Lumileds or Opto Technology) are bright, intrinsically more stable than tungsten sources, available in different wavelengths and quite economical.

Because the most precise functional maps are generated in the 600–660 nm range (see above), most investigators have used this wavelength band. However it is typical to acquire images with white or green light at some point during the experiment to visualize vascular landmarks and electrode placement. These template images can later be used for image registration, either for noise reduction or for registration with another modality (electrophysiology, fMRI; see upcoming text). Light gathered at isosbestic points of hemoglobin (e.g., 530 nm) can be used to generate blood volume images whose time course and extent are distinct from the primarily oxygen extraction images in the 600–660 nm range. Imaging at multiple wavelengths can thus provide complementary sources of physiological information.[13]

Thus far, no intraoperative imaging setup has obtained simultaneous multiwavelength images. This is not due to a lack of technology, which has been available for years. Instead, it has been difficult to implement the necessary techniques (beam splitters or filter wheel changers) in the time- and space-constrained operating room. Recent advances in compact beam-splitting apparatus may change this (see following text).

13.3.3 Detectors

OIS imaging has for some time relied on the slow-scan cooled CCD camera for detection. The greatest advantage of these systems is their low noise and large well depth, which allows for detection of very small signal changes under bright field conditions. Their disadvantages are that they are relatively slow (very few approaching video rate without significant binning) and quite costly. Given the slow nature of the hemodynamic response (onset in 200–300 ms, peak in 1–2 sec) the acquisition rate of these cameras might not seem to be a problem. However physiological and environmental noise can occur at higher frequencies, and correction for these noise sources can be better performed at higher sampling rates. In addition, slow scan CCD cameras are unlikely to detect changes any faster than the hemodynamic response (e.g., fast light scatter-mediated events).

Video rate and faster camera systems have tended to be noisy, and thus were not initially used for OIS imaging. Lower noise versions have now been implemented, and the disadvantage of noisy individual frames can be overcome by averaging sequential frames.[50] Because they are not cooled, they tend to be cheaper than slow scan CCDs. Both fast CCD and CMOS cameras are now available. The former are capable of speeds of up to 1 KHz, full frame, the latter up to 10 KHz, with noise levels and well depths acceptable for OIS imaging. Cameras of this speed are comparable with photomultiplier tubes and photodiodes in their response time, and make possible the use of voltage sensitive dyes[51] in animals. Such studies in humans await demonstration of dye safety.

13.3.4 LONI Intraoperative OIS System

The LONI human intraoperative optical imaging system (Figure 13.2) has been described in previous publications.[2,4,5,10,24,52] Briefly, we utilize a slow scan CCD camera (TE/CCD-576EFT, Princeton Instruments, Trenton, New Jersey) mounted via a custom adapter onto the video monitor port of a Zeiss operating microscope. Images are acquired through a transmission filter at 610 (605–615) nm (Corion Corp., Holliston, Massachusetts). Circular polarizing and heat filters are placed under the main objective of the operating microscope to reduce glare artifacts from the cortical surface. White light illumination is provided by the Zeiss operating microscope light

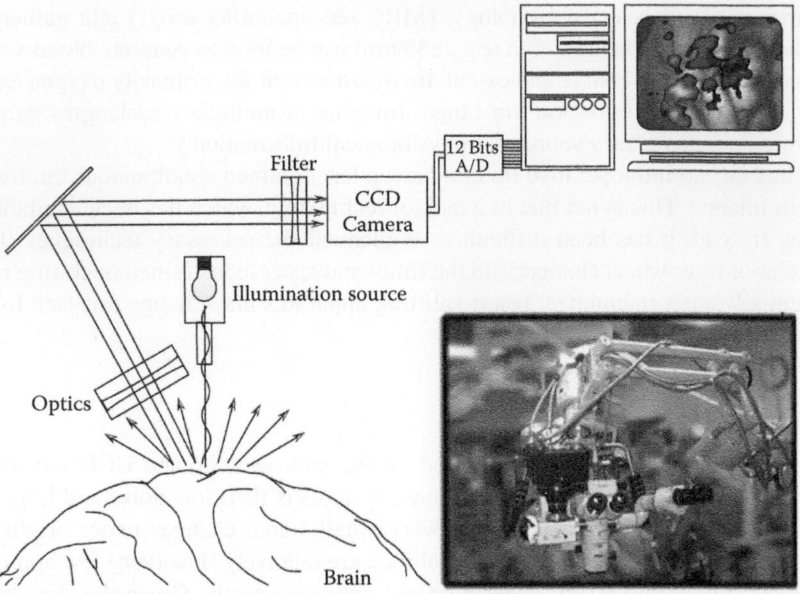

FIGURE 13.2 Schematic diagram of iOIS setup. The cortex is illuminated with white light and the reflected light is filtered at a particular wavelength of interest. Images are captured by a CCD camera during rest and during stimulation. The two conditions are compared to determine which pixels demonstrate significant changes across conditions. The window demonstrates how the CCD camera is mounted atop the operating microscope via a custom adaptor.

source, through a fiber optic illuminator. Since the microscope view is determined by the surgeon's choice of magnification and microscope placement, the experimental field of view varies, depending on lens distance from cortex and specific orientation. In general, FOV varies from 2.75 cm to 7.50 cm, resulting in pixel resolution of 110 μm to 425 μm, respectively. The spatial resolution can be increased by asking the surgeon to zoom in further on a cortical area of interest. It should be noted, however, that with increased magnification, movement of the brain due to respiration and the cardiac cycle becomes an increasing concern. The temporal resolution of acquisitions range between 200–500 ms per frame (which is sufficient to image the slowly evolving perfusion-related response) with image exposure times between 50–200 msec. The image exposure time should be kept short in order to minimize intra-image brain movement.

13.3.5 LONI Experimental Paradigms

iOIS acquisitions consist of CCD exposure of the cortex during a control state (no stimulation) followed by a subsequent stimulated state (see below). All imaging sessions are controlled via a personal computer running LabView (National Instruments, Austin, Texas). Control trials and experimental trials are interleaved. During sensorimotor mapping trials, baseline CCD images are acquired, then physiologically synchronized PC software triggers the stimulations (2 s duration). Experiments contain 10–20 stimulation trials and include an equal number of nonstimulated interleaved controls. During language mapping, patients are asked to perform verbal and nonverbal tasks in 20 s block trials with a 10 s rest interval. Paradigm blocks are repeated 4 to 8 times. All intraoperative imaging sessions are planned and timed in advance in order to minimize the duration of intraoperative imaging but also attain sufficient signal-to-noise. We have determined that the number of trials noted above seems sufficient to produce reasonable SNR (~10:1) while keeping imaging sessions as short as possible. In most cases, imaging lasts approximately 20–40 minutes. Intraoperative imaging time must be strictly controlled and surgeries should not be unnecessarily lengthened because risk of infection or anesthetic compromise increases with operative duration.

Once images have been captured, reflectance changes are estimated by performing a pixel-by-pixel ratio for each image ([image-control]/control), using the first image in each trial as a control. In order to increase the signal-to-noise ratio (SNR), multiple trials of the same task are averaged.

13.3.6 Noise Reduction

A significant obstacle in detecting the small activity-related reflectance changes in the intra-operative setting is the movement of the cortex due to cardiac and respiratory activity. This becomes an even greater challenge during awake procedures in which respiration rate is not under the direct control of the anesthesiologist. Several strategies have emerged to minimize the effect of cortical movement during imaging, including imaging through a sterile glass plate that lies atop the cortex,[1] synchronizing image acquisition with respiration and heart rate,[2] and using post-acquisition image registration.[1,10,24]

The use of a glass plate atop the cortex immobilizes the cortex by applying pressure to the cortical surface to prevent its movement. While this method may achieve immobilization, it is not be ideal because it physically interferes with the cortex and may therefore alter normal functional blood volume and cellular swelling patterns. In this respect, the latter two methods (synchronizing with respiration and heart rate and using postacquisition image registration) are preferable since they eliminate contact with the cortex being imaged.

Synchronizing image acquisition with respiration and heart rate requires monitoring of pneumogrpahic and electrocardiographic waveforms. All trials (control and experimental) should begin at the same point in time during the respiration cycle, (preferably during expiration) after which acquisition is controlled from synchronization to the cardiac cycle (500 ms post-R-wave).[4] Experimental and control trials should be collected alternately on sequential respiratory expirations. Each experimental image will have a separate control image taken from either the preceding or subsequent expiration cycle. Since data acquisition occurs at similar time points during every respiration cycle, all images are collected with the brain in a similar position, minimizing the effect of periodic brain motion. While synchronizing image acquisition with respiration and heart rate effectively minimizes movement across images, it is complicated by the need to interface with operating room anesthesiology monitors that may not be easily accessible to investigators. It also reduces temporal resolution significantly.

Postacquisition image registration utilizes automated image registration (AIR) algorithms.[53–55] When acquiring images independent of pneumographic or electrocardiographic waveforms, the precise location of the cortex within the field of view of the CCD varies from image to image. AIR can be used to realign all images in a series to a reference image either at the beginning or in the middle of the acquisition series. The realignment of images is intensity-based, using the extremely large difference in light intensities emanating from sulci and gyri to place the images into a common space. Functional changes in cortical reflectance do not interfere with this realignment algorithm since functional reflectance changes, which are <1% in magnitude, are minimal compared to the difference in reflectance from sulci and gyri (gyral light intensity levels can be up to 2000 times that of sulci). Furthermore, Woods et al.[53–55] have demonstrated (using positron emission tomography data) that simulated cortical activation sites do not interfere with realignment. Postacquisition image registration effectively compensates for movement between images and minimizes the need to interface with operating room monitors. It also allows for much higher acquisition rates.

In order to minimize vascular artifacts, the CCD camera can be focused below the cortical surface.[22] This strategy is especially useful when the area being imaged is under or near large draining veins.[22] Changing the depth of focus allows the investigator to focus on optical changes that occur only in deeper layers of cortex and to deemphasize optical signals that occur at high spatial frequencies (such as from the vasculature). Another strategy to reduce vascular artifacts is to average only images collected shortly after stimulus onset (within 1.5 s), a period of rapid oxygen consumption prior to when there are significant reflectance changes over large vessels.[56] Finally, vascular artifacts are wavelength-dependent, and therefore may be minimized by altering the wavelength at which iOIS is performed.[57]

Recently, Fourier-based algorithms have been implemented in OIS data sets.[58] They have proven especially useful in systems prone to hemodynamic noise like gyrencephalic mammals. The concept involves taking a continuous string of images (in contrast to the multiple separate trials done with the image averaging techniques described above). Repetitive stimuli are generated at a particular frequency (preferably distant from low frequency noise as well as heart rate, respiration, and vasomotion artifacts), and the dataset is temporally filtered using Fourier transform techniques. The crucial advantage is that analysis is restricted to the responses which occur at the stimulus frequency. The rejection of out-of-band signal markedly increases signal-to-noise ratio, which in turn allows for a reduction in number of stimulus repetitions, and thus total data acquisition time.[58] Fourier-based imaging techniques could prove quite useful in the time-challenged operating room setting.

13.3.7 Registration with Other Modalities: fMRI

Comparing functional magnetic resonance imaging (fMRI) signals, which are tomographic, with optical signals, which are surface signals, presents a challenge. In order to compare mapping signals from multiple modalities within a single subject, the different modalities must be warped into a common space to facilitate comparisons (Figure 13.3). The common space used is the 3D cortical model (i.e., cortical extraction) of each subject's brain, generated from high resolution T1-weighted MR scans. In order to align the fMRI activations with this cortical model, one must determine the transformation necessary to align high resolution structural scans (which are co-planar with the fMRI scans of interest) with the T1-weighted MR scans used to create the cortical model. This transformation should be a rigid-body (i.e., six parameter with 3 rotations and 3 translations) transformation since two structural scans of the same subject are being aligned. The resulting transformation is then applied to the original fMRI data to line up the functional data with the 3D cortical model (see Pouratian et al. 2001 and Cannestra et al. 2001 for examples of this approach implemented). OIS maps are projected onto the cortical model by matching sulcal landmarks on the cortical model with identical sulcal landmarks on the raw optical images. The accuracy of these projections is demonstrated in Figure 13.8

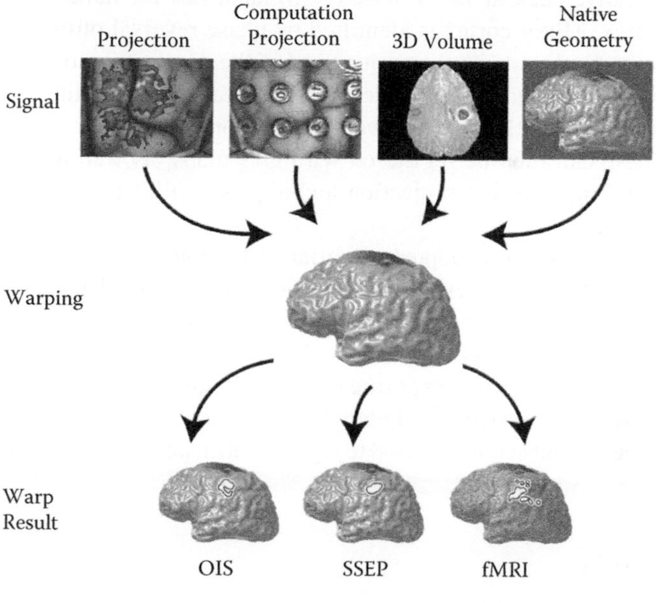

FIGURE 13.3 Inter-modality comparisons. In order to compare mapping signals across modalities, multiple modalities are warped a common space. This illustration demonstrates how each modality must be warped from its native space into a common space so that the mapping signals from each modality can be compared to each other. The common space is the cortical model of each subject's brain. Depending on the modality, different warping strategies are employed. While optical imaging is a surface imaging technique and requires projection of the data onto the cortical surface, fMRI is a tomographic technique and is warped into the common space using rigid body transformations to align the fMRI volume with the high resolution MRI used to extract the cortical surface. Electrophysiological data is warped using projection techniques similar to OIS data. The darkened area at the top of each cortical extraction demarcates tumor localization.

in the column labeled Region of Interest, which superimposes the iOIS field of view on the 3D cortical model. It is clear that sulci in the optical images are continuous with sulci on cortical extractions, confirming the success of the projections (see Figure 13.8 caption for more details).

13.3.8 Registration with Other Modalities: Electrocorticography

Somatosensory-evoked potentials (SSEPs) are recorded intraoperatively to help identify cortical sensory representations. Along with cortical electrical stimulation mapping (ESM), SSEPs are the current standard for functional localization during neurosurgical procedures.[59] Registration of SSEP maps with OIS maps is thus an important validation of intraoperative OIS. SSEPs are typically recorded using either a 20-electrode grid or an 8-electrode strip placed on the cortical surface. The median and/or ulnar nerve are stimulated using a bipolar electrode and Grass Instruments (S-12, Quincy Massachusetts) stimulator, using a 200 μsec square wave electrical pulse at 5–7 Hz for 2 sec, 15.5–17.0 mA amplitude. A maximal amplitude negative peak at 18-24 msec (N20) identifies the hand region of somatosensory cortex. Motor cortex is identified by phase reversal of the negative peak across the central sulcus, with the positive end of the dipole lying on motor cortex.[59] Images of the cortical grid or strip electrodes are taken with white or green light, which allows electrode location as well as location of vascular landmarks to be recorded. These can then be registered with OIS maps, as well as structural and functional MRI maps, using projection techniques similar to those used for OIS map registration.

The Ojemann group has adopted a similar methodology for registration of electrophysiological data.[60] However, instead of relying on sulcal landmarks for registration, they use the position of the cortical blood vessels. Preoperatively, a MR angiogram is performed and superimposed onto the subject's cortical extraction. After reflection of the dura and exposure of the brain surface, the arteries and branch points that have been superimposed onto the cortical extraction are identified on the cortical surface and subsequently used to register intraoperative maps with preoperative scans.

13.4 RESULTS

13.4.1 The Time Course of Optical Signals in Humans

The temporal profile of optical responses in humans is similar to that observed in animal models: appearing within 1 sec, peaking between 2 and 3 sec, and disappearing by 9 seconds (Figure 13.4).[2,52] This similarity suggests that analogous phenomena are being imaged in animals and humans. This also suggests that studies elucidating physiology and pathophysiology of the brain in animal models may be generalized to humans and perhaps further investigated in humans. Supporting this assertion, Cannestra et al. demonstrated that hemodynamic refractory periods, or reduced vascular response capacities with temporally close stimuli, originally observed in rats using optical imaging[61] were also observable in humans[5] (Figure 13.5). They found,

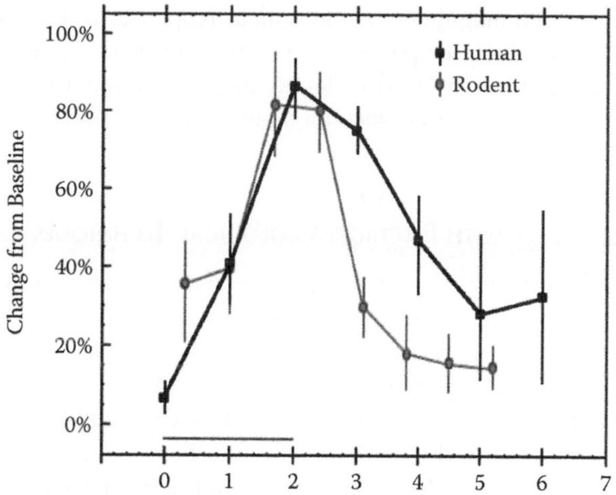

FIGURE 13.4 Rodent and human optical time courses. Comparison of the principal component time courses from optical response data in rodents and humans (Cannestra et al. 1996). Optical responses were recorded after peripheral stimulation (human: 2 s transcutaneous median or ulnar nerve stimulation, pulse interval 0.2 s, pulse duration 200 μs, 15.5–7 mA current; rodent: 0.3 s electromechanical motorized nudger, C1 whisker, 10 Hz) that began at time of 0 s. Signals appeared within 1 s, peaked between 2 and 3 s, and decayed by 6 s in both humans and rodents. Error bars indicate standard error of the mean.

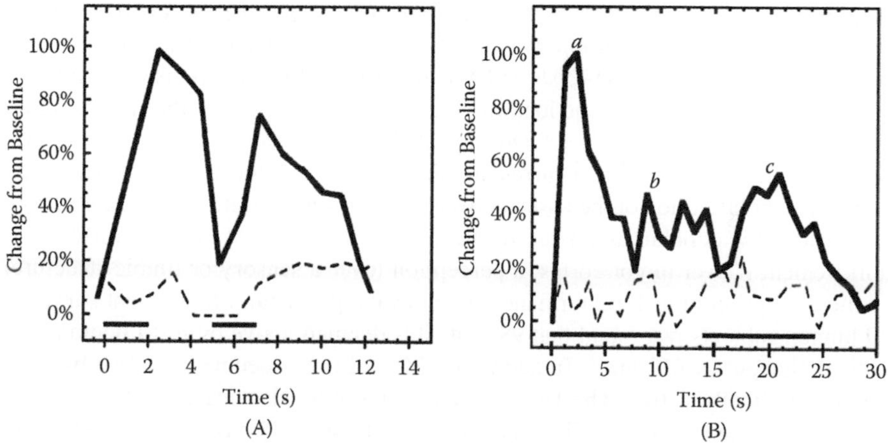

FIGURE 13.5 Hemodynamic refractory periods observed in the human cortex. Human vascular refractory period. Evidence for temporal dynamic behavior and vascular refractory periods in humans. (A) Time course of optical response to two stimuli, each 2 s in duration, separated by 3 s. The response to the second stimulus is significantly diminished. (B) Time course of optical response to two stimuli, each 10 s in duration, separated by 5 s. Once again, the response to the second stimulus appears much smaller, missing the initial peak in the optical response evident in the response to the first stimulus. Optical responses were recorded at 610 nm. Time courses are individual stimulation trials (solid line), and interleaved nonstimulated control trials (dashed line) from a single subject.

at first in rats, that the magnitude of the optical response to a stimulus presented immediately after another temporally close stimulus was significantly reduced. Subsequent investigations in humans found that delivering two sensory stimuli within seconds of one another produced a significantly diminished response to the second stimulus (Figure 13.5).

13.4.2 Correlation with Electrophysiological Techniques

The spatial correlation of perfusion-related responses and underlying electrophysiological activity is not well defined, primarily due to differing etiologies of mapping techniques and resolution differences. For example, the blood oxygenation level dependent (BOLD) fMRI signal is thought to emphasize changes in venous capillary oximetry and therefore represents a venous blood dependent indicator of brain activation.[4,24,62] In contrast, O^{15} PET studies often measure changes in blood flow to produce functional maps.[63] Optical imaging, on the other hand, produces activation maps[18,20] that are derived from both perfusion- and metabolism-related signals. Spatial maps therefore may differ across techniques.

Consistent with observations in animal studies, all the human optical studies to date indicate that the observed reflectance changes are spatially correlated with maps derived using electrophysiological techniques (somatosensory-evoked potentials (SSEPs) and electrical stimulation maps (ESM)). Stimulus-related cortical optical reflectance changes during median and ulnar nerve stimulation have been shown to co-localize with the largest SSEPs in both sensory and motor cortices[2] (Figure 13.6A–C). Perhaps more importantly, optical maps also co-localize with the current gold standard of intraoperative cortical mapping, electrical stimulation mapping or ESM (Figure 13.6D–G).[1,10,24] ESM was the first technique used for intraoperative mapping of brain function, and was first applied within the neurosurgical setting by Penfield and Boldrey.[64] During ESM, bipolar electrodes are placed on the surface of the brain and direct current applied (4 to 20 mA for 2–4 s). The technique causes a depolarization of the tissue between and around the site where the electrode is placed. This depolarization can result in the stimulation of an area, resulting in a movement if over motor cortex, a perception (over a sensory or limbic structure), or the disruption in task performance in an awake procedure. In a recent survey of 10 human subjects, we found that 98% of sites deemed active by ESM demonstrate optical changes at 610 nm.[24] In addition, ~25% of areas deemed inactive by ESM are also optically active. The latter indicates that optical maps extend beyond the regions indicated by ESM. This phenomenon of "spread" has also been observed in the rodent somatosensory "barrel" cortex in response to whisker stimulation.[65–67] Narayan et al.[18] reported optical and intravascular fluorescent dye maps overspilled regions of electrophysiologic activity (using single unit recordings) by about 20%. Optical maps encompass not only the principal barrel but also adjacent barrels. This is believed to be due to low-level neuronal activity that occurs in adjacent barrels in responses to stimulation of adjacent, nonprincipal whiskers. In humans, the "spread" phenomenon may be due, in part, to the fact that optical imaging detects both essential and secondary (i.e., active but not necessary for completion of task) cortices while ESM only detects essential areas. Alternatively, this "spread" may represent

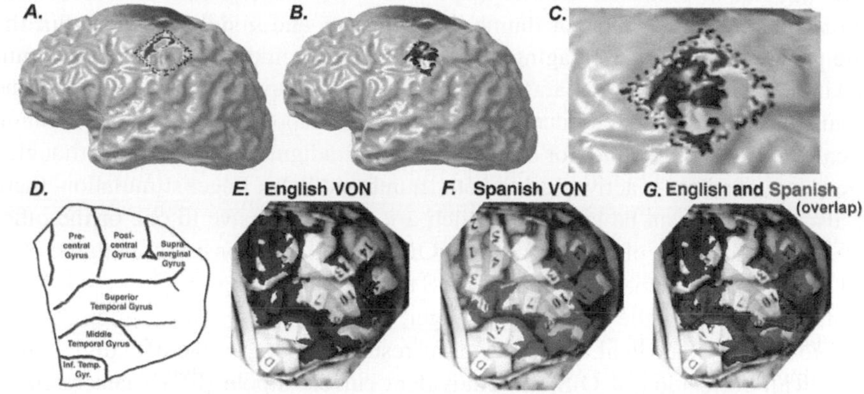

FIGURE 13.6 **(See color figure following page 224)** Colocalization of optical intrinsic signals (OIS) and electrophysiological maps. (A,B,C) Colocalization of OIS and somatosensory evoked potentials (SSEPs). (A) Cortical grid-based SSEP map of response to median nerve stimulation. Square denotes the boundary of the electrode grid. Activated region represents area bounding a +/–14 mV change from baseline. The larger activated region is over the hand region of postcentral gyrus; the smaller region is over the corresponding precentral gyrus (motor cortex). Stimulus used was a transcutaneous right median nerve stimulation (2 s duration, 200 μs square wave pulses at 5 Hz, 15 mA current). (B) Optical response to identical stimulation paradigm, thresholded at $0.5*10^{-3}$ change from baseline. (C) Superposition of optical response and SSEP, demonstrating a high degree of overlap. Both sensory and motor strips were activated under both paradigms, likely because stimulus intensity was sufficient to cause involuntary hand and finger movements. (D,E,F,G) Colocalization of optical signals and electrocortical stimulation (ESM) maps. Tags indicate critical motor and language areas (numbers) and areas in which after-discharges were observed with stimulation (letters). Blank tags demarcate areas in which no effect of cortical stimulation was observed. (D) Schematic with labels identifying gyri observed in subsequent images. (E) Optical response to an English visual object naming task (in which the patient is asked to name line drawings; VON). Note that optical activity (thresholded at 1 S.D. change from mean response) was observed over the same gyri in which critical ESM language sites were identified and no optical activity was observed over blank tags (which indicate areas not deemed essential for language processing by ESM). (F) Optical response to a Spanish VON task. (G) Comparison of English and Spanish VON. Gray region indicates areas of activation overlap between the two languages tested. iOIS maps during English and Spanish visual object naming in a single subject revealed cortical areas active during both tasks and areas specifically activated by a single language.

a nonexact co-localization of optical signals and electrophysiological activity due to imprecise physiological coupling of neuronal activity, metabolism, and perfusion.

13.4.3 OIS of Sensorimotor Cortex

iOIS studies indicate that optical maps not only co-localize with electrically active cortex (Figure 13.6A–C), but are also task-specific. Haglund et al. originally demonstrated task specificity across cortices, showing that tongue movement and naming tasks activated distinct gyri.[1] Subsequently, we demonstrated that median and ulnar nerve stimulation produced distinct maps within the same gyrus.[2] iOIS has been used to obtain distinct maps of thumb, index finger, and middle finger within the same gyrus.[5] During iOIS imaging, thumb, index, and middle fingers were stimulated individually to obtain separate cortical activation maps, mapping a subset of the homunculus. Although the borders of these maps overlapped, the areas of maximum optical changes were distinct for each activation paradigm. Furthermore, although a large extent of cortex is active during both thumb and index finger stimulation, there are also areas that can be identified which are only responsive to one or the other finger. The orientation of individual digit iOIS responses is consistent with the classical homunculus schematic, with peak iOIS responses showing a superior to inferior orientation for the middle finger, index finger, and thumb.

More recently, Sato et al. showed similar results to first through fifth digit stimulation, with correlation of OIS and equivalent current dipole (ECD) generated by preoperative magnetoencephalography (MEG) and positron emission tomography (PET).[6,7] They and others also investigated human face somatotopy,[6–8] showing responses in human consistent with maps generated in primate model systems. One study [6] also showed the potential limitations of OIS technique in pathological tissue: in a patient with arachnoid thickening over the area of interest, the highly vascularized arachnoid tissue responded nonspecifically to all peripheral stimulation, preventing localization of a functional response in surrounding tissue. However such a nonspecific response is quite easily identifiable, and might also serve as a marker for abnormal tissue.

Taken together, these studies begin to describe the morphology of perfusion related responses and cortical activity in humans. Optical reflectance imaging, combined with SSEP mapping of the sensorimotor cortex has begun to illustrate the relationship between neuronal firing, vascular and metabolic activity in human. Additionally, these experiments suggest that the development and implementation of a non-contact, repeatable, intraoperative tool for functional mapping is of practical significance, and may be used in conjunction with SSEPs to avoid eloquent brain during resection.

13.4.4 Intraoperative Maps of Language

iOIS combined with ESM mapping provides a multimodality approach for the characterization of language cortices (Figure 13.6D–G).[1] Building on previous human studies,[2,5,52] we investigated the topographical and temporal activation of language cortices within the surgical setting utilizing intraoperative OIS (iOIS) and ESM.[10] We compared disruption (ESM) and high-resolution activation (iOIS) maps, as well as paradigm specific changes observed with iOIS. This study differed from previous reports[68–70] by characterizing spatial and temporal responses both between and within Broca's and Wernicke's areas, once again demonstrating task-specificity of optical responses. iOIS was used to identify subregions of Broca's and Wernicke's by identifying different patterns of activation in each region during different language tasks.[10] These observations demonstrate response profile differences dependent upon cortical region and language task.

We also investigated language representations in a proficient bilingual patient using a visual object-naming task (Figure 13.6D–G). Our results indicate cortical areas activated by both English and Spanish (superior temporal sulcus, superior and middle temporal gyri, and parts of supramarginal gyrus). In addition, language-specific areas were identified in the supramarginal (Spanish) and precentral gyrus (English). These results suggest that cortical language representations in bilinguals may consist of both overlapping and distinct components. Furthermore, this study demonstrates the utility of iOIS in detecting topographical segregation of cognitively distinct cortices.

13.4.5 Imaging of Cortical Pathology

Intraoperative OIS is necessarily performed only in patients with significant cortical pathology requiring resection. Typically, these are epileptic foci, tumors, or arteriovenous malformations (AVM). Haglund's pioneering study[1] revealed signal differences in signal over a stimulated epileptic focus compared with surrounding normal regions, including decreased activity in the periphery of the focus, which was not seen on stimulation of normal areas and could represent an inhibitory surround. The ability to map normal (and essential) regions around a pathologic focus was also demonstrated in this original intraoperative OIS study. A more recent case study used consecutive dual-wavelength imaging to investigate the characteristics of a seizure focus in a region of motor-cortex-mediating tongue movement.[13] After ESM, which demonstrated that the region was indeed responsible for tongue movement, several spontaneous seizures occurred, and were recorded either at 530 nm (an isosbestic point sensitive to blood volume changes) or 660 nm (a point at which deoxyhemoglobin absorbtion dominates; sensitive to oxygen extraction). Contrary to conventional wisdom, blood volume changes were seen to be smaller in areal extent than oxygen extraction changes. It is possible that neurovascular coupling in a seizure focus is disregulated, with insufficient blood delivery to meet demand (small 530 nm signal), resulting in a larger than expected increase in oxygen extraction (larger 660 nm signal). However, it is not possible to rule out that the seizure focus was simply larger while 660 nm imaging took place, since imaging at 530 nm and 660 nm was sequential rather than simultaneous. Simultaneous dual-wavelength imaging would help resolve this uncertainty.

The Haglund group has also performed imaging of acute seizure interventions in the operating room.[12] In a series of patients with a neocortical or mesial temporal seizure focus, both optical and electrophysiological indices of excitability were decreased 20 minutes after infusion with either furosemide or mannitol. Electrophysiologically, interictal spike rate decreased significantly, and afterdischarge threshold increased. Optically, the area of response to a subthreshold electrical stimulation (a stimulation to the seizure focus that did not evoke after-discharges) was significantly smaller after treatment with either furosemide or mannitol. This study shows the potential of intraoperative OIS to follow dynamic changes in cortical activity in response to treatment.

Finally, Haglund et al. have investigated the characteristics of brain tumors during resection surgery, using injection of the biocompatible dye indocyanine green

(ICG; absorption maxima between 790 and 805 nm).[14] In tumors whose borders were defined by intraoperative ultrasound, the retention time of ICG was found to be significantly greater in both high and low grade tumors than in the normal periphery. However, there was considerable overlap in retention times between the three types of tissue. Intriguingly, one tumor which was noncontrast-enhancing on MRI (suggestive of a low-grade lesion) was found to have a prolonged dye retention time more consistent with a high-grade tumor, which was confirmed on pathology. The possibility of such dye-enhanced OIS for investigation of tumor margins was also investigated, but there was again considerable overlap of increased signal within both abnormal and normal structures. ICG binds plasma proteins and thus reveals either blood volume or plasma protein extravasation. Dyes more specific for tumor tissue (perhaps a chromophore bound to a tumor antigen) would allow better use of the high spatial resolution of OIS, and could facilitate the clearance of tumor margins. The fact that OIS is not a tomographic or 3D technique would not preclude this dye-based approach, as sequential imaging of the tumor cavity surface could be taken as resection advanced.

13.4.6 Comparison with BOLD fMRI

The LONI group examined the relationship between different hemodynamic mapping signals measured by iOIS and blood-oxygen level-dependent (BOLD) fMRI to better elucidate the temporal and spatial characteristics of perfusion related cortical signals in humans.[3,4] Current theory holds that the first functional or metabolic change to occur following neuronal activation is likely a brief burst of oxidative metabolism at the site of neuronal activity.[17,20,23] This burst of oxidative metabolism is believed to produce a local increase in deoxyhemoglobin concentration, which is very tightly spatially colocalized with site of neuronal activity and responsible for the negative (darkening) phase of the 610 nm optical signal and the "initial dip" in the BOLD fMRI response. Note that this initial increase in deoxyhemoglobin concentrations remains controversial since not all groups have observed it.[71] Following this brief burst of oxidative metabolism, a slower onset (approximately 1 s poststimulation) increase in functional perfusion begins, recruiting increased oxy-hemoglobin and "washing out" increases in deoxyhemoglobin concentration.[18,20,72] This increase in oxyhemoglobin and decrease in deoxyhemoglobin probably mediates the positive (brightening) 610 nm OIS and positive BOLD signals. The increased functional perfusion may peak earlier in gyral tissue (which contains arterial and capillary components) and later in the sulcal venous structures (schematized in Figure 13.7).

To test this possibility, sensorimotor cortex was studied during 110 Hz index finger vibration in eight human subjects undergoing neurosurgical procedures for removal of intracranial pathology (tumors or arteriovenous malformations, which are pathological shunts between arterial and venous systems). The same experimental parameters were used across modalities. All three modalities were coregistered in 3D.

Different spatial patterns and temporal response profiles were observed across modalities (Figure 13.8A).[4] Although overall fMRI, optical, and SSEP activities were observed with very similar spatial distributions, fMRI and the negative 610 nm optical response were separated by statistically significant distances, and temporally,

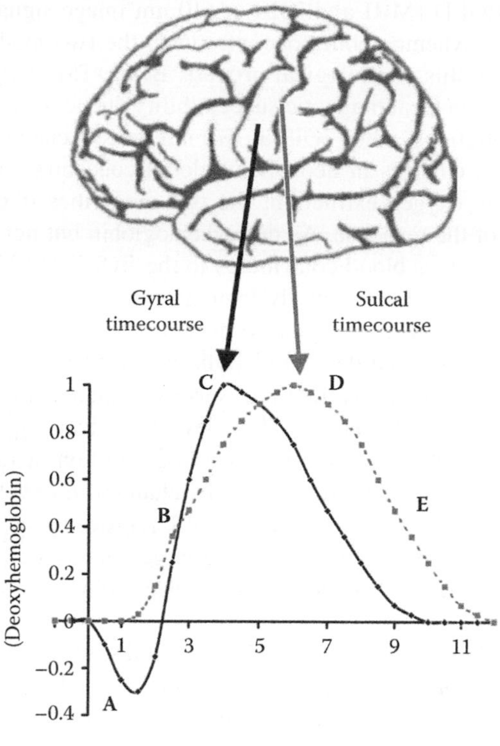

FIGURE 13.7 Schematic model of hemoglobin oximetry changes that occur following functional activation. The time course of hemoglobin changes may be different in gyri, which contain the capillary beds, and sulci, which contain large venous structures. (A) Following functional activation, there may be a brief increase in deoxyhemoglobin concentrations in the cortex, corresponding to an immediate burst in oxidative metabolism preceding functional perfusion changes. This phase is often referred to as the initial dip in the blood-oxygen level dependent fMRI signal or the first phase of the 610 nm optical response. (B) Within seconds, functional recruitment of blood volume and flow reverse the initial increase in deoxyhemoglobin, "washing out" and eventually decreasing deoxyhemoglobin concentrations below baseline. (C,D) The functional blood volume and flow changes may peak in the capillary beds and cortex before peaking in the sulci, or in the large venous structures, generating the BOLD signal and the second phase of the 610 nm optical response. (E) Eventually, functional changes subside and the brain returns to baseline. Note: Time courses displayed are hypothetical and are illustrated to explain a possible model of hemoglobin oximetry changes with functional activation.

the total fMRI response was delayed by 2–3 s relative to the optical signals. The temporal profile of the responses became more similar when pixels in the same location from each modality were examined. In contrast to fMRI, SSEP and optical activity showed greater spatial correspondence. The optical map colocalized with the SSEP map region corresponding to a 13 mV potential change (within the 0.5 cm interpolation distance required for SSEPs). This colocalization pattern is consistent with close SSEP/optical spatial coupling observed in previous experiments.[2,5]

Although both BOLD fMRI and iOIS at 610 nm image signals that arise from changes in local deoxyhemoglobin concentrations, the two modalities image distinct consequences of this physiological process. BOLD fMRI signals are believed to arise from changes in local magnetic susceptibility due to local changes in deoxyhemoglobin concentrations, while iOIS at 610 nm detects changes in cortical light reflectance related to changes in deoxyhemoglobin concentrations. Although both signals are related to oxygen extraction, the two modalities likely emphasize different components of the response. As deoxyhemoglobin but not oxyhemoglobin is paramagnetic, only venous blood contributes to the BOLD fMRI signal.[73] Because fMRI is tomographic, it collects equally from gyri and sulci, and surface and deep tissue. Meanwhile, iOIS has poor signal to noise in the sulci, given the presence of surface arteries and veins, and decreased light penetration. As a result iOIS likely emphasizes changes occurring on the gyral surface, with a consequent de-emphasis of sulcal (prominently vascular) signal. This may be an advantage, since venules may drain blood from multiple cortical areas. If a large portion of fMRI signal comes from venous structures (those most affected by changes in oxygen extraction), this may effectively reduce its spatial resolution. iOIS, registering signal primarily from the gyri, may be less affected by changes in large venous structures, which may explain the greater correlation it achieves with electrophysiological measures compared with fMRI.

In another study comparing BOLD fMRI and iOIS, we investigated the temporal/spatial correlation of the positive BOLD signal with the positive 610 nm OIS response.[3] Our prior study compared the early negative 610 nm OIS response with fMRI maps using short (2 s) somatosensory stimuli.[4] Most studies of fMRI, however, utilize a block design (30–60 s) in order to obtain a steady BOLD signal. We examined the same motor tongue task in both modalities. This task was chosen because (1) it activates primary motor and sensory cortices which are often associated with more robust and less complex hemodynamic responses,[62] and (2) motor tongue activates the inferior-lateral aspect of the pre- and postcentral gyri and the central sulcus, which are easily accessible neurosurgically for optical imaging experiments.

In all subjects, robust fMRI activations ($p < .001$, correlation with a box-car reference function convolved with a hemodynamic function with a 6 s lag to peak, 84 timepoints per subject, 5 subjects) were identified which centered on the lateral inferior aspect of the central sulcus and extended into adjacent the pre- and postcentral gyri, adjacent to the gyrus containing the ESM tongue localization. OIS and fMRI maps co-localized although the spatial extent of optical responses was larger ($p = .03$) and contained more gyral components (Figure 13.8B). In most cases, the fMRI signal was a subset (48%) of the OIS signal. The time course of the fMRI and OIS signals were similar: appearing within 2.5 s and peaking within 6 s of task onset. Although many physiological processes may contribute to the positive 610 nm response, most of this response is due to a washout of deoxyhemoglobin by oxyhemoglobin. This study, therefore, supports the theory that BOLD signals represent a similar phenomenon. The inconsistencies between OIS and fMRI maps may be attributed to cell swelling and scattering that also contribute to OIS and/or fMRI sensitivity. While OIS spatial resolution is on the order of microns, fMRI spatial

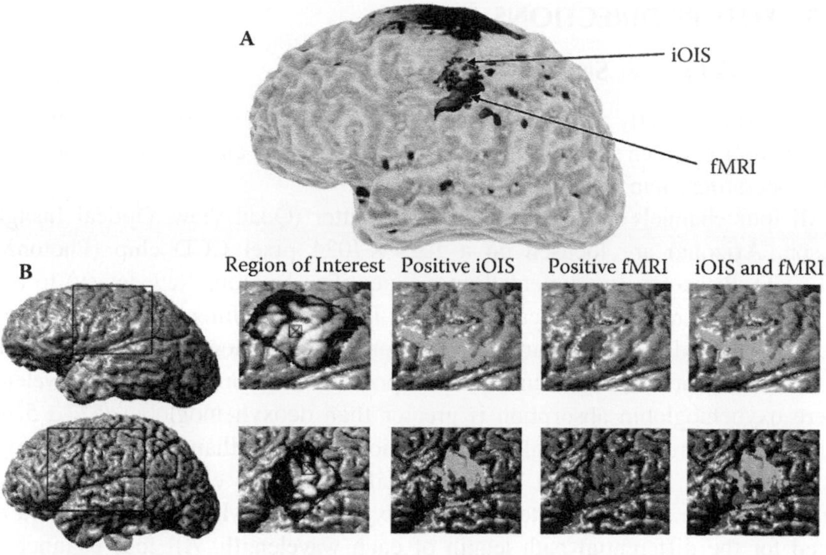

FIGURE 13.8 **(See color insert following page 224)** Comparison of negative 610 nm OIS and BOLD fMRI signals in response to 2 s somatosensory stimulus. (A) Sensori-motor cortex was studied using identical stimulation parameters for fMRI and OIS: 2 s 110 Hz index finger vibration in eight human subjects. The signals occurred adjacent to one another: the center of mass of the signals was separated by ~17 mm. While OIS signals were gyral, fMRI signals centered more superiorly and within sulci, consistent with a venous origin of BOLD fMRI signals. The gyral localization of the iOIS signal is consistent with optical signals emanating from capillary plexuses within the cortex. The BOLD fMRI signal is displayed as a 3D activation both at the surface of the brain and deep within the brain. The fMRI signal has been thresholded for display: all voxels with a $p < .01$ for the correlation of the signal with a predetermined hemodynamic function are displayed. (B) Comparison of positive 610 nm response with BOLD fMRI using a block paradigm tongue motor task (20 s blocks). Using both modalities, signals were identified near the inferior-lateral aspect of the central sulcus. The positive 610 nm response and the BOLD fMRI signal demonstrate high correlation overlapped spatially (last column). This data supports the notion that the two signals share similar etiologies, associated with the washout of deoxyhemoglobin by in functional recruitment of cerebral blood flow and excess oxyhemoglobin.

resolution is on the order of millimeters. Consequently, OIS may be able to detect small changes occurring within the gyri while similar changes occurring within the gyri of the larger fMRI voxel may be "drowned out" and undetectable.[62]

By comparing positive BOLD signals and positive 610 nm OIS signals within subject in humans, we have demonstrated the high likelihood that these signals share a common etiology. These findings support the common theory that positive BOLD signals are due to a washout of deoxyhemoglobin by oxyhemoglobin following functional activation. This work also demonstrates that fMRI signals may not precisely co-localize with electrophysiological or optical maps, instead emphasizing changes in venous structures in adjacent sulci.

13.5 FUTURE DIRECTIONS

13.5.1 2D Optical Spectroscopy

Our group has recently developed a method for 2D optical spectroscopy using image splitting optics, which should be time- and space-efficient enough to be implemented in the operating room[74] (Figure 13.9).

All four channels of a four-way beam splitter (Quad-View, Optical Insitghts, Tucson, Arizona) are focused on a 1024 × 1024 pixel CCD chip (PhotonMax EM-CCD, Princeton Instruments/Roper Scientific, Trenton, New Jersey) to allow 512 × 512 pixel resolution at each channel. Interference filters are placed in each channel, selected to pass a particular portion of the hemoglobin spectrum. There is one channel at a hemoglobin isosbestic point (570 ± 5 nm), one at a wavelength where oxyhemoglobin absorption is greater than deoxyhemoglobin (577 ± 5 nm), and two at which deoxyhemoglobin absorption is greater than that of oxyhemoglobin (560 ± 5 nm; 610 ± 5 nm). Species-specific phantoms consisting of blood and intralipid[75,76] are used to create reference oxy- and deoxyhemoglobin spectra corrected for the differential path length of each wavelength. All four channels are acquired simultaneously during a typical OIS experiment, and multiple trials are averaged as described previously. Next, all four channels are registered using an algorithm similar to Automated Image Registration,[54,55] namely, normalized cross-correlation of intensity as a similarity measure and Newton-type maximization to cross-correlate the channels across frames. Additionally, because image contrast

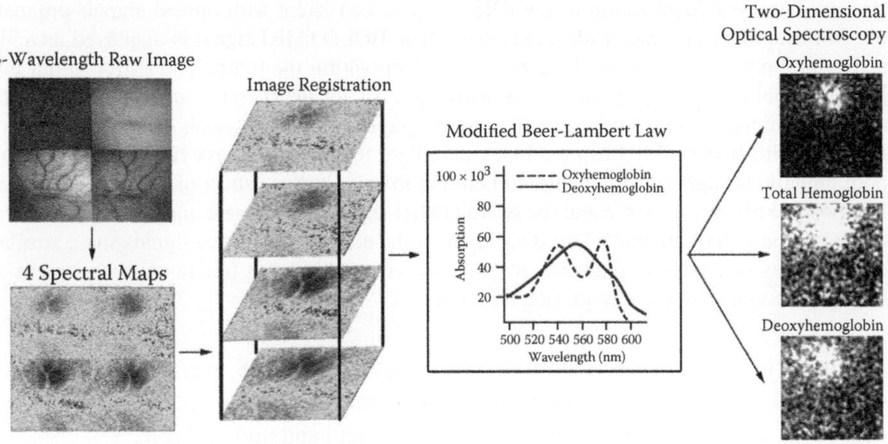

FIGURE 13.9 2D optical spectroscopy (2DOS). Images are collected simultaneously at four wavelengths (560 nm, 570 nm, 577 nm, 610 nm) using a four-way beam splitter. Images are spatially registered, and grayscale changes are normalized. Each registered pixel thus has four grayscale values (one for each wavelength) which change over time. For each pixel at each time point, least squares analysis is used to fit the measured absorbance changes to changes in oxy- and deoxyhemoglobin concentration using a modified form of the Beer–Lambert law. Oxy- and deoxyhemoglobin absorption coefficients for each wavelength are derived from an in vitro phantom, thus accounting for wavelength-dependent path differences in the brain.

is wavelength-dependent, we apply gray-level transformations to equalize contrast between channels. Finally, a modified form of the Beer–Lambert law is used to compute the fractional change in oxy-, deoxy-, and total hemoglobin concentration from baseline[74] to generate specific spatiotemporal maps for each moiety.

We have validated this approach in rat and mouse, and are currently using the 2DOS apparatus in the operating room. The advantage of the beam-splitting system is that it allows simultaneous collection of all four wavelengths in a relatively small package that can be attached to an operating microscope. Other techniques for 2D optical spectroscopy use filter wheels, which necessarily involve a small delay between image acquisitions, and also occupy more space. The acquisition of functional maps specific to each hemoglobin moiety, as well as blood volume, should allow detailed investigation of physiology in intact and pathological cortex. Co-registration with flowmetric technique such as laser doppler or laser speckle flowmetry (LDF, LSF) would allow computation of the cerebral metabolic rate of oxygen consumption (CMRO2), which could be particularly interesting in areas of pathology such as tumors, seizure foci, and arteriovenous malformations.

13.5.2 Other Areas of Inquiry

There is no reason why flavoprotein intrinsic fluorescence imaging[47] could not be implemented in the operating room, as it involves no change to the experimental apparatus except for the addition of interference filters to the reflected light path. Laser doppler flowmetry has been used intraoperatively,[77] and as mentioned previously, could provide a useful complement to spectroscopic measurements. As laser speckle flowmetry[78] uses less focused laser power than LDF, it should be safe for use in the operating room; indeed, LSF has been used to measure retinal blood flow in humans, and power deposition is within acceptable American National Standards Institute (ANSI) limits.[79]

One disadvantage of iOIS is that it is a 2D technique, sampling light within a few hundred microns of the cortical surface. Optical coherence tomography (OCT), a light-based interferometric technique used routinely in ophthalmology, allows optical sectioning of tissue to a depth of approximately 1.5 mm, and micron spatial resolution.[80] The potential utility of OCT for neurosurgery has been recognized, especially regarding investigation of tumor margins and neuroendoscopy.[81,82] However, to date, investigations have taken place in ex vivo tumor or brain specimens. The fact that OCT is tomographic and, unlike current iterations of OIS, can present real-time data, is a particular advantage.

13.6 CONCLUSIONS

13.6.1 Intraoperative Optical Imaging

(iOIS) offers a unique opportunity to image perfusion and metabolism-related responses in humans with both high spatial and temporal resolution. Several studies have already demonstrated the feasibility of the technique and its compatibility with the operating room environment. Comparison of iOIS signals with other modalities using the techniques described can potentially contribute great insight into both physiology and the functional organization of the brain. We close with some observations about the clinical and research utility of iOIS, moving forward.

13.6.2 Clinical Utility

iOIS offers several potential advantages over the current gold standard for intraoperative mapping, ESM. The spatial resolution of iOIS (~200 μm) is much greater than that of ESM. Resection within 1 cm of essential areas identified by ESM increases the likelihood of postoperative deficits,[11] suggesting the limit of resolution of ESM is on the order of 1 cm. This resolution is limited and does not ensure resection of as much pathological tissue as possible. iOIS can potentially provide a more detailed functional map of the exposed cortex, with consequently more precise resections.

The iOIS setup requires no modification of the traditional operating room configuration and does not require any major modification to neurosurgical techniques, as has been necessary for implementation of intraoperative MR (iMR). Furthermore, the camera used for optical imaging can be mounted atop a second operating microscope so that it can be moved into and out of the sterile operating field as necessary, minimizing surgical interference. iOIS offers the additional advantage that it does not perturb the brain during surgery. Since iOIS only detects changes in light reflectance, it does not require any contact with potentially normal tissue that the surgeon may not want to disturb. Moreover, optical imaging equipment is affordable since it only requires three major components: camera, light source, and a computer (note that the same equipment can be used for both animal and human imaging). Other intraoperative functional imaging techniques, like iMR, may on the other hand be prohibitive since the investigator must not only incur the expense of the MRI scanner but modify the entire surgical environment to be magnet compatible.

The greatest obstacle to wholesale clinical adoption of iOIS is the difficulty in generating reliable real-time intraoperative maps. Most iOIS image processing takes place off line, and registration with 3D datasets, such as structural and functional MRI, has been exclusively offline thus far. The result is that OIS data is not immediately available to the neurosurgeon. Another issue is the variability of OIS data, which can be considerable, especially with anesthetic changes. The first problem should be soluble with current computing technology and image registration algorithms,[54,55] The second will likely depend on coregistration of OIS data with electrophysiological (ESM, SSEP, MEG) and other imaging (fMRI, PET, angiography) data.

13.6.3 Research Utility

Intraoperative OIS is unique in allowing direct access to the human brain, in both awake and anesthetized states. Perhaps the most obvious utility of iOIS is continued investigation of the correlation of OIS and fMRI. The advent of intraoperative fMRI makes this comparison possible without much of the warping necessary to compare preoperative and operative imaging. The initial dip on fMRI remains controversial, and the rigorous comparison of optical and MR signals in humans may be one of the only ways to put the controversy to rest. A more recent issue in the fMRI literature involves the possible anatomical and functional significance of slow oscillations in BOLD signal.[84] Continuous operative iOIS imaging, in the awake and anesthetized state, should be able to resolve these oscillations at higher spatiotemporal resolution than that allowed by fMRI.

Another important theme is the basic mechanics of neurovascular coupling in human. Most investigation of neurovascular coupling is performed in lissencephalic animals one to two orders of magnitude smaller than humans. Significant differences exist even between mouse and rat in hemodynamic response to both physiological and pathological stimuli.[74,85,86] Yet other aspects of neurovascular coupling, such as response timing, appear highly conserved between rodents and human[52] (Figure 13.4). Any insights gained in animal model systems will need to be verified in humans. Of course, any difference found between human and other species would also be worthy of investigation. The ability to image relative changes in hemoglobin concentration as well as blood flow, mitochondrial redox state, and CMRO2 with the spatiotemporal resolution of iOIS, should provide such information in ways that PET or fMRI cannot.

A third opportunity afforded by intraoperative OIS is to compare the basic characteristics of normal and pathological tissue at high spatial resolution. Intraoperative OIS, especially enhanced with etiologically specific tools such as spectroscopy, intrinsic fluorescence, and possibly dye imaging, should allow development of tissue signatures which, in addition to being used acutely in the OR, could be used in the validation of animal systems.

Fourth, the fact that many operative cases involve both the awake and anesthetized state allows comparison of these states and their physiology. This is more difficult to do with fMRI and PET due to ethical constraints; by definition a patient in the operating room needs to be there and thus ethical constraints on anesthesia are less relevant. Comparatively little is known about basic cortical metabolic variables during different types of anesthesia compared with the awake state, let alone the response of cortical networks to stimulation.

Finally, the opportunity of having an awake patient on the operating table allows for higher level testing using intrinsic signals than afforded in any other situation. We have already shown that study of language function is possible;[24] other higher-order cortical functions such as visual processing, face recognition, and frontal executive function might be accessible, given a favorable surgical approach, perhaps in combination with intraoperative fMRI.

The greatest obstacle to the research applications of iOIS is lack of space and time. Clinical imperatives dictate that research imaging during surgery be kept under about 40 minutes. Given the variable nature of the hemodynamic response and lack of anesthetic control, such a time window can be insufficient for collection of good quality maps. Another major obstacle is that the individuality of each surgery—approach, pathology, anesthetic variables, etc.—makes intergroup comparison quite difficult. Despite these obstacles, the advantages of iOIS make it a compelling method whose potential has yet to be fully realized.

ACKNOWLEDGMENTS

Thanks go to the staff of the Laboratory of Neuro Imaging, without whom this work would not have been possible. This work was supported, in part, by the National Institute of Mental Health (MH 52083) and the Training Program in Neuroimaging (MH19950). KCB is supported by the National Institute of Neurological Diseases and Stroke (K08NS059072).

REFERENCES

1. Haglund, M. M., Ojemann, G. A., and Hochman, D. W. Optical imaging of epileptiform and functional activity in human cerebral cortex. *Nature* **358**, 668–671 (1992).
2. Toga, A. W., Cannestra, A. F., and Black, K. L. The temporal/spatial evolution of optical signals in human cortex. *Cereb Cortex* **5**, 561–565 (1995).
3. Pouratian, N. et al. Spatial/temporal correlation of BOLD and optical intrinsic signals in humans. *Magnetic Resonance in Medicine* **47**, 766–776 (2002).
4. Cannestra, A. F. et al. Temporal spatial differences observed by functional MRI and human intraoperative optical imaging. *Cereb Cortex* **11**, 773–782 (2001).
5. Cannestra, A. F., Pouratian, N., Shomer, M. H., and Toga, A. W. Refractory periods observed by intrinsic signal and fluorescent dye imaging. *J Neurophysiol* **80**, 1522–1532 (1998).
6. Sato, K. et al. Functional representation of the finger and face in the human somatosensory cortex: Intraoperative intrinsic optical imaging. *NeuroImage* **25**, 1292–1301 (2005).
7. Sato, K. et al. Intraoperative intrinsic optical imaging of neuronal activity from subdivisions of the human primary somatosensory cortex. *Cereb Cortex* **12**, 269–280 (2002).
8. Schwartz, T. H., Chen, L. M., Friedman, R. M., Spencer, D. D., and Roe, A. W. Intraoperative optical imaging of human face cortical topography: A case study. *Neuroreport* **15**, 1527–1531 (2004).
9. Nariai, T. et al. Imaging of somatotopic representation of sensory cortex with intrinsic optical signals as guides for brain tumor surgery. *J Neurosurg* **103**, 414–423 (2005).
10. Cannestra, A. F. et al. Temporal and topographical characterization of language cortices using intraoperative optical intrinsic signals. *NeuroImage* **12**, 41–54 (2000).
11. Haglund, M. M., Berger, M. S., Shamseldin, M., Lettich, E., and Ojemann, G. A. Cortical localization of temporal lobe language sites in patients with gliomas. *Neurosurgery* **34**, 567 (1994).
12. Haglund, M. M. and Hochman, D. W. Furosemide and mannitol suppression of epileptic activity in the human brain. *J Neurophysiol* **94**, 907–918 (2005).
13. Haglund, M. M. and Hochman, D. W. Optical imaging of epileptiform activity in human neocortex. *Epilepsia* **45**, 43–47 (2004).
14. Haglund, M. M., Berger, M. S., and Hochman, D. W. Enhanced optical imaging of human gliomas and tumor margins. *Neurosurgery* **38**, 308–317 (1996).
15. Hillman, E. M. C. Optical brain imaging in vivo: Techniques and applications from animal to man. *J Biomed Optics* **12**, 051402–051428 (2007).
16. Obrig, H. and Villringer, A. Beyond the visible—imaging the human brain with light. *J Cereb Blood Flow Metab* **23**, 1–18 (2003).
17. Fox, P. T. and Raichle, M. E. Focal physiologic uncoupling of cerebral blood flow and oxidative metabolism during somatosensory stimulation in human subjects. *Proc Natl Acad Sci U S A* **83**, 1140–1144 (1986).

18. Narayan, S., Esfahani, P., Blood, A., Sikkens, L., and Toga, A. Functional increases in cerebral blood volume over somatosensory cortex. *J Cereb Blood Flow Metab* **15**, 754–765 (1995).
19. Malonek, D. and Grinvald, A. Interactions between electrical activity and cortical microcirculation revealed by imaging spectroscopy: Implications for functional brain mapping. *Science* **272**, 551–554 (1996).
20. Frostig, R. D., Lieke, E. E., Ts'o, D. Y., and Grinvald, A. Cortical functional architecture and local coupling between neuronal activity and the microcirculation revealed by in vivo high-resolution optical imaging of intrinsic signals. *Proc Natl Acad Sci U S A* **87**, 6082–6086 (1990).
21. Grinvald, A., Frostig, R., Siegel, R., and Bartfeld, E. High-resolution optical imaging of functional brain architecture in the awake monkey. *Proc Natl Acad Sci U S A* **88**, 11559–11563 (1991).
22. Hodge, C. J., Stevens, R. T., Newman, H., Merola, J., and Chu, C. Identification of functioning cortex using cortical optical imaging. *Neurosurgery* **41**, 1137–1145 (1997).
23. Vanzetta, I. and Grinvald, A. Increased cortical oxidative metabolism due to sensory stimulation: Implications for functional brain imaging. *Science* **286**, 1555–1558 (1999).
24. Pouratian, N. et al. Optical imaging of bilingual cortical representations: Case report. *J Neurosurg* **93**, 676–681 (2000).
25. Jobsis, F. F., Keizer, J. H., LaManna, J. C., and Rosenthal, M. Reflectance spectrophotometry of cytochrome aa3 in vivo. *J Appl Physiol* **43**, 858–872 (1977).
26. LaManna, J. C., Sick, T. J., Pikarsky, S. M., and Rosenthal, M. Detection of an oxidizable fraction of cytochrome oxidase in intact rat brain. *Am J Physiol Cell Physiol* **253**, C477–C483 (1987).
27. Heekeren, H. R. et al. Noninvasive assessment of changes in cytochrome-c oxidase oxidation in human subjects during visual stimulation. **19**, 592–603 (1999).
28. Mayhew, J. et al. Spectroscopic analysis of changes in remitted illumination: The response to increased neural activity in brain. *NeuroImage* **10**, 304–326 (1999).
29. Hill, D. K. and Keynes, R. D. Opacity changes in stimulated nerve. *J Physiol* **108**, 278–281 (1949).
30. MacVicar, B. and Hochman, D. Imaging of synaptically evoked intrinsic optical signals in hippocampal slices. *J Neurosci* **11**, 1458–1469 (1991).
31. Holthoff, K. and Witte, O. Intrinsic optical signals in rat neocortical slices measured with near-infrared dark-field microscopy reveal changes in extracellular space. *J Neurosci* **16**, 2740–2749 (1996).
32. Snow, R. W., Taylor, C. P., and Dudek, F. E. Electrophysiological and optical changes in slices of rat hippocampus during spreading depression. *J Neurophysiol* **50**, 561–572 (1983).
33. Gouras, P. Spreading Depression of Activity in Amphibian Retina. *Am J Physiol* **195**, 28–32 (1958).
34. Stepnoski, R. et al. Noninvasive detection of changes in membrane potential in cultured neurons by light scattering. *Proc Natl Acad Sci U S A* **88**, 9382–9386 (1991).
35. Cohen, L. B. and Keynes, R. D. Changes in light scattering associated with the action potential in crab nerves. *J Physiol* **212**, 259–275 (1971).
36. Lipton, P. Effects of membrane depolarization on light scattering by cerebral cortical slices. *J Physiol* **231**, 365–383 (1973).
37. Fayuk, D., Aitken, P. G., Somjen, G. G., and Turner, D. A. Two different mechanisms underlie reversible, intrinsic optical signals in rat hippocampal slices. *J Neurophysiol* **87**, 1924–1937 (2002).
38. Rector, D. M., Poe, G. R., Kristensen, M. P., and Harper, R. M. Light scattering changes follow evoked potentials from hippocampal schaeffer collateral stimulation. *J Neurophysiol* **78**, 1707–1713 (1997).

39. Gratton, G. and Fabiani, M. The event-related optical signal: A new tool for studying brain function. *Int J Psychophysiol* **42**, 109–121 (2001).
40. Tse, C.-Y. et al. Imaging cortical dynamics of language processing with the event-related optical signal. *Proc Natl Acad Sci U S A* **104**, 17157–17162 (2007).
41. Lipton, P. Effects of membrane depolarization on nicotinamide nucleotide fluorescence in brain slices. *Biochem J* **136**, 999–1009 (1973).
42. Jobsis, F. F., O'Connor, M., Vitale, A., and Vreman, H. Intracellular redox changes in functioning cerebral cortex. I. Metabolic effects of epileptiform activity. *J Neurophysiol* **34**, 735–749 (1971).
43. Rosenthal, M. and Jobsis, F. F. Intracellular redox changes in functioning cerebral cortex. II. Effects of direct cortical stimulation. *J Neurophysiol* **34**, 750–762 (1971).
44. Brennan, A. M., Connor, J. A., and Shuttleworth, C. W. NAD(P)H fluorescence transients after synaptic activity in brain slices: Predominant role of mitochondrial function. *J Cereb Blood Flow Metab* **26**, 1389–1406 (2006).
45. Shuttleworth, C. W., Brennan, A. M., and Connor, J. A. NAD(P)H Fluorescence imaging of postsynaptic neuronal activation in murine hippocampal slices. *J. Neurosci.* **23**, 3196–3208 (2003).
46. Reinert, K. C., Dunbar, R. L., Gao, W., Chen, G., and Ebner, T. J. Flavoprotein autofluorescence imaging of neuronal activation in the cerebellar cortex in vivo. *J Neurophysiol* **92**, 199–211 (2004).
47. Shibuki, K. et al. Dynamic imaging of somatosensory cortical activity in the rat visualized by flavoprotein autofluorescence. *J Physiol* **549**, 919–927 (2003).
48. Husson, T. R., Mallik, A. K., Zhang, J. X., and Issa, N. P. Functional imaging of primary visual cortex using flavoprotein autofluorescence. *J Neurosci* **27**, 8665–8675 (2007).
49. Tohmi, M., Kitaura, H., Komagata, S., Kudoh, M., and Shibuki, K. Enduring critical period plasticity visualized by transcranial flavoprotein imaging in mouse primary visual cortex. *J Neurosci* **26**, 11775–11785 (2006).
50. Shtoyerman, E., Arieli, A., Slovin, H., Vanzetta, I., and Grinvald, A. Long-term optical imaging and spectroscopy reveal mechanisms underlying the intrinsic signal and stability of cortical maps in V1 of behaving monkeys. *J. Neurosci.* **20**, 8111–8121 (2000).
51. Ferezou, I., Bolea, S., and Petersen, C. C. H. Visualizing the cortical representation of whisker touch: Voltage-sensitive dye imaging in freely moving mice. *Neuron* **50**, 617–629 (2006).
52. Cannestra, A. F., Blood, A. J., Black, K. L., and Toga, A. W. The evolution of optical signals in human and rodent cortex. *Neuroimage* **3**, 202–208 (1996).
53. Woods, R. P., Cherry, S. R., and Mazziotta, J. C. Rapid automated algorithm for aligning and reslicing PET images. *J Comput Assist Tomogr* **16**, 620–633 (1992).
54. Woods, R. P., Grafton, S. T., Holmes, C. J., Cherry, S. R., and Mazziotta, J. C. Automated image registration I. General methods and intrasubject, intramodality validation. *J Comput Assist Tomogr* **22**, 139–152 (1998).
55. Woods, R. P., Grafton, S. T., Watson, J. D., Sicotte, N. L., and Mazziotta, J. C. Automated Image Registration II. Intersubject validation of linear and nonlinear models. *J Comput Assist Tomogr* **22**, 153–165 (1998).
56. Chen-Bee, C. H. et al. Visualizing and quantifying evoked cortical activity assessed with intrinsic signal imaging. *Journal of Neuroscience Methods* **97**, 157–173 (2000).
57. Shoham, D. et al. Imaging cortical dynamics at high spatial and temporal resolution with novel blue voltage-sensitive dyes. *Neuron* **24**, 791–802 (1999).
58. Kalatsky, V. A. and Stryker, M. P. New paradigm for optical imaging: Temporally encoded maps of intrinsic signal. *Neuron* **38**, 529–545 (2003).
59. Nuwer, M. R. et al. Topographic mapping of somatosensory evoked potentials helps identify motor cortex more quickly in the operating room. *Brain Topogr* **5**, 53–58 (1992).
60. Corina, D. P. et al. Correspondences between language cortex identified by cortical stimulation mapping and fMRI. *Neuroimage* **11** (2000).

61. Blood, A. J., Narayan, S. M., and Toga, A. W. Stimulus parameters influence characteristics of optical intrinsic signal responses in somatosensory cortex. *J Cereb Blood Flow Metab* **15**, 1109–1121 (1995).
62. Cohen, M. S. and Bookheimer, S. Y. Localization of brain function using magnetic resonance imaging. *Trends Neurosci* **17** (1994).
63. Phelps, M. and Mazziotta, J. Positron emission tomography: Human brain function and biochemistry. *Science* **228**, 799–809 (1985).
64. Penfield, W. and Boldrey, E. Somatic motor and sensory representation in the cerebral cortex of man as studied by electrical stimulation. *Brain* **60**, 389 (1937).
65. Chen-Bee, C. H., and Frostig, R. D. Variability and interhemispheric asymmetry of single-whisker functional representations in rat barrel cortex. *J Neurophysiol* **76**, 884–894 (1996).
66. Masino, S. A. and Frostig, R. D. Quantitative long-term imaging of the functional representation of a whisker in rat barrel cortex. *Proc Natl Acad Sci U S A* **93**, 4942–4947 (1996).
67. Godde, B., Hilger, T., von Seelen, W., Berkefeld, T., and Dinse, H. Optical imaging of rat somatosensory cortex reveals representational overlap as topographic principle. *Neuroreport* **7**, 24–28 (1995).
68. Buckner, R. et al. Functional anatomical studies of explicit and implicit memory retrieval tasks. *J. Neurosci.* **15**, 12–29 (1995).
69. Malow, B. A. et al. Cortical stimulation elicits regional distinctions in auditory and visual naming. *Epilepsia* **37**, 245–252 (1996).
70. Ojemann, G., Ojemann, J., Lettich, E., and Berger, M. Cortical language localization in left, dominant hemisphere. An electrical stimulation mapping investigation in 117 patients. *J Neurosurg* **71**, 316–326 (1989).
71. Buxton, R. B. The elusive initial dip. *NeuroImage* **13**, 953–958 (2001).
72. Ngai, A. C., Ko, K. R., Morii, S., and Winn, H. R. Effect of sciatic nerve stimulation on pial arterioles in rats. *Am J Physiol Heart Circ Physiol* **254**, H133–H139 (1988).
73. Ogawa, S. et al. Functional brain mapping by blood oxygenation level-dependent contrast magnetic resonance imaging. A comparison of signal characteristics with a biophysical model. *Biophys J* **64**, 803–812 (1993).
74. Prakash, N. et al. Temporal profiles and 2-dimensional oxy-, deoxy-, and total-hemoglobin somatosensory maps in rat versus mouse cortex. *NeuroImage* **37**, S27–S36 (2007).
75. Sheth, S. A. et al. Columnar specificity of microvascular oxygenation and volume responses: Implications for functional brain mapping. *J Neurosci* **24**, 634–641 (2004).
76. Sato, C., Nemoto, M., and Tamura, M. Reassessment of activity-related optical signals in somatosensory cortex by an algorithm with wavelength-dependent path length. *Jpn J Physiol* **52**, 301–312 (2002).
77. Arbit, E., DiResta, G. R., Bedford, R. F., Shah, N. K., and Galicich, J. H. Intraoperative measurement of cerebral and tumor blood flow with laser doppler flowmetry. *Neurosurgery* **24**, 166–170 (1989).
78. Dunn, A. K., Bolay, H., Moskowitz, M. A., and Boas, D. A. Dynamic imaging of cerebral blood flow using laser speckle. *J Cereb Blood Flow Metab* **21**, 195–201 (2001).
79. Nagahara, M., Tamaki, Y., Araie, M., and Eguchi, S. Effects of scleral buckling and encircling procedures on human optic nerve head and retinochoroidal circulation. *Br J Ophthalmol* **84**, 31–36 (2000).
80. Boppart, S. A. Optical coherence tomography: Technology and applications for neuroimaging. *Psychophysiology* **40**, 529–541 (2003).
81. Böhringer, H. J., Lankenau, E., Rohde, V., Hüttmann, G., and Giese, A. Optical coherence tomography for experimental neuroendoscopy. *Minim Invasive Neurosurg* **49**, 269–275 (2006).
82. Böhringer, H. J. et al. Time-domain and spectral-domain optical coherence tomography in the analysis of brain tumor tissue. *Lasers Surg Med* **38**, 588–597 (2006).

83. Maurer, C. R. et al. Investigation of intraoperative brain deformation using a 1.5T interventional MR system: Preliminary results. *IEEE Trans Med Imaging* **17**, 817 (1998).
84. Fox, M. D. and Raichle, M. E. Spontaneous fluctuations in brain activity observed with functional magnetic resonance imaging. *Nat Rev Neurosci* **8**, 700–711 (2007).
85. Brennan, K. C. et al. Distinct vascular conduction with cortical spreading depression. *J Neurophysiol* **97**, 4143–4151 (2007).
86. Ayata, C. et al. Pronounced Hypoperfusion During Spreading Depression in Mouse Cortex. *J Cereb Blood Flow Metab* **24**, 1172–1182 (2004).
87. Filosa, J. A. and Blanco, V. M. Neurovascular coupling in the mammalian brain. *Exp Physiol* **92**, 641–646 (2007).

14 Noninvasive Imaging of Cerebral Activation with Diffuse Optical Tomography

Theodore J. Huppert, Maria Angela Franceschini, and David A. Boas

CONTENTS

14.1 INTRODUCTION

For several centuries, optical methods have provided powerful tools to measure an assortment of physiological variables. However, the use of diffuse optical light for physiological measurements and, in particular, for noninvasive monitoring, has a somewhat shorter history. Unlike traditional ray optics as used in light microscopes, the field of diffuse optics pertains to the propagation of light through highly scattering samples, such as biological tissue. In this diffuse regime, the photons can no longer be assumed to travel in a ballistic-like path through the sample, and instead the equations of light propagation must be replaced by stochastic models to account for multiple scattering events and the numerous direction changes experienced by these photons as they migrate through scattering mediums. On one hand, diffuse optical techniques allow measurements through much thicker samples than are possible with traditional optics, including the ability to record noninvasive measurements from the human brain, which is the subject of this chapter. However, on the other hand, there are penalties to be paid for this ability and the progress of these techniques has faced many technical and methodological challenges.

The slowed development of diffuse optical methods was a result of two primary obstacles. The first obstacle was sensitivity—a need to discover a low-absorbing wavelength range where light can penetrate into tissue and to develop sufficiently sensitive detectors to monitor light that exits after traveling relatively large distances through tissue. As it turns out, the near-infrared wavelength range between roughly 600 and 950 nm is relatively poorly absorbed by biological tissue. Within this wavelength region, often referred to as the "near-infrared window," light can travel through up to several centimeters of tissue because of low intrinsic optical absorption of biological tissue. Moreover, although most components of biological tissue (e.g., proteins and water) have low absorption in these colors of light, oxy-hemoglobin and deoxy-hemoglobin (HbO_2 and Hb, respectively) are the dominant absorbers within this wavelength range. Thus, this spectral region is ideally suited for monitoring changes in these two hemoglobin chromophores, which can be achieved at relatively high signal-to-noise levels.

In 1977, F.F. Jöbsis[1] first demonstrated the use of near-infrared light to noninvasively measure changes in oxy- and deoxy-hemoglobin levels within the brain. This work employed what is now called near-infrared spectroscopy (NIRS), a point-source measurement of absorption changes within tissue. In the early years following Jöbsis's initial work, many groups followed to examine a variety of biologically relevant parameters in tissue, including the brain.[2,3] Much of this early research, during the 1980s and early 1990s, focused on the use of diffuse near-infrared light for measuring cerebral hemoglobin oxygen saturation within neonate and adult humans utilizing a small number of point-source measurements.[4–8] It was to be several years before actual spatially resolved imaging with diffuse light (diffuse optical tomography, or DOT) would appear. For this, diffuse optical researchers needed to overcome a second obstacle—the need for a better understanding of how light propagates through highly scattering (diffusive) tissue. Today, although still an active area of research, considerable work has appeared on the theory of light propagation through scattering media in recent years (for a review see Reference 9), and this work, along with phantom studies, has made imaging with diffuse light a reality.

Over the past 30 years since Jöbsis's initial work, the number and diversity of optical studies has grown tremendously. Much of this growth can be attributed to increased availability of optical instruments and the recognition of several advantages to NIRS imaging, which complement other existing neuroimaging modalities, including functional magnetic resonance imaging (fMRI) with additional temporal and spectroscopic resolutions. In addition, optical experiments are typically less costly to perform than conventional functional MRI. In general, optical instruments are also smaller, less expensive to purchase, and require little to no maintenance costs. A further attraction of NIRS is that many instruments are portable, which allows experimental procedures to be conducted in a more controllable environment than the noisy and confining scanners used in MRI and several other neuroimaging modalities. This has allowed the NIRS technique to study subject populations that have been traditionally difficult to study with fMRI including infants,[10–12] children,[13–15] and the elderly.[16–18] This feature of optical systems also has allowed optical measurements in a wider range of study types such as those requiring movement,[19] long-term monitoring,[20] or studies that are sensitive to experimental conditions (e.g., noise or the potentially claustrophobia-inducing nature of the MR scanner).

The advantages of optical imaging must, of course, be weighed against the limitations of these techniques and, in particular, the current lack of anatomical imaging for localization purposes and the relatively shallow penetration depth (particularly during brain monitoring). In the remainder of this chapter, we review the basic principles of NIRS and DOT and discuss these limitations and work being done to improve these methods. We will discuss example instrumentation implementations, and introduce techniques for the analysis of optical brain studies. Finally, we summarize with a discussion of the future directions for functional optical brain imaging.

14.2 THEORY OF OPTICAL IMAGING

In comparison to traditional ray optics, the theory of diffuse optics requires several important modifications in order to account for the effects of multiple scattering events and the propagation of light through thick biological samples. Briefly, diffuse optical techniques are based on the measurement of the absorption and scattering of near-infrared light. Low levels of light are sent into tissue while detectors measure the amount of light exiting at positions away from the point of entry. The absorption and scattering properties of the underlying tissue determines the amount and distribution of the light which exits the tissue. The total absorption through such a sample can be quantified using a modification of the Beer–Lambert law adjusted for the increased distance light travels due to multiple scattering and geometry effects. The modified Beer–Lambert law (MBLL) is an empirical description of optical attenuation in a highly scattering medium,[4,21] and is given by the expression

$$OD = -\log\frac{I}{I_o} = \varepsilon CLB + G \tag{14.1}$$

where OD is the optical density (also referred to as absorption), I_o is the incident light intensity, I is the detected light intensity, ε is the extinction coefficient of the chromophore, C is the concentration of a particular chromophore (e.g., an absorbing compound such as hemoglobin). L is the linear distance between where the light enters the tissue and where the detected light exits the tissue, and B is a pathlength factor that multiplies L to account for the increased effective distance traveled by light due to scattering. Finally, G is a factor that accounts for the measurement geometry and light lost at the boundaries of the sample. By convention, optical density is expressed in log base e. The calculation of the concentration of chromophores in the sample region (e.g., hemoglobin changes in the brain) requires a value for the factor B, which must include both a differential pathlength factor (DPF) and partial volume factor for each wavelength.[8,22,23] The DPF is a scaling factor transforming the source-detector separation into a measure of the effective pathlength through the tissue, which increases due to multiple scattering events. While the DPF can be measured by using frequency domain and time domain optical systems,[24–26] the complexity and cost of such systems often prohibits DPF measurement for each subject prior to functional recording. Alternatively, one can calibrate the DPF against the cardiac-induced arterial pulsation,[27] but this requires an accurate measurement of the cardiac pulsation within the optical signal, which is not always achievable. When using a continuous wave optical system, one most often has to rely on previously tabulated DPF values.[24,28] However, this approach suffers from uncertainties in the DPF, which may vary between individuals and tissue composition, and may change with age or anatomical changes.[24,29] Since the path traveled by light through cortical tissue varies with the wavelength-dependent scattering properties of the head and head layers, errors in the DPF correction used in MBLL can give rise to cross-talk errors in the separation of oxy- and deoxy-hemoglobin estimates. However, this cross-talk may be minimized by an optimal choice of wavelength pairs as discussed in several recent studies.[30–32]

The MBLL assumes that changes in chromophore concentrations are spatially uniform over the measurement sampling volume. Thus, application of the MBLL must also account for potential partial volume factors in the value of B. For example, during functional brain imaging, changes in optical properties result from localized changes in the blood oxygenation and blood volume within the brain. This is most clearly understood by considering the geometry involved in noninvasive measurements. In an adult human, the first layer of tissue is 0.5–1 cm of scalp, followed by 0.5–1 cm of skull, followed by a thin layer (0–2 mm) of cerebral spinal fluid, and finally the gray and white matter of the brain. Functional activation changes primarily occur in the cortex, the outermost layer of the brain, which are therefore up to 2 cm below the source and detector. Thus, light traveling from the source to the detector will at best encounter the functional change over only a very small portion of the optical sampling volume, even for a spatially large activation. Such local changes violate the MBLL assumption of a global change and, thereby, introduce an error reminiscent of partial volume averaging. Typical partial volume corrections are often assumed to be approximately a factor of 40 to 60,[30] but vary between different brain regions, individuals, and functional conditions

In most functional brain studies, one is generally interested in dynamic measurements of optical absorption changes rather than on the absolute quantification of total absorption. According to the MBLL (Equation 14.1), a change in the chromophore concentration results in changes to the intensity of light passing through the sample. For such differential measurements, the extinction coefficient ε, and distance L remain constant. In addition, it is often be assumed that B and G are also constant. This is generally true for reasonably small changes in absorption compared to total absorption from background tissue. For measurements of changes in light intensity, equation 1 can be rewritten as

$$\Delta OD = -\log \frac{I_{Final}}{I_{Initial}} = \varepsilon \Delta CLB \tag{14.2}$$

where $\Delta OD = OD_{Final} - OD_{Initial}$ is the change in optical density, $I_{Initial}$ and I_{final} are the measured intensities before and after the concentration change, and ΔC is the change in concentration. While in principle, the measurement of absolute light attenuation would allow for the quantification of baseline levels of hemoglobin, in practice this would require precise calibration of the probe and instrument components. In particular with regard to measurements on the brain, where the coupling of light sources to the scalp, hair, and various other experimental factors make this calibration extremely difficult, absolute quantification of absorption based on the intensity of light is not currently possible. Because ΔOD is calculated from normalized light intensities, most of the calibration factors determined by the power of the instrument lasers, efficiency of fiber optics and photon detectors, and similar instrumentation issues can be ignored, and thus the voltage of the instrument photon detectors can be used in place of the light fluence (assuming that the detector voltage is linear with the level of incident light).

When multiple species of absorbing chromophores are present—as is the case of brain activation studies where both oxy- and deoxy-hemoglobin concentrations are expected to change—the MBLL is expanded to consider the contributions of each chromophore species. The total changes in optical density are expressed as a linear combination of the contributions from each change in species concentration weighted by the respective molar extinction coefficient.

$$\Delta OD^{\lambda} = \left(\varepsilon^{\lambda}_{\mathrm{HbO_2}} \Delta\left[\mathrm{HbO_2}\right] + \varepsilon^{\lambda}_{\mathrm{Hb}} \Delta\left[\mathrm{Hb}\right]\right) B^{\lambda} L, \tag{14.3}$$

where λ indicates a particular wavelength. Equation 14.3 explicitly accounts for independent concentration changes in oxy-hemoglobin ($\Delta[\mathrm{HbO_2}]$) and deoxy-hemoglobin ($\Delta[\mathrm{Hb}]$). By measuring ΔOD at two or more wavelengths and using the known extinction coefficients of oxy-hemoglobin ($\varepsilon_{\mathrm{HbO2}}$) and deoxy-hemoglobin ($\varepsilon_{\mathrm{Hb}}$) at those wavelengths, we can then determine the concentration changes of oxy-hemoglobin and deoxy-hemoglobin by solving the system of linear equations defined in Equation 14.3. For the common example of optical measurements at two distinct wavelengths (λ_1 and λ_2), the concentration change in oxy- and deoxy-hemoglobin can be given by

$$\Delta[\mathrm{Hb}] = \frac{\varepsilon_{HbO_2}^{\lambda 2} \dfrac{\Delta OD^{\lambda 1}}{B^{\lambda 1}} - \varepsilon_{HbO_2}^{\lambda 1} \dfrac{\Delta OD^{\lambda 2}}{B^{\lambda 2}}}{\left(\varepsilon_{Hb}^{\lambda 1} \cdot \varepsilon_{HbO_2}^{\lambda 2} - \varepsilon_{Hb}^{\lambda 2} \cdot \varepsilon_{HbO_2}^{\lambda 1}\right) L}. \tag{14.4}$$

$$\Delta[\mathrm{HbO_2}] = \frac{\varepsilon_{Hb}^{\lambda 1} \dfrac{\Delta OD^{\lambda 2}}{B^{\lambda 2}} - \varepsilon_{Hb}^{\lambda 2} \dfrac{\Delta OD^{\lambda 1}}{B^{\lambda 1}}}{\left(\varepsilon_{Hb}^{\lambda 1} \cdot \varepsilon_{HbO_2}^{\lambda 2} - \varepsilon_{Hb}^{\lambda 2} \cdot \varepsilon_{HbO_2}^{\lambda 1}\right) L}$$

The generalization of this inverse formula for more than two wavelengths can be found in Arridge et al.[21] Figure 14.1 plots the extinction coefficients for oxy- and deoxy-hemoglobin versus wavelength as measured by Cope et al.[5,21,33] The other chromophores of significance in tissue in this wavelength range are water, lipids, and cytochrome oxidase. We do not consider these chromophores in this chapter, as their contribution in general is an order of magnitude less significant than hemoglobin, and are not easily measured without use of six or more wavelengths.

14.2.1 Photon Diffusion in Tissue

While the MBLL has been shown to work well in highly scattering media with uniform properties, it has known deficiencies in media with more complex structures (e.g., the layered structure of the scalp, skull, and brain),[23] and it does not provide a framework for reconstructing images. In order to form images, one must model where the light has gone within the tissue. For light propagation through samples that are much thicker than several times the scattering length of light (around 0.1 mm for tissue), which is the case of diffuse imaging of the human brain, a diffuse approximation is well justified. Analogous to the expression from Fick's second law to

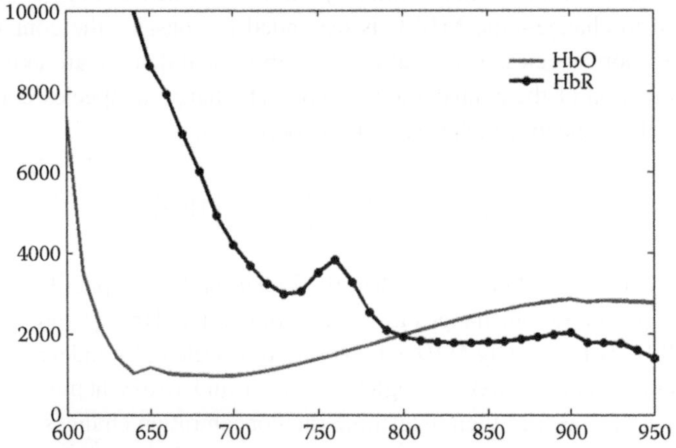

FIGURE 14.1 Absorption spectrum of oxy-hemoglobin (HbO) and deoxy-hemoglobin (Hb). Below 600 nm the absorption increases by more than a factor of 10 while above 950 nm water absorption increases significantly.

describe a random-walk of macroscopic particles, the photon diffusion approximation is given by the equation,[34–37]

$$-D\nabla^2\Phi(\mathbf{r},t)+v\mu_a\Phi(\mathbf{r},t)+\frac{\partial\Phi(\mathbf{r},t)}{\partial t}=vS(\mathbf{r},t) \tag{14.5}$$

$\Phi(\mathbf{r},t)$ is the photon fluence at position r and time t. The photon fluence is defined as the number of photons intersecting a differential area and is proportional to the intensity of light. $S(\mathbf{r},t)$ is the source distribution of photons. $D = v/(3\mu_s')$ is the photon diffusion coefficient,[38,39] μ_s' is the reduced scattering coefficient, μ_a is the absorption coefficient, and v is the speed of light in the medium. For a combination of the hemoglobin chromophores, absorption coefficient is defined by the spectrally weighted contribution of each absorbing species, i.e.,

$$\mu_a=\varepsilon_{HbO_2}\left[HbO_2\right]+\varepsilon_{Hb}\left[Hb\right]. \tag{14.6}$$

Equation 14.5 accurately models the migration of light through highly scattering media, provided that the probability of scattering is much greater than the absorption probability. Solutions of the photon diffusion equation can be used to predict the photon fluence detected for typical diffuse measurements.[37]

Assuming only a small perturbation in absorption, the analytic solution to Equation 14.5 for measurements of changes in optical fluence is only known for a limited number of simple geometries. One of the simplest of these geometries is the semi-infinite homogenous geometry, which is defined by a medium that has only one surface. In practice, this geometry approximates many objects whose boundaries are located sufficiently far enough away from the imaging volume. Assuming that hemoglobin concentration changes are both global and small, the solution of the photon diffusion equation for a semi-infinite medium is

$$\Delta OD=\frac{1}{2}\sqrt{\left(\frac{3\mu_s'}{\mu_A^{Initial}}\right)}\left(1-\frac{1}{1+L\sqrt{\left(3\mu_S'^{Initial}\mu_A^{Initial}\right)}}\right)\Delta\mu_A L. \tag{14.7}$$

The solution of the photon diffusion equation for the representative semi-infinite tissue geometry (Equation 14.7) tells us that the modified Beer–Lambert law is reasonable for tissues with spatially uniform optical properties when the chromophore concentration does not change significantly (i.e., Δ[X]/[X]<< 1). Equation 14.7 also explicitly defines the pathlength factor B for a semi-infinite medium, which was introduced in Equation 14.3, as

$$B=\frac{1}{2}\sqrt{\left(\frac{3\mu_s'}{\mu_A^{Initial}}\right)}\left(1-\frac{1}{1+L\sqrt{\left(3\mu_S'^{Initial}\mu_A^{Initial}\right)}}\right) \tag{14.8}$$

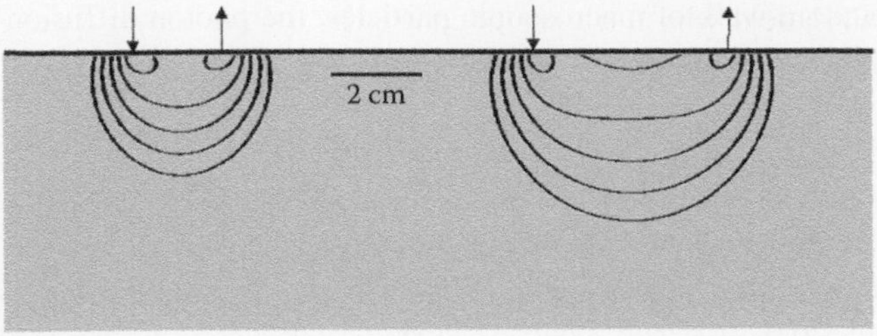

FIGURE 14.2 Photon migration sensitivity profile shown for a semi-infinite medium for a 2 cm and 4 cm source-detector separation. Contour lines are shown every half order of magnitude.

This shows that *B* depends on tissue scattering, initial chromophore concentration, extinction coefficient (and thus *B* is wavelength dependent), and optode separation. In practice, the validity of the assumption that *B* is independent of μ_a and *L* has often been ignored since *B* is in general empirically determined and/or the changes in μ_a are typically small.

In Figure 14.2, we show the sensitivity profile for a simple reflectance geometry with a continuous-wave (CW) light source using typical tissue optical properties of $\mu_s' = 10$ cm^{-1} and $\mu_a = 0.1$ cm^{-1}. The most important point to notice is that the spatial sensitivity profile is not localized but covers a large volume and therefore intrinsically has relatively poor spatial resolution. Second, notice that as the separation between the source and detector increases, the penetration depth also increases. This fact can be used to probe deeper into the brain and possibly even to distinguish superficial scalp from deeper brain signals.

14.2.2 Theoretical Optical Sensitivity to the Brain

The diffusion approximation to the optical forward problem as described in the previous section may be used to model the propagation of light through tissue. By using numerical solvers to Equation 14.5 for finite element meshes or by using Monte Carlo techniques,[28,35,36,40–42] we can simulate the propagation of photons through tissue and obtain a spatial sensitivity map for a given grid of source and detector measurement combinations. For example, in our Monte Carlo simulations, individual photon trajectories are traced by sampling appropriate probability distributions for scattering events.[41] After 10^6–10^8 photons are traced, quantities such as the fraction of photons reaching a particular detector or the spatial sampling of the tissue can be determined. In our Monte Carlo model, the tissue is a 3D volume with optical properties assigned to each voxel, and source and detectors can be placed at any voxel. Photons are propagated until they exit the tissue, and the total path length in each tissue type is recorded for all photons reaching a detector. Only recently has the computational power become widely available for the practical use of Monte Carlo simulations (one simulation typically takes 3–4 hours on current high-end desktop systems).

Figure 14.3 shows an anatomical MRI of a human head, segmented into five tissue types (air, scalp, skull, cerebral spinal fluid, and gray/white matter; see Reference 43),

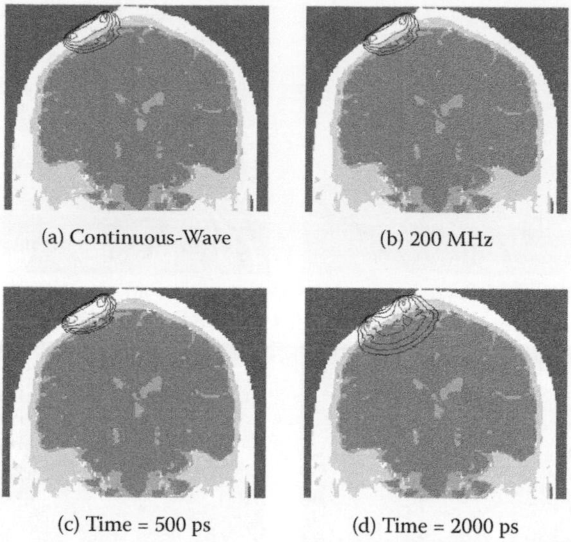

FIGURE 14.3 Contour plots of the photon migration sensitivity profile in a 3D human head as determined from Monte Carlo simulations. Details provided in the text.

with a contour overlay indicating the photon migration spatial sensitivity profile for (a) continuous-wave, (b) 200 MHz modulation, and (c,d) pulsed measurements. One contour line is shown for each half order of magnitude (10 dB) signal loss, and the contours end after 3 orders of magnitude in loss (60 dB). For 3D Monte Carlo simulation, we assumed that $\mu_s' = 10$ cm^{-1} and $\mu_a = 0.4$ cm^{-1} for the scalp and skull, $\mu_s' = 0.1$ cm^{-1} and $\mu_a = 0.01$ cm^{-1} for the CSF, and $\mu_s' = 12.5$ cm^{-1} and $\mu_a = 0.25$ cm^{-1} for the gray/white matter. Note how the contours extend several millimeters into the brain tissue, indicating sensitivity to changes in cortical optical properties. The depth penetration difference between the continuous-wave and 200 MHz measurements is difficult discern. A ratio of the two sensitivity profiles (not shown) shows that the 200 MHz profile is shifted slightly toward the surface. The time-domain sensitivity profiles suggest the possibility of obtaining greater penetration depths in the head from measurements made at longer delay times.

14.2.3 Diffuse Optical Imaging and Tomography

It is a relatively simple step, conceptually speaking at least, to move from the simple NIRS point-measurement to a spatially resolved imaging of these same variables. In a typical diffuse optical imaging study of brain function, a grid of light sources and light detectors is positioned on the head of a subject as pictured in Figure 14.4. Often this is done using fiber optics where one end of these fibers, which we refer to as an optode, is secured into a head gear device, and the other end of the fiber is connected to the imaging instrument. The arrangement of the optodes on the head determines the depth-sensitivity of the measurements to underlying cerebral changes, which is a function of the distance between the source and detector pair as shown in Figure 14.2.

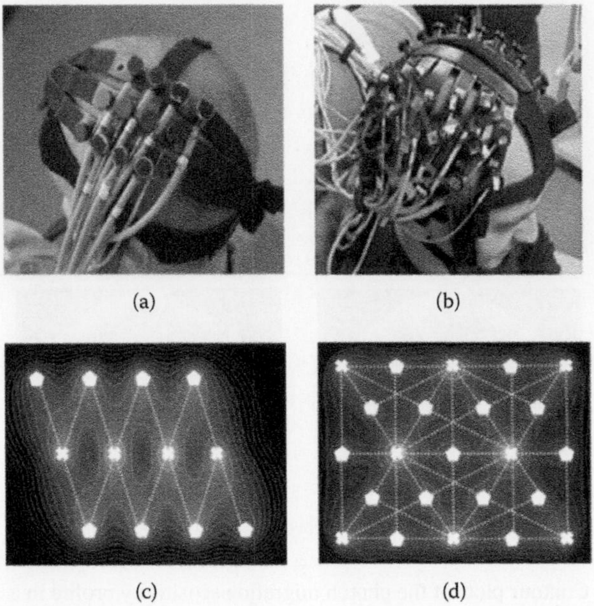

FIGURE 14.4 Probe arrangement for NIRS and DOT experiments. Panels (a) and (c) show a nearest-neighbor NIRS probe. Panels (b) and (d) show a tomographic-style (DOT) probe. The sensitivity profiles for these two probes are shown in panels (c) and (d) with contour lines at 5 dB intervals based on a semi-infinite homogenous ($\mu_a = 0.1$ cm^{-1} and $\mu_s' = 10$ cm^{-1}) slab geometry.

In addition, the arrangement of optodes will also dictate the lateral spatial resolution. In the case of true tomographic imaging (DOT), a dense array of optodes and overlapping measurement combinations of the sample volume are used to provide uniform spatial coverage and achieve the best spatial resolution.[31,44,45] However, optical tomography is generally more experimentally difficult than less dense array probes because of the number of optodes involved—requiring additional setup time for adjustment of the probe and quality assurance of the signals—and the range of light intensities that need to be recorded at the detector optodes. In a tomographic probe as shown in Figure 14.4b, light from sources located various distances away must be recorded by a single detector. Since the intensity of light reaching the surface decays exponentially as the detector moves away from the source, several orders of magnitude in fluence will be reduced between a near and far source-detector spacing. For example, there is approximately a 50-fold intensity difference between light from a 1.5 cm and 3 cm spacing.[46] The collection of overlapping measurements requires either detectors with a high dynamic range (e.g., Reference 47) or source-switching schemes where only a fraction of the sources are turned on at any one time (e.g., References 45,48).

In general, optical imaging—as opposed to tomography—is more common in brain activation studies. Unlike optical tomography, diffuse optical imaging does not require overlapping sets of measurements and, thus, the design and implementation of these experiments is somewhat easier with a compromise of less uniform spatial resolution and lower resolution. Optical imaging typically uses a grid of optodes arranged to collect nearest-neighbor combinations of sources and detectors. However, a nearest-neighbor probe geometry (Figure 14.4a) may often contain of areas of low measurement sensitivity. In particular, optical measurements are most sensitive in an area immediately between the source and detector positions and are less sensitive to changes occurring immediately beneath these optode positions as shown in Figure 14.4. The choice between a point-source measurement, imaging, or true tomography probe design depends on the application and specific hypothesis of the experiment. There are several practical trade-offs to consider when designing this probe, not the least of which is the additional setup time required to position and adjust the probe on the head of a subject.

Whether optical measurements are collected from a true tomography or imaging geometry, the image reconstruction theory and procedure is very similar (reviewed in References 9,49). In order to generate an image of spatial variations in Hb and HbO_2, in principle, all we need to do is to measure the photon fluence at multiple source and detector positions. Each measurement samples a different volume of underlying tissue and can be modeled using photon migration theory discussed in the previous sections of this chapter. This set of measurements can then be back-projected to generate the image of underlying concentration changes. The general solution of the photon diffusion equation at the detector position r_d for a medium with spatially varying absorption is

$$\Phi(\mathbf{r}_d) = \Phi_{incident}(\mathbf{r}_d) - \int \Phi(\mathbf{r}) \frac{v}{D} \left(\varepsilon_{HbO_2} \Delta[\mathrm{HbO_2}] + \varepsilon_{Hb} \Delta[\mathrm{Hb}]\right) G(\mathbf{r}, \mathbf{r}_d) d\mathbf{r}. \quad (14.9)$$

For reflectance in a semi-infinite medium, the solutions for $\Phi_{incident}$ and $G(\mathbf{r}, \mathbf{r}_d)$ can be found in Farrell.[50] Equation 14.9 is an implicit equation for the measured fluence as $\Phi(\mathbf{r})$ appears in the integral on the right-hand side. One way to solve this equation uses the first Born approximation, which assumes that $\Phi(\mathbf{r}) = \Phi_{incident} + \Phi_{sc}$.[51] In this approximation, the measured fluence is given by

$$\Phi_{sc}(\mathbf{r}_d) = -\int \Phi_{incident}(\mathbf{r}) \frac{v}{D} \left(\varepsilon_{HbO_2} \Delta[\mathrm{HbO_2}] + \varepsilon_{Hb} \Delta[\mathrm{Hb}]\right) G(\mathbf{r}, \mathbf{r}_d) d\mathbf{r}. \quad (14.10)$$

By measuring $\Phi_{sc}(\mathbf{r}_d)$ using multiple source and detector positions and multiple wavelengths, we can invert the integral equation to obtain images of $\Delta[HbO_2]$ and $\Delta[Hb]$. Equation 10 defines the contribution of each position in the underlying volume to the measurement made between a particular source and detector pairing and defines the measurement sensitivity for each measurement on the optode grid. The process of image reconstruction from such measurements involves the inversion of Equation 14.10 in order to solve for the changes in oxy- and deoxy-hemoglobin at each position in the sample volume.

For small changes in absorption, the optical trajectories may be assumed to remain constant and thus the hemoglobin terms may be moved outside of the integral in Equation 14.10. Many imaging procedures relay on reducing Equation 14.10

to a matrix formulation by rewriting this spatial integral as a discrete sum over the volume elements (voxels) in the desired image, i.e., Y = A·x where Y is the vector of measurements (i.e., $\Phi_{sc}(\mathbf{r}_d)$), x is the vector of image voxels, and A is the transformation matrix obtained from the integrand of Equation 14.10. In the case of fewer measurements than unknowns, the linear inverse problem is underdetermined and, thus, cannot be uniquely solved.[49] In addition, this inverse problem is ill-posed, which means that multiple equivalent solutions to the inverse problem may exist. The optical inverse problem requires regularization. The usual form of the regularized optical inverse problem is given by the (regularized) Moore–Penrose generalized inverse[9]

$$x = -A^T\left(AA^T + \lambda I\right)^{-1} \cdot Y, \qquad (11)$$

where I is the identity matrix, and $\lambda = \alpha\ \max(AA^T)$ is termed the Tikhonov regularization parameter. Y is the optical density measurements and x is the estimated absorption changes in the image. Note that a weighted back-projection ($x = -s \cdot A^T \cdot Y$) is an approximation to Equation 14.10 in that the matrix product $A \cdot A^T$ approaches an identity matrix

Figure 14.5 illustrates the method of reconstructing an image with diffuse light. The process is conceptually similar to that of x-ray computed tomography, which is included for comparison. In the case of x-ray tomography, a collimated beam transilluminates the tissue, maintaining high spatial resolution, and the attenuation of the beam is measured. The measured attenuation is then back-projected into the tissue

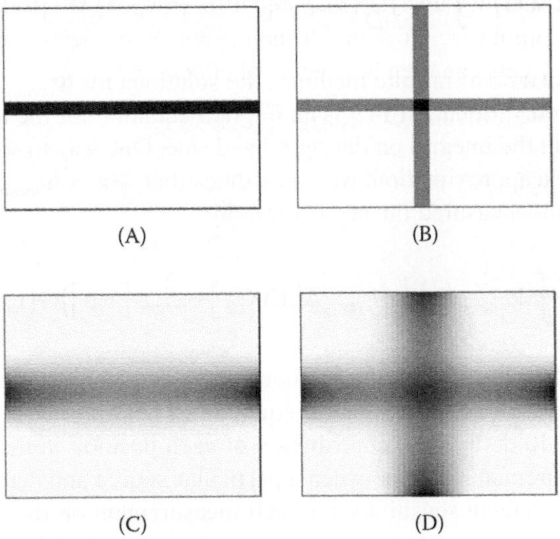

FIGURE 14.5 Comparison of x-ray computed tomography (A,B) and diffuse optical tomography (C,D) of a square absorbing object in the center of the outer square. X-ray tomography has greater resolution as indicated by the sharper edges of the reconstructed image. Diffuse optical tomography reconstructs the object but with reduced resolution because of the diffusive nature of near-infrared photon migration in tissue.

along the axis of the x-ray. For a simple object positioned in the center of the plane, we see that a single x-ray projection sharply defines the lateral boundaries of the object, but does not localize the object along the x-ray propagation axis. Through the combination of additional measurements at different angles with respect to the object, it is possible to obtain a high-resolution image of the x-ray-attenuating object within the tissue. Figure 14.5B shows the result from two orthogonal projections. The artifacts near the boundaries are reduced with additional projections. Figures 14.5C,D show the results obtained with diffuse light by following the same back-projection procedure. The striking difference obtained with a single projection is that the boundaries of the object are not sharply defined, and the object appears to be near the sources and detectors where the sensitivity of the measurement is greatest. The first observation is an intrinsic limitation of imaging with diffuse light, while the second observation is an artifact of the back-projection procedure. Utilizing an additional orthogonal projection (Figure 14.5D), we see better localization of the absorbing object. We note that while the back-projection method provides quantitative images with x-ray CT, it is only an approximation for diffuse optical tomography. This difference results from the fact that the measured diffuse light has sampled a volume of tissue as opposed to the measured x-ray that samples only along a line through the sample. The back-projection method, therefore, does not produce an image that optimally fits the experimental data. More sophisticated optimization is needed to obtain quantitative image reconstruction with diffuse light.[49]

14.3 INSTRUMENTATION FOR BRAIN STUDIES

For brain imaging experiments, several technical solutions exist for implementation of NIRS and DOT, including time domain (TD), frequency domain (FD), and continuous-wave (CW) systems. The principle difference between these methods is the characteristics of the incident light used and subsequently the type of detectors. TD systems[2,9,52–54] introduce into tissue extremely short (picosecond) incident pulses of light. These narrow pulses are broadened and attenuated by the various tissue layers such as skin, skull, cerebrospinal fluid (CSF), and brain. A TD system detects the temporal distribution of photons as they leave the tissue, and the shape of this distribution provides information about tissue scattering and absorption values. The collection of time-of-flight information from TD systems allows for the discrimination absorption changes from different depths. For example, deeper absorption changes affect the tail of the temporal distribution function representing the photons that have traveled for the longest duration in the tissue. By modeling this temporal point-spread function, TD methods can be used to increase sensitivity to deeper changes.[55] In principle, TD systems can theoretically obtain the highest spatial resolution and can accurately determine absorption and scattering. However, drawbacks of TD measurements include long acquisition times to achieve reasonable signal-to-noise ratios, a need to mechanically stabilize the instrument, and the large dimensions and high cost of the necessary ultrafast lasers.[9] In addition, TD methods require acquisition of data from multiple temporal windows in order to reconstruct the temporal distribution of light exiting the tissue; thus, the overall acquisition rate is inversely related to the number of time-gates measured and consequently the depth resolution. Although TD instruments have advanced in recent years,[56] their cost and complexity has still generally limited their widespread use.

Another implementation of NIRS uses frequency domain (FD) signals.[57–60] In FD systems the light source is on continuously, but amplitude-modulated at frequencies on the order of tens to hundreds of megahertz. Information about the absorption and scattering properties of tissue are obtained by recording amplitude decay and phase shift (delay) of the detected signal with respect to the incident light.[61] State-of-the-art FD systems cover a wide range of applications for clinical use, and can achieve considerably higher temporal resolution than TD systems. FD instruments are typically less expensive to build than TD systems, but are still more so than CW systems, and they require more careful engineering and optimization to eliminate noise, ground loops, and RF-transmitter effects. The detection of FD light requires sensors with high dynamic range, and such systems typically require photo multiplier tubes (PMT) detectors. In recent years, FD systems have been developed commercially (e.g., the Imagent system from ISS Inc.), and this has allowed these methods to be more readily accessible.

Finally, in CW systems,[62–64] light sources emit light continuously as FD systems do, but at constant amplitude or modulated at frequencies not higher than a few tens of kilohertz. In CW setups, only the amplitude of the light is detected. Thus, the primary shortcoming of a CW system compared to TD and FD systems is the inability to uniquely quantify the effects of light scattering and absorption.[65] While CW imaging techniques have the potential to provide quantitative images of hemodynamics changes during brain activation, quantitative imaging of baseline brain states is likely to only be possible with FD and TD methods, as they provide both time-delay (e.g., optical pathlength) and amplitude information. However, CW technology can be engineered with relatively inexpensive and widely available components; a hospital pulse-oximeter being an example for the small dimensions achievable for a CW NIRS instrument. There are currently a number of commercially available such systems (Hamamatsu: the NIRO 500, and the INVOS 3100 systems; NIRx Medical technologies: the Dynot system; Hitachi Medical Corporation: the ETG Optical Topography System; and Techen: the NIRS2 and CW5 systems). The commercialization of these instruments has made this technology more available to a variety of researchers. It is likely that CW technology will quickly find widespread use among brain researchers because of its relatively low cost, portability, and ease of implementation and use compared to FD and TD systems, while still displaying sensitivity to cerebral hemodynamic features. Arguably, a CW imaging system will also achieve the best signal-to-noise ratio at image frame rates faster than 1 Hz and up to several hundred Hz. In the following two sections, we will describe two examples of basic instrument design of CW NIRS and DOT systems from our lab.

14.3.1 NIRS Instrument

A basic NIRS system delivers light of preferably two or more wavelengths to a single position on the tissue, and then collects and detects the diffusely reemitted light at one or more locations. Because the light must travel several centimeters through the scalp and skull to sample the adult human brain, it is necessary that light sources be chosen with sufficient power, and in the wavelength range between 600 and 950 nm where light absorption is minimized. This typically requires the use of laser diodes although filtered white light sources have also been successfully used in adult humans.[27] For safe, long-term tissue exposure to laser light, the optical power incident on the tissue must be no more than 4 mW per 1 mm.[2,66] To gain sensitivity to the brain in the adult human, detectors must be placed at least 2.5 cm from the source (less for babies, with their thinner scalps and skulls), as the depth of sensitivity is roughly proportional to the source-detector separation. With such separations, the amount of light reaching the detector is on the order of 10 pico-Watts (an attenuation of 9 decades). To obtain a decent signal-to-noise ratio at desired bandwidths of 1 Hz or greater, therefore, it is necessary to use high-sensitivity detectors: photo-multiplier tubes (PMT's) or avalanche photodiodes (APDs), or CCD cameras.

Figure 14.6 shows a photograph of our CW NIRS system. The sources in this system are two low-power laser diodes emitting light at discrete wavelengths, typically 690 nm (Sanyo, DL7140-201) and 830 nm (Hitachi, HL8325G). These lasers are powered by a stabilized current, intensity-modulated by an approximately 5 kHz square wave at a 50% duty-cycle. Both diodes are driven at the same frequency, but phase shifted by 90° with respect to one another. This phase encoding—known as an in-phase/quadrature-phase (IQ) demodulation circuit—allows simultaneous laser operation as well as separation of the contributions of each source to a given detector's signal. The system has four separate detector modules (Hamamatsu C5460-01), which are optically and electrically isolated from each other. Each module consists of a silicon avalanche photodiode (APD) with a built-in high-speed current-to-voltage amplifier, and temperature-compensation. They achieve a gain of typically 10^8 V/W with a noise equivalent power of 0.04 pW/Hz resulting from the high detector sensitivity. The incoming signal at a given detector is composed of both source colors,

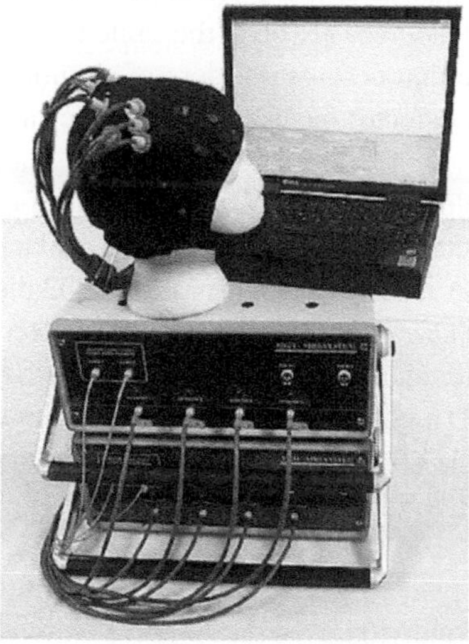

FIGURE 14.6 Photograph of the NIRS system and cap used to hold the fiber optics on the head.

which are separated by synchronous (lock-in) detection using the two source signals as a reference. The output of the decoding portion of the IQ circuit includes two signal components, corresponding to the two laser wavelengths. These two components are low-pass filtered at 20 Hz and digitized by a computer at 200 samples per second.

14.3.2 Imaging Instrumentation

Conceptually, an imaging instrument is a simple extension of a NIRS system to include more sources and detectors. Adding more detectors is straightforward as each detector operates independently without interaction with the other detectors. The sources, on the other hand, all introduce light into the tissue, and detectors can only detect the total amount of light present at a given time. Thus, in order to properly assign light received at a detector to each individual source, the source light must be encoded. Several encoding strategies exist: time-sharing, time-encoding, and frequency-encoding. Time-sharing is the easiest to implement. In this scheme, each source is turned on, one at a time, long enough for the detectors to acquire a decent signal (typically 10–100 ms). Large numbers of sources gives rise to slow image frame rates and temporal skew between sources. Time-encoding overcomes the temporal skew problem by switching between sources at a much faster rate (100 μs or faster) but integrating the signal from each source over several switch cycles (e.g., the same 10–100 ms), resulting in roughly the same image frame rate as with time-sharing. Finally, frequency encoding turns all sources on at the same time but modulates the intensity of each one at a slightly different frequency. The individual source signals are then discriminated at the detector by employing simple analog or digital band-pass filters at the frequencies of source modulation. While the duty cycle in this scheme is maximized, the effective dynamic range of the system is limited by the fact that all sources are on at the same time. That is, it can be difficult to detect a distant source that is several orders of magnitude weaker than a source close-by given that each detector has a fixed dynamic range (typically 3–4 orders of magnitude).

Figure 14.7 shows a block diagram and photograph of our DOT imaging system with 32 lasers and 32 detectors. The detectors are avalanche photodiodes (APDs, Hamamatsu C5460-01). A master clock generates the 32 distinct carrier frequencies between 4.0 kHz and 7.4 kHz in approximately 200 Hz steps. These frequencies are then used to drive the individual lasers with current stabilized square-wave modulation. Following each APD module is a band-pass filter, cut-on frequency of ~500 Hz to reduce 1/f noise and the 60 Hz room light signal, and a cut-off frequency of ~10 kHz to reduce the third harmonics of the square-wave signals. After the band-pass filter is a programmable gain stage to match the signal levels with the acquisition level on the analog-to-digital converter within the computer. Each detector is digitized at ~40 kHz and the individual source signals separated using an IQ demodulation scheme against the carrier frequency of each laser.

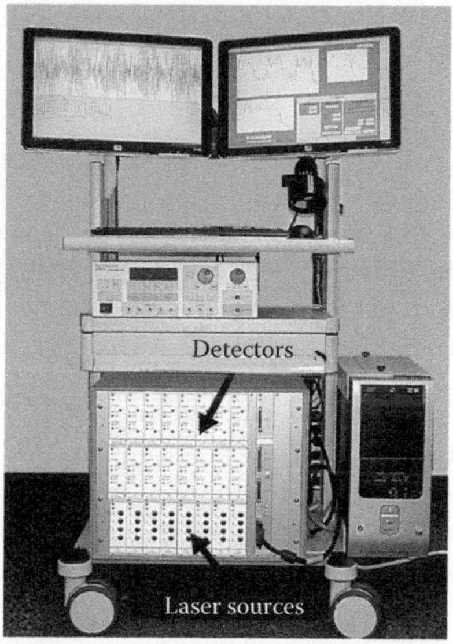

FIGURE 14.7 Photograph of the optical tomography system. Details provided in the text.

14.4 EXPERIMENTAL DESIGN OF OPTICAL STUDIES

In the reminder of this chapter, we will discuss applications of NIRS and DOT instruments. Over the past several years near-IR optical methods have been applied to study a variety of cerebral regions including the visual,[67–69] auditory,[70–72] and somatosensory cortices.[73,74] Other areas of investigation have included the motor,[75–78] prefrontal/cognitive,[79–82] and language systems.[83] In addition to its use in brain mapping experiments, optical imaging is beginning to be applied to study several clinical problems, in part motivated by the portability and bedside applicability of optical instruments. Clinically oriented applications have included the prevention and treatment of seizures[84–89] and Alzheimer's disease,[18,90,91] and provided a potential tool to monitor stroke rehabilitation.[92–99] NIRS has also been applied to psychiatric health concerns such as depression[79,100–104] and schizophrenia.[104–109]

Several groups have also investigated the possibility of using optical methods to measure the cytochrome oxidase (cyt-ox) redox state.[1,110] Cyt-ox oxidation is a marker of intracellular energy metabolism, and it has been suggested as a possible parameter to assess the functional state of the brain.[27,111–114] These optical measurements are usually conducted over a wide spectral range (extending up to 500 nm), because the absorption contributions of cyt-ox are an order of magnitude (or more) smaller than those of hemoglobin.

Lastly, a few groups have also worked on using diffuse light to directly measure neuronal activity, rather than indirectly via hemoglobin. For example, it has been shown in vitro that neuronal activity is associated with an increase in light scattering, induced by a change in the index of refraction of the neuronal membranes.[115–117] Following up on this work, near-infrared optical methods have been successfully used to detect such light-scattering changes in vivo in adult humans.[118–120] This fast optical signal appears to show the same time course as the electrophysiological response measured with EEG/electrophysiological techniques.

Conceptually, the design of an optical imaging brain experiment study is similar to many other imaging modalities (particularly functional magnetic resonance imaging; fMRI). There are two primary things to consider in designing an optical study: the expected characteristics of the activation signal and the specific hypothesis that will be addressed by the study. More so than in fMRI experiments, optical studies generally begin with some foreknowledge of the location and extent of the expected activation signals. It is particularly important when designing an optical study to consider the expected depth of the activation signal as this determines the type of optical method used and the arrangement of the head probe. Although, in general optical measurements are limited to the outer layers (approximately 5–8 mm) of the brain, this depth sensitivity may be adjusted based on the source-to-detector separation distance and on the characteristics of the underlying tissue (see Figure 14.2). For example, in the visual cortex, where both bone and venous blood vessel densities are high, optical penetration is lower and the signal strength at a given distance may be lower requiring closer spacing to achieve a given ratio of signal-to-noise. In contrast, cognitive studies of brain activity in the frontal lobe may need to consider the proximity to sinus cavities and underlying cerebral spinal fluid layers, which results in a lower attenuation of optical signals, and allow for larger optode separation distances at similar signal-to-noise ratios. These factors should be noted when designing an optical probe for such a study.

A second consideration in the design of optical experiments is the specific hypothesis that will be addressed. In comparison to the majority of fMRI based studies, which are based on the amplitude of the evoked response, optical measurements have several unique characteristics that are factors in designing an experimental hypothesis. First, optical signals generally have high temporal resolutions (often far exceeding the typically slow dynamics of the hemodynamic activation response). This characteristic allows experimenters to develop hypotheses based on the temporal dynamics of the evoked signal. Examples of this approach include examination of differences in the early or late phases of the evoked response between subject populations or tasks (e.g., see Reference 121). A second unique characteristic of optical measurements is the added spectroscopic information. This allows hypotheses based on relative changes, for example, the relationship of oxy- and deoxy-hemoglobin. This information may be used to (at least qualitatively) infer the influence of flow verses oxygen metabolism effects (e.g., see References 122,123). In contrast, for the majority of fMRI studies, hypotheses are formulated to test for differences in the amplitude or spatial characteristics of the response rather then these timing differences. Although this can also be done in optical imaging, due to the generally low spatial resolution and depth sensitivity of optical measurements, the amplitude of optically measured evoked response may be distorted by partial volume errors.[31] Thus, particular care must be taken to spatial maps between conditions or in drawing conclusions about the amplitude of evoked responses, since conditional differences in the depth or location of the activation may offer alternative explanations of such results. As discussed in the prior section, the design of the optical experiment—for example, the choice between a NIRS versus a DOT (tomography) study—will determine the sensitivity, reproducibility, and accuracy of the results.

A final unique consideration to the design of optical studies is the potential contamination from background physiological signals. Since optical measurements are made through the superficial layers of skin on the head, these measurements are sensitive to systemic fluctuations. Further discussion of the sources of this noise and analysis methods for reducing their impact will be discussed later in this chapter. However, in designing an optical study, it is important to also consider the possibility for systemic effects of the stimulation task. For example, does heart rate change when the subject performs the task? In addition, we have observed that systemic fluctuations may occasionally be influenced by the design of the optical experiment. For example, blocked-design experiments, which present stimuli at regular time intervals, may exhibit coupling between physiological oscillations (Mayer waves) and stimulus presentation. The presence of these systemic oscillations may depend on subject posture, breathing rate, or simply reflect individual variations in physiology.[124] Additionally, the performance of difficult tasks, such as motor activity, may change this physiology and produce systemic evoked responses, which may be confused with global brain activity—for example, if subjects unintentionally hold their breath while performing a brief finger-tapping exercise or the task evokes a startled response (i.e., raising heart rate). In Figure 14.8, we show an example of stimulus-evoked increases in heart rate that were recorded during a 5-s finger-tapping exercise. This heart-rate increase, which is roughly a 0.5% change, has approximately the same timing and duration as the expected cerebral hemodynamic signals, and

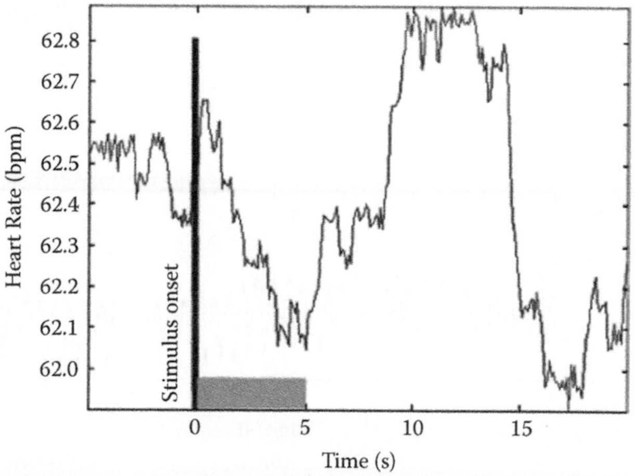

FIGURE 14.8 Example of a systemic "functional" response recorded from a finger clip pulse-oximeter during a 2-s finger-tapping task. The pulse-rate was recorded on the stationary hand during the motor task. In addition to the cortical activity involved in this task, systemic reactions related to the difficulty of the task may also confound analysis of NIRS signals. This figure shows the average heart-rate response from 37 trials from a single subject.

although these changes are smaller than the focal brain activation changes—usually on the order of a few percent change in optical density—these observations highlight one of the difficulties in optical measurements and the ongoing need for advancement in data analysis methods to deal with these specific aspects of optical.

Aside from the filtering and analysis methods, which will be discussed later in this chapter as methods to reduce the impact of this physiology, in experiments where systemic physiology is believed to pose a confounding problem, it is advisable to collect additional information about systemic changes. For example, heart rate, respiration, and blood pressure can be measured using auxiliary devices to characterize systemic changes. Alternatively, NIRS measurements from short source-detector separations can be used to measure local fluctuations in the skin layers above the brain activation region.

In the following sections, we will present examples of both NIRS and DOT experiments to emphasize these experimental considerations.

14.4.1 Design Example 1: NIRS Baseline Recording

Many efforts have been made to better understand background physiological fluctuations with hopes of improving the data analysis methods to remove these nuisance variables from functional brain mapping studies. Figure 14.9 shows the observed amplitude modulations from a single subject during two different baseline conditions

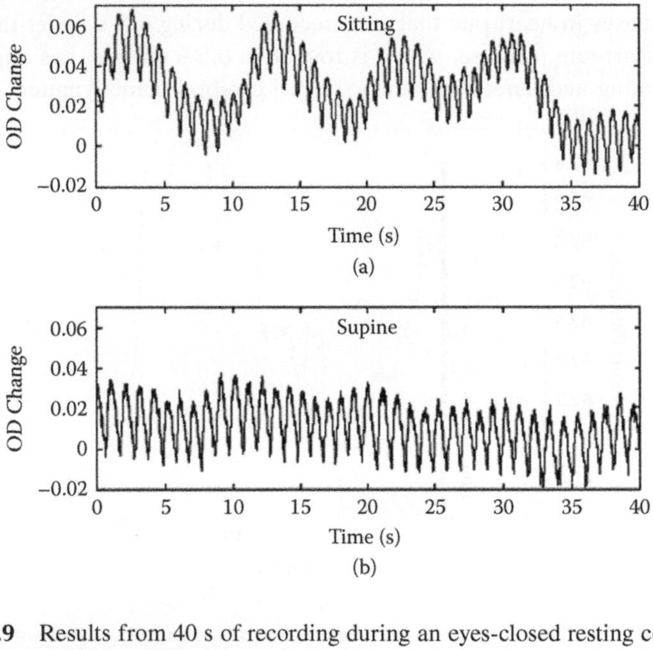

FIGURE 14.9 Results from 40 s of recording during an eyes-closed resting condition in a subject while in a sitting position (top) and supine (bottom). Optical density changes observed from the 830-nm laser are shown. In both traces, heart pulsations are clearly evident. In the sitting case, a second, high-amplitude periodicity of ~0.1 Hz is also evident, while it is essentially absent in the supine position (see text).

(in optical density units). The top trace—with data gathered at 830 nm from approximately location C3 in the international 10–20 system—shows 40 s of eyes-closed baseline while the subject was seated upright. The bottom trace is from the same subject, again eyes closed baseline but while lying supine. Heart pulsations are clear in both records, while in the upright case an additional high-amplitude oscillation of ~0.1 Hz appears. The frequency of this periodicity corresponds to the Mayer wave—a systemic blood pressure oscillation that is more prominent when standing or sitting than when lying down.[27,125] In recent years, several studies of resting state (baseline) physiological fluctuations have turned away from considering these fluctuations as simply noise and have begun to investigate information about cerebral physiology within these baseline signals. For example, Franceschini et al.[124] have used optical measurements of resting physiology to examine asymmetry between hemispheres. In addition, several groups have begun to use NIRS to explore normal vascular physiology, such as vasomotion or autonomic regulation.[124,126–128] We believe that these approaches will be continually developed in future studies and offer great promise for the use of optical imaging in a clinical setting.

14.4.2 Design Example 2: Functional NIRS

While studies of spontaneous physiology are beginning to appear, to date the majority of NIRS and DOT studies have examined evoked hemodynamic responses to cued functional tasks. In Figure 14.10, we show an example of an evoked hemodynamic response to a train of pulses (4 Hz for 1-s duration) measured using optical imaging during mild electrical stimulation of the median nerve. For the MBLL

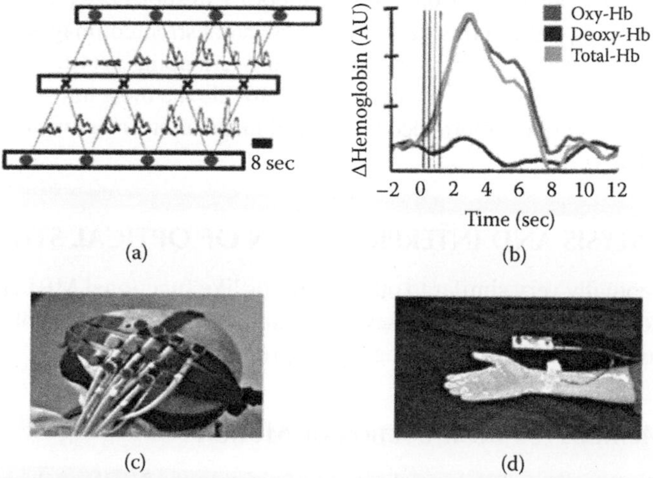

FIGURE 14.10 The hemodynamic response to train of pulses (4 Hz for a 1-s duration) to the median nerve measured using NIRS. Panel (a) and (b) show the spatial distribution of hemodynamic responses and region-of-interest average ($p < 0.05$ defined for oxy-hemoglobin). Panel (c) and (d) show the NIRS probe arrangement and setup for the median nerve stimulation.

analysis we assumed that the differential path-length factor was equal to 6 at each wavelength and a partial volume correction of 60 was used. Figure 14.10a shows the spatial distribution of oxy- (red), deoxy- (blue), and total-hemoglobin (green) changes based on the optical measurements from the probe shown. Figure 14.10b shows the evoked response to a single stimulation and a train of stimuli from region-of-interest analysis ($p < 0.05$) on seven subjects. These responses show the typical hyperemic (increased blood flow) response that is often observed for functional responses. This response corresponds to the typical "positive BOLD" signal observed in fMRI. In this response, both oxy-hemoglobin and total-hemoglobin are observed to increase due to increased regional blood flow and volume in the activated brain region. These changes are the result of neurovascular coupling— the action of neurons and neurotransmittors to dilate blood vessels around the brain activation site (reviewed in Reference 129). Increases in arterial blood flow causes deoxy-hemoglobin levels to decrease—a phenomena termed the *washout effect*. Increased neuronal activity is also associated with increased cerebral oxygen metabolism ($CMRO_2$). Although an increase in $CMRO_2$ removes more oxygen from the blood—converting oxy-hemoglobin into deoxy-hemoglobin—the increases in blood flow generally overcompensate for this effect, and the net result is a usual increase in blood oxygen saturation (for a review see Reference 130). Figure 14.10 shows this typical hyperemic response from a median nerve electrical stimulation study.

14.4.3 Design Example 3: Tomographic Optical Imaging

In Figure 14.11, we show results from a functional brain study using a tomographic probe. In this study by Joseph et al.,[45] an overlapping (tomographic) optical probe was used to record optical signals from the motor cortex in subjects as they preformed a brief finger-tapping task. In this figure, we show reconstructed images from the first nearest-neighbor, second-nearest-neighbor, and overlapping sets of measurements. The overlapping (tomographic) measurements provide the most uniform spatial coverage and most accurately reproduce the spatial maps of brain activation when compared to fMRI.

14.5 ANALYSIS AND INTERPRETATION OF OPTICAL STUDIES

While conceptually very similar to other methods like functional MRI and EEG, the analysis of optical signals requires several additional considerations of some of the unique features of this data, including signal processing.

14.5.1 Methods for the Reduction of Motion

Subject motion can be a detrimental source of noise in the NIRS analysis. While the best opportunities to deal with such motion artifacts occur during the data acquisition and reflect subject compliance, experimental design, and the experimenter's expertise in positioning and securing the head probe, motion artifacts will undoubtedly still affect many experiments. Motion is particularly common in dealing with infant or animal subjects, where subject compliance is more difficult to control. In

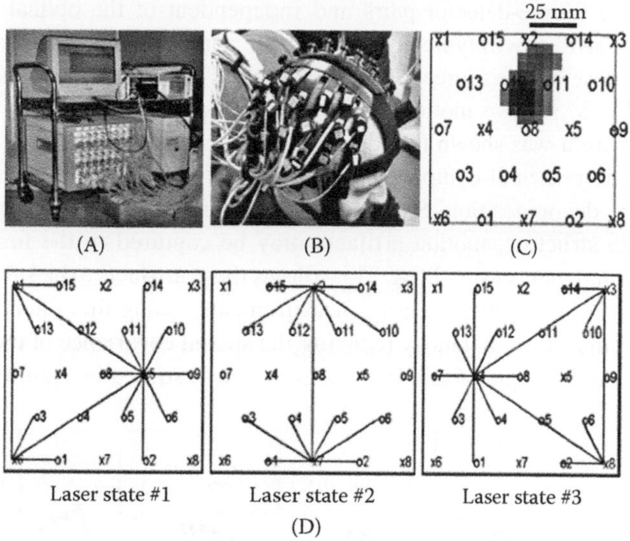

FIGURE 14.11 In this figure, we show a time-multiplexing scheme for of our optical tomography system (panel A). The system switches between three combinations of light source and detectors arranged on the head (panels B and D). This provides a more uniform spatial sensitivity and can be used to improve image reconstruction. Panel C shows the results of a brain activation pattern from a 2-s motor task from a human volunteer.

general, motion artifacts appear as large jumps in the amplitude of optical signals and are often several orders of magnitude larger then the measurement variance expected from other physiological fluctuations within the head including activation. This feature can be used to help detect regions affected by motion artifacts by simple visual inspection or outlier analysis (for example, by examination of the studentized residual error of the data after functional analysis or using a similar regression analysis model). Subject movement may also move the head probe causing a change in the optode couplings to the head. This can be a major issue since it requires a renormalization of the measured signal or, in a worse case, could mean that measurements represent a different brain region. These errors can greatly affect the quantitative accuracy of optical experiments. Once identified, temporal regions affected by motion artifacts can be excluded from further analysis by either removing these time points from the model used for the functional analysis, or by excluding the entire data file or subject. Filtering methods such as autoregressive models or adaptive filtering methods such as Weiner or Kalman filters have also been suggested for reducing the impact of motion artifacts.[131,132] Since many of the subsequent response estimation, filtering, and other signal processing methods will not work properly if motion is present in the data, motion correction is often done as a early step in data analysis. However, it is important to have an objective criterion for removing data from a study.

Since motion artifacts arise from the physical displacement of the optical probe from the surface of the subject's head, motion artifacts are statistically covariant across multiple source-detector pairs and independent of the optical wavelength. Such a motion artifact will typically demonstrate a large change of signal in all or at least a large subset of the source-detector channels. The use of principal component analysis (PCA) to remove motion artifacts was first demonstrated by Wilcox and coworkers where it was shown to reduce motion artifacts in a study of infant brain function.[133] The principal components of the data covariance are calculated using singular value decomposition. Since motion artifacts will typically have a highly covariant data structure, motion artifacts may be captured in the first (strongest) eigenvector components using PCA. This allows these artifacts to be subtracted from the time-series of the optical density measurements. Using this filter, motion artifacts can be removed from data by reducing the spatial covariance of the set of measurements.[133] An example of this PCA filter is demonstrated in Figure 14.12. Here,

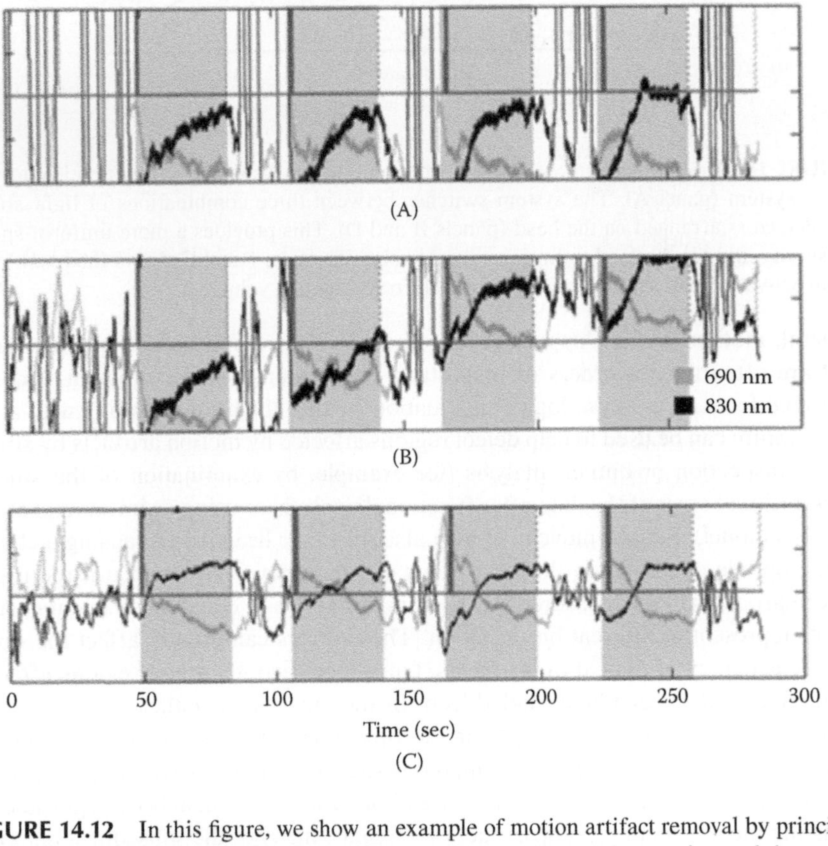

FIGURE 14.12 In this figure, we show an example of motion artifact removal by principal component (PCA) filtering. The raw data shown in panel (A, top) is experimental data measured in the inferior temporal cortex in an infant. This data was dominated by several motion artifacts that resulted from the movement of the probe on the head's surface. These motions artifacts were highly covariant across the entire probe. In panels (B) and (C) are shown the filtered data after the removal of the first and second principal components. After the reduction of the covariance in the data by the removal of these components, the motion artifacts were greatly reduced. The red boxes indicate the presentation of stimuli.

we show an example data set taken as part of a NIRS study on object recognition in infants.[133] The raw data, shown in Figure 14.12A (top) demonstrates very strong motion artifacts, which dominate the optical signals. In addition, these motion artifacts were covariant across the majority of the NIRS probe, making analysis of this data virtually impossible. Using this PCA filter to remove the first and second principal components, the motion artifact is greatly reduced (shown in panels B and C).

This motion correction algorithm exploits the property of a motion artifact to have an intense and correlated effect over a large area of the NIRS probe. This is a general characteristic of motion, but not entirely encompassing of all motion artifacts. This algorithm works best for reducing motion that affects areas much larger than the functional region. For a small number of source-detector pairs or a probe with a spatial coverage comparable to the extent of the functional region, principal component analysis can negatively affect the estimate of the hemodynamic response. To ensure that this does not occur, statistical effects analysis can be used to estimate the number of components that maximize the probability of functional activity.

14.6 METHODS FOR THE REDUCTION OF PHYSIOLOGICAL NOISE

Once optical signals have been checked and corrected for motion artifacts, there will still be several other sources of noise in the data. Physiological systemic noise arising from the cardiac cycle, respiration, systemic blood pressure, and Mayer wave fluctuations are a common component of signal interference in NIRS measurements.[31,119,120,124,127,134,135] This is enhanced by the oversensitivity of NIRS to superficial tissue. In fact, the absence of cardiac and other fluctuations can be a sign that data quality is low (perhaps due to poor coupling of the optical probe with the head) and often indicates that functional analysis will not be possible. Background physiological signals have been explored and may reveal interesting details about vascular physiology.[136] However, in most studies, attempts are made to reduce background physiological signals, and several techniques will be discussed in the following sections.

14.6.1 Linear Filtering Techniques

Band-pass filtering can provide a method for effectively removing many sources of noise in the data, such as high-frequency instrument noise or low frequency signal drifts. This can also be used to remove physiological noise such as low-frequency blood pressure oscillations, which are generally found between 0.08 and 0.12 Hz or cardiac pulsation (around 1 Hz). However, some physiology such as respiration (around 0.25 Hz) cannot be removed using frequency filtering methods because their frequency content overlaps with that of functional signals. Band-pass filtering must be used carefully to avoid frequencies associated with the functional hemodynamic response, which depend on the experimental design.

14.6.2 Principal Component Analysis

Principal component analysis has also been used in optical studies to remove systemic physiology based on the spatial and temporal structure of these signals as first described by Zhang and coworkers.[135] Similar to motion artifacts, systemic fluctuations are covariant between the NIRS measurements from different regions of the head. Thus, the reduction of this covariance can be used to filter physiology of systemic origin from the measurements. In comparison to the motion correction, for this PCA filter, the spatial covariance may be calculated from a separate baseline (rest state) data set, rather than from the functional data itself. By using a separate baseline time course to derive the spatial covariance associated with the systemic physiology, this allows better distinction from the evoked hemodynamic functional response. As this systemic signal interference causes different variation in oxy- and deoxy-hemoglobin, the spatial covariance should calculated independently for each hemoglobin species or from each optical wavelength.

In Figure 14.13, we demonstrate how the reduction of spatial covariance allows the better localization hemodynamic activation associated with a functional motor task by the removal of global, systemic changes. This data, previously described in Franceschini and coworkers,[124] demonstrates that, in addition to the regional cerebral

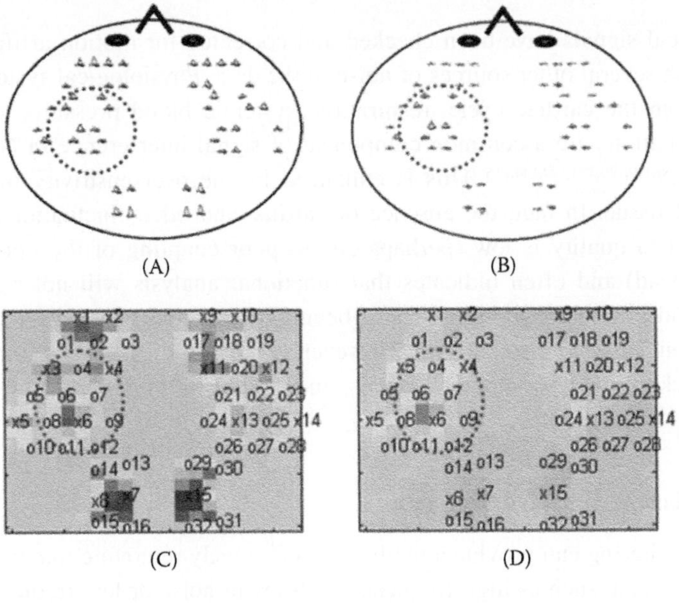

FIGURE 14.13 NIRS measurements are often sensitive to systemic fluctuations arising from blood pressure changes, respiration, or the cardiac cycle. As a complication, this physiology may change during the performance of intense stimulus tasks, such as motor activity. As shown in the first column (left), these changes can result in a systemic response giving the appearance of global "functional activation." Using the PCA filter described in the text to remove this systemic effect by reducing this covariance, the activation region is localized to the motor area (right column). The back-projection images of oxy-hemoglobin changes, shown across the bottom row, further demonstrate the localization of the response after filtering.

changes, functional tasks can result in global, systemic changes as well. During the finger-tapping task, as demonstrated in Figure 14.9, changes in heart rate, respiration, and blood pressure resulted in a task-associated, global hemodynamic response. These changes result in the appearance of a "functional activation" over nearly the entire head of the subject, as shown in Figure 14.13A. Using the PCA filter to project the strongest principal components calculated from a separate baseline recording, the resulting filtered response was localized to the contralateral motor cortex.

14.6.3 Estimation of Functional Signals in NIRS

Since the continuous-wave NIRS technique is capable of measuring only hemodynamic changes and unable to report baseline values, the most common use of this technique is to assess changes in hemoglobin levels in response to functional stimulation. In general, this is done after motion correction and possible physiological noise reduction steps have occurred.

In comparison to typical MRI studies, optical measurements have more spectroscopic information due to the sensitivity to both oxy- and deoxy-hemoglobin. Since optical measurements provide information about both oxy- and deoxy-hemoglobin species, the analysis and interpretation of these signals is inherently multidimensional—including time, space, and spectroscopic (oxy/deoxy-hemoglobin) degrees-of-freedom. Deciding which of these parameters or hemoglobin species is most informative in analysis and in testing a hypothesis is not clear. Although in practice, oxy-hemoglobin changes are most often considered because these are often larger and more robust changes compared to deoxy-hemoglobin, the measurement of both signals offers the possibility to investigate more hypothesis then with fMRI studies. For example, it may be important to not only consider the magnitude of one individual signal, but also the relative changes of these signals. Thus, it may be important to consider how the measured hemodynamic changes originate according to the underlying vascular structure and are driven by the indirect effects of local oxygen metabolism and vascular dilation in response to brain "activation."[130] Several models of the vascular system (reviewed in Reference 130) have been proposed for interpretation of fMRI[130,137–140] and optical[141,142] data. However, it is still unclear how best to use these models in the analysis of such data. These issues have not yet been fully addressed and must be considerations in future NIRS and DOT studies. Such nonlinear analysis methods based on vascular models have started to be used in direct analysis of fMRI signals[140,143,144] and may be more feasible for optical signals, which have lower numbers of spatial measurements (less computational load) and higher temporal resolution.

14.6.4 Linear Model of Hemodynamic Change

Most analysis of functional hemodynamic changes in neuroimaging techniques such as NIRS or fMRI is based on an assumption of a linear model of hemodynamic changes. In the linear model, the effects of hemodynamic variations are additive with the individual components summing linearly to give the measured total signal change. For example, the hemodynamic responses associated with multiple types of stimuli or overlapping responses are assumed to add to give the overall measured signal. In this model, the observations of hemodynamic changes over time can be expressed as a series of linear equations, described by the equation

$$\mathbf{Y} = \mathbf{U}_{\text{stimulus}} \otimes \beta + e \quad (14.13a)$$

$$\mathbf{Y} = \mathbf{G} \cdot \beta + e \quad (14.13b)$$

In this notation, Y is the recorded time-series of observations of a hemodynamic variable. β represents the parameter vector, for which we wish to solve. The operator ⊗ represents convolution of the regression variable (i.e., the timing of stimuli) with the impulse response to that regressor (β) (i.e., the hemodynamic response). In the matrix form of this equation (14.13b), the matrix G is typically termed the *design matrix*. When multiplied by the impulse response functions (β), this matrix performs the convolution of the regression parameters and the response functions. The effects of individual responses are summed to reproduce the observations. Lastly, the final term in Equation 14.13, e is the error in each measurement. This is generally assumed to be a normally distributed, zero-mean, random noise.

In unconstrained estimation (deconvolution), a separate value for β is estimated for each time point in the hemodynamic response. In this case, β represents an impulse response function that convolves for the stimulus vector and defines the entire time course of the evoked response. This is often referred to as a finite impulse response (FIR) model of the activation. The full estimation of the evoked response makes no assumptions about the shape of the response, which can be advantageous if the hypothesis of the experiment concerns the timing of the signal or differences in the timing between conditions or subjects. The drawback of this approach is that each time-point of the response is estimated individually—resulting in many degrees-of-freedom to the functional model. An alternative approach is to use a defined canonical temporal basis to constrain the shape of the evoked response[145,146,126,142,147,148]. This approach is similar to that used in SPM[149,150] or AFNI[151] software packages commonly used in fMRI analysis. The advantage of this approach is to reduce the degrees-of-freedom of the model and to constrain the functional response with prior information about its expected temporal shape. This may improve detection of functional signals, and this approach shows promise for future work.

In order to estimate the mean effect of each functional response, we must perform an inversion of Equation 14.13 in order to estimate β. The minimum ordinary least-squares solution to this linear equation is provided by the equation

$$\hat{\beta} = \left(\mathbf{G}^{t} \cdot \mathbf{G}\right)^{-1} \cdot \mathbf{G}^{t} \cdot \mathbf{Y} \quad (14.14)$$

The superscript t represents the transpose matrix operation, G is the design matrix[43] and Y is the measurement vector. The efficiency of this inverse equation is optimized by the careful experimental design of the timing of stimulus presentation.[43,152–154] Inappropriate stimulus timing can result in cross-talk between components of β or create an ill-posed structure to the matrix G, which prevents inversion or introduces numerical imprecision. In addition, the nonlinear effects of short interstimulus intervals should also be considered, which may effect the assumption of the linear model.[155]

14.6.5 HOMER—A Graphical Interface for Functional Optical Signals

In order to analyze our functional NIRS and DOT experiments, we have developed a Matlab-based graphic interface called HOMER for hemodynamic optically measured evoked response. The HOMER program is based on flexible data architecture that allows data to be imported with limited preparatory processing from virtually any currently existing commercial NIRS system. The analysis of this data can be tailored to allow the users complete versatility to specify the details of their experiment, including NIRS probe geometry and layout, optical wavelength selection, stimulus design, and other instrumental and acquisition parameters. The structure of the HOMER program is based on three levels of data processing. (1) At the first level, individual data files are represented. This level contains the data for a single experimental run, for example, a single acquisition scan. Scan-level processing allows the data from a single experimental run to be interactively visualized, time-series data filtered, and analyzed for functional responses. For a typical experiment, multiple such data files will be recorded within a single experimental session. (2) The session level represents the second organizing level in HOMER. This level includes the joint processing of one or more individual data files for a single experimental subject, recorded during a single session. At the session level, response averages and effects analysis can be calculated from multiple data runs. Statistical tests can be preformed to investigate the significance of functional changes (i.e., the f-test and t-test). In addition, images of the changes in optical absorption or hemoglobin concentrations can be reconstructed from these session averages. Session-level data should have consistent optical probe placement and geometry to allow analysis between scans. (3) At the final level of processing, multisession or group information is represented. This allows group averaging from multiple experimental subjects or multiple sessions from the same subject. Data processing at this level is allowed for region-of-interest comparisons to avoid intersubject registration issues. At all these three levels, data analysis, visualization, and data export are provided to Matlab or ASCII formats.

As shown in Figure 14.14, the layout of the HOMER program is based around an interactive graphical display of the NIRS or DOT probe, shown in the upper right. The user specifies this probe geometry within the data file imported into HOMER as described in the text. By selecting source (displayed as "x") or detector ("o") positions on this probe layout, the user navigates through the display of their data. Data shown in Figure 14.14 are from Reference 156.

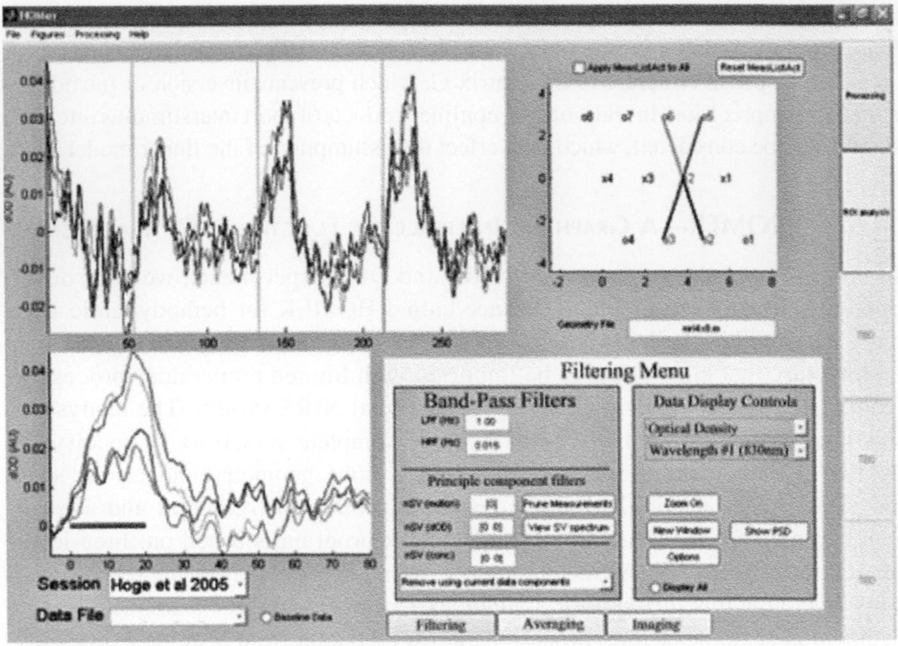

FIGURE 14.14 The layout of the HOMER program is based around an interactive graphical display of the NIRS probe, shown in the upper right (B). The user specifies this probe geometry within the data file imported into HOMER. By selecting source (displayed as "x") or detector ("o") positions on this probe layout, the user navigates through the display of their data. The original data is shown in panel (A) and the average evoked response is shown in panel (C). The data was recorded during a 20-s finger-tapping task. Data shown is from a single run of one of the subjects.

14.7 MULTIMODAL AND VALIDATION STUDIES

In the early 1990s, functional magnetic resonance imaging methods began to appear, providing whole-brain imaging of blood flow[157,158] and the blood oxygen level dependent (BOLD) signals[159–163] associated with changes in blood flow and oxygen metabolism within the brain. Shortly after this development, NIRS was shown to have sensitivity similar to fMRI in a series of comparison studies.[164–167] Optical measurements of brain activation arise from similar hemodynamic contrast as functional MRI and thus, these two signals can be used to cross-validate one another and to explore assumptions about these technologies, such as vascular sensitivity. Concurrent multimodal measurements can also be used to explore the effects of intersubject and intertrial variation in the evoked hemodynamic response, since both modalities measure common physiological fluctuations at the same time, which can be used to separate measurement and physiological noise.

A first step toward cross-validating fMRI and optical techniques is to examine simultaneously acquired time courses from spatially coregistered locations. To date a number of groups have accomplished this as recently reviewed by Steinbrink and

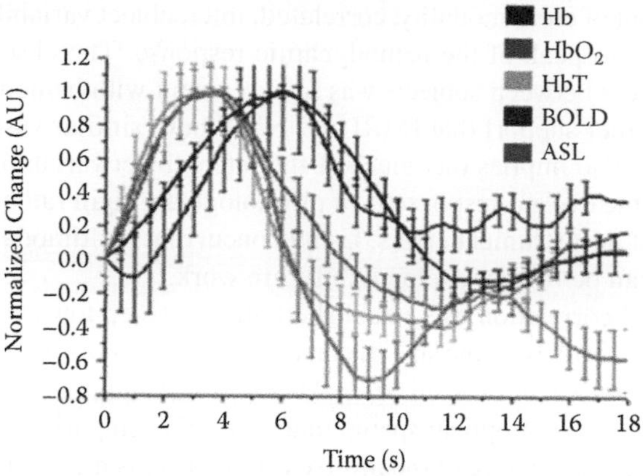

FIGURE 14.15 Simultaneous NIRS and fMRI measurements of blood oxygen level dependant (BOLD) and blood flow using arterial spin labeling (ASL) were recorded during a brief 2-s finger-tapping task using a rapid event-related experimental design. The temporal dynamics of the NIRS measure of deoxy-hemoglobin is highly correlated to the fMRI BOLD signal in agreement with the theoretical basis of the BOLD signal. Likewise, the oxy- and total-hemoglobin signals are slightly earlier than the venous weighted BOLD and Hb signals and are more correlated to the blood flow (ASL) signal. (From Huppert, T.J. et al. *NeuroImage* **29**, 368–382, 2006. With permission.)

colleagues.[168] Figure 14.15 shows an example of a comparison between optical and fMRI signals from Huppert et al.[169] In this study, we preformed simultaneous near-infrared spectroscopy (NIRS) along with BOLD (blood oxygen level dependant) and blood flow measured using ASL (arterial spin labeling)-based fMRI during an event-related motor activity in human subjects in order to compare the temporal dynamics of the hemodynamic responses recorded in each method. The functional MRI BOLD signal represents changes in blood oxygenation and deoxy-hemoglobin concentration,[170] and thus is expected to correlate to the optical measurement of deoxy-hemoglobin. Both are venous weighted measurements. Likewise, blood flow measured by ASL is an arterial weighted measurement[171] and expected to correlate to the arterial weighted oxy-hemoglobin changes. Multimodal measurements allow us to examine the validity of the biophysical models underlying each modality and, as a result, gain greater insight into the hemodynamic responses to neuronal activation. Concurrent measurements also allow us to examine physiological variability in the evoked response. In Figure 14.15, we show that the fMRI measured BOLD response is more correlated with the NIRS measure of deoxy-hemoglobin ($R = 0.98$; $p < 10^{-20}$) than with oxy-hemoglobin ($R = 0.71$) or total hemoglobin ($R = 0.53$). This result was predicted from the theoretical grounds of the BOLD response and is in agreement with several previous works.[168,172–175] This data has also allowed us to examine more detailed measurement models of the fMRI signal and comment on the roles of the oxygen saturation and blood volume contributions to the BOLD response. In addition, we found high correlation between the NIRS measured total hemoglobin and ASL measured cerebral blood flow ($R = 0.91$; $p < 10^{-10}$) and oxy-hemoglobin with flow ($R = 0.83$; $p < 10^{-5}$) as predicted by the biophysical models. Finally, we noted a significant amount of cross modality, correlated, intersubject variability in amplitude change, and time-to-peak of the hemodynamic response. The observed covariance in these parameters between subjects was in agreement with hemodynamic models, and provided further support that fMRI and NIRS have similar vascular sensitivity. This observation also implies that much of the intersubject variability in the temporal dynamics of the evoked response is of physiological origin rather then representing measurement or instrument errors. Using concurrent multimodal measurements, this variability can be further explored in future work.

Once temporal correlation was established, the next step toward combining MRI and optical signals is to examine spatial correspondences in the diffuse optical image and the fMRI. However, as pointed out in the discussion of optical image reconstruction, the reconstruction of optical spatial maps for this comparison can be challenging due to the ill-poised nature of the inverse problem. This makes direct comparison of optical and MR images difficult because care must be taken to account for possible artifacts of the reconstruction process. Recently, Toronov et al.[172] and Joseph et al.[45] both described spatial comparisons between fMRI and optical. In the work by Joseph et al., tomographic imaging was used to improve spatial resolutions of the optical data.[45] For the most part, these spatial comparisons are qualitative, since the optical inverse problem cannot be strictly solved. To work around this problem, Sassaroli et al.,[176] Huppert et al.,[177] and Toronov et al.[173] proposed to use the calculated optical measurement sensitivity to weight the fMRI signal, thereby creating a forward projection of the high-spatial resolution MR image onto the optical source-detector measurement space. This approach allowed optical and MRI signals to be quantitatively compared by testing the MR image as a possible solution to the optical image problem.

In Figure 14.16, we show a spatial comparison between optical and fMRI. Using a model of light propagation through the tissue, which was estimated from the anatomical MRI images from each of the subjects, we calculated a weighted average of the underlying fMRI data that represented the optical measurements expected based on the fMRI images. We used this approach to compare the optical and fMRI spatial and temporal maps, and to examine the consistency of the fMRI data as a solution to the optical imaging problem. Like our earlier temporal comparison, we found statistically better spatial correlation between the fMRI-BOLD response and the optically measured deoxy-hemoglobin ($R = 0.71$; $p = 1 \times 10^{-7}$) than between the BOLD and oxy- or total hemoglobin measures ($R = 0.38$; $p = 0.04$ | 0.37; $p = 0.05$, respectively). Similarly, we found that the correlation between the ASL measured blood flow and optically measured total and oxy-hemoglobin is stronger ($R = 0.73$; $p = 5 \times 10^{-6}$ and $R = 0.71$; $p= 9 \times 10^{-6}$, respectively) than the flow to deoxy-hemoglobin spatial correlation ($R = 0.26$; $p = 0.10$).

One of the current limitations to optical methods is the process of image reconstruction and the mathematically ill-posed inverse problem. Several groups have demonstrated that is it possible to anatomical information provided by the MRI to obtain more accurate estimates of where the light has sampled within the head, using the Monte Carlo method described earlier.[56,178–180] To see how this synergy develops,

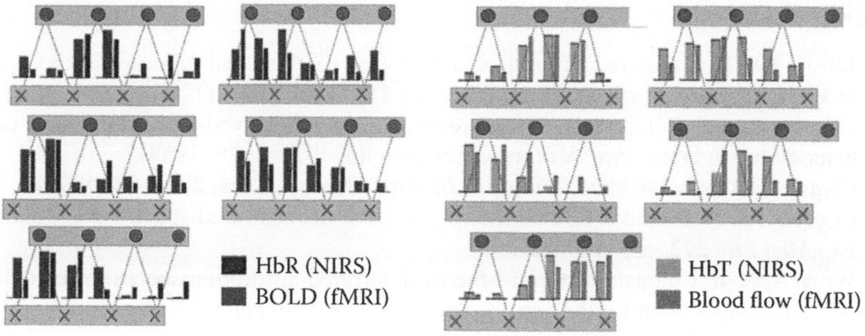

(a) Study I-BOLD/Hb

(b) Study II-ASL/HbT

FIGURE 14.16 The spatial profiles of optical and fMRI signals are compared for 10 subjects. Monte Carlo simulations (described in the text) were used to generate the sensitivity profiles for each optical measurement pair and were used to generate spatially weighted projections of the fMRI signals. In panel (a), the mean amplitude of the fMRI BOLD (blood oxygen level dependent) and optical Hb signals for each of these source-detector projections, showing good spatial agreement between the two modalities. In panel (b), the fMRI ASL (arterial spin labeling) measure of blood flow and the optical total-hemoglobin signals are shown.

consider the following sequence of events. First, we acquire anatomical MRI scans that (1) help localize our optical probes with respect to the underlying brain anatomy, and (2) can be used to segment the head into a variety of tissue types (typically, air, scalp, skull, cerebral spinal fluid, and white and gray matter). The segmented head can then be used for Monte Carlo studies in photon diffusion to determine the DOT instrument's sensitivity to various tissue types and depths. Knowledge of the underlying tissue structure can also be used to limit the location of the reconstructed functional activation to the cortex of the brain.[181] These estimates can also be used as a model of the optical changes, which would allow more accurate quantitation of the Hb and HbO_2 changes during functional challenge. In this way, combined MRI and optical recordings have the potential to provide more changes found there—something that is impossible with either technique alone.[142]

14.8 SUMMARY

The rapidly growing literature from the past 30 years clearly demonstrates the ability of near-infrared spectroscopy to noninvasively measure cerebral hemodynamic changes in response to stimuli. Furthermore, the literature is growing in support of the unique ability of NIRS to measure metabolic (cytochrome oxidase) and neuronal (fast optical scattering changes) signals. NIRS is unique as a neuro-monitoring/imaging technique as it is the only method which can potentially measure hemodynamic, metabolism, and neuronal signals simultaneously. Furthermore, the technology is relatively inexpensive and portable, allowing studies of brain activation in populations not easily studied in the past, such as babies and children, and freely behaving paradigms.

REFERENCES

1. Jobsis, F.F. Noninvasive, infrared monitoring of cerebral and myocardial oxygen sufficiency and circulatory parameters. *Science* **198**, 1264–1267 (1977).
2. Chance, B. et al. Comparison of time-resolved and -unresolved measurements of deoxyhemoglobin in brain. *Proc Natl Acad Sci USA* **85**, 4971–4975 (1988).
3. Chance, B. Optical method. *Annu Rev Biophys Biophys Chem* **20**, 1–28 (1991).
4. Cope, M. et al. Methods of quantitating cerebral near infrared spectroscopy data. *Adv Exp Med Biol* **222**, 183–189 (1988).
5. Wray, S. et al. Characterization of the near infrared absorption spectra of cytochrome aa3 and haemoglobin for the non-invasive monitoring of cerebral oxygenation. *Biochim Biophys Acta* **933**, 184–192 (1988).
6. McCormick, P.W. et al. Noninvasive cerebral optical spectroscopy for monitoring cerebral oxygen delivery and hemodynamics. *Crit Care Med* **19**, 89–97 (1991).
7. Benaron, D.A. et al. Noninvasive methods for estimating in vivo oxygenation. *Clin Pediatr (Phila)* **31**, 258–273 (1992).
8. Elwell, C.E. et al. Quantification of adult cerebral hemodynamics by near-infrared spectroscopy. *J Appl Physiol* **77**, 2753–2760 (1994).
9. Gibson, A.P. et al. Recent advances in diffuse optical imaging. *Phys Med Biol* **50**, R1–43 (2005).
10. von Siebenthal, K. et al. Near-infrared spectroscopy in newborn infants. *Brain Dev* **14**, 135–143 (1992).
11. Elwell, C.E. et al. Measurement of CMRO2 in neonates undergoing intensive care using near infrared spectroscopy. *Adv Exp Med Biol* **566**, 263–268 (2005).
12. Delpy, D.T. et al. Cerebral monitoring in newborn infants by magnetic resonance and near infrared spectroscopy. *Scand J Clin Lab Invest Suppl* **188**, 9–17 (1987).
13. Weber, P. et al. Cerebral hemodynamic changes in response to an executive function task in children with attention-deficit hyperactivity disorder measured by near-infrared spectroscopy. *J Dev Behav Pediatr* **26**, 105–111 (2005).
14. Soul, J.S. and du Plessis, A.J. New technologies in pediatric neurology. Near-infrared spectroscopy. *Semin Pediatr Neurol* **6**, 101–110 (1999).
15. Adelson, P.D. et al. The use of near infrared spectroscopy (NIRS) in children after traumatic brain injury: A preliminary report. *Acta Neurochirurgica* **71**, 250–254 (1998).
16. Mehagnoul-Schipper, D.J. et al. Reproducibility of orthostatic changes in cerebral oxygenation in healthy subjects aged 70 years or older. *Clin Physiol* **21**, 77–84 (2001).
17. Kwee, I.L. and Nakada, T. Dorsolateral prefrontal lobe activation declines significantly with age—Functional NIRS study. *J Neurol* **250**, 525–529 (2003).
18. Hock, C. et al. Near infrared spectroscopy in the diagnosis of Alzheimer's disease. *Ann N Y Acad Sci* **777**, 22–29 (1996).
19. Miyai, I. et al. Cortical mapping of gait in humans: A near-infrared spectroscopic topography study. *NeuroImage* **14**, 1186–1192 (2001).
20. Igawa, M. et al. Activation of visual cortex in REM sleep measured by 24-channel NIRS imaging. *Psychiatry Clin Neurosci* **55**, 187–188 (2001).
21. Arridge, S.R. et al. Reconstruction methods for infra-red absorption imaging. *Proc. SPIE* **1431**, 204–215 (1991).
22. Sassaroli, A. and Fantini, S. Comment on the modified Beer-Lambert law for scattering media. *Phys Med Biol* **49**, N255–257 (2004).
23. Kocsis, L. et al. The modified Beer-Lambert law revisited. *Phys Med Biol* **51**, N91–98 (2006).
24. Duncan, A. et al. Optical pathlength measurements on adult head, calf and forearm and the head of the newborn infant using phase resolved optical spectroscopy. *Phys Med Biol* **40**, 295–304 (1995).

25. van der Zee, P. et al. Experimentally measured optical pathlengths for the adult head, calf and forearm and the head of the newborn infant as a function of inter optode spacing. *Adv Exp Med Biol* **316**, 143–153 (1992).
26. Fantini, S. et al. Non-invasive optical monitoring of the newborn piglet brain using continuous-wave and frequency-domain spectroscopy. *Phys Med Biol* **44**, 1543–1563 (1999).
27. Kohl, M. et al. Determination of the wavelength dependence of the differential pathlength factor from near-infrared pulse signals. *Phys Med Biol* **43**, 1771–1782 (1998).
28. Hiraoka, M. et al. A Monte Carlo investigation of optical pathlength in inhomogeneous tissue and its application to near-infrared spectroscopy. *Phys Med Biol* **38**, 1859–1876 (1993).
29. Yoshitani, K. et al. Effects of hemoglobin concentration, skull thickness, and the area of the cerebrospinal fluid layer on near-infrared spectroscopy measurements. *Anesthesiology* **106**, 458–462 (2007).
30. Strangman, G. et al. Factors affecting the accuracy of near-infrared spectroscopy concentration calculations for focal changes in oxygenation parameters. *NeuroImage* **18**, 865–879 (2003).
31. Boas, D.A. et al. Diffuse optical imaging of brain activation: Approaches to optimizing image sensitivity, resolution, and accuracy. *NeuroImage* **23(Suppl. 1)**, S275–288 (2004).
32. Sato, H. et al. Practicality of wavelength selection to improve signal-to-noise ratio in near-infrared spectroscopy. *NeuroImage* **21**, 1554–1562 (2004).
33. Matcher, S.J. et al. Performance comparison of several published tissue near-infrared spectroscopy algorithms. *Anal Biochem* **227**, 54–68 (1995).
34. Ishimaru, A. *Wave Propagation and Scattering in Random Media*, Academic Press, New York, 1978.
35. Flock, S.T. et al. Monte Carlo modeling of light propagation in highly scattering tissue—I: Model predictions and comparison with diffusion theory. *IEEE Trans Biomed Eng* **36**, 1162–1168 (1989).
36. Flock, S.T. et al. Monte Carlo modeling of light propagation in highly scattering tissues—II: Comparison with measurements in phantoms. *IEEE Trans Biomed Eng* **36**, 1169–1173 (1989).
37. Haskell, R.C. et al. Boundary conditions for the diffusion equation in radiative transfer. *J Opt Soc Am A Opt Image Sci Vis* **11**, 2727–2741 (1994).
38. Furutsu, K. and Yamada, Y. Diffusion approximation for a dissipative random medium and the applications. *Phys Rev E Stat Phys Plasmas Fluids Relat Interdiscip Topics* **50**, 3634–3640 (1994).
39. Durduran, T. et al. Does the photon-diffusion coefficient depend on absorption? *J Opt Soc Am A Opt Image Sci Vis* **14**, 3358–3365 (1997).
40. Fukui, Y. et al. Monte Carlo prediction of near-infrared light propagation in realistic adult and neonatal head models. *Appl Opt* **42**, 2881–2887 (2003).
41. Boas, D. et al. Three dimensional Monte Carlo code for photon migration through complex heterogeneous media including the adult human head. *Opt. Express* **10**, 159–170 (2002).
42. Wang, L. et al. MCML—Monte Carlo modeling of light transport in multi-layered tissues. *Comput Methods Programs Biomed* **47**, 131–146 (1995).
43. Dale, A.M. Optimal experimental design for event-related fMRI. *Hum Brain Mapp* **8**, 109–114 (1999).
44. Boas, D.A. et al. Improving the diffuse optical imaging spatial resolution of the cerebral hemodynamic response to brain activation in humans. *Opt Lett* **29**, 1506–1508 (2004).
45. Joseph, D.K. et al. Design and validation of a time division multiplexed continuous wave diffuse optical tomography (DOT) system optimized for brain function imaging. *Appl Opt* **45**, 814208151 (2006).
46. Custo, A. et al. Effective scattering coefficient of the cerebral spinal fluid in adult head models for diffuse optical imaging. *Appl Opt* **45**, 4747–4755 (2006).

47. Zeff, B.W. et al. Retinotopic mapping of adult human visual cortex with high-density diffuse optical tomography. *Proc Natl Acad Sci USA* **104**, 12169–12174 (2007).
48. Barbour, R.L. et al. Optical tomographic imaging of dynamic features of dense-scattering media. *J Opt Soc Am A Opt Image Sci Vis* **18**, 3018–3036 (2001).
49. Arridge, S.R. Optical tomography in medical imaging. *Inverse Problems* **15**, 14–93 (1999).
50. Farrell, T.J. et al. A diffusion theory model of spatially resolved, steady-state diffuse reflectance for the noninvasive determination of tissue optical properties in vivo. *Med Phys* **19**, 879–888 (1992).
51. Natterer, F. *The Mathematics of Computerized Tomography*. Society for Industrial and Applied Mathematics, Philadelphia, 2001.
52. Benaron, D.A. and Stevenson, D.K. Optical time-of-flight and absorbance imaging of biologic media. *Science* **259**, 1463–1466 (1993).
53. Cubeddu, R. et al. Time-resolved imaging on a realistic tissue phantom: mu(s)' and mu(a) images versus time-integrated images. *Appl Opt* **35**, 4533–4540 (1996).
54. Grosenick, D. et al. Time-resolved imaging of solid phantoms for optical mammography. *Appl Opt* **36**, 221–231 (1997).
55. Selb, J. et al. Improved sensitivity to cerebral hemodynamics during brain activation with a time-gated optical system: Analytical model and experimental validation. *J Biomed Opt* **10**, 11013 (2005).
56. Ntziachristos, V. et al. MRI-guided diffuse optical spectroscopy of malignant and benign breast lesions. *Neoplasia* New York, **4**, 347–354 (2002).
57. Gratton, E. et al. Measurements of scattering and absorption changes in muscle and brain. *Philos Trans R Soc Lond B Biol Sci* **352**, 727–735 (1997).
58. Jiang, H.B. et al. Simultaneous Reconstruction of Optical-Absorption and Scattering Maps in Turbid Media from near-Infrared Frequency-Domain Data. *Opt Lett* **20**, 2128–2130 (1995).
59. Pogue, B.W. and Patterson, M.S. Frequency-domain optical absorption spectroscopy of finite tissue volumes using diffusion theory. *Phys Med Biol* **39**, 1157–1180 (1994).
60. Pogue, B. et al. Instrumentation and design of a frequency-domain diffuse optical tomography imager for breast cancer detection. *Optics Express* **1**, 391 (1997).
61. Chance, B. et al. Phase measurement of light absorption and scatter in human tissue. *Rev Scientific Instruments* **69**, 3457–3481 (1998).
62. Nioka, S. et al. Human brain functional imaging with reflectance CWS. *Oxygen Transport to Tissue Xix* **428**, 237–242 (1997).
63. Siegel, A.M. et al. Design and evaluation of a continuous-wave diffuse optical tomography system. *Opt Express* **4**, 287–298 (1999).
64. Maki, A. et al. Spatial and temporal analysis of human motor activity using noninvasive NIR topography. *Med Phys* **22**, 1997–2005 (1995).
65. Arridge, S.R. and Lionheart, W.R.B. Nonuniqueness in diffusion-based optical tomography. *Opt Lett* **23**, 882–884 (1998).
66. American National Standard for Save Use of Lasers. In *ANSI Z136.1-1993*, American National Standards Institute, New York, 1993.
67. Heekeren, H.R. et al. Cerebral haemoglobin oxygenation during sustained visual stimulation—A near-infrared spectroscopy study. *Philos Trans R Soc Lond B Biol Sci* **352**, 743–750 (1997).
68. Meek, J.H. et al. Regional changes in cerebral haemodynamics as a result of a visual stimulus measured by near infrared spectroscopy. *Proc Biol Sci* **261**, 351–356 (1995).
69. Ruben, J. et al. Haemoglobin oxygenation changes during visual stimulation in the occipital cortex. *Adv Exp Med Biol* **428**, 181–187 (1997).
70. Sakatani, K. et al. Cerebral blood oxygenation changes induced by auditory stimulation in newborn infants measured by near infrared spectroscopy. *Early Hum Dev* **55**, 229–236 (1999).

71. Chen, S. et al. Auditory-evoked cerebral oxygenation changes in hypoxic-ischemic encephalopathy of newborn infants monitored by near infrared spectroscopy. *Early Hum Dev* **67**, 113–121 (2002).
72. Zaramella, P. et al. Brain auditory activation measured by near-infrared spectroscopy (NIRS) in neonates. *Pediatr Res* **49**, 213–219 (2001).
73. Franceschini, M.A. et al. Hemodynamic evoked response of the sensorimotor cortex measured noninvasively with near-infrared optical imaging. *Psychophysiology* **40**, 548–560 (2003).
74. Obrig, H. et al. Cerebral oxygenation changes during motor and somatosensory stimulation in humans, as measured by near-infrared spectroscopy. *Adv Exp Med Biol* **388**, 219–224 (1996).
75. Obrig, H. et al. Cerebral oxygenation changes in response to motor stimulation. *J Appl Physiol* **81**, 1174–1183 (1996).
76. Colier, W.N. et al. Human motor-cortex oxygenation changes induced by cyclic coupled movements of hand and foot. *Exp Brain Res* **129**, 457–461 (1999).
77. Hirth, C. et al. Non-invasive functional mapping of the human motor cortex using near-infrared spectroscopy. *Neuroreport* **7**, 1977–1981 (1996).
78. Kleinschmidt, A. et al. Simultaneous recording of cerebral blood oxygenation changes during human brain activation by magnetic resonance imaging and near-infrared spectroscopy. *J Cereb Blood Flow Metab* **16**, 817–826 (1996).
79. Kameyama, M. et al. Frontal lobe function in bipolar disorder: A multichannel near-infrared spectroscopy study. *NeuroImage* **29**, 172–184 (2006).
80. Nagamitsu, S. et al. Prefrontal cerebral blood volume patterns while playing video games-A near-infrared spectroscopy study. *Brain Dev* (2006).
81. Okamoto, M. et al. Prefrontal activity during taste encoding: An fNIRS study. *NeuroImage* 47(2), 220–232 (2006).
82. Sumitani, S. et al. Hemodynamic changes in the prefrontal cortex during mental works as measured by multi channel near-infrared spectroscopy (NIRS). *J Med Invest* **52(Suppl.)**, 302–303 (2005).
83. Sato, H. et al. Temporal cortex activation during speech recognition: An optical topography study. *Cognition* **73**, B55–66 (1999).
84. Adelson, P.D. et al. Noninvasive continuous monitoring of cerebral oxygenation periictally using near-infrared spectroscopy: A preliminary report. *Epilepsia* **40**, 1484–1489 (1999).
85. Sokol, D.K. et al. Near infrared spectroscopy (NIRS) distinguishes seizure types. *Seizure* **9**, 323–327 (2000).
86. Steinhoff, B.J. et al. Ictal near infrared spectroscopy in temporal lobe epilepsy: A pilot study. *Seizure* **5**, 97–101 (1996).
87. Watanabe, E. et al. Noninvasive cerebral blood volume measurement during seizures using multichannel near infrared spectroscopic topography. *J Biomed Opt* **5**, 287–290 (2000).
88. Buchheim, K. et al. Decrease in haemoglobin oxygenation during absence seizures in adult humans. *Neurosci Lett* **354**, 119–122 (2004).
89. Watanabe, E. et al. Focus diagnosis of epilepsy using near-infrared spectroscopy. *Epilepsia* **43(Suppl. 9)**, 50–55 (2002).
90. Fallgatter, A.J. et al. Loss of functional hemispheric asymmetry in Alzheimer's dementia assessed with near-infrared spectroscopy. *Brain Res Cogn Brain Res* **6**, 67–72 (1997).
91. Hanlon, E.B. et al. Near-infrared fluorescence spectroscopy detects Alzheimer's disease in vitro. *Photochem Photobiol* **70**, 236–242 (1999).
92. Liebert, A. et al. Bed-side assessment of cerebral perfusion in stroke patients based on optical monitoring of a dye bolus by time-resolved diffuse reflectance. *NeuroImage* **24**, 426–435 (2005).
93. Orihashi, K. et al. Near-infrared spectroscopy for monitoring cerebral ischemia during selective cerebral perfusion. *Eur J Cardiothorac Surg* **26**, 907–911 (2004).

94. Terborg, C. et al. Bedside assessment of cerebral perfusion reductions in patients with acute ischaemic stroke by near-infrared spectroscopy and indocyanine green. *J Neurol Neurosurg Psychiatry* **75**, 38–42 (2004).
95. Thiagarajah, J.R. et al. Noninvasive early detection of brain edema in mice by near-infrared light scattering. *J Neurosci Res* **80**, 293–299 (2005).
96. Chen, W.G. et al. Hemodynamic assessment of ischemic stroke with near-infrared spectroscopy. *Space Med Med Eng (Beijing)* **13**, 84–89 (2000).
97. Nemoto, E.M. et al. Clinical experience with cerebral oximetry in stroke and cardiac arrest. *Crit Care Med* **28**, 1052–1054 (2000).
98. Saitou, H. et al. Cerebral blood volume and oxygenation among poststroke hemiplegic patients: Effects of 13 rehabilitation tasks measured by near-infrared spectroscopy. *Arch Phys Med Rehabil* **81**, 1348–1356 (2000).
99. Vernieri, F. et al. Near infrared spectroscopy and transcranial Doppler in monohemispheric stroke. *Eur Neurol* **41**, 159–162 (1999).
100. Eschweiler, G.W. et al. Left prefrontal activation predicts therapeutic effects of repetitive transcranial magnetic stimulation (rTMS) in major depression. *Psychiatry Res* **99**, 161–172 (2000).
101. Matsuo, K. et al. Alteration of hemoglobin oxygenation in the frontal region in elderly depressed patients as measured by near-infrared spectroscopy. *J Neuropsychiatry Clin Neurosci* **12**, 465–471 (2000).
102. Okada, F. et al. Dominance of the "nondominant" hemisphere in depression. *J Affect Disord* **37**, 13–21 (1996).
103. Matsuo, K. et al. Decreased cerebral haemodynamic response to cognitive and physiological tasks in mood disorders as shown by near-infrared spectroscopy. *Psychol Med* **32**, 1029–1037 (2002).
104. Suto, T. et al. Multichannel near-infrared spectroscopy in depression and schizophrenia: Cognitive brain activation study. *Biol Psychiatry* **55**, 501–511 (2004).
105. Fallgatter, A.J. and Strik, W.K. Reduced frontal functional asymmetry in schizophrenia during a cued continuous performance test assessed with near-infrared spectroscopy. *Schizophr Bull* **26**, 913–919 (2000).
106. Okada, F. et al. Impaired interhemispheric integration in brain oxygenation and hemodynamics in schizophrenia. *Eur Arch Psychiatry Clin Neurosci* **244**, 17–25 (1994).
107. Kubota, Y. et al. Prefrontal activation during verbal fluency tests in schizophrenia—A near-infrared spectroscopy (NIRS) study. *Schizophr Res* **77**, 65–73 (2005).
108. Shinba, T. et al. Near-infrared spectroscopy analysis of frontal lobe dysfunction in schizophrenia. *Biol Psychiatry* **55**, 154–164 (2004).
109. Watanabe, A. and Kato, T. Cerebrovascular response to cognitive tasks in patients with schizophrenia measured by near-infrared spectroscopy. *Schizophr Bull* **30**, 435–444 (2004).
110. Wyatt, J.S. et al. Quantification of cerebral oxygenation and haemodynamics in sick newborn infants by near infrared spectrophotometry. *Lancet* **2**, 1063–1066 (1986).
111. Chance, B. and Williams, G.R. The respiratory chain and oxidative phosphorylation. *Adv Enzymol Relat Subj Biochem* **17**, 65–134 (1956).
112. Lockwood, A.H. et al. Effects of acetazolamide and electrical stimulation on cerebral oxidative metabolism as indicated by the cytochrome oxidase redox state. *Brain Res* **308**, 9–14 (1984).
113. Wong-Riley, M.T. Cytochrome oxidase: An endogenous metabolic marker for neuronal activity. *Trends Neurosci* **12**, 94–101 (1989).
114. Wobst, P. et al. Linear aspects of changes in deoxygenated hemoglobin concentration and cytochrome oxidase oxidation during brain activation. *NeuroImage* **13**, 520–530 (2001).
115. Salzberg, B.M. and Obaid, A.L. Optical studies of the secretory event at vertebrate nerve terminals. *J Exp Biol* **139**, 195–231 (1988).

116. Ogawa, S. et al. An approach to probe some neural systems interaction by functional MRI at neural time scale down to milliseconds. *Proc Natl Acad Sci USA* **97**, 11026–11031 (2000).
117. Rector, D.M. et al. A focusing image probe for assessing neural activity in vivo. *J Neurosci Methods* **91**, 135–145 (1999).
118. Gratton, G. et al. Shades of gray matter: Noninvasive optical images of human brain responses during visual stimulation. *Psychophysiology* **32**, 505–509 (1995).
119. Toronov, V. et al. Near-infrared study of fluctuations in cerebral hemodynamics during rest and motor stimulation: Temporal analysis and spatial mapping. *Med Phys* **27**, 801–815 (2000).
120. Obrig, H. et al. Spontaneous low frequency oscillations of cerebral hemodynamics and metabolism in human adults. *NeuroImage* **12**, 623–639 (2000).
121. Tanosaki, M. et al. Variation of temporal characteristics in human cerebral hemodynamic responses to electric median nerve stimulation: A near-infrared spectroscopic study. *Neurosci Lett* **316**, 75–78 (2001).
122. Fallgatter, A.J. et al. Prefrontal hypooxygenation during language processing assessed with near-infrared spectroscopy. *Neuropsychobiology* **37**, 215–218 (1998).
123. Okada, F. et al. Region-dependent asymmetrical or symmetrical variations in the oxygenation and hemodynamics of the brain due to different mental stimuli. *Brain Res Cogn Brain Res* **2**, 215–219 (1995).
124. Franceschini, M.A. et al. Diffuse optical imaging of the whole head. *J Biomed Opt* **11**, 054007 (2006).
125. Taylor, J.A. et al. Low-frequency arterial pressure fluctuations do not reflect sympathetic outflow: Gender and age differences. *Am J Physiol* **274**, H1194–1201 (1998).
126. Diamond, S.G. et al. Physiological system identification with the Kalman filter in diffuse optical tomography. *Med Image Comput Comput Assist Interv Int Conf Med Image Comput Comput Assist Interv* **8**, 649–656 (2005).
127. Diamond, S.G. et al. Dynamic physiological modeling for functional diffuse optical tomography. *NeuroImage* **30**, 88–101 (2006).
128. von Siebenthal, K. et al. Cyclical fluctuations in blood pressure, heart rate and cerebral blood volume in preterm infants. *Brain Dev* **21**, 529–534 (1999).
129. Edvinsson, L. and Krause, D.N. *Cerebral Blood Flow and Metabolism*, (Lippincott Williamns and Wilkins, 2002).
130. Buxton, R.B. et al. Modeling the hemodynamic response to brain activation. *NeuroImage* **23 Suppl 1**, S220–233 (2004).
131. Matthews, F. et al. Hemodynamics for Brain-Computer Interfaces. *IEEE Signal Processing Magazine* **January**, 87–94 (2008).
132. Izzetoglu, M. et al. Motion artifact cancellation in NIR spectroscopy using Wiener filtering. *IEEE Trans Biomed Eng* **52**, 934–938 (2005).
133. Wilcox, T. et al. Using near-infrared spectroscopy to assess neural activation during object processing in infants. *J Biomed Opt* **10**, 11010 (2005).
134. Strangman, G. et al. Non-invasive neuroimaging using near-infrared light. *Biol Psychiatry* **52**, 679–693 (2002).
135. Zhang, Y. et al. Eigenvector-based spatial filtering for reduction of physiological interference in diffuse optical imaging. *J Biomed Opt* **10**, 11014 (2005).
136. Baillet, S. et al. Combined MEG and EEG source imaging by minimization of mutual information. *IEEE Trans Biomed Eng* **46**, 522–534 (1999).
137. Buxton, R.B. et al. Dynamics of blood flow and oxygenation changes during brain activation: The balloon model. *Magn Reson Med* **39**, 855–864 (1998).
138. Obata, T. et al. Discrepancies between BOLD and flow dynamics in primary and supplementary motor areas: Application of the balloon model to the interpretation of BOLD transients. *NeuroImage* **21**, 144–153 (2004).

139. Mandeville, J.B. et al. Evidence of a cerebrovascular postarteriole windkessel with delayed compliance. *J Cereb Blood Flow Metab* **19**, 679–689. (1999).
140. Deneux, T. and Faugeras, O. Using nonlinear models in fMRI data analysis: Model selection and activation detection. *NeuroImage* (2006).
141. Boas, D.A. et al. Can the cerebral metabolic rate of oxygen be estimated with near-infrared spectroscopy? *Phys Med Biol* **48**, 2405–2418 (2003).
142. Huppert, T.J. Harvard University (PhD Thesis) (2007).
143. Riera, J. et al. Fusing EEG and fMRI based on a bottom-up model: Inferring activation and effective connectivity in neural masses. *Philos Trans R Soc Lond B Biol Sci* **360**, 1025–1041 (2005).
144. Wan, X. et al. The neural basis of the hemodynamic response nonlinearity in human primary visual cortex: Implications for neurovascular coupling mechanism. *NeuroImage* **32**, 616–625 (2006).
145. Ciftci, K. et al. Constraining the general linear model for sensible hemodynamic response function waveforms. *Medical and biological engineering and computing* (2008).
146. Jasdzewski, G. et al. Differences in the hemodynamic response to event-related motor and visual paradigms as measured by near-infrared spectroscopy. *NeuroImage* **20**, 479–488 (2003).
147. Zhang, Y. et al. A haemodynamic response function model in spatio-temporal diffuse optical tomography. *Phys Med Biol* **50**, 4625–4644 (2005).
148. Cohen-Adad, J. et al. Activation detection in diffuse optical imaging by means of the general linear model. *Med Image Anal* **11**, 616–629 (2007).
149. Friston, K.J. et al. Characterizing evoked hemodynamics with fMRI. *NeuroImage* **2**, 157–165 (1995).
150. Friston, K.J. et al. Event-related fMRI: Characterizing differential responses. *NeuroImage* **7**, 30–40 (1998).
151. Cox, R.W. AFNI: Software for analysis and visualization of functional magnetic resonance neuroimages. *Comput Biomed Res* **29**, 162–173 (1996).
152. Josephs, O. and Henson, R.N. Event-related functional magnetic resonance imaging: Modelling, inference and optimization. *Philos Trans R Soc Lond B Biol Sci* **354**, 1215–1228 (1999).
153. Friston, K.J. Experimental Design and Statistical Issues. 33–58 (2000).
154. Wager, T.D. and Nichols, T.E. Optimization of experimental design in fMRI: A general framework using a genetic algorithm. *NeuroImage* **18**, 293–309 (2003).
155. Cannestra, A.F. et al. Refractory periods observed by intrinsic signal and fluorescent dye imaging. *J Neurophysiol* **80**, 1522–1532 (1998).
156. Hoge, R.D. et al. Simultaneous recording of task-induced changes in blood oxygenation, volume, and flow using diffuse optical imaging and arterial spin-labeling MRI. *NeuroImage* **25**, 701–707 (2005).
157. Belliveau, J.W. et al. Magnetic resonance imaging mapping of brain function. Human visual cortex. *Invest Radiol* **27 Suppl 2**, S59–65 (1992).
158. Kim, S.G. Quantification of relative cerebral blood flow change by flow-sensitive alternating inversion recovery (FAIR) technique: Application to functional mapping. *Magn Reson Med* **34**, 293–301 (1995).
159. Belliveau, J.W. et al. Functional mapping of the human visual cortex by magnetic resonance imaging. *Science* **254**, 716–719 (1991).
160. Kwong, K.K. et al. Dynamic magnetic resonance imaging of human brain activity during primary sensory stimulation. *Proc Natl Acad Sci USA* **89**, 5675–5679 (1992).
161. Ogawa, S. et al. Intrinsic signal changes accompanying sensory stimulation: Functional brain mapping with magnetic resonance imaging. *Proc Natl Acad Sci USA* **89**, 5951–5955 (1992).

162. Frahm, J. et al. Dynamic MR imaging of human brain oxygenation during rest and photic stimulation. *J Magn Reson Imaging* **2**, 501–505 (1992).
163. Bandettini, P.A. et al. Time course EPI of human brain function during task activation. *Magn Reson Med* **25**, 390–397 (1992).
164. Hoshi, Y. and Tamura, M. Dynamic multichannel near-infrared optical imaging of human brain activity. *J Appl Physiol* **75**, 1842–1846 (1993).
165. Villringer, A. et al. Near infrared spectroscopy (NIRS): A new tool to study hemodynamic changes during activation of brain function in human adults. *Neurosci Lett* **154**, 101–104 (1993).
166. Okada, F. et al. Gender- and handedness-related differences of forebrain oxygenation and hemodynamics. *Brain Res* **601**, 337–342 (1993).
167. Boas, D.A. et al. Scattering and wavelength transduction of diffuse photon density waves. *Physical Review. E. Statistical Physics, Plasmas, Fluids, and Related Interdisciplinary Topics* **47**, R2999–R3002 (1993).
168. Steinbrink, J. et al. Illuminating the BOLD signal: Combined fMRI-fNIRS studies. *Magn Reson Imaging* **24**, 495–505 (2006).
169. Huppert, T.J. et al. A temporal comparison of BOLD, ASL, and NIRS hemodynamic responses to motor stimuli in adult humans. *NeuroImage* **29**, 368–382 (2006).
170. Norris, D.G. Principles of magnetic resonance assessment of brain function. *J Magn Reson Imaging* **23**, 794–807 (2006).
171. Buxton, R.B. Quantifying CBF with arterial spin labeling. *J Magn Reson Imaging* **22**, 723–726 (2005).
172. Toronov, V. et al. Investigation of human brain hemodynamics by simultaneous near-infrared spectroscopy and functional magnetic resonance imaging. *Med Phys* **28**, 521–527 (2001).
173. Toronov, V.Y. et al. A spatial and temporal comparison of hemodynamic signals measured using optical and functional magnetic resonance imaging during activation in the human primary visual cortex. *NeuroImage* 34(3), 1136–1148 (2007).
174. MacIntosh, B.J. et al. Transient hemodynamics during a breath hold challenge in a two part functional imaging study with simultaneous near-infrared spectroscopy in adult humans. *NeuroImage* **20**, 1246–1252 (2003).
175. Toronov, V. et al. The roles of changes in deoxyhemoglobin concentration and regional cerebral blood volume in the fMRI BOLD signal. *NeuroImage* **19**, 1521–1531 (2003).
176. Sassaroli, A. et al. Spatially weighted BOLD signal for comparison of functional magnetic resonance imaging and near-infrared imaging of the brain. *NeuroImage* **33**, 505–514 (2006).
177. Huppert, T. et al. A quantitative spatial comparison of diffuse optical imaging with BOLD- and ASL-based fMRI. *J Biomed Opt* **11**, 064018 (2006).
178. Barbour, R.L. et al. MRI-guided optical tomography: Prospectives and computation for a new imaging method. *IEEE Comput. Sci. Eng* **2**, 63–77 (1995).
179. Pogue, B.W. and Paulsen, K.D. High-resolution near-infrared tomographic imaging of the rat cranium by use of a priori magnetic resonance imaging structural information. *Opt Lett* **23**, 1716–1718 (1998).
180. Intes, X. et al. Diffuse optical tomography with physiological and spatial a priori constraints. *Phys Med Biol* **49**, N155–163 (2004).
181. Boas, D.A. and Dale, A.M. Simulation study of magnetic resonance imaging-guided cortically constrained diffuse optical tomography of human brain function. *Appl Opt* **44**, 1957–1968 (2005).

15 Fast Optical Signals
Principles, Methods, and Experimental Results

Gabriele Gratton and Monica Fabiani

CONTENTS

15.1 INTRODUCTION

The purpose of this chapter is to describe the use of optical methods to record, in a noninvasive fashion, intrinsic signals related to neuronal activity (i.e., fast optical signals). Fast signals have been used to study non-invasively the time course of activity in localized cortical areas,[1–2] and, more recently, from the entire cortical surface.[3] In this chapter, we will first present a brief introduction to noninvasive functional imaging, and describe different approaches that can be used in this area of research. We will then describe the rationale and procedures related to the use of optical methods to create functional images of brain activity. We will then discuss examples of applications of fast optical imaging. We will conclude the chapter by discussing some advantages and limitations of the technique.

15.1.1 Fast Optical Signals and the Event-Related Optical Signal (EROS)

Fast optical signals are recently developed imaging methods whose purpose is to provide spatio-temporal maps of the transmission of near-infrared (NIR) light through the intact human brain. With this method, NIR light is shone through the scalp and skull, and changes in either the time taken by the photons to move through the head tissues[1] and/or in the amount of light reaching the detector[4] are observed. These two parameters (i.e., the photons' time-of-flight and the light intensity) are influenced by the scattering and absorption properties of the tissue. Of particular importance here is the observation that the scattering properties of brain tissue vary concurrently with neural activity in the tissue.[5–7] Because optical measures are relatively localized (to volumes of just a few mm in diameter), it is possible to use these methods to derive estimates of the time course of activity in specific brain areas. The recording of fast optical signals can be time-locked to particular events (such as stimuli or responses), hence the name event-related optical signals, or EROS. The EROS technique has both advantages and limitations that will be described later in this chapter.

15.1.2 Why Non-invasive Functional Brain Imaging?

The past 20 years have seen a phenomenal expansion of noninvasive functional brain imaging.[8] One of the major advantages of neuroimaging is the possibility of measuring in vivo physiological signals from various brain areas in normal human subjects. This possibility is of great significance for scientists and clinicians alike. Of course, there are also limitations attached to the use of these measures: at present, noninvasive imaging techniques only allow us to measure the cumulative physiological response of macroscopic areas of the brain (thus summarizing the activity of thousands of neurons at a time), and cannot provide data about the activity of individual neurons. Further, experimental lesion studies are not possible, and it is therefore difficult to ascertain causal links between the activity observed in various brain areas and behavioral and physiological outcomes, although the use of transcranial magnetic stimulation (TMS[9]) may allow us to narrow down or experimentally test some of these links.

15.1.3 Types of Functional Imaging Methods

Several types of noninvasive functional methods are now available. These methods differ on a number of dimensions, most notably the type of signal that is imaged to obtain estimates of brain activity. Most of the experimental work in functional neuroimaging is based on imaging either hemodynamic/metabolic or neuronal signals (although other types of functional imaging signals, such as receptor–ligand distribution, have also been used[10]). Optical methods are unique in that they can be used to visualize both hemodynamic *and* neuronal signals.[6–11] In the next section we will describe some general principles of each of these two types of signals.

15.1.3.1 Hemodynamic/Metabolic Methods

This class of techniques aims at measuring changes in the concentration of metabolically significant substances during the course of an experiment. These changes in concentration can be inferred directly (e.g., by labeling the substance using radioactive, magnetic, or optical markers) or indirectly (e.g., by studying the effects of a target substance, such as deoxy-hemoglobin, on the magnetic resonance properties of the tissue where the substance occurs naturally). Techniques that allow us to measure hemodynamic/metabolic effects include positron emission tomography (PET[8]), functional magnetic resonance imaging (fMRI[12]), and near-infrared spectroscopy (NIRS[13]). In particular, fMRI has become widely used because of its excellent spatial resolution, its easily attainable coupling with anatomical imaging (structural MRI), and the wide diffusion of the required hardware (as MR scanners are available in most hospitals because of their clinical use). However, most fMRI data do not provide absolute estimates of the concentration of oxy- and deoxy-hemoglobin, but only a derived measure indexing their relative concentration (the blood-oxygenation-level-dependent, or BOLD, signal). Quantitative estimates can instead be obtained using optical spectroscopic techniques,[13–14] although this requires solving several practical problems (see chapter on NIRS in this volume).

These hemodynamic/metabolic changes are of interest to neuroscientists and psychologists because they allow us to study not only metabolism and blood circulation in the brain, but also, indirectly, neuronal activity. In fact, it has been known for a long time that blood flow and metabolism increase in brain areas that are active during a particular task (e.g., Grinvald et al.[15]). Thus, hemodynamic/metabolic neuroimaging is based on the assumption that, by studying changes in these parameters, one can infer whether a particular brain area is involved in the neural activity associated with a particular task. Although this is a reasonable assumption in many cases, it also points to a limitation of these techniques as measures of neuronal activity. Namely, the inference of the occurrence of neuronal activity from hemodynamic/metabolic data is indirect, being mediated by *neurovascular coupling*.[11,16] This term refers to the set of biochemical and biophysical steps that occur in between the onset of neuronal activity and the physiological (hemodynamic) events that follow it, and that are ultimately measured. Neurovascular coupling complicates the measures for two reasons: (1) it introduces an inherent delay in the physiological variable that is measured (e.g., change in the concentration of deoxy-hemoglobin) with respect to the phenomenon that is inferred (i.e., neuronal activity), which, in turn, limits the temporal resolution of the method, and (2) neurovascular coupling may not be constant, but may in fact vary as a function of the state of the organism, of the brain area from which the measures are taken, and of individual and group variables.[17]

15.1.3.2 Neuronal Methods

Another type of signal used in functional neuroimaging is related to the movement of ions through and around the cell membranes, which occurs during neuronal activity. There are several methods that can be used to record ion-movement noninvasively. Some methods rely on the changes in electro-magnetic fields that can be detected from sensors located outside the head. Examples of these techniques are the electroencephalogram (EEG[18]), evoked and event-related potentials (EPs and ERPs[19]), and magnetoencephalography (MEG[20]). Note, however, that differently from single- and multiple-units activity measures obtained with implanted micro-electrodes, noninvasive electromagnetic measures taken at a distance (e.g., from the surface of the head) cannot provide information about the activity of individual neurons, but need to rely on the spatial and temporal summation of the fields generated by individual neurons. Spatial summation only occurs if the individual neurons are oriented in a consistent direction (with respect to their dendritic field–axon axis), generating an "open-field" configuration.[21] Thus, only a portion of the neuronal activity generated in the brain can be studied with these methods. In the case of the cerebral cortex, electromagnetic methods are most sensitive to the activity of pyramidal neurons (the output cells of the cortex), because of the size, prominence, and orientation of their dendritic trees. They are also most sensitive to the postsynaptic activity measurable in the dendrites rather than to the all-or-none activity typical of axons, because of the comparatively large size of the dendritic fields and the longer duration of postsynaptic potentials. Therefore, the effects observed with these methods differ from those observed with single-unit methods (which do not have a directional bias and are based mostly on action potential measures in axons and/or in the vicinity of cell bodies), but are more similar to field potentials recorded invasively. Further, the spatial resolution of general purpose 3D-reconstruction algorithms for electromagnetic measures, allowing investigators to reconstruct the distribution of the activity inside the head from surface measures, is limited[22] and/or needs to rely on assumptions that are often difficult to verify.[23] Optical methods, reviewed in the next section, provide a complementary approach for studying neuronal and hemodynamic activity and their coupling.

15.2 PRINCIPLES OF FAST OPTICAL IMAGING

15.2.1 The Scattering Signal

This book reviews techniques that use optical methods for studying brain function. Several optical signals can be used. This chapter focuses on signals that are temporally *concurrent* with neuronal changes. Changes of this type have been reported elsewhere in this book. They can generally be described in terms of changes of the scattering coefficient of neural tissue when the tissue is active. These changes have been demonstrated in individual axons,[5] brain slices,[6] and depth optical recording in animals.[7] For instance, Rector et al.[7] developed a methodology for recording changes in the transparency of rat hippocampus or other deep brain structures in vivo. They showed that stimulation of the Schaeffer's collateral elicits a change in the transparency of the CA1 field whose latency is consistent with that of the evoked potential from CA1 recorded with electrophysiological methods. They attributed these effects to changes in the scattering properties of this area, presumably associated with the migration of ions through the neuronal membrane. Similarly, the same group reported fast and localized intrinsic optical signals from the barrel cortex of rats in response to single-whisker mechanical stimulation.[24]

Although there is strong evidence for the existence of scattering changes concurrent with electrically recorded neuronal activity, the exact causes of these changes are still under investigation. Two phenomena are considered as potential causes: (a) changes in neuronal membrane conformation related to the difference in potential between intra- and extra-cellular compartments;[25] and (b) volumetric changes in the intra- and extra-cellular compartments due to the movement of ions and associated water molecules during cell depolarization or hyperpolarization. Recent data suggest that the former mechanism is probably strictly related to birefringence changes, whereas the scattering changes are more likely related to the volumetric phenomena.[26]

15.2.2 Non-invasive Measurement of Optical Properties

15.2.2.1 Continuous vs. Time-Resolved Measures

Although scattering changes accompanying neuronal activity have been demonstrated for quite some time using invasive techniques, only recently methods have become available for their non-invasive measurement. Because of the high absorption by hemoglobin and water at other wavelengths, noninvasive measurements of the human brain are restricted to the near-infrared range. Further, the scattering-based functional changes are relatively small, and require appropriate instrumentation for their measurement. Two types of measures have been used: Measures of delay in the flight of NIR photons through active areas (*delay measures*[1]), and measures of the amount of light moving through active areas (*intensity measures*[4]). In heterogeneous surface-bound media such as the head, both delay and intensity measures are sensitive to both absorption and scattering changes, although in a different manner.[1,27] In addition, these measures differ in terms of spatial resolution, relative sensitivity to deep and superficial events, and type of noise or artifact contamination.

Intensity measures are simpler to obtain and model than photon-delay measures. The simplest recording apparatus (the *optode*) consists of a source of NIR light, which can be connected to the head using an optic fiber, and a light measurement instrument, which can also be connected to the head using an optic fiber. Several types of light sources can be used, including incandescent lamps, light emitting diodes (LEDs), laser diodes, and regular lasers. Several types of detectors can also be used, including light-sensitive diodes, photo-multiplier tubes, CCD-cameras, and so on. Irrespective of the sources and detectors used, it is important that environmental light sources be controlled or excluded. This can be achieved either by insulating the detector instrument (and the head) from environmental light sources (e.g., Steinbrink et al.[4]) or, preferably, by labeling the instrumental light source in a special manner. This can be achieved, for example, by modulating the light source at a particular frequency (e.g., Wolf et al.[28]).

Photon-delay measures are more complex, because the delays to be measured (and their changes) are of the order of picoseconds or fractions thereof. The measurement requires that the light-source vary over time. This can be achieved by making the light-sources either pulsate or oscillate in intensity at a very high frequency. The latter measures are more convenient in terms of cost.

Time-resolved measures are used in NIR spectroscopy because they make the computation of the absolute concentration of oxy- and deoxy-hemoglobin easier.[29] However, if the main interest of the investigation is in detecting changes (such as those related to neuronal events), this particular advantage of time-resolved measures may be less important. Another differentiation between measures obtained with continuous light (i.e., measures of tissue transparency, usually labeled *intensity measures*) and time-resolved measures (i.e., measures of modulation and delay of the light waves or pulses when passing through the head tissue) is their relative sensitivity (or lack thereof) to environmental light, movement artifacts, and to activity in superficial (i.e., skin and skull) compared to deep (i.e., intracranial) head layers. Continuous measures cannot distinguish between light coming from the instrumental source and light coming from other environmental sources. Modulated or pulsated sources make this distinction possible because they produce light that changes with a particular frequency and/or time course. Time-of-flight measures are relatively insensitive to movements, which may cause changes in the coupling between the optodes and the head, and may have strong effects on intensity measures.[30] Finally, time-of-flight measures are more sensitive to deep effects[27,31–32] than intensity measures.

15.2.2.2 Penetration and 3D Reconstruction

Noninvasive measures are obtained from instruments placed on the surface of the head. A central issue for this research is to determine which parts of the brain are responsible for the effects that are observed. Photons injected in a particular point of the head (a scattering medium) propagate randomly through the tissue. However, because a large number of photons are always involved, the propagation of the photon population can in principle be expressed mathematically, using equations that describe the statistical volume explored by each source-detector pair. Any change in the measurement parameters can then be probabilistically ascribed to this volume. These equations are relatively simple for surface-bound homogenous media. In this case, photons moving between a particular source and detector (both located on the surface of the medium at a certain distance from each other) are most likely to be contained within a volume shaped as a "curved spindle" whose extremes are located at the source and detector (see Figure 15.1). The spatial gradients defining this spindle (which is important in defining the spatial resolution of the technique) depend on various factors, including the parameter (intensity or photon delay) used for the measurement. In fact, these gradients are sharper for measures of photon delay than for measures of the amount of light transmitted (continuous measures). Thus, photon delay measures possess a higher spatial resolution. In addition, they may have effects

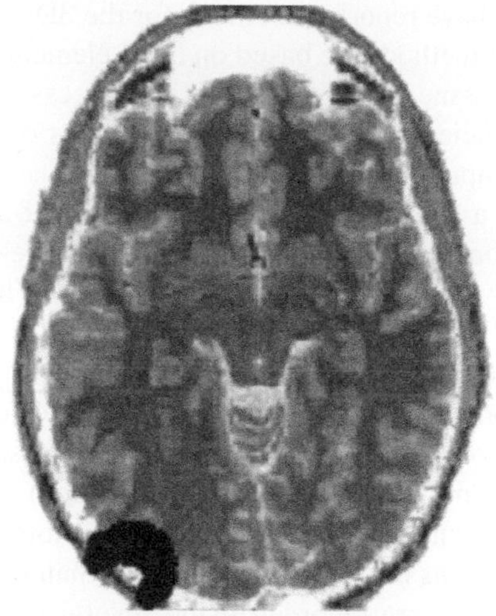

FIGURE 15.1 Schematic representation of the volume in the brain that is explored by a particular source-detector pair located in left occipital areas. (From Gratton, G. and Fabiani, M. *Int. J. Psychophysiol.*, 42, 109–121, 2001. With permission.)

that are of opposite sign for very superficial events (occurring in the skin) rather than in deep layers (occurring in the brain[27]).

An important consideration is that head tissues vary considerably with respect to their optical properties (scattering and absorption). For instance, bone has a relatively high scattering coefficient but a low absorption coefficient compared to the gray matter. The cerebrospinal fluid is almost transparent to light, but the subarachnoid space in which it is contained is traversed by a dense web of blood vessels. Thus, within this space, NIR light can travel without much resistance in a direction perpendicular to the surface, but can only travel very little in a direction parallel to the surface, as it will be absorbed. For this reason, the subarachnoid space has little influence on the transmission of light through the head.[32] The opposite occurs with the white matter, which has a very high scattering coefficient, and whose effect is to reflect most of the light reaching it. The heterogeneity of the optical properties of the head tissues has the consequence of deforming the volume investigated by a particular source-detector pair.[27,32] The effects of this distortions appear to be relatively moderate (a few mm) at source-detector distances of less than 5 cm,[33] but could potentially be larger at longer source-detector distances.

The maximum depth at which fast (scattering-dominant) optical effects can be studied noninvasively is largely determined by the source-detector distance. For homogenous media, the maximum depth is between half and a quarter of the source-detector distance.[34] However, the sensitivity to phenomena occurring at different depths varies depending on the type of measure used. Measures of amount of light transmission are more sensitive to superficial phenomena, whereas measures of photon-delay are more sensitive to deep phenomena.[27,32,34] This may lead to discrepancies in the results obtained with these two sets of measures.

Various investigators have reported methods for the 3D reconstruction of photon migration data.[35–43] Most methods are based on finite element models of the propagation of photons through tissue, and use an iterative process to determine the scattering and absorption coefficients of various compartments of the tissue itself. These models are therefore computation-intensive.

We have so far used a simpler approach[44] consisting of approximating the path between a particular source-detector pair using a spindle-shaped volume (to be scaled in size depending on the source-detector distance). This approach introduces approximations in the 3D reconstruction of the order of several mm, but may be adequate for many practical applications and is computationally very simple. Gratton et al.[33] reported data comparing estimates of the depth of regions of brain activity obtained using this approach and fMRI. The study was based on manipulating the eccentricity of visual stimuli, which influences the depth of the region of medial occipital cortex where activity should be observed (i.e., more eccentric visual stimuli are processed in deeper areas of the calcarine fissure than more central visual stimuli). Optical imaging data were recorded using several source-detector distances, and the distance at which the largest effect was observed was used to estimate the depth of the optical effect (estimated as half of this distance). Functional MRI data were also recorded from the same subjects, using the same paradigm. Results indicate that the two estimates differ, on average, by less than 1 mm. This provides support for the simplified approach to the estimation of the depth of the optical effects (and therefore to 3D reconstruction) described above.

15.2.2.3 Data Collection and Artifact Correction

There are at present several companies that produce recording instruments suitable for recording fast optical data. These instruments are capable of collecting measures of the intensity and/or delay of the photons reaching the detector at a very fast pace (up to more than 1 kHz). Multiple-channel systems are available (up to 64 sources and 32 detectors). The use of multiple channels may involve time-multiplexing to distinguish the signals associated with different source-detector pairs. Time-multiplexing may reduce the temporal resolution of the instrument. All systems are designed to be movable (and are therefore potentially portable) and use low-energy nonionizing radiation, which makes them safe and suitable for extended and/or repeated recordings with the same subject. Also, most systems allow for an external event channel or at least for external triggering, so as to enable synchronization with the system producing the stimulation and/or recording the subject response. This allows for the recording of event-related brain activity (EROS).

In addition to the fast neuronal signals, the raw data obtained with optical recording systems (and in particular the intensity data) carry a number of other signals. These signals are related to (a) arterial and capillary pulse (about 1 Hz); (b) respiration (about 0.25 Hz); and (c) pressure waves in the blood vessels (about 0.1 Hz). These signals (which can be of interest in a number of cases) are considerably larger than the neuronal signal and can in fact obscure it. Thus, they should be treated as artifacts if the research focuses on the recording of neuronal signals. These artifacts are quite evident for intensity data, but less so for time delay data, which are less sensitive to superficial phenomena (and therefore less sensitive to capillary pulse). Some of these artifacts (as well as slow drifts related to heating or cooling of the apparatus) can be eliminated using appropriate high-pass filters. The pulse artifact is more problematic, because its frequency is closer to that of the neuronal signals. However, this artifact (which is ubiquitous in intensity recordings) generates very characteristic sawtooth waves. The shape regularity of this artifact can be used to devise algorithms for its correction. Gratton and Corballis[45] proposed one such method, which is now used in several laboratories.

Other types of artifacts are due to movements of the subjects or of the optical apparatus. These movements produce large signals, due to variations in the interface between the measurement instruments (e.g., optic fibers) and the head, which are temporally well localized and can be discarded using automatic procedures.[30] In addition, these artifacts are most evident in intensity measures, but produce very limited effects on photon delay measures, most likely because the latter reflect the differential proportion of photons with long and short diffusive paths within the head, rather than their total amount. In Figure 15.2 we report an example of the head gear (a modified motorcycle helmet) that we use to hold the source and recording fiber into place, so that the coupling between the optic fibers, and the head is firmly in place.

15.2.2.4 Signal Processing and Statistical Analysis

The recording apparatus described above yields time series that can be time-locked to particular events (i.e., to stimuli and/or to the subject's responses to the stimuli). These time series can be processed with the standard methodologies used for analyzing other types of time series, such as those obtained with ERPs or fMRI recordings. A simple way of extracting the fast optical signal is by averaging the time-locked waveforms obtained for particular stimuli or response conditions. More complex analysis procedures can also be used, such as cross correlation, wavelet analysis, combined time-frequency analysis, and so on. In addition, low- and high-pass filtering can be used to improve the signal-to-noise ratio.[46] Statistics about the reliability of the observed effects can then be computed either within or across subjects, using

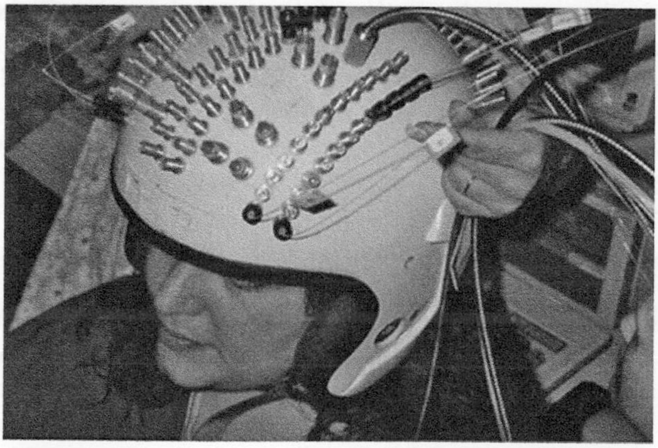

FIGURE 15.2 Example of head gear (a modified motorcycle helmet) that is used to hold in place the source and detector fibers.

standard statistical methods. Fabiani et al.[47] have shown that nonparametric statistics, based on bootstrap methods, can also be useful to determine the reliability of effects.

As mentioned above, data from a single source-detector pair (or channel) refer to a curved spindle-shaped volume with vertices at the source and detector. These data, therefore, provide an estimate of the origin of the signal. Given the variability in cortical anatomy across individuals, it is necessary to record from multiple surface locations to make sure to "hit" the active area in a particular subject, as well as to refer the functional data to the underlying structural anatomy of the individual (see Whalen et al.[48]). Whenever multiple recordings are obtained, it is important to establish that effects are not due to chance, as the probability of capitalizing on chance of course increases with the number of locations. This can be achieved in several ways:

1. Bonferroni-corrected confidence intervals or p-values can be computed. Our current software uses a more refined version of this correction procedure, based on the application of Gaussian field theory to the computation of number of independent comparisons.[49] This, in turn, is used to compute p-values corrected for multiple comparisons.
2. An alternative approach is to first compute an omnibus statistics, across all locations, and then, if the omnibus effect is significant, to determine the exact location at which the effect is maximum. Bootstrap methods may be useful for this second step.[47]
3. Finally, a multivariate approach can be used to deal with the problem of multiple comparisons, or at least to minimize the number of locations investigated.[50]

When collapsing data across subjects, it is important to consider that recordings from the same surface location (identified using scalp landmarks) in different subjects may not correspond to the same functional cortical area in all subjects. The earliest studies using optical imaging data by and large ignored this issue, which may however lead to a marked reduction of the signal-to-noise ratio (because active locations from one subject may be averaged with inactive locations from another subject). This problem is not unique to optical imaging, as it is typical of any brain imaging technique with a spatial resolution higher than a few cm. Several investigators have proposed a number of tools for realigning and rescaling brain and cortical areas in different subjects. In principle, all of these tools can also be applied to EROS data. However, this requires that (1) the exact locations used for each source and detectors be recorded within a fiducial space (this can be achieved by using a magnetic 3D digitizer, such as the Polhemus[7] 3D digitizer system), (2) the same fiducial locations (e.g., nasion and preauricular points) be marked on structural MR images (e.g., by using vitamin E pills or other markers), (3) a method is used to co-register the digitized data about the location of the sources and detectors and the subject's head (see Whalen et al.[48]), (4) a 3D reconstruction method be used to determine the volume from which data are recorded from, separately for each individual subject (see description above), (5) data from different source-detector pairs whose investigated volume overlap at least in part be combined to improve spatial

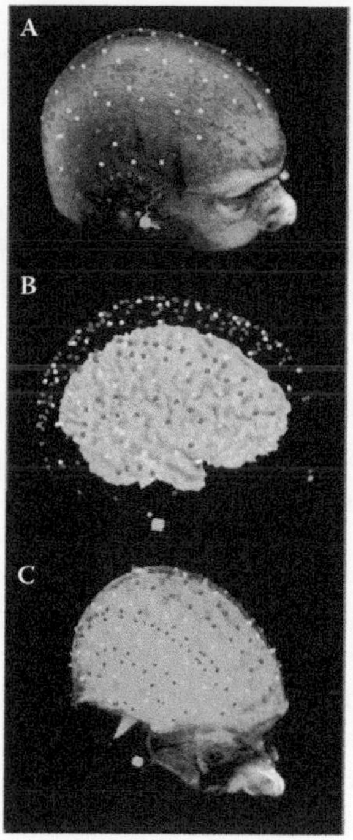

FIGURE 15.3 **(See color insert following page 224)** (A) Digitized sources (red dots) and detectors (yellow dots) overlaid on a subject's structural MRI; (B) Sources and detector projected onto the brain with scalp and skull removed; (C) Overlapping "spindle" volumes used for 3D reconstruction, emanating from source-detector pairs. Green dots in all plots show locations of digitized fiducial points (also identified by markers in the MR).

resolution and signal quality (an example of this approach is given by the pi-detector proposed by Wolf et al.[28]), and (6) the 3D reconstructed data from different subjects be aligned and rescaled using one of the procedures described in the literature (the simplest of which is the Talairach transformation[51]). Once these transformations have been obtained, data from different subjects can be pooled together, and between-subjects statistics can be computed. Figure 15.3 shows digitized sources and detectors overlaid on a subject's structural MRI (A), projected onto the brain with scalp and skull removed (B) and showing overlapping "spindle" volumes used for 3D reconstruction (C).

15.3 EXPERIMENTAL RESULTS

There are currently more than 30 published reports on peer-reviewed journals, coming from eight different laboratories of original studies investigating fast optical signals. All but one[52] reported the observation of a fast optical signal, although the methods used in different studies varied significantly. The majority of these studies has been conducted in our laboratory, and has used photon-delay measures as the main dependent variable. Our major goal has been to show that the fast optical signal can be used to study noninvasively the time course of activity in selected cortical areas, although our recent work has extended the measurement to progressively larger portions of the cortex, and most recently to the entire cortical surface.[3] To support this claim, we have run experiments demonstrating the following:

A. Rapid changes in the optical signal can be recorded immediately after stimulation (within 100 ms or less) or immediately before a motor response. This is essential to demonstrate that a fast optical signal occurs. The short latency of the signal makes it very unlikely that it can be due to hemodynamic effects, which take several hundred milliseconds to develop.[53]
B. Similarly, when recording fast signals with NIR light of two wavelengths, such as 690 and 830 nm, which are located at opposite sides of the isosbestic point for oxy- and deoxy-hemoglobin, the EROS data show effects in the same direction.[54] This again indicates that the signal is due to scattering changes (and therefore of neuronal origin) rather than to rapid deoxygenation phenomena (such as the initial dip reported with high-field fMRI by Ugurbil and colleagues[55]—and in exposed cortex optical imaging studies—e.g., Malonek and Grinvald[53]) as the latter should result in effects of opposite sign at these wavelengths.
C. These changes only occur if a stimulus is used, which is appropriate for eliciting activity in that particular area of cortex, whereas other stimuli, even within the same modality but not expected to activate that particular cortical region do not elicit a response. (For instance, when stimulating the right visual field effects should be visible in the left occipital cortex.) This is important to demonstrate that the phenomenon is specific to particular cortical regions and not a general or artifactual response (such as one due to movements, global arousal phenomena, etc.).
D. These changes are temporally overlapping with other *neuronal* signals, such as ERPs, recorded concurrently from the same subjects and conditions.[56–58] This is important if one intends to use these measures to study the time course of neuronal activity. The comparison can also help narrow down the time intervals of interest (much in the same way that regions of interest can be established in the spatial domain—see point E).
E. The fast optical effects are localized to brain areas where *hemodynamic* measures with high spatial resolution (such as fMRI) indicate that neuronal activity occurs.[59–60] This is important if these measures are used as indices of local neuronal activity in selected cortical areas.
F. Similar effects are recorded from different areas, when different stimulus or response modalities are used. This is important to demonstrate that fast optical signals are not the property of one particular cortical area, but are a general property of the cortex. This, in turn, allows us to generalize our studies to different domains.

We will now describe some of the findings obtained with EROS in various modalities and paradigms. Figures 15.4–15.7 show examples of data obtained in these studies.

15.3.1 Visual Modality

A large number of studies on the fast signal have been conducted in the visual modality. With one exception,[52] these studies indicate that a fast optical signal can be recorded in response to visual stimuli. Recordings have been made from both medial and more lateral occipital areas. The first study[1] was based on the reversal of one of four grids placed in different quadrants of the visual field. In this study, the reversals occurred every 500 ms. The original study was based on a small sample size (N = 3),* but the results were replicated in a subsequent study using a larger sample size (N = 8, Gratton and Fabiani[61]). A comparison of the results of these two studies is presented in Figure 15.4. Both studies indicated the existence of a fast optical effect consisting of an increase in the photon delay that occurred approximately 100 ms after stimulation (grid reversal). This effect was localized, in that it occurred at locations placed on the contralateral hemisphere to where the stimulation occurs. Further, stimulation of upper quadrants resulted in larger effects at lower locations in the occipital cortex (and vice versa), in a manner consistent with the well-known representation of the visual field in medial occipital cortex. The location of the response of our original study was found to correspond closely with that of the hemodynamic response measured with fMRI, whereas the time course was similar to that of the visual evoked potential (VEP[59]).

Our 1995 study, as well as the subsequent 2003 replication shown in Figure 15.4, were based on a fixed interstimulus interval (ISI). It is therefore theoretically possible that the observed response could be due to a time-locked response to some previous stimulation. However, two subsequent studies rule out this possible confound. In one of these studies[11] (see also Gratton et al.[54]) we systematically varied the ISI from 100 to 1000 ms; in each case, a response (characterized by an increase in photon delay) was observed with a latency between 60 and 100 ms from stimulation. In fact, the latency of the response increased slightly with the ISI. The same study also indicated that the fast optical response increased in amplitude with ISI, in a manner consistent with that of the concurrently recorded VEP. Increases in phase delay with similar peak latency (60–100 ms) were also reported in several other studies.[33,62–63] In all cases, when the stimulus was unilateral, this fast response was only observed in the hemisphere contralateral to stimulation, thus confirming the spatial localization of the response. Further, we found that the more lateral the area stimulated, the larger the source-detector distance at which the response (i.e., an increase in photon delay) was observed.[33] This is consistent with the idea that lateral stimuli elicit activity in deeper cortical areas than foveal stimuli, consistent both with fMRI findings obtained in the same study and with the well-known functional organization of medial occipital cortex. The latency of the optical effects is also generally consistent with what is known about the visual system. For instance, typically stimuli with abrupt onset tend to elicit activity with a shorter latency (about 60–80 ms) than isoluminant alternating stimuli (about 80–100 ms). This is also generally consistent with what is observed for VEPs. In some of our studies the recording was extended to

* The small sample size was due to the necessity of repeating the measurement approximately 12 times per subject to create a functional *map* with a 1-channel (1 source, 1 detector) system.

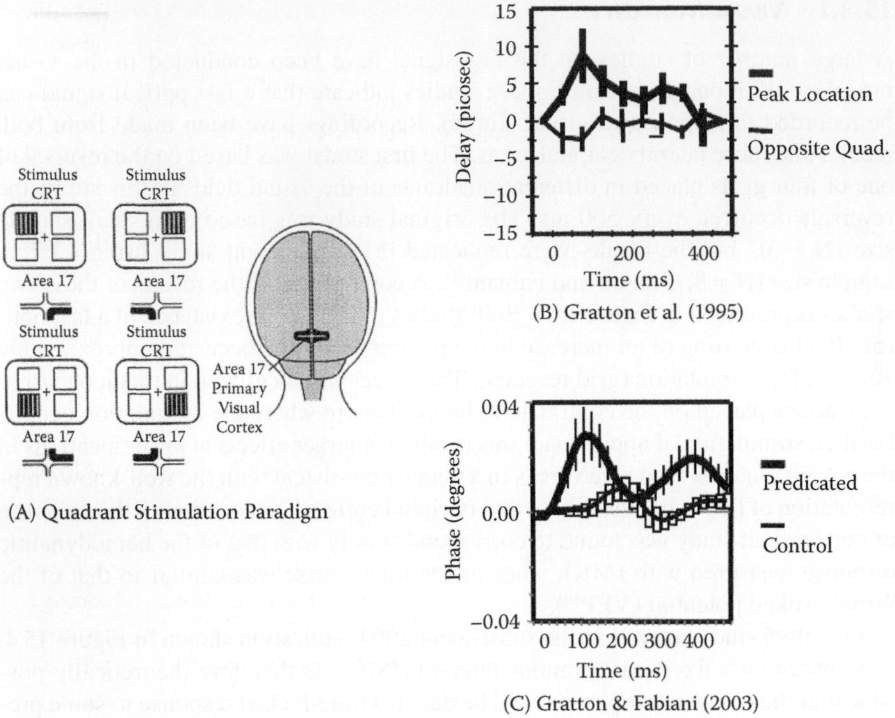

FIGURE 15.4 EROS recorded from medial occipital areas during two visual stimulation experiments using the same paradigm.[1,61] (A) Diagram of the stimulation conditions. The stimulation consisted of reversals of black-and-white grids occurring every 500 ms; only one quadrant of the visual field was stimulated during a trial block, and the stimulated quadrant was rotated across blocks. The head diagram represents the predicted location of the response, based on a contralateral, inverted representation of the visual field in medial occipital areas. (Modified from Gratton, G., and Fabiani, M. *Psychophysiology,* 40, 561, 2003. With permission.) (B,C) EROS effects in the two experiments. In both experiments, the thicker lines refer to the EROS activity recorded from predicted locations for each quadrant stimulation conditions, and the thinner lines refer to the EROS activity recorded from the same locations when the other quadrants were stimulated. ([B] From Gratton, G. et al. *Psychophysiology*, 32, 505, 1995. [C] From Gratton, G., and Fabiani, M. *Psychophysiology,* 40, 561, 2003. Both with permission.) The second experiment (lower panel) was based on a larger data sample and had a higher sampling rate than the first experiment (upper panel).

more lateral occipital areas. Fast optical responses could also be recorded from these areas with a latency similar or only slightly longer than in more medial areas.[50,59,64] Finally, several studies showed optical responses with much longer latencies (200–300 ms[1,61,63,65]). These responses were often ipsilateral or bilateral, perhaps reflecting subsequent feedback processes from other cortical or subcortical areas.

In summary, a number of studies indicate the presence of a fast optical response to visual stimuli in medial occipital areas, characterized by an increase in phase delay with a latency of 60–100 ms from stimulation. The localization of this response is consistent with the well-known representation of the visual field in medial occipital cortex, and the temporal properties of the signal are consistent with the time course of neuronal signals, such as the VEP.

15.3.2 Auditory Modality

Several studies have been completed in the auditory modality by our lab and others.[57–58,66–69] Most of these studies are based on passive auditory oddball paradigms. In these studies, the subjects read books (or perform other tasks) while series of tones are presented to them via headphones. The tones vary along some dimension. For instance, in our first paper,[66] a tone could be either long (75 ms duration) or short (25 ms duration). The long duration tone occurred 80% of the times, and was labeled the "standard" stimulus. The short duration tone occurred 20% of the times, and was labeled the "deviant" stimulus. The ERPs elicited in this paradigm have been the subject of extended investigation,[70] and are typically characterized by two types of activities. First, there are some responses that are elicited by all stimuli (although their amplitude is larger for stimuli with higher overall intensity). They are labeled exogenous potentials, the most evident of which is the N1 (which is probably itself the sum of several components[71]). Second, there are some potentials that are typically generated only by deviant stimuli; among them is the Mismatch-Negativity (MMN[70]), an ERP component with a peak latency of approximately 120–200 ms. There is some uncertainty about whether the N1 and MMN components are generated in the same or in different cortical areas. The Rinne et al.'s study[66] was set up to show (1) that fast optical effects could be recorded from auditory areas, (2) that different optical responses with characteristics similar to those of the N1 and the MMN could be obtained, and (3) whether these two fast optical responses were generated in the same or different areas. The results confirmed the first two hypotheses, and showed that the optical response to deviant stimuli (corresponding to the electrical MMN) was systematically recorded approximately 1 cm below the location where the response to the high-energy standard stimuli (corresponding to the electrical N1) was maximal. The response to the deviant stimuli had a latency of 160–180 ms, whereas the response to the standard stimuli had a latency of 100 ms. These latencies closely correspond to the latencies of the electrical MMN and N1, respectively. Both responses consisted of an increase in the photon delay parameter. The results of subsequent work by our lab and others[57–58,67–69] confirmed and extended those of Rinne's work,[66] and also indicated that the responses to the deviant stimuli occurred later (160–180 ms) and at more ventral locations (1 cm) with respect the responses to the standard stimuli (whose peak latency was 100 ms). Further, anatomical 3D reconstruction of these subsequent studies indicated that the optical responses occurred in superior or medial temporal cortex.

In summary, the data from the auditory modality show that fast optical responses with latencies consistent with those of auditory evoked potentials can be recorded from temporal cortex. Several phenomena can be distinguished on the basis of their locations, time course, and response to experimental manipulations.

15.3.3 Somatosensory Modality

Fast optical imaging in the somatosensory modality has been conducted in several labs, including a study by Steinbrink and colleagues,[4] one by Franceschini and Boaz,[72] and one by Maclin and colleagues in our lab.[56] Steinbrink's work was based on a different technique than that used by our group. They used a continuous light source (a lamp), instead of a radio-frequency modulated light source. The measures obtained were changes in the amount of light transmitted from the source to the detector in the period immediately following somatosensory stimulation. Stimulations were brief low-intensity electric shocks delivered to either arm (median nerve stimulation). Continuous measures were obtained from parietal locations. Rapid (60–160 ms from stimulation) changes were observed when the contralateral side was stimulated, but not when the ipsilateral side was stimulated. These data confirm that a fast optical signal can be recorded noninvasively, and that this signal is localized (at least to the contralateral hemisphere). Similar results were obtained by Franceschini and Boaz,[72] and by Maclin et al.[56] In the latter study, however, we observed a faster response with a latency of approximately 20 ms, in addition to the 100–200 ms response reported by the other studies. Both of these responses are consistent with those obtained with other technologies, such as ERPs.[56]

15.3.4 Motor Modality

We have conducted two studies investigating fast optical signals in the motor modality. One of them was an early study[27] that involved unilateral hand and/or foot tapping at a frequency of 0.8 Hz. In this study, there was no synchronization between the subjects' movements and the optical recordings, and a single-channel system was used. The analysis was conducted in the frequency domain, and showed significant oscillations of the optical signal (photon delay) at the frequency of tapping and its harmonics. These oscillations were larger on the side contralateral to the tapping limb.

In a subsequent study,[73] we recorded fast optical imaging data during a choice reaction time task. This experiment was based on a synchronized multichannel system, which allowed us to generate maps of brain activity at different times with respect to the response. Some of these maps are shown in Figure 15.5. They show that a fast optical signal (consisting of an increase in phase delay, analogous to those observed in the visual and auditory modalities) can be observed in the last 100 ms before the button press on the side contralateral to the responding hand. The time course of the optical response is very similar to that of the Lateralized Readiness Potential (LRP), an electrical measure used to index response preparation.[74] Similar data have been also obtained by Morren et al.[75] and Franceschini and Boas.[72] Morren and colleagues[75] showed that a fast optical signal shortly preceding the response can be observed using either photon delay or intensity as dependent variables, although the phase effects were observed in a much smaller number of channels than the intensity data, presumably reflecting their higher spatial resolution.

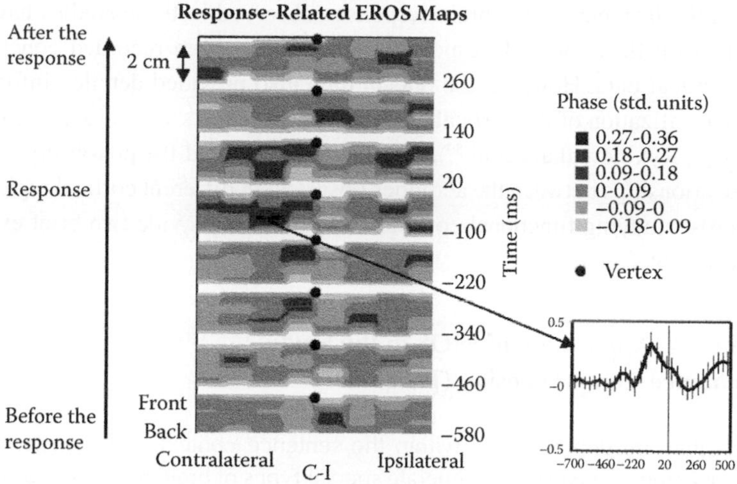

FIGURE 15.5 Maps of the optical activity preceding and following a motor response. The maps are presented as if viewed from the top of the head. The brain area contralateral to the responding hand is presented on the left, and that ipsilateral to the responding hand is presented on the right. The numbers beside the maps refer to the relative time (in ms) with respect to the response (negative values refer to time before the response). The small panel on the right shows the time course of the response (with standard-error bars) for the location of largest activity. (From DeSoto, M. C. et al., *J. Cogn. Neurosci.,* 13, 523, 2001. With permission.)

These data provide further support to the generality of the fast optical signal as a measure of localized cortical activity. They also indicate that different methodologies can be used to record fast optical signals.

15.3.5 Cognitive Studies

Many of the studies reported in the previous sections were conducted to assess the feasibility of using fast optical signals as a noninvasive index of neuronal function. As such, they used relatively simple sensory stimulation. However, there have been an increasing number of experiments using fast optical signals to investigate a variety of cognitive functions. For instance, EROS has been used to investigate attention effects in early visual cortex,[62] visual memory in occipital areas,[63,65] conflict conditions in motor cortex,[73] sensory memory processes in the superior temporal gyrus,[57–58,66–69] and inferior frontal gyrus,[67–69] working memory load effects in the middle and superior frontal gyri,[76] executive function in prefrontal cortex,[3,50,77] and sentence processing in superior temporal and inferior frontal gyrus.[78] The general purpose of these studies has been to investigate variations in the time course of activity in these cortical regions as a function of various experimental manipulations. By and large, the latencies of the brain responses observed in these studies have been consistent with those of ERP responses, which were often recorded concurrently with the optical data. However, the optical data also provided detailed information about the localization of the cortical areas where the effects were observed. In some studies (e.g., Rykhlevsakaia et al.[50]), the data also provided the possibility of studying the relationships between the activities observed in different cortical regions—in other words, studying functional connectivity. We now provide two brief examples of recently published data.

Example 1: Application of EROS to the Study of Sentence Comprehension (Tse et al.[78])

Words that are unpredictable within the sentence context, either in terms of their meaning or their syntax, generate special types of brain responses, as identified with ERP measures. Specifically, Kutas and Hillyard[79] showed that semantically anomalous words (e.g., the last word of the sentence "the pizza is too hot to cry") elicit a particular ERP component with a latency of 400 ms from word presentation, which they labeled the "N400." This component was later shown to be elicited by any word providing unpredictable information about the semantic meaning of a particular sentence, and has been considered to be a measure of lexical access (for reviews see[19,80]). Although the N400 has been studied extensively, the exact location(s) in the brain where it is generated is still under some debate. As EROS can provide information about both the temporal and spatial properties of brain activity, we applied it to the analysis of anomalous sentences.[78] Specifically we compared the brain activity elicited by semantically anomalous words (such as the word "cry" in the example above) to that elicited by predictable words in the same sentence positions (such as the word "eat" appearing in the same position), or to words that were syntactically anomalous, although semantically appropriate (these are known to produce a different ERP response, labeled the "P600"[80]). We recorded optical activity from an extended montage covering all of the perisylvian structures in the left hemisphere (an area traditionally linked to both language production and language comprehension). The results (some of which are shown in Figure 15.6) showed that semantically anomalous words elicited a set of specific brain activities in both the superior temporal gyrus and the inferior frontal cortex. These activities were not observed for semantically appropriate words. One of these optical responses occurred with a latency of approximately 400 ms in the superior temporal gyrus. This activity was correlated across subjects with their electrical N400s, which was recorded simultaneously with the optical response. Syntactically anomalous words elicited a different pattern of activities, including activity with a somewhat longer latency in more dorsal areas of the superior temporal cortex. In this case the optical response also correlated both in time and in amplitude (across subjects) with the electrical scalp activity.

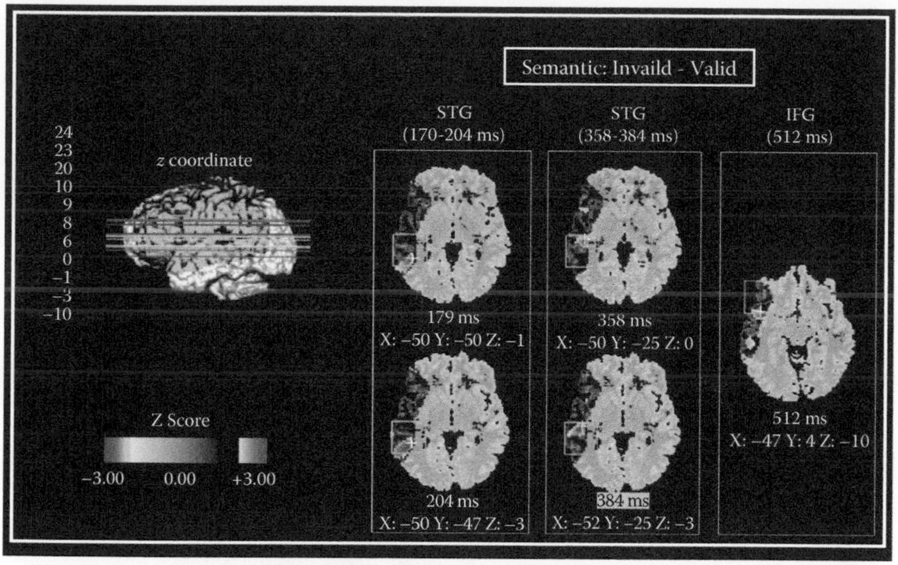

FIGURE 15.6 **(See color insert following page 224)** Statistical maps of EROS data (modified from Tse, C.-Y. et al., Imaging the cortical dynamics of language processing with the event-related optical signal, *Proc. Natl. Acad. Sci. USA*, 104, 17157, 2007. With permission). *Left panel*: *z* (top–bottom) Talairach coordinates for the slices presented in the right panel, and color scale corresponding to the *Z* score of the functional data presented in the right panel (the functional activations shown are significant). *Right panel*: Axial slices with significant increases in optical signal for semantic anomalies. Green boxes indicate the locations of the 4 cm cubes defining each ROI. The white cross within the ROI indicates the location of peak activity. The dark gray shading shows the area of cortex interrogated by the recording montage. Significant EROS activities were typically found at similar locations across more than five axial slices (1.25 mm thickness for each slice) and across adjacent time points (25.6 ms sampling rate). EROS activities at similar locations and time points were regarded as spatio-temporal clusters (white labels above each slice). Some of the slices with the most significant peak activity for each spatio-temporal cluster are shown here. The latencies of the peak optical activities are indicated in yellow below each slice. The activity in the superior temporal gyrus (STG) at 384 ms in the semantic condition (corresponding to the electrical N400) is highlighted.

Example 2: Application of EROS to the Analysis of Frontal Activity in an Oddball Task (Low et al.[76])

The human brain is fine-tuned to analyze departures from regularities in the environment. For instance, when a series of identical tones are presented, the occurrence of a different tone elicits a very special brain response. However, when the tone series is unattended, this brain response is simpler: ERP studies have shown that this situation elicits a component labeled the Mismatch Negativity (MMN[70]), typically followed by another brain response labeled P3a. Optical imaging studies, some of which conducted in our laboratory, have also found that deviant auditory stimuli generate a response in superior temporal cortex followed by activation in inferior prefrontal cortex.[58,66–69] The brain response, however, is enhanced when the auditory stimuli are relevant to the subject's task, and are therefore attended. In this case, ERP studies indicate the presence of a large response, labeled the P3 or P300 component (for a review see[19]). The brain origin of this scalp-recorded component is still under some debate, and it is likely that multiple structures may contribute to its generation. In one recently published study, we[76] examined frontal activity elicited by attended and unattended deviant auditory stimuli with EROS. Some of the results of this study are presented in Figure 15.7. They show two different types of responses for attended and unattended deviant items. Whereas a positive response is observed in right dorsolateral prefrontal cortex for task-relevant deviant items, the same area does not respond in the same way to unattended deviant items. The optical prefrontal activity elicited by attended deviant items is simultaneous with early parts of the electrical P300 response. This is consistent with various modeling efforts that suggest that this region may

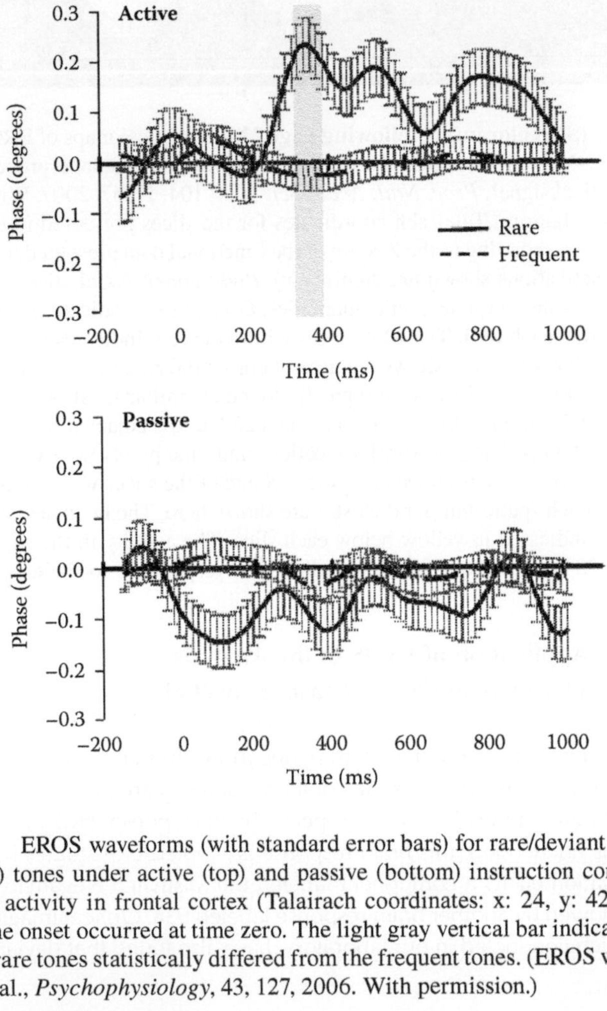

FIGURE 15.7 EROS waveforms (with standard error bars) for rare/deviant (solid) and frequent (dashed) tones under active (top) and passive (bottom) instruction conditions for the voxel of peak activity in frontal cortex (Talairach coordinates: x: 24, y: 42) in an oddball paradigm. Tone onset occurred at time zero. The light gray vertical bar indicates the window for which the rare tones statistically differed from the frequent tones. (EROS waveforms from Low, K. A. et al., *Psychophysiology*, 43, 127, 2006. With permission.)

be related to the frontal P3, a subcomponents that is typically observed slightly before the posterior, parietal aspects of the P300. The prefrontal region where the EROS response was observed is also consistent with fMRI activity recorded in similar paradigms.[81]

15.4 DISCUSSION

15.4.1 Summary of Empirical Data

In this chapter we have reviewed evidence indicating that it is possible to record fast optical signals using noninvasive procedures. Fast optical signals have been recorded in different modalities (including visual, auditory, somatosensory, and motor) in the context of simple sensory and motor paradigms, as well as more complex cognitive tasks. They are likely related to scattering changes associated with neuronal activity. In general, these responses are characterized by an increase in photon delay and a reduction in light transmission in active cortical areas.

15.4.2 Relationship with Other Techniques

Several studies show that the time course of the fast optical signal is similar to the event-related electrical activity recorded in the same conditions. Also, several studies show that the optical signal is localized in a manner consistent with the known functional anatomy of the cortical areas involved, and is also consistent with fMRI data recorded in similar conditions. These data provide support for the idea that fast optical imaging can be used to study the time course of activity in localized cortical areas.

15.4.3 Unresolved Issues and Future Research Directions

Several issues remain to be addressed. First, it has been hypothesized that the fast optical signal (or EROS) is due to changes in the scattering coefficient in active cortical areas determined by the movement of ions across the neuronal membrane. The research conducted using depth recordings with miniaturized optical instruments or implanted optic fibers[7] is clearly relevant for this purpose, as is the work by various groups on in vitro preparations.[6,25] However, additional investigation on the physiological basis of the signal is needed to determine the exact relationships between physical events in the brain and the noninvasive measures. Further, it would be very useful to determine whether these relationships hold equally for different types of neurons (i.e., pyramidal cells versus interneurons) and/or different neurotransmitter systems.

Second, more research is needed to determine which methods are most suitable for recording fast optical signals. Both photon delay and intensity signals have been used. We have carried out some comparisons between these two methods,[54] but a more systematic analysis of the relationships between these two measures is needed. It is also possible that a combination of the two measures may be useful to increase the signal-to-noise ratio.

Third, although substantial advancements have been made in the area of 3D reconstruction, a single method still needs to obtain general acceptance. This area of research will gain from the developments of easy-to-use, general purpose analysis software. Comparisons with fMRI data, such as that carried out by[11,60] can provide validation to the various proposed methods.

Fourth, so far fast optical signals appear to be recordable non-invasively from relatively shallow areas (3 cm or so from the surface of the head), but the penetration to deeper structures is severely limited. However, the penetration may be improved by increases in signal-to-noise ratio and by other technological advancements (such as the use of the "pi detector,"[28] the use of photon delay measures compared to intensity measures, or by varying the modulation frequency[82]).

Fifth, raw data are at present noisy, often requiring data collection from a large number of trials and subjects. Several tools can be used to increase the signal-to-noise ratio, including instrumental improvements (such as the use of more powerful light sources and reduced time multiplexing) and analytical methods (such as the use of appropriate filtering methods,[46] wavelet analysis, cross-correlation and auto-regression methods,[50] independent component analysis,[75] etc.).

15.4.4 Conclusions

Fast optical imaging is a promising new tool for studying brain function, and in particular the time course of activity in selected cortical areas. Advantages of the technique include noninvasivity, portability, good combination of spatial and temporal resolution, relatively low cost, and easy integration with other techniques, such as fMRI, electrophysiological methods, and NIR spectroscopy. Main limitations are its reduced penetration and, at least at present, its low signal-to-noise ratio. The technique appears particularly well suited to provide a bridge between hemodynamic and neuronal methods, and can be used profitably both in the study of brain physiology and in cognitive neuroscience. In particular, in combination with anatomical methods such as diffusion tensor imaging (for a review see Rykhlevskaia et al.[83]) or structural MRI, fast optical signals recorded from the entire cortical surface can be used to test hypotheses about changes in structural and functional connectivity, as the temporal order of activation of various brain areas can be modeled together with anatomical information to provide constraints on models of cognitive processing.[50,77]

ACKNOWLEDGMENTS

The work presented in this chapter was supported in part by NIMH grant MH ROI 80182 to G. Dratton and by NIA grant AGROI 21887 to M. Fabiani.

REFERENCES

1. Gratton, G. et al., Shades of gray matter: Noninvasive optical images of human brain responses during visual stimulation, *Psychophysiology*, 32, 505, 1995.
2. Gratton, G. and Fabiani, M., Shedding light on brain function: The event-related optical signal, *Trends Cogn. Sci.*, 5, 357, 2001.
3. Agran J. et al., When the rules keep changing: The timing of activation of task-general and task-specific brain regions involved in preparation, submitted.

4. Steinbrink, J. et al., Somatosensory evoked fast optical intensity changes detected non-invasively in the adult human head, *Neurosci. Lett.*, 291, 105, 2000.
5. Cohen, L. B., Changes in neuron structure during action potential propagation and synaptic transmission, *Physiol. Rev.,* 53, 373, 1972.
6. Frostig, R. D. et al. Cortical functional architecture and local coupling between neuronal activity and the microcirculation revealed by in-vivo high resolution optical imaging of intrinsic signals, *Proc. Natl. Acad. Sci. USA,* 87, 6082, 1990.
7. Rector, D. M. et al., Light scattering changes follow evoked potentials from hippocampal Schaeffer collateral stimulation, *J. Neurophysiol.,* 78, 1707, 1997.
8. Raichle, M. E., Visualizing the mind, *Sci. Amer.,* 58, April 1994.
9. Pascual-Leone, A. et al., Transcranial magnetic stimulation in cognitive neuroscience—virtual lesion, chronometry, and functional connectivity, *Curr Opin Neurobiol.*, 10, 232, 2000.
10. Toga, A. W. and Mazziotta, J. C. (Eds.), *Brain Mapping: The Methods*, Academic Press, San Diego, 1996.
11. Gratton, G. et al., Comparison of neuronal and hemodynamic measure of the brain response to visual stimulation: An optical imaging study, *Hum. Brain Mapping,* 13, 13, 2001.
12. Ogawa, S. et al. Intrinsic signal changes accompanying sensory stimulation: Functional brain mapping with magnetic resonance imaging, *Proc. Natl. Acad. Sci. USA*, 89, 5951, 1992.
13. Villringer, A. and Chance, B., Non-invasive optical spectroscopy and imaging of human brain function, *Trend. Neurosci.,* 20, 435, 1997.
14. Jobsis, F. F., Noninvasive, infrared monitoring of cerebral and myocardial oxygen sufficiency and circulatory parameters, *Science,* 198, 1264, 1977.
15. Grinvald, A. et al., Functional architecture of cortex revealed by optical imaging of intrinsic signals, *Nature,* 324, 361, 1986.
16. Villringer, A. and Dirnagl, U., Coupling of brain activity and cerebral blood flow: Basis of functional neuroimaging, *Cerebrovasc. Brain Metab. Rev.,* 7, 240, 1995.
17. Miller, K. L. et al., Non-linear transport dynamics of the cerebral blood flow response, *Hum. Brain Mapping*, 13, 1, 2001.
18. Pizzagalli, D. A., Electroencephalography and high-density electrophysiological source localization, in *Handbook of Psychophysiology* (3rd edition), J. Cacioppo, L. Tassinary, and G. Berntson (Eds.), Cambridge University Press, New York, 2007, p. 56.
19. Fabiani, M. et al., Event related brain potentials, in *Handbook of Psychophysiology* (3rd edition), J. Cacioppo, L. Tassinary, and G. Berntson (Eds.), Cambridge University Press, New York, 2007, p. 85.
20. Hari, R. et al., Timing of human cortical functions during cognition: Role of MEG. *Trends Cogn. Sci.*, 4, 455, 2000.
21. Allison, T. et al., The central nervous system, in *Psychophysiology: Systems, Processes, and Applications*, Coles, M. G. H., Porges, S. W. and Donchin, E. (Eds.), Guilford, New York, 1986, p. 5.
22. Pascual-Marqui, R. D. et al., Low resolution electromagnetic tomography: A new method for localizing electrical activity in the brain, *Int. J. Psychophysiol.*, 18, 49, 1994.
23. Dale, A. M. et al., Dynamic statistical parametric mapping: Combining fMRI and MEG for high-resolution imaging of cortical activity, *Neuron,* 26, 55, 2000.
24. Rector, D. M. et al., Spatio-temporal mapping of rat whisker barrels with fast scattered light signals, *Neuroimage*, 26, 619, 2005.
25. Stepnoski, R. A. et al., Noninvasive detection of changes in membrane potential in cultured neurons by light scattering, *Proc. Natl. Acad. Sci. USA*, 88, 9382, 1991.
26. Foust, A.J., and Rector, D.M., Optically teasing apart neural swelling and depolarization. *Neuroscience*, 145, 887, 2007.
27. Gratton, G. et al., Rapid changes of optical parameters in the human brain during a tapping task, *J. Cogn. Neurosci.,* 7, 446, 1995.

28. Wolf, U. et al., Detecting cerebral functional slow and fast signals by frequency-domain near-infrared spectroscopy using two different sensors, *OSA Biomed. Top. Meeting, Tech. Dig.*, 427, 2000.
29. Gratton, E. et al., The possibility of a near-infrared optical imaging system using frequency-domain methods, *Proc. III Int. Conf. Peace through Mind/Brain Sci.*, 183, 1990.
30. Maclin, E. et al., Amelioration of movement artifacts in slow and fast optical recordings, presented at the *Cogn. Neurosci. Soc.*, San Francisco, 2008.
31. Okada, E. et al., Theoretical and experimental investigation of near-infrared light propagation in a model of the adult head, *Appl. Opt.*, 36, 21, 1997.
32. Firbank, M. et al., A theoretical study of the signal contribution of regions of the adult head to near-infrared spectroscopy studies of visual evoked responses, *Neuroimage*, 8, 69, 1998.
33. Gratton, G. et al., Toward non-invasive 3-D imaging of the time course of cortical activity: Investigation of the depth of the event-related optical signal (EROS), *NeuroImage*, 11, 491, 2000.
34. Gratton, G. et al., Feasibility of intracranial near-infrared optical scanning, *Psychophysiology*, 31, 211, 1994.
35. Alfano, R. R. et al., Advances in optical imaging of biomedical media, *Ann. New York Acad. Sci.*, 820, 248, 1997.
36. Arridge, S. R. and Hebden, J. C. (1997), Optical imaging in medicine: II. Modeling and reconstruction, *Phys. Med. Biol.*, 42, 841, 1997.
37. Arridge, S. R., and Schweiger, M., Image reconstruction in optical tomography, *Phil. Trans. Royal Soc. London—Series B: Biol. Sci.*, 352, 717, 1997.
38. Benaron, D. A. et al., Non-recursive linear algorithms for optical imaging in diffusive media, *Adv. Exper. Med. Biol.*, 361, 215, 1994.
39. Chang, J. et al., Optical imaging of anatomical maps derived from magnetic resonance images using time-independent optical sources, *IEEE Trans. Med. Imaging*, 16, 68, 1997.
40. Franceschini, M. A. et al., On-line optical imaging of the human brain with 160-ms temporal resolution, *Opt. Expr.*, 6, 49, 2000.
41. Jiang, H. et al., Optical image reconstruction using DC data: Simulations and experiments, *Phys. Med. Biol.*, 41, 1483, 1996.
42. Paulsen, K. D. and Jiang, H., Spatially varying optical property reconstruction using a finite element diffusion equation approximation, *Med. Phys.*, 22, 691, 1995.
43. Zhu, W. et al., Iterative total least-squares image reconstruction algorithm for optical tomography by the conjugate gradient method, *J. Opt. Soc. Amer. A—Opt. Image Sci.*, 14, 799, 1997.
44. Gratton, G., AOpt-cont@ and AOpt-3D@: A software suite for the analysis and 3D reconstruction of the event-related optical signal (EROS). *Psychophysiology*, 37, S44, 2000.
45. Gratton, G. and Corballis, P. M., Removing the heart from the brain: Compensation for the pulse artifact in the photon migration signal, *Psychophysiology,* 32, 292, 1995.
46. Maclin, E. et al., Optimum filtering for EROS measurements, *Psychophysiology*, 40, 542, 2003.
47. Fabiani, M. et al., Bootstrap assessment of the reliability of maxima in surface maps of brain activity of individual subjects derived with electrophysiological and optical methods, *Beh. Res. Meth., Instr., Comp.*, 30, 78, 1998.
48. Whalen, C., Maclin, E.L., Fabiani, M., and Gratton, G. (2008). Validation of the method for coregistering scalp recording locations with 3D structural MR images. *Human Brain Mapping*, 29(11), 1288–1301.
49. Kiebel, S. J. et al., Robust smoothness estimation in statistical parametric maps using standardized residuals from the general linear model, *NeuroImage*, 10, 756, 1999.
50. Rykhlevskaia, E. et al. Lagged covariance structure models for studying functional connectivity in the brain, *NeuroImage*, 30, 1203, 2006.

51. Talairach, J., and Tournoux, P., *Co-Planar Stereotaxic Atlas of the Human Brain: Three-Dimensional Proportion System: An Approach to Cerebral Imaging*, Thieme, New York, 1988.
52. Steinbrink, J. et al., The fast optical signal—robust or elusive when non-invasively measured in the human adult? *Neuroimage*, 26, 996, 2005.
53. Malonek, D. and Grinvald, A., The spatial and temporal relationship between cortical electrical activity and responses of the microcirculation during sensory stimulation: implications for optical, PET, and MR functional brain imaging. *Science,* 272, 551–554, 1996.
54. Gratton, G. et al., Effects of measurement method, wavelength, and source-detector distance on the fast optical signal, *NeuroImage*, 32, 1576, 2006.
55. Menon, R. S. et al., BOLD based functional MRI at 4 Tesla includes a capillary bed contribution: Echo-planar imaging correlates with previous optical imaging using intrinsic signals, *Magn. Reson. Med.*, 33, 453, 1995.
56. Maclin, E. L. et al., The Event Related Optical Signal (EROS) to electrical stimulation of the median nerve, *NeuroImage*, 21,1798, 2004.
57. Fabiani, M. et al., Reduced suppression or labile memory? Mechanisms of inefficient filtering of irrelevant information in older adults, *J. Cogn. Neurosci.*, 18, 637, 2006.
58. Sable, J. J. et al., Optical imaging of perceptual grouping in human auditory cortex, *Eur. J. Neurosci.*. 25, 298, 2007.
59. Gratton, G. et al., Fast and localized event-related optical signals (EROS) in the human occipital cortex: Comparison with the visual evoked potential and fMRI, *NeuroImage*, 6, 168, 1997.
60. Zhang, X. et al., The study of cerebral hemodynamic and neuronal response to visual stimulation using simultaneous NIR optical tomography and BOLD fMRI in humans, in *Proc. SPIE Vol. 5686, Photonic Therapeutics and Diagnostics*, K. E. Bartels, L. S. Bass, W.T. de Riese, K. W. Gregory, H. Hirschberg, A. Katzir, N. Kollias, S. J. Madsen, R. S. Malek, K. M. McNally-Heintzelman, L. P. Tate, Jr., E. A. Trowers, B. J.-F. Wong (Eds.), 2005, p. 566.
61. Gratton, G., and Fabiani, M., The event related optical signal (EROS) in visual cortex: Replicability, consistency, localization and resolution, *Psychophysiology,* 40, 561, 2003.
62. Gratton, G., Attention and probability effects in the human occipital cortex: An optical imaging study, *NeuroReport*, 8, 1749, 1997.
63. Gratton, G. et al., Memory-driven processing in human medial occipital cortex: An event-related optical signal (EROS) study, *Psychophysiology,* 38, 348, 1998.
64. Gratton, G. et al., Time course of activation of human occipital cortex measured with the event-related optical signal (EROS), *Biomedical Optics 2006 Technical Digest* (Optical Society of America, Washington, DC, 2006), MD4, 2006.
65. Fabiani, M. et al., Multiple visual memory phenomena in a memory search task, *Psychophysiology*, 40, 472, 2003.
66. Rinne, T. et al., Scalp-recorded optical signals make sound processing from the auditory cortex visible, *NeuroImage*, 10, 620, 1999.
67. Tse, C. Y. et al., Event-related optical imaging reveals the temporal dynamics of right temporal and frontal cortex activation in pre-attentive change detection, *Neuroimage*, 29, 314, 2006.
68. Tse, C. Y., and Penney, T. B., Preattentive change detection using the event-related optical signal, *IEEE Eng. Med. Biol. Mag.*, 26, 52, 2007.
69. Tse, C. Y., and Penney, T. B., On the functional role of temporal and frontal cortex activation in passive detection of auditory deviance, *Neuroimage*, 2008.
70. Näätänen, R., Mismatch negativity (MMN): Perspectives for application, *Int. J. Psychophysiol.,* 37, 3, 2000.

71. Näätänen, R., and Picton, T., The N1 wave of the human electric and magnetic response to sound: A review and an analysis of the component structure, *Psychophysiology*, 24, 375, 1987.
72. Franceschini, M. A. and Boas, D. A., Noninvasive measurement of neuronal activity with near-infrared optical imaging, *Neuroimage*, 21, 372, 2004.
73. DeSoto, M. C. et al. When in doubt, do it both ways: Brain evidence of the simultaneous activation of conflicting responses in a spatial Stroop task, *J. Cogn. Neurosci.*, 13, 523, 2001.
74. Gratton, G. et al. Pre- and poststimulus activation of response channels: A psychophysiological analysis, *J. Exp. Psychol.: Hum Perc. Perf.*, 11, 331, 1988.
75. Morren, G. et al., Detection of fast neuronal signals in the motor cortex from functional near infrared spectroscopy measurements using independent component analysis, *Med. Biol. Eng. Comput.*, 42, 92, 2004.
76. Low, K. A. et al., Fast optical imaging of frontal cortex during active and passive oddball tasks, *Psychophysiology*, 43, 127, 2006.
77. Gratton, G., Rykhlevskaia, E., Nee, E., Leaver, E., and Fabiani, M. (in press). Does white matter matter? Spatiotemporal dynamics of task switching in aging, *J. Cogn. Neurosci.*
78. Tse, C.-Y. et al., Imaging the cortical dynamics of language processing with the event-related Optical Signal, *Proc. Natl. Acad. Sci. USA*, 104, 17157, 2007.
79. Kutas, M. and Hillyard, S. A., Reading senseless sentences: Brain potentials reflect semantic incongruity, *Science*, 207, 203, 1980.
80. Kutas, M. et al., Language, in *Handbook of Psychophysiology* (3rd edition), J. Cacioppo, L. Tassinary, and G. Berntson (Eds.), Cambridge University Press, New York, 2007, p. 555.
81. McCarthy, G. et al., Infrequent events transiently activate human prefrontal and parietal cortex as measured by functional MRI, *J. Neurophysiol.*, 77, 1630, 1997.
82. Maclin, E. L. et al., Improving the signal-to-noise ratio of Event Related Optical Signals (EROS) by manipulating wavelength and modulation frequency, *IEEE EMBM*, 26, 47, 2007.
83. Rykhlevskaia, E. I. et al., Combining structural and functional neuroimaging data for studying brain connectivity: A review, *Psychophysiology*, 45, 173, 2008.
84. Gratton, G. and Fabiani, M., The event-related optical signal: A new tool for studying brain function. *Int. J. Psychophysiol.*, 42, 109–121, 2001.

Index

F

L

M

N

O

P

Q

R

S

T

V

Z

Printed and bound by CPI Group (UK) Ltd, Croydon, CR0 4YY
23/10/2024
01778266-0005